Applications of Diamond-like Carbon Coatings

Edited by Abdul Wasy Zia

Institute of Mechanical Process and Energy Engineering (IMPEE),
School of Engineering and Physical Sciences,
Heriot-Watt University,
Edinburgh, United Kingdom

WILEY

Published by John Wiley & Sons, Inc., Hoboken, New Jersey.

Published simultaneously in Canada.

Library of Congress Cataloging-in-Publication Data

Names: Zia, Abdul Wasy, editor.
Title: Applications of diamond-like carbon coatings / edited by Abdul Wasy Zia.
Description: Hoboken, New Jersey : John Wiley & Sons Inc., [2025] | Includes bibliographical references and index.
Identifiers: LCCN 2024047186 | ISBN 9781394189113 (hardback) | ISBN 9781394189120 (adobe pdf) | ISBN 9781394189137 (epub)
Subjects: LCSH: Diamond thin films. | Protective coatings.
Classification: LCC TP873.5.D5 A77 2025 | DDC 667/.9--dc23/eng/20250224
LC record available at https://lccn.loc.gov/2024047186

Cover Image: ©Springer Nature
Cover Design: WILEY

Set in 9.5/12.5pt STIXTwoText by Lumina Datamatics

Contents

List of Contributors *x*
Preface *xiv*

Section I Diamond-like Carbon Coatings *1*

1 Introduction of Diamond-like Carbon Coatings *3*
Naveen Joshi, Jagdish Narayan, Roger J. Narayan
1.1 Introduction *3*
1.2 DLC Coatings for Mechanical and Transportation Applications *6*
1.2.1 DLC Coatings for Lubricants *6*
1.2.2 DLC Coatings for Pipelines, Tubes, and Molds *7*
1.3 DLC Coatings for Medical Applications *8*
1.3.1 DLC Coatings for Orthopedics *10*
1.3.2 DLC Coatings for Antimicrobial Performance *10*
1.3.3 DLC Coatings for Cardiovascular Implants *11*
1.4 DLC Coatings for Electrical and Optical Devices *12*
1.4.1 DLC Coatings for MEMS *12*
1.4.2 DLC Coatings for Nonvolatile Storage, Low-K Dielectrics, and Field Emitters *13*
1.5 DLC Coatings for Optical Applications *14*
1.6 Emerging Applications of DLC Coatings *15*
1.6.1 DLC Coatings for Energy Storage Applications *15*
1.6.2 DLC Coatings for Polymers and Textiles *17*
1.6.3 Optimizing DLC Coatings Through the Utilization of Artificial Intelligence Tools *17*
1.7 Limitations of DLC Coatings *18*
1.8 Q-Carbon as a Superior Coating Material *19*
1.9 Summary *22*

2 Deposition Methods of Diamond-like Carbon Coatings *31*
Pranay Kalakonda, Jagdish Narayan, Roger J. Narayan
2.1 Introduction *31*
2.2 Classification of Carbon Coatings *34*
2.3 Deposition Methods *36*
2.3.1 Physical Vapor Deposition *37*
2.3.1.1 Ion Beam Deposition *37*
2.3.1.2 Mass Selected Ion Beam Deposition *38*
2.3.1.3 Cathodic Arc Deposition *39*
2.3.1.4 Sputtering *42*
2.3.1.5 Ion Beam-assisted Deposition *43*
2.3.1.6 Pulsed Laser Deposition *44*
2.3.2 Chemical Vapor Deposition *48*
2.4 Deposition Mechanisms *50*
2.5 Influence of Deposition Parameters *53*
2.6 Properties and Analysis Methods *54*
2.7 Applications *56*
2.8 Conclusions *57*

3 Properties of Diamond-like Carbon Coatings *71*
Muhammad Abdul Basit, Ramsha Zafar, Nawal Haider
3.1 Introduction to Diamond-like Carbon Coating *72*
3.2 General Categorization of DLC Coatings *72*
3.3 Preparation of DLC Coatings *74*
3.4 Characterization Techniques of DLC Coatings *74*
3.5 Properties of DLC Coatings *76*
3.5.1 Physical Properties of DLC Coatings *76*
3.5.1.1 Density *77*
3.5.1.2 Surface-free Energy *77*
3.5.1.3 Thermal Conductivity *79*
3.5.2 Chemical Properties of DLC Coatings *80*
3.5.2.1 Reactivity and Chemical Stability *81*
3.5.3 Anticorrosion Properties of DLC Coatings *82*
3.5.4 Biocompatibility of DLC Coatings *83*
3.5.5 Opto-electrical Properties of DLC Coatings *84*
3.5.5.1 Electrical Resistance *84*
3.5.5.2 Bandgap *86*
3.5.5.3 Refractive Index *86*
3.5.6 Tribological Properties of DLC Coatings *87*
3.5.6.1 Friction Coefficient *88*
3.5.6.2 Wear Resistance *91*
3.5.7 Adhesion of DLC Coatings to Substrates *92*

3.5.8 Mechanical Properties of DLC Coatings *94*
3.5.8.1 Hardness and Young's Modulus *94*
3.5.8.2 Fatigue Strength of DLC Coatings *96*
3.6 Improving the Properties of DLC Coatings *96*
3.6.1 Modification Through Alloy Addition *97*
3.6.2 Modification Through Multilayer and Hybrid Coatings *98*
3.7 Applications and Uses of DLC Coatings *100*
3.8 Summary and Future Outlook *100*

Section II Diamond-like Carbon Coatings for Mechanical and Transportation Applications *107*

4 Diamond-like Carbon Coatings for Tools and Molds *109*
Farooq Ahmad, Raies Abdullah, Jamile Mohammadi Moradian
4.1 Introduction *109*
4.2 Hot Forging Die Tool Failure Mechanisms *111*
4.3 Surface Treatments in Die Tools and Molds *111*
4.4 Deposition of DLC Coatings *113*
4.4.1 Physical Vapor Deposition *113*
4.4.2 Chemical Vapor Deposition *114*
4.4.3 Hybrid Coating Techniques *115*
4.5 Structural Properties of DLC Coatings *115*
4.6 Mechanical and Tribological Properties of DLC Coatings *117*
4.7 Applications of DLC Coatings in Die Tools and Molds *118*
4.8 Challenges and Future Perspective of DLC Coatings *121*

5 Diamond-like Carbon Coatings for Solid Lubrication: Production Techniques, Properties, and Applications *129*
Deepak Kumar, Virendra Pratap Singh, Himanshu Rai
5.1 Introduction *130*
5.2 Classifications of DLC Coatings *131*
5.2.1 Hydrogenated Dip-layer Capacitors (a-DLC) *132*
5.2.2 DLC Coatings Devoid of Hydrogen (ta-C) *132*
5.2.3 DLC Coatings Doped With Nitrogen (a-C:N) *132*
5.2.4 Fluorinated DLC coatings (a-C:F) *133*
5.2.5 Silicon-doped DLC coatings (a-C:Si) *133*
5.2.6 Anhydrous Carbon Nitride Amorphous (a-CN:H) *133*
5.3 Deposition Methods of DLC Coatings *133*
5.3.1 Sputtering *133*
5.3.1.1 Direct Current Magnetron Sputtering *133*
5.3.1.2 Sputtering at Radio Frequency *134*
5.3.2 Ion Beam Deposition *135*

5.3.3 Cathodic Arc *135*
5.3.4 Pulsed Laser Deposition *136*
5.3.5 Chemical Vapor Deposition *137*
5.3.6 Physical Vapor Deposition *138*
5.4 Tribology of DLC Coatings *140*
5.4.1 Effect of Surface Roughness on Wear and Friction *140*
5.4.2 Effect of Environment on Wear and Friction *141*
5.4.3 Effect of Substrate on Wear and Friction *142*
5.5 Modification of DLC Coatings for Reduced Wear and Friction *143*
5.5.1 DLC Coatings With Interlayers *144*
5.5.2 Metal-doped DLC *144*
5.5.3 Cr-based Interlayers DLC Coatings *145*
5.6 Applications of DLC Coatings *145*
5.7 Summary and Future Directions *146*

6 Diamond-like Carbon Coatings and Green Lubricants *155*
Irfan Nadeem, Muhammad Shahid Arshad, Marko Polajnar
6.1 Introduction *155*
6.2 DLC Coatings in Sustainable Tribology: Navigating Green Engineering Concepts *158*
6.2.1 Biodegradable Oils *159*
6.2.2 Low-SAPS Lubricants *160*
6.2.3 Fatty Acids *161*
6.2.4 Nanoparticles as Lubricant Additives *162*
6.2.5 Ionic Liquids *167*
6.2.6 Water-based Lubricants and Polyhydric Alcohols *169*
6.3 Role and Importance of Doping Elements on DLC Lubricating Properties *171*
6.3.1 Influence of Dopants on the Surface Energy of DLC Coatings *171*
6.3.2 Influence of Dopants on the Wetting Properties of DLC Coatings *174*
6.4 Interactions of DLC Coatings with Green Additives *176*
6.5 Lubrication Behavior of DLC Coatings with Base Lubricants *179*
6.6 Conclusion *182*

Section III Diamond-like Carbon Coatings for Medical Applications *191*

7 Diamond-like Carbon Coatings for Artificial Implants *193*
Benedict Rothammer, Max Marian
7.1 Introduction *193*
7.2 Requirements for DLC Coatings When Employed to Artificial Implants *195*

7.2.1 Coating-Substrate Adhesion *196*
7.2.2 Biological Compatibility *198*
7.3 Application for Load-bearing Implants *198*
7.3.1 Biotribological Studies on the Model Level *201*
7.3.2 Biotribological Studies on the Component Level *213*
7.4 Application for Dental Implants and Braces/Retainers *216*
7.4.1 Dental Implants *216*
7.4.2 Braces/Retainers *219*
7.5 Concluding Remarks *222*

8 Diamond-like Carbon Coatings for Orthopedic Implants and Surgical Tools *245*
Witold Kaczorowski, Krzysztof Jastrzębski, Piotr Niedzielski
8.1 Introduction *245*
8.2 Orthopedic Substrates and DLC Coating Techniques *246*
8.3 Carbon Coatings as Part of Multilayer Systems or as a Base for Doping *248*
8.4 Properties of DLC Coatings and Their Use in Orthopedics and Surgical Instruments *249*
8.5 Examples of Applications of Carbon Coatings in Orthopedics and on Surgical Instruments *261*
8.6 Conclusion *263*

Section IV Diamond-like Carbon Coatings for Electrical and Optical Devices *277*

9 Diamond-like Carbon Coatings for Electric Storage Batteries *279*
Muhammad Ihsan-Ul-Haq
9.1 Introduction *279*
9.2 Applications of DLC Coatings in Secondary Batteries *280*
9.2.1 DLC Coatings on Alkali Metal Anodes *282*
9.2.2 DLC Coatings on Carbon or Alloy-based Anodes *283*
9.2.3 DLC Coatings on Cathodes *284*
9.2.4 DLC Coatings on the Separator *286*
9.3 Conclusions and Future Research *287*

10 DLC Coatings for Optical and Optoelectronic Applications *293*
Mümin Mehmet Koç, Naim Aslan
10.1 Introduction *293*
10.2 Production and Characterization Methods *295*
10.3 Optical and Optoelectronic Properties *299*

10.4 Effect of Production Methods/Parameters on DLC Film Properties *301*
10.5 DLC Films in Optical and Optoelectronic Applications *303*
10.6 Conclusion *307*

Section V Emerging Applications of Diamond-like Carbon Coatings *313*

11 Diamond-like Carbon Coatings for Polymers and Textiles *315*
Witold Kaczorowski
11.1 Introduction *315*
11.2 Polymer Substrates and Techniques for Developing DLC Coatings on Them *317*
11.3 Properties of DLC Coatings for Which They Find Application in Polymer Modification *318*
11.4 Application Areas for Carbon Coatings for Polymers and Textiles *331*
11.5 Conclusion *335*

12 Diamond-like Carbon Coating Applications in MEMS *345*
Muhammad Usman, Hafiz Mahmood Ilahi, Muhammad Umar Farooq, Mudassar Rauf, Muhammad Asim, Sarmad Ali Khan
12.1 Introduction *346*
12.2 Mechanical Properties of DLC Coatings *348*
12.3 DLC Applications for Sensors *351*
12.3.1 DLC Applications for Force Sensors *351*
12.3.2 DLC Applications for Temperature Sensors *354*
12.3.3 DLC Applications for Piezoresistive Sensor *355*
12.3.4 DLC Applications for Strain Sensors *355*
12.3.5 DLC Applications for Pressure Sensors *356*
12.4 DLC Applications for Nanoactuators *357*
12.5 Industrial Processing *358*
12.6 Applications of DLC Coatings Structural Integrity *359*
12.7 DLC Applications for MEMS Tribology *359*
12.7.1 Nanofretting *360*
12.7.2 Nanodot Contacts *361*
12.7.3 Lubricant Life Assessment *363*
12.8 DLC Coatings System and Performance *363*
12.8.1 Hydrogenated and Non-hydrogenated DLC Coating *363*
12.8.2 Role of Substrate Material *364*
12.8.3 Wear at Microscopic Scale *365*
12.8.4 Static Friction *366*
12.9 Conclusion *366*

13 Data-centric Process, Property, and Performance of Carbon Coatings *373*
Muhammad Ikhlaq, Abdul Wasy Zia
13.1 Introduction *373*
13.2 Data-centric Manufacturing Process of Carbon Coatings *375*
13.3 Data-centric Virtual Synthesis and Structural Design of Carbon Coatings *380*
13.4 Understanding the Properties of Carbon Coatings With Data-centric Approaches *383*
13.5 Data-centric Performance Assessment of Carbon Coatings *387*
13.6 Applications of Carbon Coating in Enhancing the Heat Transfer Processes *397*
13.6.1 Condensation Heat Transfer *398*
13.6.2 Boiling Heat Transfer *402*
13.7 Conclusion *405*

14 Circularity and Techno-economic Analysis of Carbon-coated Products *413*
Ashmal James, Abdul Wasy Zia
14.1 Introduction *413*
14.2 Potential Reprocessing Methods for Carbon De-coating *415*
14.3 Decision Matrix of Carbon De-coating Methods *417*
14.3.1 Process Robustness for De-coating *417*
14.3.2 Process Efficiency to De-coat Carbon-coated Components *418*
14.3.3 De-coating Processes and Base Material Safety *418*
14.3.4 Sustainability Aspects of De-coating Processes *418*
14.3.5 Technology Readiness Level of De-coating Processes *419*
14.3.6 Cost *419*
14.4 Technical Analysis of Carbon De-coating Methods *419*
14.4.1 Robustness of De-coating Processes *420*
14.4.2 Circularity Index for De-coating Processes *421*
14.4.3 Safety Factor Index for De-coating Processes *421*
14.4.4 Technology Readiness Levels of De-coating Processes *422*
14.5 Economic Analysis of Carbon De-coating Methods *422*
14.5.1 Capital Cost *422*
14.5.2 Running Cost *422*
14.6 Case Study – Carbon De-coating from Cutting Tools *423*
14.7 Conclusion *426*

Index *429*

List of Contributors

Raies Abdullah
School of Materials Science and Engineering
Jiangsu University
Zhenjiang
China

Farooq Ahmad
School of Materials Science and Engineering
Jiangsu University
Zhenjiang
China

Muhammad Shahid Arshad
Laboratory for Tribology and Interface Nanotechnology
University of Ljubljana
Ljubljana
Slovenia

Department for Nanostructured Materials
K7
Jožef Stefan Institute
Ljubljana
Slovenia

Muhammad Asim
School of Professional Education & Executive Development
The Hong Kong Polytechnic University
Kowloon
Hong Kong

Naim Aslan
Faculty of Engineering
Department of Metallurgical and Materials Engineering
Munzur University
Tunceli
Turkiye

Rare Earth Elements Application and Research Center
Munzur University
Tunceli
Turkiye

Muhammad Abdul Basit
Department of Materials Science and Engineering
Institute of Space Technology
Islamabad
Pakistan

Muhammad Umar Farooq
The Sargent Centre for Process Systems Engineering
Department of Chemical Engineering
University College London
London
UK

Nawal Haider
Department of Materials Science and Engineering
Institute of Space Technology
Islamabad
Pakistan

Muhammad Ihsan-Ul-Haq
Energy Storage and Distributed Resources Division
Lawrence Berkeley National Laboratory
Berkeley, CA
USA

Interdisciplinary Research Center for Sustainable Energy Systems (IRC-SES)
King Fahd University of Petroleum and Minerals
Dhahran
Saudi Arabia

Materials Science and Engineering Department
King Fahd University of Petroleum and Minerals
Dhahran
Saudi Arabia

Muhammad Ikhlaq
School of Engineering
Newcastle University
Newcastle upon Tyne
UK

Dyson Institute of Engineering and Technology
Malmesbury
UK

Hafiz Mahmood Ilahi
Department of Mechanical Engineering
City University of Hong Kong
Hong Kong SAR
China

Ashmal James
Institute of Mechanical Process and Energy Engineering (IMPEE)
School of Engineering and Physical Sciences
Heriot-Watt University
Edinburgh
UK

Krzysztof Jastrzębski
Division of Biomedical Engineering and Functional Materials
Institute of Material Science and Engineering
Lodz University of Technology
Lodz
Poland

Naveen Joshi
Department of Materials Science and Engineering
Centennial Campus
North Carolina State University
Raleigh, NC
USA

Witold Kaczorowski
Division of Biomedical Engineering and Functional Materials
Institute of Material Science and Engineering
Lodz University of Technology
Lodz
Poland

Pranay Kalakonda
Department of Materials Science and Engineering
North Carolina State University
Raleigh, NC
USA

Sarmad Ali Khan
Department of Industrial and Manufacturing Engineering
University of Engineering and Technology
Lahore
Pakistan

Mümin Mehmet Koç
Department of Medical Services and Techniques
School of Medical Service
Kırklareli University
Kırklareli
Turkiye

Department of Physics
Faculty of Science and Art
Kırklareli University
Kırklareli
Turkiye

Deepak Kumar
Department of Mechanical Engineering
Carnegie Mellon University
Pittsburgh, PA
USA

Department of Materials Design and Innovation
University at Buffalo
New York, NY
USA

Max Marian
Department of Mechanical and Metallurgical Engineering
School of Engineering
Pontificia Universidad Católica de Chile
Macul
Región Metropolitana
Chile

Institute of Machine Design and Tribology (IMKT)
Leibniz University Hannover
Garbsen
Germany

Jamile Mohammadi Moradian
School of Materials Science and Engineering
Jiangsu University
Zhenjiang
China

Irfan Nadeem
Laboratory for Tribology and Interface Nanotechnology
University of Ljubljana
Ljubljana
Slovenia

Jagdish Narayan
Department of Materials Science and Engineering
Centennial Campus
North Carolina State University
Raleigh, NC
USA

Roger J. Narayan
Department of Materials Science and Engineering
Centennial Campus
North Carolina State University
Raleigh, NC
USA

Joint Department of Biomedical Engineering
Centennial Campus
North Carolina State University and UNC Chapel Hill
Raleigh, NC
USA

Piotr Niedzielski
Division of Biomedical Engineering and Functional Materials
Institute of Material Science and Engineering
Lodz University of Technology
Lodz
Poland

Marko Polajnar
Laboratory for Tribology and Interface Nanotechnology
University of Ljubljana
Ljubljana
Slovenia

Himanshu Rai
Department of Control Engineering
Faculty of Electrical Engineering
Czech Technical University in Prague
Prague
Czech Republic

Mudassar Rauf
School of Mechanical and Electrical Engineering
Wenzhou University
Wenzhou
China

Benedict Rothammer
Engineering Design
Friedrich-Alexander-Universität Erlangen-Nuremberg (FAU)
Erlangen
Germany

Virendra Pratap Singh
Department of Mechanical Engineering
IES College of Technology
Bhopal
India

Muhammad Usman
Department of Mechanical Engineering
City University of Hong Kong
Hong Kong SAR
China

Ramsha Zafar
Department of Materials Science and Engineering
Institute of Space Technology
Islamabad
Pakistan

Abdul Wasy Zia
Institute of Mechanical Process and Energy Engineering (IMPEE)
School of Engineering and Physical Sciences
Heriot-Watt University
Edinburgh
UK

Preface

Carbon coatings were reported in 1954 and exhibit graphite-like behaviors since diamond-like carbon (DLC) coatings were reported around 1970. DLC coatings grasped rapid recognition in the industry and were widely adopted for mechanical applications as DLC coatings exhibit high hardness, lower friction coefficient, and excellent antiwear properties. DLC coatings have progressed through a range of deposition techniques like physical vapor deposition, plasma-enhanced chemical vapor deposition, etc.; coating architectures such as monolayer, multilayer, doped; nanocrystalline, amorphous, or granular morphologies; and have been used for a wider range of industrial applications including automotive, mechanical components and cutting tools, healthcare products, renewable energy, computing and electronics, plastics, textiles sectors, etc.

This book presents current advancements in the applications of DLC coatings. Thirty-three subject experts have contributed 14 chapters to this book, which presents 5 major sections on DLC coatings. The first section introduces DLC coatings, details on deposition methods, and their properties. The second section presents DLC applications for mechanical and transportation with emphasis at tools and molds, solid lubricants, and green tribology. The third section presents DLC applications for medical sector such as artificial implants and surgical tools. The fourth section presents DLC applications in electrical and optical devices such as energy storage batteries and optoelectronic products. New DLC markets are emerging with industrial transformation. Therefore, the last section presents new trends in DLC applications such as DLC coatings for polymer and textiles, micro-electromechanical systems. The last two chapters highlight the significance of the data-centric process–property–performance relationship of DLC coatings and the circularity of DLC-coated products back to cradle-to-cradle product life cycle after the de-coating process in aspiration of a circular economy for green and sustainable ecology.

The book contents are suitable for new researchers in DLC to subject experts and would be a useful source for teaching, research, policymaking, and industrial audiences associated with DLC coatings.

Editor
Abdul Wasy Zia, PhD
Edinburgh

Section I

Diamond-like Carbon Coatings

1

Introduction of Diamond-like Carbon Coatings

Naveen Joshi[1], *Jagdish Narayan*[1], *Roger J. Narayan*[1,2]

[1] *Department of Materials Science and Engineering, Centennial Campus, North Carolina State University, Raleigh, NC, USA*
[2] *Joint Department of Biomedical Engineering, Centennial Campus, North Carolina State University, Raleigh, NC, USA*

CONTENTS

1.1 Introduction, 3
1.2 DLC Coatings for Mechanical and Transportation Applications, 6
 1.2.1 DLC Coatings for Lubricants, 6
 1.2.2 DLC Coatings for Pipelines, Tubes, and Molds, 7
1.3 DLC Coatings for Medical Applications, 8
 1.3.1 DLC Coatings for Orthopedics, 10
 1.3.2 DLC Coatings for Antimicrobial Performance, 10
 1.3.3 DLC Coatings for Cardiovascular Implants, 11
1.4 DLC Coatings for Electrical and Optical Devices, 12
 1.4.1 DLC Coatings for MEMS, 12
 1.4.2 DLC Coatings for Nonvolatile Storage, Low-K Dielectrics, and Field Emitters, 13
1.5 DLC Coatings for Optical Applications, 14
1.6 Emerging Applications of DLC Coatings, 15
 1.6.1 DLC Coatings for Energy Storage Applications, 15
 1.6.2 DLC Coatings for Polymers and Textiles, 17
 1.6.3 Optimizing DLC Coatings Through the Utilization of Artificial Intelligence Tools, 17
1.7 Limitations of DLC Coatings, 18
1.8 Q-Carbon as a Superior Coating Material, 19
1.9 Summary, 22

1.1 Introduction

Carbon materials have attracted immense attention due to their exceptional mechanical strength and unique functionalities [1]. Among carbon materials, diamond-like carbon (DLC) has been extensively examined for its versatile structure and

Applications of Diamond-like Carbon Coatings, First Edition. Abdul Wasy Zia.

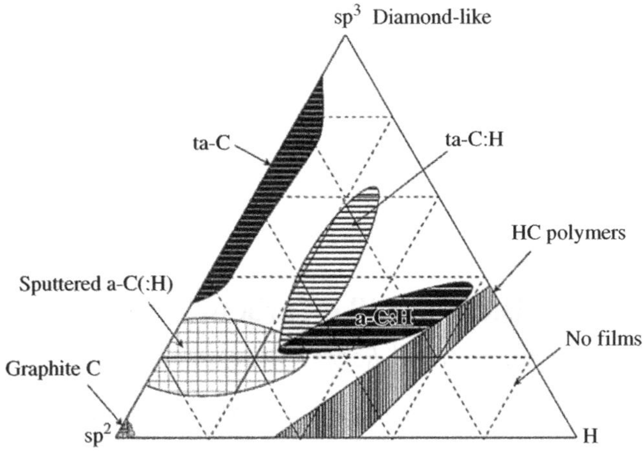

Figure 1.1 Ternary phase diagram that describes various parameters of DLC materials. *Source:* Reproduced from [6] / with permission of ELSEVIER.

multifunctional capabilities [2]. The term DLC is generally utilized to define a type of carbon-based materials containing a mixture of (graphitic) sp^2- and (diamond) sp^3-hybridized phases. DLC can contain 0–50% hydrogen atoms depending on the deposition methods employed [3]. However, the DLC coatings can be formed without hydrogen incorporation. Narayan et al. showed the formation of hydrogen-free DLC films with high sp^3 bonding and superior mechanical properties by pulsed laser deposition (PLD) [4]. The ternary phase diagram of DLC structures, as proposed by Robertson, is shown in Figure 1.1 [5]. Regions designated as a-C, ta-C, and a-C:H refer to pure carbon, tetrahedral amorphous carbon, and hydrogenated carbon with the corresponding extent of hydrogenation, respectively. Due to the variable fractions of sp^2- and sp^3-hybridized states, DLC structures show excellent mechanical strength, resistance to wear, resistance to corrosion, reduced friction coefficient, and unique functional properties [2, 4]. Thus, DLC materials have been a popular choice of coating materials for applications in mechanical and transportation industries, biomedical implants and devices, storage devices, and optical products. Some of the significant properties of different classes of carbon materials are summarized in Table 1.1 [6].

Several approaches have been developed for processing DLC films, such as both chemical vapor deposition (CVD) and physical vapor deposition (PVD) techniques. Among PVD techniques, ion beam deposition, PLD, and magnetron sputtering are widely explored for the deposition of DLC films [7]. Ion beam deposition facilitates the deposition of high-quality coatings at near-room temperature. However, the deposition rate is too low; the substrates require complex manipulation methods for uniform deposition, limiting the practical application of this approach [8]. Sputtering is employed as another popular technique to deposit DLC films with variable sp^2 content. Even so, the low ion energies associated with

Table 1.1 Important physical characteristics of carbon materials currently in use.

	Mass density ($g\ cm^{-3}$)	Hydrogen content (at. %)	Number density ($1\ cm^{-3}$)	Hard/soft
Diamond	3.51	0	0.29	Superhard
Graphite	2.25	0	0.19	Soft
ta-C	3.2	1	0.27	Superhard
a-C:H	2.3	11	0.21	Hard
Polystyrene	1.05	50	0.16	Soft

Source: Reproduced from [6] / with permission of ELSEVIER.

sputter deposition result in the poor mechanical strength of the coatings, making them unsuitable for high-strength applications [9]. On the other hand, PLD yields smooth DLC coatings with high sp^3 content, but it has not yet been developed for the large-scale manufacturing of DLC, and it is often challenging to produce uniform coatings using this technique [10]. Among the CVD techniques, the plasma-enhanced chemical vapor deposition (PECVD) process has been commonly utilized for the growth of DLC films with high uniformity, conformal coverage, and reasonable deposition rates [11]. The PECVD also enables the growth of strain-free and adherent DLC coatings with high levels of mechanical strength and resistance to wear and corrosion [11–13]. Thus, it is highly preferred for the large-scale manufacturing of DLC coatings. A schematic showing the PECVD chamber used in the growth of DLC films is provided in Figure 1.2 [13].

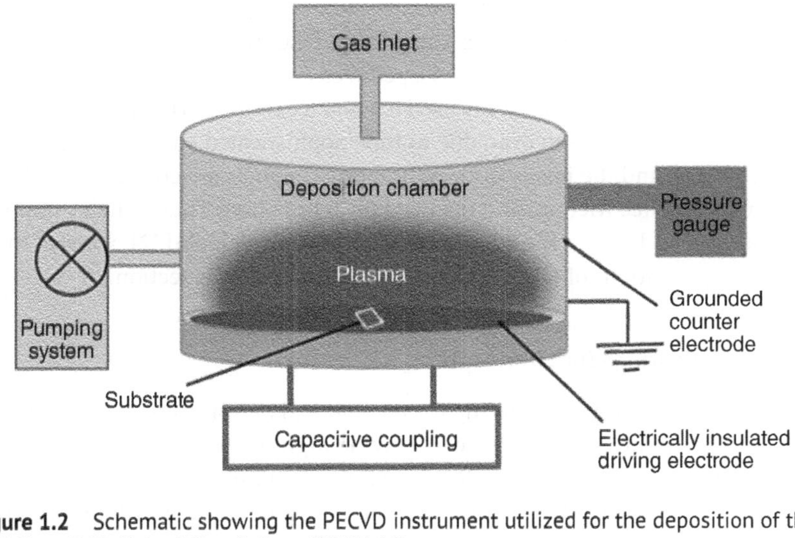

Figure 1.2 Schematic showing the PECVD instrument utilized for the deposition of the DLC films [13] / John Wiley & Sons/CC BY 4.0.

DLC structures are doped with several elements and compounds to improve their functionality [14]. Metals and semiconductors such as Mo, Cu, Cr, Ag, Ti, Ni, Al, W, and Si and compounds such as ZnO and W_2C have been reported to have doped to improve the performance of DLC coatings [13–17]. Among them, silicon (Si) is highly preferred as a dopant material, as it is known to reduce the internal stresses in the coating, enhance the adhesion of the coating to the underlying substrate, and improve the thermal stability of DLC structures [13, 18, 19]. Si-incorporated DLC (Si-DLC) is also shown to have a low friction coefficient, improved resistance to wear, and improved resistance to corrosion [18]. In addition, Yang et al. noted that Si-DLC coatings are biocompatible and inhibit microbial activity, expounding their applications to implantable medical devices [13]. This chapter gives an overview of several doped and undoped DLC coatings and their applications in mechanical and transportation, biomedical devices, electrical devices, and optical products. The limitations of DLC coatings are discussed briefly, and quenched carbon (Q-carbon) is introduced as an emerging, superior alternative for coating materials that are currently in use.

1.2 DLC Coatings for Mechanical and Transportation Applications

The unique combination of high hardness, reduced friction coefficient, resistance to wear, and resistance to corrosion in DLC materials have made them an unambiguous choice for protective coatings in the mechanical and transportation industry [2, 6]. Since their discovery in the 1950s, DLC coatings were first investigated for their unique scratch resistance properties [20]. In 1973, Aisenberg and Chabot reported that the cutting characteristics of paper cutting blades may be significantly improved by coating them with DLC films [21]. The wear tests revealed that the coefficient of friction is reduced on the coated blades. Since then, DLC coatings have been commonly used to enhance the performance of machines and mechanical tools. Moreover, due to their sophisticated structural properties, gears, bearings, and the inside of the automotive engines are coated with DLC to reduce wear, reduce friction, and the need for lubrication. Thus, it helps improve fuel efficiency and reduce the maintenance costs of the vehicle [22]. Some of these structural applications of DLC coatings are discussed in this section.

1.2.1 DLC Coatings for Lubricants

DLC is a popular choice of coating for mechanical and transportation applications, as it is a material that is able to provide high resistance to corrosion, low friction, and high hardness under dry sliding conditions with a coefficient friction as low as 0.01–0.5 based on the environmental conditions the DLC coating is subjected to [5, 23]. For instance, hydrogenated coatings perform

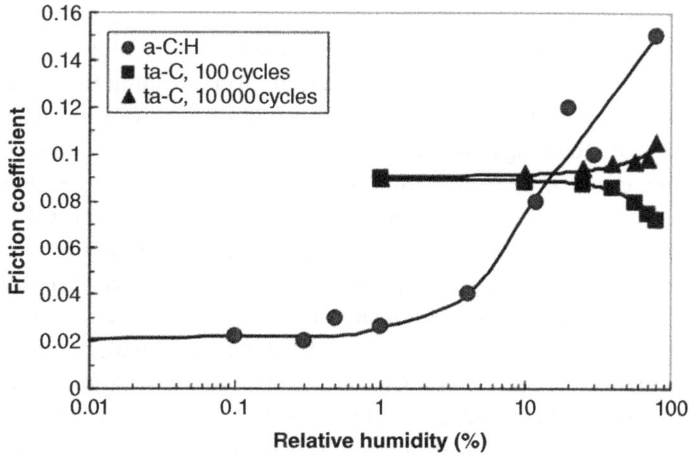

Figure 1.3 Changes in frictional coefficient with relative humidity for a-C:H and ta-C coatings. *Source:* Reproduced from [25] / with permission of IOP Publishing.

better in an inert environment. In contrast, hydrogen-free DLC is preferred in humid environments where a friction coefficient as low as 0.1 can be achieved for an extended period [5, 24]. The variation of the coefficient of friction with relative humidity for hydrogenated and nonhydrogenated DLC coatings is shown in Figure 1.3 [25]. Another class of DLC coatings, which is referred to as near-frictionless carbon, is associated with the lowest possible friction coefficient of any known material (~0.005) due to the passivation of the contact surface achieved by hydrogenated carbon atoms in DLC [25–27]. As such, the hydrogen content and chemical bonding in the coatings have a significant impact on the coefficient of friction. However, fabricating a single type of DLC to achieve low friction in both dry and humid environments is quite challenging. To address this issue, a new class of DLC coatings called diamond-like nanocomposite (DLN) coatings was developed [28, 29]. The structure of DLN materials consists of two amorphous interpenetrating networks with minimal bonding between them. Mutual stabilization within these networks reduces the coating residual stress, improves adhesion, and inhibits graphitic carbon growth at elevated temperatures [25, 28–30]. Furthermore, DLN provides a low friction coefficient and can adapt to dry and humid environments, expounding the mechanical applications of DLC coatings [25, 29].

1.2.2 DLC Coatings for Pipelines, Tubes, and Molds

In addition to a low coefficient of friction, DLC coatings show low wear rates ranging from 10^{-7} to 10^{-9} $mm^3\,Nm^{-1}$ [31]. However, specific test conditions (e.g. temperature, relative humidity, and dopant materials) are known to influence the wear properties of DLC coatings. For instance, the water vapor condensed on the contact zone in humid conditions can inhibit the dry lubricating effect of DLC and prevent the formation of transferred layers, leading to an increase in wear [31, 32]. Similarly, working temperature has a significant impact on the wear rates of DLC coatings. Gharam et al. have shown that the wear rate increases from a value of 0.5×10^{-5} mm^3 Nm^{-1} to a value of 8.5×10^{-5} mm^3 Nm^{-1} from room temperature to 100 °C and then gradually decreases to 400 °C [33]. Thus, several strategies, such as element doping, surface implantation, and surface thermal treatments, have been employed to overcome these challenges. Among them, doping is found to be a promising technique to enhance the mechanical and transportation properties of DLC [34]. DLC structures have been doped with various elements such as N, Mo, Cu, Cr, Ag, Ti, Ni, Al, W, and Si and various compounds such as W_2C and Mo_2C [14–18, 34]. One of the studies showed that co-doping DLC structures with Ti and Ni significantly reduced the internal stresses in the film, improving the durability of the coatings [35]. In another study, Wei et al. have shown that the metal interlayers of Cr, Ti, and Al in DLC led to a significant enhancement in the adhesion of these coatings by stabilizing the sp^2 clusters and thus decreasing the graphitization in the DLC structures [36]. Damasceno et al. observed a similar effect in the incorporation of Si into DLC [18]. Additionally, Kim and Lee's work showed that Si-DLC films resulted in a significant reduction in wear rate and a reduction in friction coefficient, leading to improved tribological properties in the film [37]. Figure 1.4 shows the variation of wear rate in Si-DLC films with varied concentrations of Si [38]. The sample with the highest Si content showed a significant reduction in the wear rate as well as an increase in the elastic modulus of the film. In addition, co-doping N/Si and O/Si in DLC was shown to elevate the thermostability of these structures [39]. Furthermore, Si prefers to create sp^3 bonding within DLC, increasing the sp^3/sp^2 ratio in the films. The increase in the sp^3 content in the films stabilizes the DLC structure, enhancing the strength of these coatings [40]. Thus, Si has been a popular choice of dopant materials to enhance the multifunctional capabilities of DLC coatings for mechanical applications [18, 19, 37–40].

1.3 DLC Coatings for Medical Applications

DLC is one of the oldest and most popular forms of carbon-based materials to be employed for biomedical applications [41]. In the 1990s, Thomson et al. reported that DLC coatings are biocompatible [42]. Since then, they have been widely studied for various biomedical applications due to their versatile

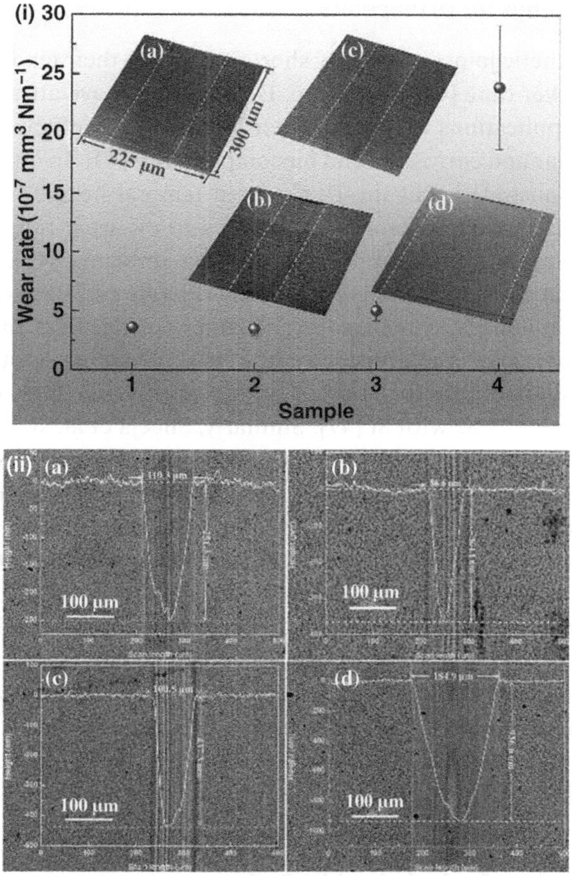

Figure 1.4 (i) 3D and (ii) 2D morphologies and profiles of wear rates for Si-DLC films with varied concentrations of silicon. *Source:* Reproduced from [38] / with permission of ELSEVIER.

properties. In addition to biocompatibility, DLC materials exhibit high resistance to corrosion, improved tribological properties, antimicrobial activity, and anti-inflammatory action, making them an ideal choice for biomedical implants and devices [43]. DLC films can be deposited on several types of substrates at relatively low deposition temperatures; this characteristic makes this class of carbon-based materials technologically and commercially important. Thus, it has been a popular choice of coating for orthopedic, cardiovascular, and dental implants [41–43].

1.3.1 DLC Coatings for Orthopedics

The life of prosthetic joints is severely shortened due to their tendency to wear and corrosion over time [44]. Therefore, DLC coatings have attracted interest in orthopedic applications because of their low friction coefficient, improved resistance to wear and corrosion, and biocompatibility [43]. In one of the studies, it was demonstrated that the DLC-coated femoral head Ti-6Al-4V alloy showed enhanced wear resistance as compared to the wear rate of uncoated stainless steel, alumina, titanium, and zirconia implants [45]. In another study, Tiainen et al. demonstrated that the wear rate of DLC-coated metal-polyethylene hip joints was reduced by 10^6 times smaller than that of metal-polyethylene joints or metal-metal joints [46]. Dowling group reported improved biocompatibility in DLC-coated hip joints with high sp^3 content, achieved by doping DLC with Si [47]. Similarly, Sheeja et al. showed that the corrosion resistance and hardness of Co-Cr-Mo alloy (orthopedic implant material) was significantly increased by coating it with Si-DLC [48]. In another study performed by Mitura and others, DLC-coated orthopedic screws showed no signs of corrosion or inflammation even after 52 weeks [49]. On the contrary, there are also reports on the failure of DLC-coated orthopedic implants. DLC-coated knee joint implants showed excessive wear and spallation when paired with ultrahigh molecular weight polyethylene [50]. In a similar study, Taeger et al. found that the DLC-coated Ti-6Al-V femoral head failed faster than the alumina femoral head, indicating a need for the development of a biomaterial with enhanced functionalities having superior wear, toughness, and adhesion [51].

1.3.2 DLC Coatings for Antimicrobial Performance

Microbial biofilms are ubiquitously increasing the risk of medical device infection [52]. As such, treatment related to medical device infections often requires the removal and replacement of devices, increasing healthcare costs and mortality [52]. DLC coating acts as a simple yet powerful antimicrobial layer by preventing the adhesion of microbes to the surface of the implants [46]. Yang et al. demonstrated that oxygen plasma-treated Si-DLC is hydrophilic, exhibits a lower negative zeta potential value, and showed antifungal behavior against *Candida albicans* (Figure 1.5) [13]. In another study, Robertson and others reported that Ge-doped DLC significantly reduced the formation of *Pseudomonas aeruginosa* biofilms on SS316 substrates [53]. Bociaga et al. opine that the antimicrobial response of Si-DLC is promising but needs further investigation [54]. A recent report on the antibiofilm characteristics of prosthetic meshes coated with several metal-containing DLC thin films showed significant antimicrobial activity against five microbial species. They propose that

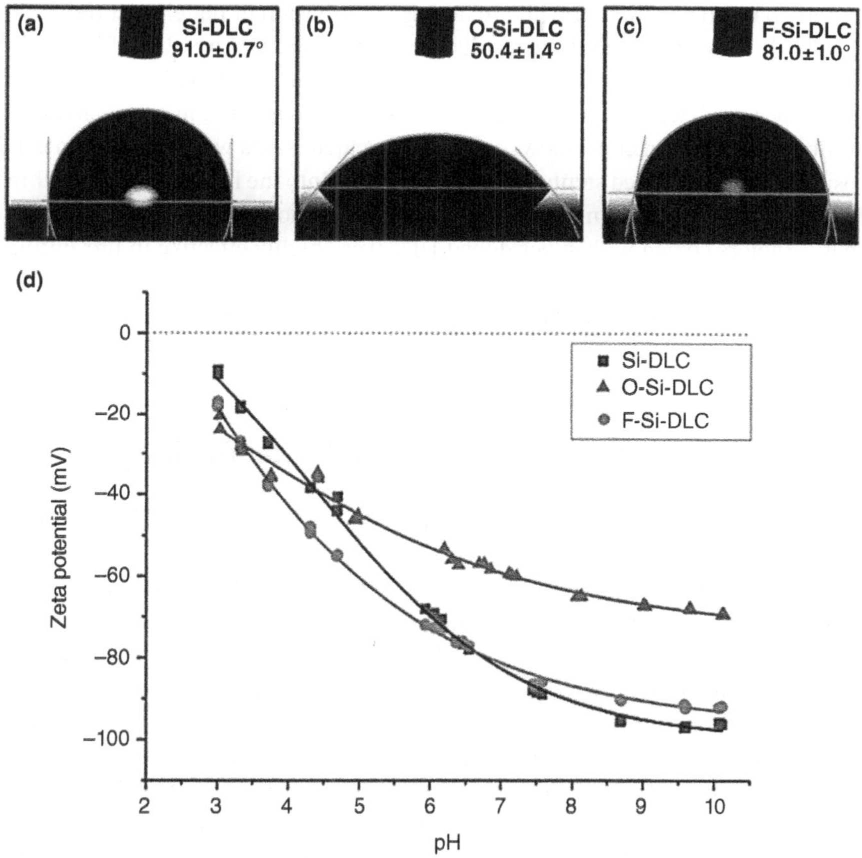

Figure 1.5 The water contact angle values for (a) Si-DLC, (b) oxygen plasma-treated Si-DLC (O-Si-DLC), (c) fluorine-terminated Si-DLC (F-Si-DLC), and (d) zeta potential values of Si-DLC, O-Si-DLC, and F-Si-DLC coatings versus the electrolyte solution pH values [13] / John Wiley & Sons/CC BY 4.0.

modified DLC coatings are a suitable choice of biomedical coatings because of their biocompatibility, mechanical strength, and antimicrobial activity [55].

1.3.3 DLC Coatings for Cardiovascular Implants

One of the primary requirements of a cardiovascular implant is that it should not activate plasma enzymes in order to prevent thrombosis. As such, early studies on DLC-coated heart implants showed a tendency to prevent platelet activation and coagulation of blood [56]. Cheng et al. have reported improved hemocompatibility in Ti-doped DLC due to its hydrophobicity and smooth surface [57]. In another study, researchers have shown that incorporating Si into DLC improves endothelial cell attachment while reducing the adhesion of platelets at the same time [58]. They showed that the results were even better when Si-DLC films were treated with fluorine. All these findings showed that functionalized DLC coatings may be utilized for cardiovascular devices such as stents and heart valves. In addition, Gutensohn et al. showed that DLC-coated stents were resistant to corrosion, while noncoated stents released metal ions into the human plasma within four days of implantation [59]. In another study, Choi and others have shown that catheters coated with silver-doped DLC were highly effective in preventing bacterial infection while preventing the adhesion of platelets [60]. They believe that a greater focus should be placed on employing DLC coatings on cardiovascular devices owing to their multifunctional capabilities.

1.4 DLC Coatings for Electrical and Optical Devices

Due to their unconventional and excellent photoelectronic performance, broadband transparency, and ease of synthesis, DLC materials are among the emerging coating materials for electrical and optical devices [61]. The optical bandgap of DLC materials is governed by the amount and the size of sp^2 graphitic clusters because the sp^2 structure has one π bond [62]. As such, the bandgap value of DLC may be suitably tuned by controlling the sp^2/sp^3 content in the films. Similarly, the electrical conductivity of DLC structures may be increased by several orders of magnitude by increasing the sp^2 content and through the incorporation of metals within the films [61, 63]. Due to this precise manipulation of their optical and electrical functionalities, DLC materials are being evaluated for their applications in devices such as micro-electromechanical systems (MEMS), batteries, field emitters, and antireflective coatings in infrared (IR) devices [61].

1.4.1 DLC Coatings for MEMS

MEMS technology consists of miniaturized devices, including electronic and mechanical components grown through microfabrication techniques [25]. It is widely explored as a feasible option to automate devices for various applications [7, 25]. Owing to their tunable electronic properties and the ability for wafer-scale integration of electric circuits, Si-based materials are a popular choice for the MEMS devices currently in use [25]. However, the fabrication of Si microelectronic systems is not economical and requires state-of-the-art facilities to ensure their proper growth and integration of circuits. Moreover, the poor mechanical strength and low fracture toughness of Si make this process even more challenging [64]. This phenomenon has led to the search for new and efficient material systems for use in MEMS. As such, polymers (e.g. polymethylmethacrylate and polydimethylsiloxane) are being explored due to their suitability [7, 25]. Among the prospective candidates for MEMS devices, diamond and DLC are the most promising due to their unusual mechanical properties and tunable electrical and optical properties that are otherwise not possible in Si-based materials [25]. Due to their very high tensile and fracture strength, Young's modulus, increased resistance to wear, and reduced coefficient of friction, DLC coatings are deemed suitable for high-frequency MEMS applications. In addition, the low coefficient of thermal expansion and very high thermal conductivity of DLC make them suitable candidates for packaging devices and thermal actuators that require efficient heat dissipation [25]. Moreover, the chemical inertness, mechanical strength, biocompatibility, and hydrophobicity of DLC make them a popular choice for coating materials for applications in bio-MEMS, such as biosensing and implantable medical devices [65]. Thus, DLC can be exploited as a promising structural material to provide unique features that are otherwise not possible with the existing materials and also as a coating material in order to enhance the functional properties of MEMS devices [25].

1.4.2 DLC Coatings for Nonvolatile Storage, Low-K Dielectrics, and Field Emitters

The tunable electrical properties achieved by controlling the sp^2-hybridized states in DLC structures have enabled their applications in nonvolatile storage devices [7, 62]. As such, the high density of electronic trap states in DLC that are otherwise detrimental for electronic devices is being explored to create nonvolatile digital information storage devices [66]. Gerstner et al. observed a reduction in resistance as well as an increase in the capacitance of DLC structures after a negative bias was applied. They showed that the effects are reversible after the application of a positive bias [67]. They describe a two-state conduction mechanism where electron hopping and Pool–Frenkel conduction were observed in the high resistance states and low resistance states, respectively. Memory devices with high retention times as high as one year were designed by defining the binary states corresponding to the low and high resistance states. Another promising application of DLC coatings is its potential use as a low-K material that can be used for back end of the line circuits to upgrade their performance [61, 68]. Even though the dielectric constant of DLC materials is significantly higher (5.6) than SiO_2 (4.0) that is currently used, Grill et al. have shown that the dielectric constant of DLC can be modified to lower values (2.7–3.8) than SiO_2 by controlling the deposition conditions [68]. Furthermore, the outstanding mechanical properties, good adhesion, and high thermal stability of DLC coatings make them futuristic candidates for low-K dielectrics [25, 61]. Since the 1990s, DLC

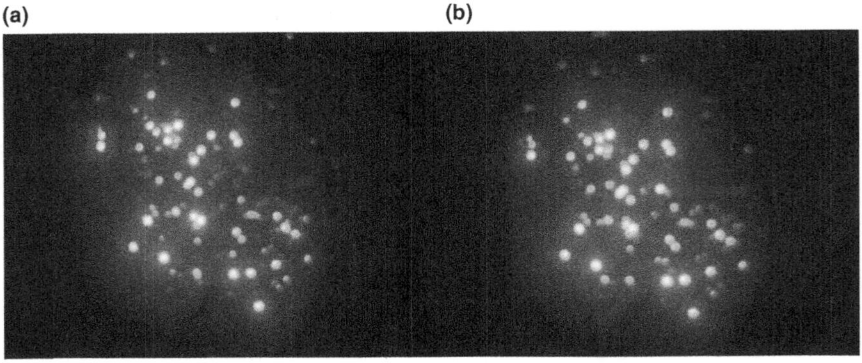

Figure 1.6 The emission images obtained from the patterned DLC coating: (a) initial image and (b) image after three hours. *Source:* Reprinted with permission from Ref. [73].

coatings have been considered for field emitter applications and have been shown to enhance the field emission performance of traditional Spindt-type emission tips [69]. Oh et al. have shown that the nitrogen-incorporated DLC coatings reduced the turn-on voltage by 45% and simultaneously increased the emission current (from a value of 160 μA to a value of 1520 μA) [70]. Furthermore, the functionalization of the DLC-coated tips by plasma etching in hydrogen and oxygen also demonstrated a significant reduction in threshold voltage and an increase in the emission current [71]. More recently, the Narayan Research Group has shown that the field emission properties of Q-carbon (a new allotrope of carbon) outweigh the performance of DLC-coated filed emitters with a turn-on electric field as low as ~2.38 V μm^{-1} and a substantial emission current density of ~33 μA cm^{-2}, which was associated with (a) the sp^2/sp^3 mixture in Q-carbon and (b) a large local field enhancement because of the unique diamond tetrahedral microstructure and the local geometry that is absent in DLC coatings [72]. Figure 1.6 illustrates the temporal evolution of field emission in DLC coatings, revealing a notable decline in both the total emission current and the emitting area over a three-hour period [73].

1.5 DLC Coatings for Optical Applications

DLC is one of the most popular antireflective coatings for IR optics that are made from ZnSe, ZnS, or Ge because of its IR transparency [74]. DLC coatings are also utilized to protect ZnS IR windows when it is used with a GeC intermediate layer [75]. He et al. have demonstrated that DLC protective coatings can enhance durability and also improve the functionalities of organic photoconductors [76].

The antiwear properties and low deposition temperatures of DLC coatings make them suitable for protection against abrasion caused by polycarbonate sunglasses. Thermal imaging systems use DLC coatings as a barrier against deterioration of the surface of Al mirrors [76]. In addition to their uses as protective optical coatings, DLC is also being explored for the fabrication of optical components [61]. Soares et al. have demonstrated the facile patterning of DLC with hard masks, such as SiO_2 and Al, by etching in oxygen plasma [77]. In conjunction with IR transparency, this expands the applications of DLC coatings for the recording of IR diffractive optical components that exhibit excellent pattern quality and precise surface control. DLC coatings are also being explored as candidates for photovoltaic applications. A high optical absorption coefficient obtained by increasing the sp^3 fraction in DLC structures was exploited for absorbing a significant amount of solar energy [78]. As such, a photovoltaic effect was observed in Si-DLC coatings, where the reverse current rose by three orders of magnitude on exposure of the structures to AM1 light (when compared to the undoped DLC coatings) [78].

1.6 Emerging Applications of DLC Coatings

DLC materials have been the number one choice of coatings for cutting tools, automotive parts, and biomedical implants due to their outstanding strength as well as resistance to corrosion. In this regard, efforts involving DLC coatings have mostly focused on enhancing their mechanical properties in the recent past. However, due to their multifunctional capabilities and extended product life, DLC coatings are being explored for various applications such as energy storage, innovative smart textiles, and polymers [6, 7]. In addition, recent advancements in machine learning and materials informatics have revolutionized data-driven materials development and viable manufacturing of technologically and commercially important materials with enhanced performance and durability. This approach has enabled the virtual optimization of DLC coatings to improve their structural and functional properties by saving cost and energy resources [7]. Thus, materials informatics has a strong potential to aid the sustainable production of DLC coatings by uplifting their capabilities [79].

1.6.1 DLC Coatings for Energy Storage Applications

Carbon materials are given significant attention for energy storage applications because of their structural and electrochemical stability, electrical conductivity, and durability [80]. Among them, DLC is being explored for use in Li-ion batteries (LIBs) due to its lithium-ion trapping characteristics and electronic conductivity [81]. The electrical properties of DLC can be suitably tuned by modifying the deposition conditions, post-processing treatments, and concentration of the doping elements.

Due to their superior structural stability, DLC coatings are known to resist the dendrite growth in LIB and provide a three-dimensional conducting environment for Li^+ ions [81]. As such, Cho et al. have demonstrated the increased capacity of up to 50 cycles in DLC-coated cathodes, whereas uncoated cathodes failed after 7 cycles [82]. In another study, Zhang et al. showed that the cyclic efficiency associated with lithium electrodes increased by up to 60% after 50 cycles when coated with DLC [83]. Figure 1.7 presents a comparative analysis of the cyclic performance of as-fabricated and laser-annealed carbon-based electrodes with varying sp^2 content, indicating enhanced capacity in the laser-annealed electrodes [84]. DLC coatings have also been shown to lower the contact resistance of current collectors, enhancing the performance of $LiFePO_4$ (LFP) composite cathodes. Onsurd et al. reported a considerable increase in the rate capability up to 89% in LFP cathodes with an area capacity of 1.7 mAh cm^{-2} upon cycling at 5C [85]. Thus, DLC coatings have shown promising features for battery applications, particularly for LIBs. Even so, there is a need for more statistical data to understand the potential use of DLC coatings for energy storage applications and to ensure consistent and comparable results across different studies [81].

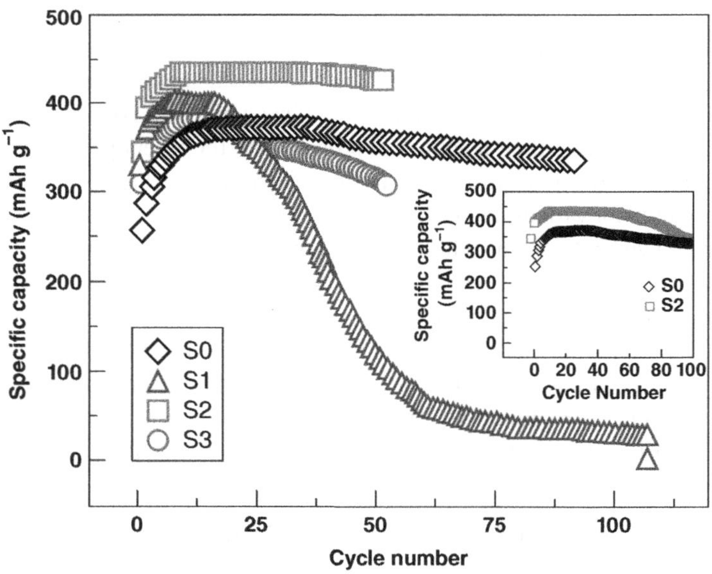

Figure 1.7 Cyclic characteristics of the reference electrode (S0) and laser-annealed carbon electrodes (S2) at the optimized parameters for the anodes in LIB. The results indicate enhanced capacity in the annealed sample when compared to the reference sample in LIB. S1 and S3 are laser-treated samples before optimization. The inset indicates the specific capacity of the optimized laser-annealed samples and reference electrode for 100 cycles. *Source:* Reproduced from [84] / with permission of ELSEVIER.

1.6.2 DLC Coatings for Polymers and Textiles

Smart textiles have revolutionized the field of the textile industry due to their ability to sense and react to external stimuli [86]. Several Kevlar and chromic materials, as well as polymer composites, are used to fabricate smart textiles [86, 87]. Specifically, carbon-based materials are being explored as a popular choice for next-generation smart textiles due to their peculiar properties. Among them, DLC coatings are particularly preferred due to their intrinsic strength, flexibility, and toughness, contributing to innovations in textile industries [88]. The superhydrophobic properties, flame resistance, and antimicrobial properties of DLC materials make them suitable candidates for next-generation smart textiles, self-cleaning windows, and applications that necessitate a reduction of drag in fluid flow [88]. Cortese et al. demonstrated the development of DLC-coated cotton textiles that exhibited superhydrophobic properties and used it as a membrane for filtration applications [89]. Additionally, DLC-coated cotton has been shown to exhibit a significant improvement in flame resistance, with a nearly twofold increase in resistance to flame propagation compared to pristine cotton [90]. Furthermore, Juknius et al. reported the development of DLC/silver nanocomposite thin film deposited on synthetic silk textiles for smart bandage applications [91]. They showed that incorporating DLC into the composite increased the efficiency of Ag^+ ion into aqueous media, enhancing the antimicrobial efficiency of the bandage by up to 50% compared to silver composite without DLC. As such, the emerging applications of DLC materials in polymers and textiles hold immense potential for transforming industries such as healthcare, energy, and consumer goods. As research and development continue to advance, DLC coatings are anticipated to play a vital role in shaping the future of smart materials. Figure 1.8 illustrates the potential smart applications of DLC coatings in textiles [88].

1.6.3 Optimizing DLC Coatings Through the Utilization of Artificial Intelligence Tools

DLC coatings are highly preferred for several applications because of their versatile, multifunctional capabilities. It should be noted that the deposition of high-quality DLC coatings is primarily controlled by the growth kinetics and balancing the energy per unit carbon atom/ion [4]. Moreover, the optimization of DLC production is contingent upon deposition methodology, technological specifications, and system configurations, which can entail over 50 combinations of chamber pressure, bias voltage, gas flow rate, deposition temperature, and other variables, enabling a need for a systematic approach [79]. The field has evolved through four generations: (i) experimental optimization of parameter-to-property relationships; (ii) multiparameter optimization using hybrid experimental methods and statistical approaches;

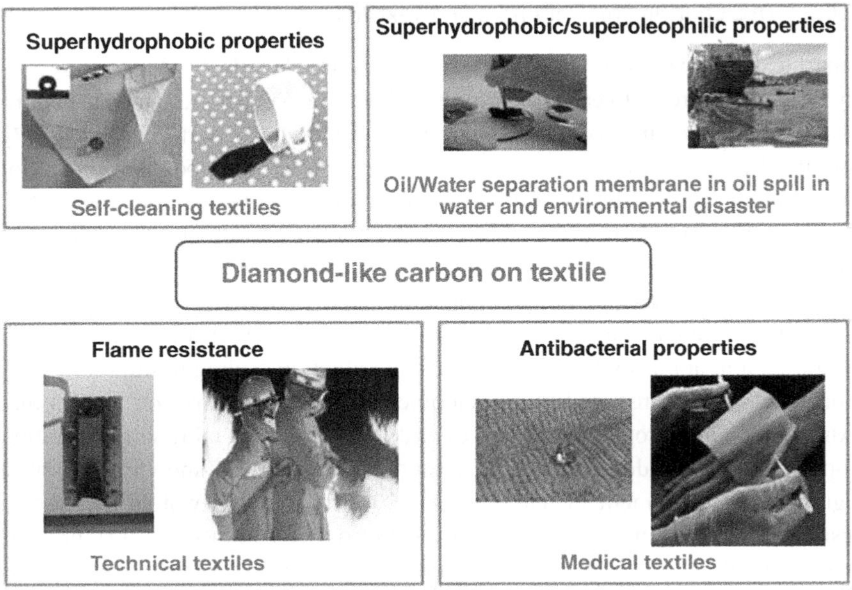

Figure 1.8 Potential innovative applications for DLC-modified textiles [88].

(iii) integration of statistical methods with artificial intelligence (AI) approaches; and (iv) AI-driven multiparameter and multiproperty optimization with predictive capabilities [92]. The current generation is expected to revolutionize DLC development, addressing the challenges of emerging markets and complex environments such as robots, fusion reactors, electric vehicles, and extraterrestrial vehicles. However, the limited data points generated by the experimental studies for developing the DLC modeling techniques are insufficient for training the AI models [79, 92]. Virtual data sets are being created by augmenting sparse experimental data through interpolation and extrapolation to address this issue. This approach necessitates comprehensive experimental studies covering a wide range of parameters to generate a substantial and reliable data set for model verification [79, 93]. A comprehensive data set would enable the development of accurate AI models that can predict DLC properties and behavior with high reliability [93].

1.7 Limitations of DLC Coatings

DLC materials are an unambiguous choice for coating materials due to their exceptional mechanical strength, corrosion resistance, and multifunctional capabilities. Because of their unique combination of structural and functional properties, DLC materials are a popular choice of coatings in various sectors, including biomedical implants, electrical and optical devices, energy storage, next-generation textiles, and smart polymers. However, current fabrication techniques have limitations, including difficulty reproducing precise structures, toxicity, and defects [2, 7]. Furthermore, high-temperature annealing processes and decreased mechanical strength at elevated temperatures reduce the practicality and lifespan of DLC coatings [94]. Therefore, there is a pressing need for a novel coating material that offers a unique combination of economic viability, reproducibility, mechanical strength, chemical inertness, biocompatibility, and enhanced functionalities that surpass the performance of DLC coatings.

1.8 Q-Carbon as a Superior Coating Material

Q-carbon is a newly discovered allotrope of carbon that exhibits unusual structural and functional characteristics [95]. Q-carbon is comprised of randomly packed tetrahedra; it exhibits a very high sp^3/sp^2 ratio at its interface, making it the hardest known material to date [95, 96]. The Raman spectra and the microstructure of Q-carbon nanostructures are provided in Figure 1.9 [96]. Because of its unique structure, Q-carbon exhibits a higher number density of covalently bonded carbon atoms, making it up to 70% higher in hardness than diamond. Thus, it provides ultrahard coatings, which are harder, tougher, and more adherent than diamond and DLC structures [97]. As such, the Narayan Research Group at NC State University, Raleigh, USA, has developed a process for growing Q-carbon/diamond composite protective coatings that can be applied to practical substrates like sapphire, Si, glass, steels, and WC-Co, offering prospective opportunities for superhard coatings [97, 98–102]. In addition to its high hardness and toughness, Joshi et al. demonstrated that Q-carbon is biocompatible with superior antimicrobial properties [102]. They have shown that Si-incorporated Q-carbon inhibited the formation of biofilms up to 80%, making it a promising material for biomedical applications. Q-carbon is conventionally grown by the PLD and subsequent pulsed laser annealing of DLC thin film with a nanosecond pulsed excimer laser source [97, 98]. Recently, we have shown that Q-carbon can be grown with wafer-scale integration on substrates as big as 6" in diameter, enabling the industrial-scale production of Q-carbon with controlled thickness and properties [103].

In addition to its fascinating structural properties, Q-carbon demonstrates robust room-temperature ferromagnetism; it exhibits magnetization as high as 20 emu g^{-1} and Tc of ~757 K, making it an excellent candidate for targeted drug delivery applications [104]. When compared to DLC coatings, Q-carbon shows superior field emission potential due to its unique sp^2/sp^3 mixture in the emitter, as well as the presence of a semimetallic alpha-carbon matrix in the material [72].

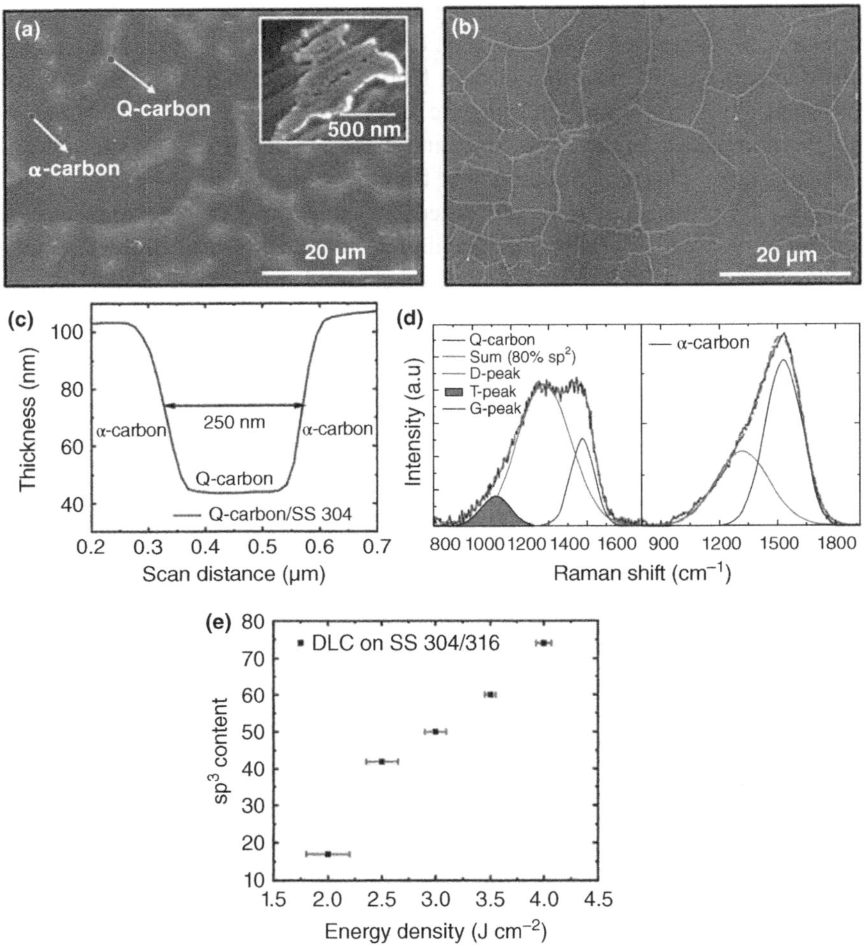

Figure 1.9 (a) Scanning electron microscope (SEM) image that shows the formation of Q-carbon on 304 SS; inset shows a high-resolution image of Q-carbon in alpha-matrix, (b) SEM image of Q-carbon, (c) surface roughness data for Q-carbon, (d) Raman spectroscopy data for Q- and alpha-carbon, and (e) relationship between the sp^3 content of the DLC and the laser energy density utilized during deposition. *Source:* Reproduced from [96] / with permission of ELSEVIER.

Q-carbon-based field emitters display exceptional performance, featuring (i) hysteresis-free emission, (ii) high emission stability, (iii) high current density operation without defect generation, and (iv) long-term reliability over successive emission measurements [72]. Unlike DLC, Q-carbon is highly resistant to radiation damage, withstanding ion radiation that induces extreme atomic displacements and electronic excitations [72, 105]. The Q-carbon anodes have shown

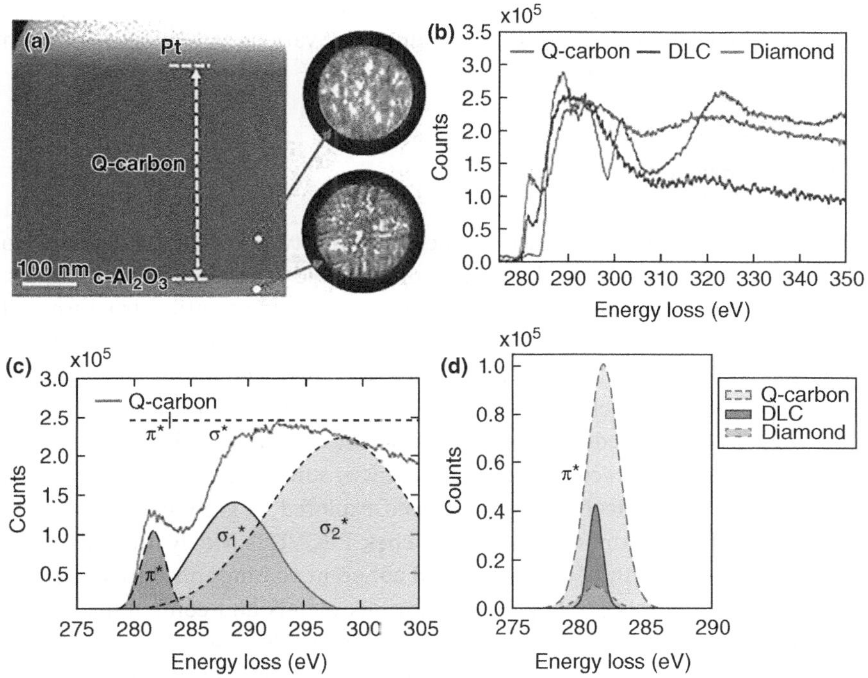

Figure 1.10 (a) High-angle annular dark-field image of Q-carbon on c-sapphire with the Ronchigram pattern, which demonstrates the amorphous and crystalline nature of Q-carbon and c-sapphire, respectively, (b) EELS results from Q-carbon, as-deposited DLC and diamond, (c) representative fitting routine obtained from Q-carbon EEL spectrum, and (d) the fitted π^* peaks that show an increase in the density of the electronic states in Q-carbon when compared to DLC and diamond. *Source:* Reproduced from [104] / with permission of American Chemical Society.

superior power, performance, and cyclability in lithium as well as sodium-ion batteries. This characteristic is attributed to the higher trapping efficiency of Li and Na ions, increasing from 1/6 in graphite to 1/2 in Q-carbon with its unique diamond tetrahedral structure [84, 106]. Figure 1.10 presents a comparative analysis of the electron energy loss spectra (EELS) of diamond, DLC, and Q-carbon, revealing an unexpected enhancement in the π^* region (EELS) of Q-carbon [104]. It is attributed to an increase in unpaired spin electron density, which is 80% higher than in DLC. This phenomenon is associated with the remarkable magnetic and optoelectronic properties of Q-carbon, which are absent in diamond and DLC. The unique electronic structure of Q-carbon is responsible for its fascinating characteristics, making it an ideal substitute for DLC coatings. Thus, Q-carbon is a versatile material with extraordinary properties, finding practical applications in nearly every technological domain. With its ability to overcome the limitations of existing carbon materials, we envision Q-carbon as a superior alternative to DLC coatings, poised to revolutionize various technological fields in a wide range of applications.

1.9 Summary

Carbon materials have gained prominence in the twenty-first century due to their outstanding mechanical strength and tunable functional characteristics. DLC is a type of carbon-based material known for its versatile structure and multifunctional capabilities. DLC materials contain a variable mixture of sp^2- and sp^3-hybridized C–C bonds, with or without hydrogen atoms embedded in their structure. As such, the variable structure of DLC provides excellent mechanical strength, reduced friction coefficient, biocompatibility, resistance to wear, and resistance to corrosion, making them suitable for protective coating applications. Several techniques have been explored to grow conformal DLC coatings at low cost. Among these approaches, PECVD is widely utilized for the industrial-scale growth of strain-free and adherent coatings on intricate geometries. DLC structures are doped with various metals and compounds to add new functionalities and improve their performance. Si is widely used to dope with DLC structures, as it enhances adhesion, reduces internal stresses, and improves thermal stability in the coatings. This chapter describes DLC coatings and their applications in mechanical and automotive industries, biomedical devices, MEMS, dielectrics, and optical devices. Some emerging applications in energy storage, polymers, and smart textiles are also explored. In addition, the limitations of DLC coatings are discussed, and Q-carbon is introduced as a superior alternative to DLC materials, thereby providing an overview of the current coating technology. The Q-carbon coatings with enhanced hardness, toughness, and adhesion stand to revolutionize the field of high-performance coatings and take this field to the next level.

References

1 Li, Z., Wang, L., Li, Y. et al. (2019). Carbon-based functional nanomaterials: preparation, properties, and applications. *Composites Science and Technology* 179: 10–40.

2 Wei, Q. and Narayan, J. (2000). Superhard diamond-like carbon: preparation, theory, and properties. *International Materials Reviews* 45 (4): 133–164.

3 Robertson, J. (2008). Classification of diamond-like carbons. In: *Tribology of Diamond-like Carbon Films: Fundamentals and Applications*, 13–24.

4 Krishnaswamy, J., Rengan, A., Narayan, J. et al. (1989). Thin-film deposition by a new laser ablation and plasma hybrid technique. *Applied physics Letters* 54 (24): 2455–2457.

5 Robertson, J. (2002). Diamond-like amorphous carbon. *Materials Science and Engineering: R: Reports* 37 (4–6): 129–281.

6 Bewilogua, K. and Hofmann, D. (2014). History of diamond-like carbon films—from first experiments to worldwide applications. *Surface and Coatings Technology* 242 (1–2): 214–225.

7 Zia, A.W. and Birkett, M. (2021). Deposition of diamond-like carbon coatings: conventional to non-conventional approaches for emerging markets. *Ceramics International* 47 (20): 28075–28085.

8 Kimock, F.M. and Knapp, B.J. (1993). Commercial applications of ion beam deposited diamond-like carbon (DLC) coatings. *Surface and Coatings Technology* 56 (3): 273–279.

9 Chowdhury, S., Laugier, M.T., and Rahman, I.Z. (2004). Characterization of DLC coatings deposited by rf magnetron sputtering. *Journal of Materials Processing Technology* 153: 804–810.

10 Voevodin, A.A. and Donley, M.S. (1996). Preparation of amorphous diamond-like carbon by pulsed laser deposition: a critical review. *Surface and Coatings Technology* 82 (3): 199–213.

11 Lusk, D., Gore, M., Boardman, W. et al. (2008). Thick DLC films deposited by PECVD on the internal surface of cylindrical substrates. *Diamond and Related Materials* 17 (7–10): 1613–1621.

12 Lin, Z., Lv, S.B., Yu, Z.J. et al. (2008). Effect of bias voltage on diamond-like carbon film deposited on PMMA substrate. *Surface and Coatings Technology* 202 (22–23): 5386–5389.

13 Yang, K.H., Riley, P., Rodenhausen, K.B. et al. (2022). Antifungal behavior of silicon-incorporated diamond-like carbon by tuning surface hydrophobicity with plasma treatment. *International Journal of Applied Ceramic Technology* 19 (5): 2545–2555.

14 Oral, B., Hauert, R., Müller, U. et al. (1995). Structural changes in doped aC:H films during annealing. *Diamond and related materials* 4 (4): 482–487.

15 Sun, L., Zuo, X., Guo, P. et al. (2019). Role of deposition temperature on the mechanical and tribological properties of Cu and Cr co-doped diamond-like carbon films. *Thin Solid Films* 678: 16–25.

16 Foong, Y.M., Koh, A.T.T., Lim, S.R. et al. (2012). Materials properties of ZnO/ diamond-like carbon (DLC) nanocomposite fabricated with different source of targets. *Diamond and related materials* 25: 103–110.

17 Yong, Q., Ma, G., Wang, H. et al. (2016). Influence of tungsten content on microstructure and properties of tungsten-doped graphite-like carbon films. *Journal of Materials Research* 31 (23): 3766–3776.

18 Damasceno, J.C., Camargo Jr, S.S., Freire Jr, F.L. et al. (2000). Deposition of Si-DLC films with high hardness, low stress and high deposition rates. *Surface and Coatings Technology* 133: 247–252.

19 Wu, W.J. and Hon, M.H. (1999). Thermal stability of diamond-like carbon films with added silicon. *Surface and Coatings Technology* 111 (2–3): 134–140.

20 Schmellenmeier, H.Z. (1956). Carbon layers with diamond structure. *Physical Chemistry* 205: 349–360.

21 Aisenberg, S. and Chabot, R. (1971). Ion-beam deposition of thin films of diamond-like carbon. *Journal of Applied Physics* 42 (7): 2953–2958.

22 Kolawole, F.O., Kolade, O.S., Bello, S.A. et al. (2023). The improvement of diamond-like carbon coatings for tribological and tribo-corrosion applications in automobile engines: an updated review study. *The International Journal of Advanced Manufacturing Technology* 126 (5): 2295–2322.

23 Rha, J.J., Kwon, S.C., Cho, J.R. et al. (2005). Creation of ultra-low friction and wear surfaces for micro-devices using carbon films. *Wear* 259 (1–6): 765–770.

24 Voevodin, A.A., Phelps, A.W., Zabinski, J.S. et al. (1996). Friction induced phase transformation of pulsed laser deposited diamond-like carbon. *Diamond and Related Materials* 5 (11): 1264–1269.

25 Luo, J.K., Fu, Y.Q., Le, H.R. et al. (2007). Diamond and diamond-like carbon MEMS. *Journal of Micromechanics and Microengineering* 17 (7): S147.

26 Ye, J., Okamoto, Y., and Yasuda, Y. (2008). Direct insight into near-frictionless behavior displayed by diamond-like carbon coatings in lubricants. *Tribology Letters* 29: 53–56.

27 Yu, G., Gong, Z., Jiang, B. et al. (2020). Superlubricity for hydrogenated diamond like carbon induced by thin MoS_2 and DLC layer in moist air. *Diamond and Related Materials* 102: 107668.

28 Neerinck, D.P.M.A.C.D.C.P., Persoone, P., Sercu, M. et al. (1998). Diamond-like nanocomposite coatings for low-wear and low-friction applications in humid environments. *Thin Solid Films* 317 (1–2): 402–404.

29 Neerinck, D., Persoone, P., Sercu, M. et al. (1998). Diamond-like nanocomposite coatings (aC: H/a-Si: O) for tribological applications. *Diamond and Related Materials* 7 (2–5): 468–471.

30 Yang, W.J., Choa, Y.H., Sekino, T. et al. (2003). Thermal stability evaluation of diamond-like nanocomposite coatings. *Thin Solid Films* 434 (1–2): 49–54.

31 Palokunnas, P. (2018). Diamond-like carbon coating in tribological applications. Bachelor's thesis, Tampere University of Technology, Finland.

32 Scharf, T.W. and Prasad, S.V. (2013). Solid lubricants: A review. *Journal of Materials Science* 48: 511–531.

33 Abou Gharam, A., Lukitsch, M.J., Balogh, M.P. et al. (2011). High temperature tribological behavior of W-DLC against aluminum. *Surface and Coatings Technology* 206 (7): 1905–1912.

34 Ghadai, R.K., Das, S., Kumar, D. et al. (2018). Correlation between structural and mechanical properties of silicon doped DLC thin films. *Diamond and Related Materials* 82: 25–32.

35 Zhang, S., Wu, S., Huang, T. et al. (2023). The effects of Ti/Ni doping on the friction and wear properties of DLC coatings. *Coatings* 13 (10): 1743.

36 Wei, C. and Yen, J.Y. (2007). Effect of film thickness and interlayer on the adhesion strength of diamond like carbon films on different substrates. *Diamond and Related Materials* 16 (4–7): 1325–1330.

37 Kim, H.G., Ahn, S.H., Kim, J.G. et al. (2005). Effect of Si-incorporation on wear–corrosion properties of diamond-like carbon films. *Thin Solid Films* 482 (1–2): 299–304.

38 Guo, C.Q., Lin, S.S., Gao, D. et al. (2020). Modulation of Si on microstructure and tribo-mechanical properties of hydrogen-free DLC films prepared by magnetron sputtering. *Applied Surface Science* 509: 145381.

39 Jongwannasiri, C., Moolsradoo, N., and Watanabe, S. (2015). Heat resistant properties of some elements-incorporated diamond-like carbon films and their trial applications for micro end mill coatings. *Materials Sciences and Applications* 6 (1): 9.

40 Guo, C.T. (2008). Diamond-like carbon films deposited on polycarbonates by plasma-enhanced chemical vapor deposition. *Thin Solid Films* 516 (12): 4053–4058.

41 Dearnaley, G. and Arps, J.H. (2005). Biomedical applications of diamond-like carbon (DLC) coatings: a review. *Surface and Coatings Technology* 200 (7): 2518–2524.

42 Thomson, L.A., Law, F.C., Rushton, N. et al. (1991). Biocompatibility of diamond-like carbon coating. *Biomaterials* 12 (1): 37–40.

43 Roy, R.K. and Lee, K.R. (2007). Biomedical applications of diamond-like carbon coatings: a review. *Journal of Biomedical Materials Research Part B: Applied Biomaterials: An Official Journal of the Society for Biomaterials, The Japanese Society for Biomaterials, and The Australian Society for Biomaterials and the Korean Society for Biomaterials* 83 (1): 72–84.

44 Arnholt, C.M., MacDonald, D.W., Malkani, A.L. et al. (2016). Corrosion damage and wear mechanisms in long-term retrieved CoCr femoral components for total knee arthroplasty. *Journal of Arthroplasty* 31(12): 2900–2906. doi: 10.1016/j.arth.2016.05.006.

45 Kim, D.H., Kim, H.E., Lee, K.R. et al. (2002). Characterization of diamond-like carbon films deposited on commercially pure Ti and Ti–6Al–4V. *Materials Science and Engineering: C* 22 (1): 9–14.

46 Tiainen, V.M. (2001). Amorphous carbon as a bio-mechanical coating—mechanical properties and biological applications. *Diamond and Related Materials* 10 (2): 153–160.

47 Dowling, D.P., Kola, P.V., Donnelly, K. et al. (1997). Evaluation of diamond-like carbon-coated orthopaedic implants. *Diamond and Related Materials* 6 (2–4): 390–393.

48 Sheeja, D., Tay, B.K., Lau, S.P. et al. (2001). Tribological characterisation of diamond-like carbon coatings on Co–Cr–Mo alloy for orthopaedic applications. *Surface and Coatings Technology* 146 (2–3): 410–416.

49 Mitura, E., Mitura, S., Niedzielski, P. et al. (1994). Diamond-like carbon coatings for biomedical applications. *Diamond and Related Materials* 3 (4–6): 896–898.

50 Escudeiro, A., Wimmer, M.A., Polcar, T. et al. (2015). Tribological behavior of uncoated and DLC-coated CoCr and Ti-alloys in contact with UHMWPE and PEEK counterbodies. *Tribology International* 89: 97–104.

51 Taeger, G., Podleska, L.E., Schmidt, B. et al. (2003). Comparison of diamond-like-carbon and alumina-oxide articulating with polyethylene in total hip arthroplasty. *Materialwissenschaft und Werkstofftechnik: Entwicklung, Fertigung, Prüfung, Eigenschaften und Anwendungen technischer Werkstoffe* 34 (12): 1094–1100.

52 Habash, M. and Reid, G. (1999). Microbial biofilms: their development and significance for medical device—related infections. *The Journal of Clinical Pharmacology* 39 (9): 887–898.

53 Robertson, S.N., Gibson, D., MacKay, W.G. et al. (2017). Investigation of the antimicrobial properties of modified multilayer diamond-like carbon coatings on 316 stainless steel. *Surface and Coatings Technology* 314: 72–78.

54 Bociaga, D., Sobczyk-Guzenda, A., Szymanski, W. et al. (2017). Mechanical properties, chemical analysis and evaluation of antimicrobial response of Si-DLC coatings fabricated on AISI 316 LVM substrate by a multi-target DC-RF magnetron sputtering method for potential biomedical applications. *Applied Surface Science* 417: 23–33.

55 Cazalini, E.M., Miyakawa, W., Teodoro, G.R. et al. (2017). Antimicrobial and anti-biofilm properties of polypropylene meshes coated with metal-containing DLC thin films. *Journal of Materials Science: Materials in Medicine* 28 (6): 1–10.

56 Nurdin, N., Francois, P., Moret, M. et al. (2003). Hemocompatible diamond-like carbon (DLC) surfaces. *European Cells & Materials* 5: 17–28.

57 Cheng, H.C., Chiou, S.Y., Liu, C.M. et al. (2009). Effect of plasma energy on enhancing biocompatibility and hemocompatibility of diamond-like carbon film with various titanium concentrations. *Journal of Alloys and Compounds* 477 (1–2): 931–935.

58 Ogwu, A.A., Okpalugo, T.I.T., Ali, N. et al. (2008). Endothelial cell growth on silicon modified hydrogenated amorphous carbon thin films. *Journal of Biomedical Materials Research Part B: Applied Biomaterials: An Official Journal of the Society for Biomaterials, The Japanese Society for Biomaterials, and The Australian Society for Biomaterials and the Korean Society for Biomaterials* 85 (1): 105–113.

59 Gutensohn, K., Beythien, C., Bau, J. et al. (2000). In vitro analyses of diamond-like carbon coated stents: reduction of metal ion release, platelet activation, and thrombogenicity. *Thrombosis Research* 99 (6): 577–585.

60 Choi, H.W., Dauskardt, R.H., Lee, S.C. et al. 2008. Characteristic of silver doped DLC films on surface properties and protein adsorption. *Diamond and Related Materials* 17 (3): 252–257.

61 Grill, A. (1999). Electrical and optical properties of diamond-like carbon. *Thin Solid Films* 355: 189–193.

62 Kaplan, S., Jansen, F., and Machonkin, M. (1985). Characterization of amorphous carbon-hydrogen films by solid-state nuclear magnetic resonance. *Applied Physics Letters* 47 (7): 750–753.

63 Khan, M.I., Adil, F., Majeed, S. et al. (2019). Structural, morphological, electrical and optical properties of Cu doped DLC thin films. *Materials Research Express* 6 (12): 126420.

64 Kahn, H., Tayebi, N., Ballarini, R. et al. (2000). Fracture toughness of polysilicon MEMS devices. *Sensors and Actuators A: Physical* 82 (1–3): 274–280.

65 Carlisle, J. A. (2004). Precious biosensors. *Nature Materials* 3 (10): 668–669.

66 Ray, S.C., Okpalugo, T.I.T., Pao, C.W. et al. (2005). Electronic structure and photoluminescence study of silicon doped diamond like carbon (Si: DLC) thin films. *Materials Research Bulletin* 40 (10): 1757–1764.

67 Gerstner, E.G. and McKenzie, D.R. (1998). Fabrication and characterization of novel electronic devices using tetrahedral amorphous carbon. *Diamond and Related Materials* 7 (8): 1172–1177.

68 Grill, A., Patel, V., Saenger, K.L. et al. (1996). Diamond-like carbon materials as low-k dielectrics for multilevel interconnects in ULSI. *MRS Online Proceedings Library (OPL)* 443: 155.

69 Amaratunga, G.A. and Silva, S.R.P. (1996). Nitrogen containing hydrogenated amorphous carbon for thin-film field emission cathodes. *Applied Physics Letters* 68 (18): 2529–2531.

70 Oh, M.H. (1997). Micromachined field emitter arrays for application to FED. *Smart Electronics and MEMS* 3242: 220–228.

71 Ding, M.Q., Myers, A.F., Choi, W.B. et al. (1997). Field emission from amorphous diamond coated Mo tip emitters by pulsed laser deposition. *Journal of Vacuum Science & Technology B: Microelectronics and Nanometer Structures Processing, Measurement, and Phenomena* 15 (4): 840–844.

72 Narayan, J., Bhaumik, A., Gupta, S. et al. (2018). Progress in Q-carbon and related materials with extraordinary properties. *Materials Research Letters* 6 (7): 353–364.

73 Paik, N. (2005). Field emission characteristics of diamond-like carbon (DLC) films prepared using a magnetron sputter-type negative ion source (MSNIS). *Diamond and Related Materials* 14 (9): 1556–1561.

74 Lettington, A.M. (1987). Amorphous hydrogenated carbon films. In: *Proceedings of the European Materials Research Society Symposium*, 359. Paris: Les Editions de Physique.

75 Lettington, A.H. (1998). Applications of diamond-like carbon thin films. *Carbon* 36 (5–6): 555–560.

76 He, X.M., Lee, S.T., Chan, Y.C. et al. (1998). Effect of negative rf bias on electrophotographic properties of hard diamond-like carbon films deposited on organic photoconductors. *Journal of Physics: Condensed Matter* 10 (35): 7835.

77 Soares, L.L., Lima, C.R., Cescato, L. et al. (1998). Recording of relief structures in amorphous hydrogenated carbon (aC: H) films for infrared diffractive optics. *Journal of Modern Optics* 45 (7): 1479–1486.

78 Cheah, L.K., Shi, X., Liu, E. et al. (1998). Nitrogenated tetrahedral amorphous carbon films prepared by ion-beam-assisted filtered cathodic vacuum arc technique for solar cells application. *Applied Physics Letters* 73 (17): 2473–2475.

79 Zia, A.W., Hussain, S.A., and Baig, M.M.F.A. (2022). Optimizing diamond-like carbon coatings-from experimental era to artificial intelligence. *Ceramics International* 48 (24): 36000–36011.

80 Zhai, Y., Dou, Y., Zhao, D. et al. (2011). Carbon materials for chemical capacitive energy storage. *Advanced Materials* 23 (42): 4828–4850.

81 Zia, A.W., Hussain, S.A., Rasul, S. et al. (2023). Progress in diamond-like carbon coatings for lithium-based batteries. *Journal of Energy Storage* 72: 108803.

82 Cho, S.M., Lee, J.K., and Yoon, W.Y. (2015). Electrochemical effects of silicon/diamond-like carbon layered composite on oxygen cathodes in lithium–oxygen batteries. *Electrochimica Acta* 158: 246–252.

83 Zhang, Y.J., Liu, X.Y., Bai, W.Q. et al. (2014). Magnetron sputtering amorphous carbon coatings on metallic lithium: towards promising anodes for lithium secondary batteries. *Journal of Power Sources* 266: 43–50.

84 Khosla, N., Narayan, J., Narayan, R. et al. (2023). Microstructure and defect engineering of graphite anodes by pulsed laser annealing for enhanced performance of lithium-ion batteries. *Carbon* 205: 214–225.

85 Onsrud, M., Tezel, A.O., Fotedar, S. et al. (2022). Novel carbon coating on aluminum current collectors for lithium-ion batteries. *SN Applied Sciences* 4 (8): 225.

86 Cherenack, K. and Van Pieterson, L. (2012). Smart textiles: challenges and opportunities. *Journal of Applied Physics* 112 (9): 091301.

87 Wang, S., Xuan, S., Liu, M. et al. (2017). Smart wearable Kevlar-based safeguarding electronic textile with excellent sensing performance. *Soft Matter* 13 (13): 2483–2491.

88 Caschera, D., Toro, R.G., Cortese, B. et al. (2019). Diamond-like carbon: a versatile material for developing innovative smart textiles applications. A short review. *Atti della Accademia Peloritana dei Pericolanti-Classe di Scienze Fisiche, Matematiche e Naturali* 97 (S2): 27.

89 Cortese, B., D'Amone, S., Manca, M. et al. (2008). Superhydrophobicity due to the hierarchical scale roughness of PDMS surfaces. *Langmuir* 24 (6): 2712–2718.

90 Caschera, D., Toro, R.G., Federici, F. et al. (2015). Flame retardant properties of plasma pre-treated/diamond-like carbon (DLC) coated cotton fabrics. *Cellulose* 22 (4): 2797–2809.

91 Juknius, T., Ružauskas, M., Tamulevičius, T. et al. (2016). Antimicrobial properties of diamond-like carbon/silver nanocomposite thin films deposited on textiles: towards smart bandages. *Materials* 9 (5): 371.

92 Casado, R. and Younas, M. (2015). Emerging trends and technologies in big data processing. *Concurrency and Computation: Practice and Experience* 27 (8): 2078–2091.

93 Jose, N.A., Kovalev, M., Bradford, E. et al. (2020). An accelerated approach for efficient development and scaling of new material technologies, combining flow synthesis with machine learning. Case study: Nanostructured ZnO for antibacterial coatings. In: *2020 AIChE Annual Meeting*. American Institute of Chemical Engineers.

94 Knight, J.C., Page, T.F., and Chandler, H.W. (1991). Thermal instability of the microstructure and surface mechanical properties of hydrogenated amorphous carbon films. *Surface and Coatings Technology* 49 (1–3): 519–529.

95 Narayan, J. and Bhaumik, A. (2015). Novel phase of carbon, ferromagnetism, and conversion into diamond. *Journal of Applied Physics* 118 (21): 215303.

96 Joshi, P., Gupta, S., Haque, A. et al. (2020). Fabrication of ultrahard Q-carbon nanocoatings on AISI 304 and 316 stainless steels and subsequent formation of high-quality diamond films. *Diamond and Related Materials* 104 (239): 107742.

97 Sachan, R., Gupta, S., and Narayan, J. (2019). Nonequilibrium structural evolution of Q-carbon and interfaces. *ACS Applied Materials & Interfaces* 12 (1): 1330–1338.

98 Gupta, S., Bhaumik, A., Sachan, R. et al. (2018). Structural evolution of Q-carbon and nanodiamonds. *JOM* 70: 450–455.

99 Gupta, S., Sachan, R., Bhaumik, A. et al. (2018). Enhanced mechanical properties of Q-carbon nanocomposites by nanosecond pulsed laser annealing. *Nanotechnology* 29 (45): 45LT02.

100 Narayan, J., Gupta, S., Bhaumik, A. et al. (2018). Q-carbon harder than diamond. *MRS Communications* 8 (2): 428–436.

101 Haque, A., Pant, P., and Narayan, J. (2018). Large-area diamond thin film on Q-carbon coated crystalline sapphire by HFCVD. *Journal of Crystal Growth* 504: 17–25.

102 Joshi, N., Shukla, S., Khosla, N. et al. (2024). Q-carbon as an emergent surface coating material for antimicrobial applications. *Thin Solid Films* 791 (9): 140227.

103 Riley, P.R., Joshi, P., Khosla, N. et al. (2022). Formation of Q-carbon with wafer scale integration. *Carbon* 196: 972–978.

104 Bhaumik, A., Nori, S., Sachan, R. et al. (2018). Room-temperature ferromagnetism and extraordinary hall effect in nanostructured Q-carbon: Implications for potential spintronic devices. *ACS Applied Nano Materials* 1 (2): 807–819.

105 Haque, A., Karmakar, S., Trivedi, R.K. et al. (2023). Electric-field emission mechanism in Q-carbon field emitters. *ACS Omega* 8 (10): 9307–9318.

106 Khosla, N. (2023). *Laser Processing of Q-Carbon, Diamond, and Anode and Cathode for Lithium-ion Batteries*. Raleigh, NC: North Carolina State University.

2

Deposition Methods of Diamond-like Carbon Coatings

Pranay Kalakonda[1], Jagdish Narayan[1], Roger J. Narayan[1,2]

[1] *Department of Materials Science and Engineering, Centennial Campus, North Carolina State University, Raleigh, NC, USA*
[2] *Joint Department of Biomedical Engineering, North Carolina State University and UNC Chapel Hill, Raleigh, NC, USA*

CONTENTS

2.1 Introduction, 31
2.2 Classification of Carbon Coatings, 34
2.3 Deposition Methods, 36
 2.3.1 Physical Vapor Deposition, 37
  2.3.1.1 Ion Beam Deposition, 37
  2.3.1.2 Mass Selected Ion Beam Deposition, 38
  2.3.1.3 Cathodic Arc Deposition, 39
  2.3.1.4 Sputtering, 42
  2.3.1.5 Ion Beam-assisted Deposition, 43
  2.3.1.6 Pulsed Laser Deposition, 44
 2.3.2 Chemical Vapor Deposition, 48
2.4 Deposition Mechanisms, 50
2.5 Influence of Deposition Parameters, 53
2.6 Properties and Analysis Methods, 54
2.7 Applications, 56
2.8 Conclusions, 57

2.1 Introduction

Diamond-like carbon (DLC), as the name suggests, is similar to diamond due to the fact that a significant portion of the bonds contained within the material are sp^3 hybridized in nature. It can be described broadly as a metastable state of amorphous carbon (a-C). Carbon films, in general, were first reported in 1953 by Schmellenmeier, which were fabricated using glow discharge plasma in the presence of acetylene gas [1]. This showed a substantial diamond-like composition,

Applications of Diamond-like Carbon Coatings, First Edition. Abdul Wasy Zia.

which explains the properties of higher hardness and wear resistance. These results were later corroborated by the findings of Aisenberg and Chabot [2] in 1971, and the term DLC was initially introduced. The insulating carbon films were deposited on the room temperature substrates by developing an ion beam deposition (IBD) technique, which yielded a partially crystalline film with comparable characteristics to diamond.

Multiple allotropes, such as diamond, graphite, and Buckminsterfullerene (C60), can exist with distinct atomic structures and entropy, as carbon can take on different bonding configurations like sp^3, sp^2, and sp hybridizations (Figure 2.1) [3]. This makes the formation of multiple disordered carbon structures possible. In the sp^3 bonding configuration, only the strong, directional σ-bonds are formed, as one s-orbital and three p-orbitals hybridize to form four sp^3-hybridized orbitals arranged in a tetrahedrally directed fashion. These four sp^3-hybridized orbitals form σ-bonds with the adjacent carbon atoms, as in diamond. One s-orbital and two p-orbitals hybridize to form three sp^2-hybridized orbitals arranged in a trigonally directed fashion and participate in a σ-bond formation, while the other pπ-orbital exists normal to the σ-bonding plane and forms a π-bond with the pπ-orbitals of adjacent carbon atoms, as in graphite. In the sp bonding configuration, one s-orbital and one p-orbital hybridize to form two sp-orbitals, which form σ-bonds along the x-axis, and the other two pπ-orbitals form π-bonds along the y and z axes.

Due to the strong, directional, tetrahedral σ-bonds, diamond exhibits high number density and extraordinary physical properties such as the highest hardness, highest bulk modulus, lowest thermal expansion coefficient, and chemical and electrochemical inertness; it also exhibits a wide bandgap of 5.5 eV [4]. Graphite has strong σ-bonds and nondirectional π-bonds within the layer, and only weak van der Waals bonding between the layers. Due to this configuration, it is an anisotropic metal in three dimensions and a zero bandgap material within the single

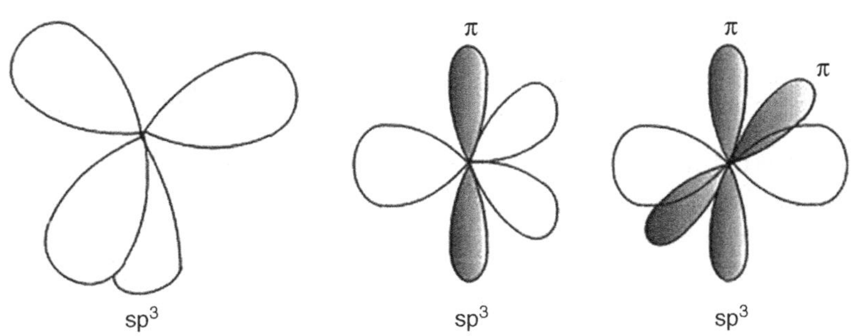

Figure 2.1 The sp^3-, sp^2-, and sp-hybridized bonding configurations. *Source:* Reproduced from [3] / ELSEVIER.

graphite plane [5, 6]. DLC exhibits extreme properties such as high hardness, Young's modulus, and chemical inertness similar to diamond [7, 8], but the material is much cheaper to produce than diamond because of the isotropic disordered nature of the DLC thin film. The molecular dynamics simulations made with the Large-scale Atomic/Molecular Massively Parallel Simulator (LAMMPS) software show the similarities between the structure of diamond and DLC materials (Figure 2.2) [9]. Figure 2.2a shows the tetrahedral arrangement of carbon atoms in the crystalline structure of diamond. Figure 2.2b shows the sp^2 and sp^3 network arrangement of carbon atoms in a DLC structure. Figures 2.2c,d show changes in the atomistic networking of carbon atoms within DLC as a function of sp^3 content, with 32.5% sp^3 content in Figure 2.2c and 71.3% sp^3 content in Figure 2.2d, respectively. As shown in Figure 2.2, the DLC materials are an intermediate state where sp^2 nanoclusters are bonded by sp^3 atoms [10].

DLC materials possess good mechanical, chemical, optical, and tribological properties, such as good chemical inertness, high hardness, high Young's modulus, biocompatibility, scratch resistance, impermeability, lower friction, wear

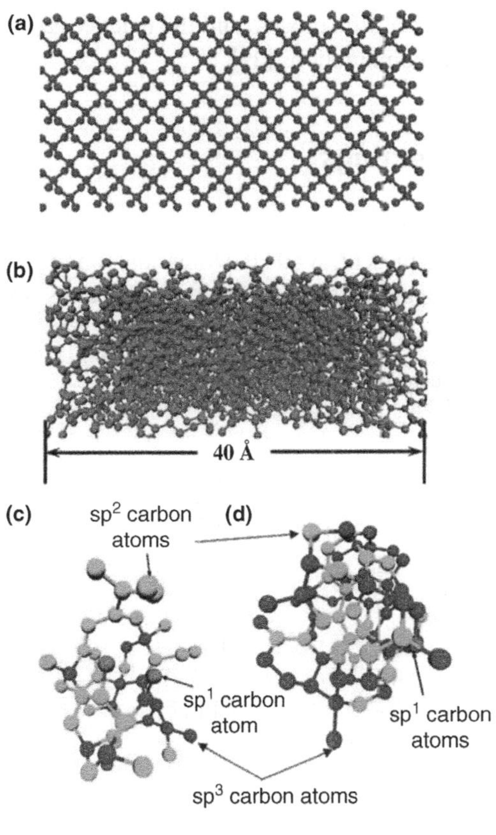

Figure 2.2 Molecular dynamics simulation of (a) crystalline structure of diamond, (b) disordered structure of DLC, (c) DLC with 32.5% sp^3 content, and (d) DLC with 71.3% sp^3 content. *Source:* [11] / ELSEVIER / CC BY 4.0.

resistance, thermal stability up to about 300 °C, and also have good antiwear resistance and antisticking properties [12–17]. Owing to these excellent properties, DLC coatings find a wide variety of applications in the aerospace and automotive industries, such as cutting tools, mechanical components, and engine components [18–20]. The materials are useful in the electronic industry in the form of microelectronic components, photoelectronic components, and sensors [21–24]; the materials are also useful as coated stents as well as surgical instruments in the biomedical industry [25].

DLC films are generally deposited by using conventional physical vapor deposition (PVD) techniques like ion beam deposition, sputtering deposition, pulsed laser deposition (PLD), cathodic arc deposition (CAD), and plasma-enhanced chemical vapor deposition (PECVD) techniques like radio frequency (RF-PECVD), microwave (MW-PECVD), and other hybrid methods [3, 11, 26, 27]. The sp^3 and sp^2 content of the deposited films largely depend on the deposition methods used and operational parameters [28] employed, which in turn govern the structure-property relationships. The operating parameters that can be optimized to regulate the properties of deposited DLC films include deposition temperature and pressure, precursor characteristics, kinetics of the process and the reactive gases, plasma power, flux, frequency, and other combinations [29].

2.2 Classification of Carbon Coatings

Referring to Figure 2.3, carbon films can be first generally classified into two categories depending on the sp^2 and sp^3 fractions present. While the films containing predominantly sp^2 bonds are classified as graphite-like carbon, the DLC films are

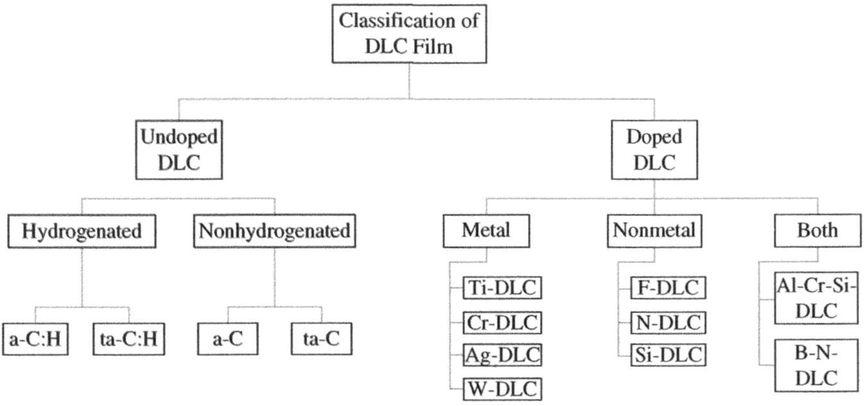

Figure 2.3 Classification of DLC films.

dominated by sp^3 bonds [30]. The sp^3, sp^2 carbon (C) bond fraction and the hydrogen (H) content in the film are the determining factors to get the desired class of DLC materials with tailored properties for suitable applications. The DLC films without any doping are generally subcategorized into four groups depending on the sp^3 C content and the H content. The films with sp^3 C content <50% and H content <5% can be called a-C, while the hydrogenated version of this type contains <50% sp^3 C and >5% H content, which can be referred to as hydrogenated amorphous carbon (a-C:H). Similarly, >50% sp^3 C and <5% H in the carbon films are associated with a harder variant of DLC, known as tetrahedral carbon (ta-C), and the hydrogenated ta-C contains about >50% sp^3 C and >5% H content. The films with much higher H (>40%) content are dominant with a linear chain structure and, therefore, classified as polymer-like carbon films [31]. Undoped DLC or pure DLC films typically have high hardness, low friction, and superior adhesion. One of the first undoped and hydrogen-free DLC films with improved uniformity and homogeneity was prepared by using a hybrid technique involving PLD simultaneously accompanied by DC plasma discharge [32]. The properties of the undoped films can be further tailored by introducing and varying the H content in the films, typically a higher H content producing softer films comparatively. Undoped DLC films find applications in the medical field as coatings on surgical instruments, implants, and biomedical devices, where low friction, good wear resistance, and enhanced biocompatibility are desired [33].

Undoped films generally contain stresses and strains, which can lead to the delamination of the thin films. These stresses and strains can be minimized by doping with metals. The pure DLC was doped with metals such as titanium (Ti), chromium (Cr), silver (Ag), tungsten (W), nickel (Ni), and iron (Fe) [34–39] and nonmetals like nitrogen (N), fluorine (F), and silicon (Si) [39, 40]. The pure DLC was also co-doped with a combination of metals and nonmetals to get hybrid coatings like aluminum–chromium–silicon-doped DLC [41] and nitrogen–boron-doped DLC [42]. Ag doping is interesting for biomedical applications due to its antibacterial properties [43, 44]. Metal doping can also impart increased hardness, enhanced thermal stability, and improved functionalities for specific applications. Ti and Cr doping resulted in an increased corrosion resistance while decreasing internal stresses; there was an exceptional increase in wear resistance by adding low amounts of copper (1.2%) and Ti (2.75%) [45–47]. Si-doped DLC showed significant antibacterial and antifungal behavior, enhanced biocompatibility, and bioactivity [48, 49]. Nitrogen-doped DLC showed increased thermal stability and decreased residual stress [33]. Pure DLC doped with a combination of metals and nonmetals creates a unique balance of hardness and wear resistance for desired applications.

A ternary phase diagram showing the C, H system was first used by Jacob and Moller to effectively describe the carbon films according to their sp^3 C, sp^2 C, and

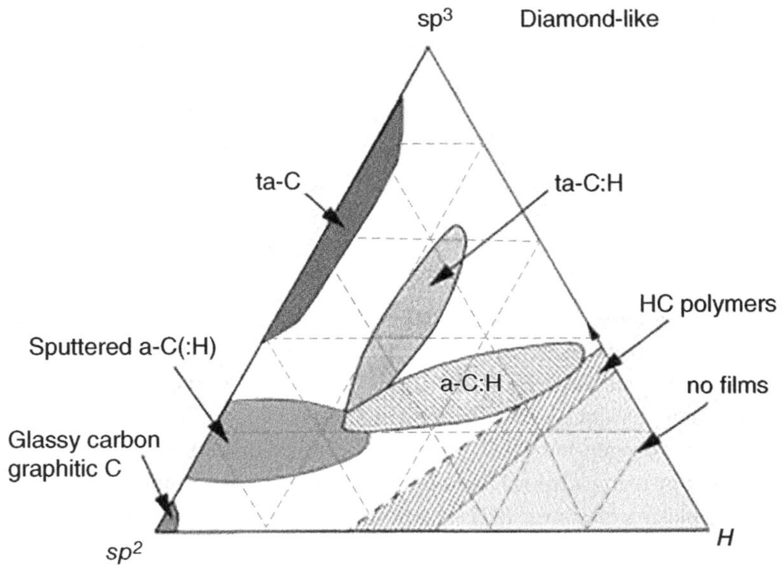

Figure 2.4 Ternary phase diagram of the C, H system. *Source:* Reproduced from [3] / with permission of ELSEVIER.

H percent (Figure 2.4). This representation makes it more straightforward to visualize the type of carbon film based on the sp^3, sp^2 C, and H content [50]. The type of carbon on the lower left-hand corner contains sp^2 bonding largely and, therefore, graphitic in nature. As we go up towards the tip of the triangle, the sp^3 content gradually increases, leading to a-Cs with sp^2 bonding predominantly and then ta-Cs with sp^3 bonding for the most part. At the summit, sp^3 content peaks; therefore, the films are diamond-like in nature. As we move from the lower left-hand corner to the lower right-hand corner, the hydrogen content in the films increases. In the right-hand corner, the limits of the triangle are defined by hydrocarbon polymers like polyethylene and polyacetylene. Past this point, there is no film formation, and only molecules form due to the absence of interconnecting C–C networks owing to the high percentage of hydrogen. The hydrogenated a-Cs and hydrogenated ta-Cs with varying hydrogen content can be mainly found in the middle of the triangle [3].

2.3 Deposition Methods

The deposition techniques for fabricating DLC films can be broadly distinguished as PVD and CVD. Any method of DLC deposition necessitates a source of carbon and a source of energy for the carbon species. An additional filtering stage may be required depending on the control of parameters like the type of species and energy distribution of that species. In the case of PVD methods, the carbon source is in the form of a pure target, which can then be evaporated thermally, ablated with the help of a laser, or sputtered using ions. These methods generally do not involve any chemical reactions during or after the growth process as opposed to the CVD methods unless the process was hybridized for the purposes of doping with other elements, consequently giving hydrogen-free DLC films. CVD methods typically involve chemical reactions that take place either at or near the substrate surface, with the carbon source being the carbon-containing gases such as hydrocarbons to produce ionized carbon species; hydrogenated DLC films are formed as a result. The energy required in both PVD and CVD methods may come from specimen bias or electrostatic acceleration. The other energy source is the momentum transfer from the collision with the energetic species either during the deposition involving simultaneous bombardment or before deposition, as in the case of sputtering [51]. The prominent methods of depositing DLC films are listed in the following sections.

2.3.1 Physical Vapor Deposition

2.3.1.1 Ion Beam Deposition

Typically, ion beam methods involve the condensation or deposition of DLC films from a beam consisting of carbon or hydrocarbon ions having energies in the range of approximately 20 eV up to a few keV. It is a physical process in the sense that the impact of these energetic ions on the growing film promotes the sp^3 bonding. In CVD methods, the sp^3 bonding is stabilized by a chemical process [52]. A carbon ion flux with energies in the vicinity of 100 eV and a narrow energy distribution of few eVs to several tens of eVs, containing a minimum number of nonenergetic or neutral species and optimally single energetic species, are considered to be the best deposition conditions [53].

The first IBD method to deposit carbon films at room temperature was employed by Aisenberg and Chabot in 1971 [2]. The resulting films were insulating and partially crystalline, with characteristics similar to those of diamond. The plasma sputtering at the ion source under an Ar atmosphere in a magnetically confined plasma created positive carbon ions; these ions are accelerated to the target electrostatically. Typically, Ar is used for creating positive C ions, and Cs is used for creating negative C ions in this approach, which results in the presence of Ar/Cs contamination. This method is called a direct IBD method, as there is no filtering or mass selection involved. The ion energy in this process was approximately 40 eV.

The conventionally used coaxial ion source is very similar to the one used in the IBD system described earlier. Figure 2.5a shows the schematic representation of this type of ion source used in the IBD system. The carbon ions are created via plasma

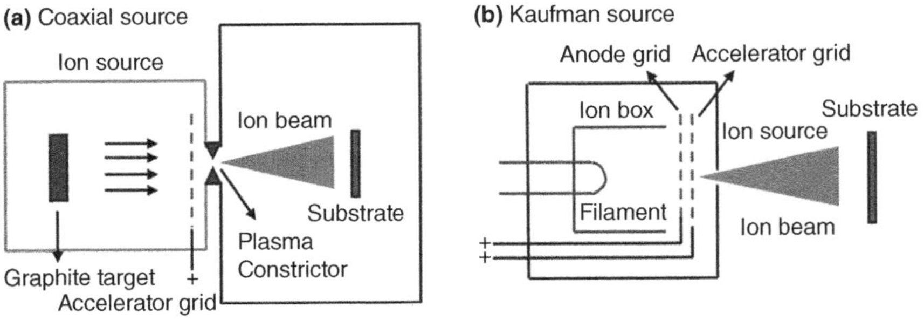

Figure 2.5 Schematics of IBD apparatus using (a) coaxial source and (b) Kaufman source.

sputtering of a graphite cathode in the ion source, which is maintained at a relatively low vacuum to allow for a large flux of ions along with unionized neutral species. As an alternative, a popular ion source called Kaufman source is illustrated in Figure 2.5b. In this type of ion source, high ionization rates are achieved by ionizing a hydrocarbon gas such as methane using electrons from a thermionic emission of cathode filament using an axial magnetic field to give rise to plasma [54, 55]. This source is also kept at a low vacuum for the same reason. The coaxial source generates ions of the form C_m^+, while the Kaufman source creates ions of the form $C_mH_n^+$ [56]. In both sources, the resulting plasma is then extracted through the grid by applying a bias and then directed toward the substrate in the high vacuum deposition chamber using an accelerator grid, thus forming the ion beam. Another important source used in the IBD systems is the cascade arc source [57], in which an intense plasma is produced under low vacuum, and it expands supersonically into the high vacuum deposition chamber. This source results in an enormous flux of ions and radicals.

2.3.1.2 Mass Selected Ion Beam Deposition

An excellent control over deposition by using a single and isotopically pure ion at a well-defined energy is highly desirable. This can be achieved in an intricate apparatus that uses magnetic filters suitable for isolating the ions based on their e/m ratio to get a single and isotopically pure ion species (usually C^+ ion), which was termed as mass selected ion beam deposition (MSIBD) and was first described by Aksenov et al. [58] for growing carbon films [59–61]. The films obtained through this method were highly amorphous, containing the highest sp^3 bonding fraction comparatively, although problems like high compressive stresses in the film were observed limiting the adhesion of the films [56]. A coaxial ion source employing a graphite target or a Kaufman source utilizing hydrocarbon gas like methane was utilized to produce carbon ions with a small spread of 1–10 eV in the ion energies. A bias is applied to accelerate the ions to get a sharply defined ion energy in the range of 5–40 keV and passed through a magnetic mass analyzer for e/m selection.

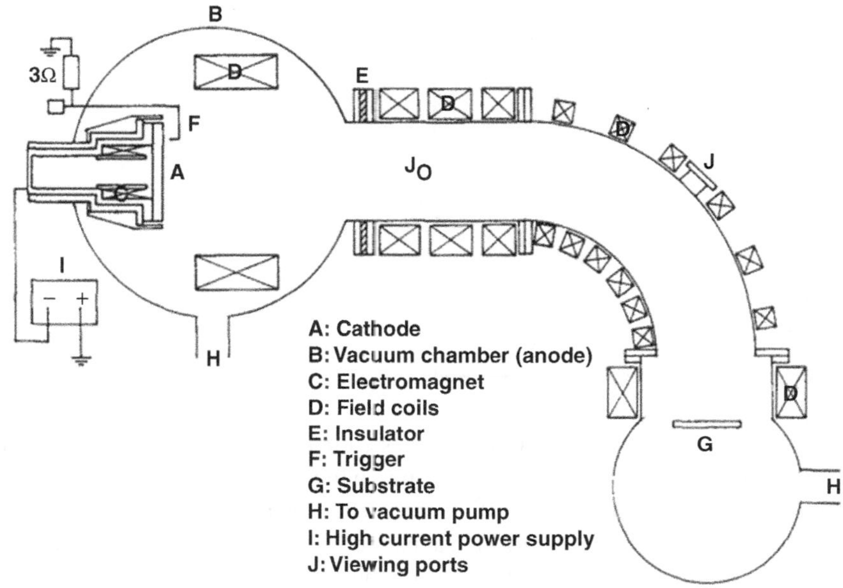

Figure 2.6 Mass selected ion beam deposition apparatus. *Source:* Reproduced from [62] / with permission of ELSEVIER.

Unionized neutrals, clusters, carbon particulates, and impurities are filtered out, and the beam will diverge at low voltages due to Coulombic repulsion. An electrostatic lens is used to decelerate the ions to get the desired ion energy and then directed onto the substrate to form ta-C film in the deposition chamber at vacuum on the order of 10^{-6} Torr and can reach 10^{-10} to 10^{-8} Torr in ultrahigh vacuum systems. For depositing on the insulating substrates, a low energy electron flood gun can be utilized for neutralization. By using an alternating beam of another ion source of the desired species, doped DLC films can be formed [59]. This can also be achieved by employing a dual-beam system consisting of two ion sources. One major drawback of the MSIBD process is the low rate of the deposition process, on the order of 10^{-3} Angstrom/s. The detailed schematic of the MSIBD system is depicted in Figure 2.6, which uses carbon arc as the ion source for higher rates.

2.3.1.3 Cathodic Arc Deposition

CAD is yet another subtle variant of ion beam methods in which a particularly energetic plasma is generated with a high level of ion density in the order of 10^{13} cm^{-3} by using a low voltage and high current power supply to initiate an arc in a high vacuum environment by striking a graphite cathode using a small carbon trigger electrode and withdrawing the trigger [62–68]. A simplified configuration of the cathodic arc setup is provided in Figure 2.7.

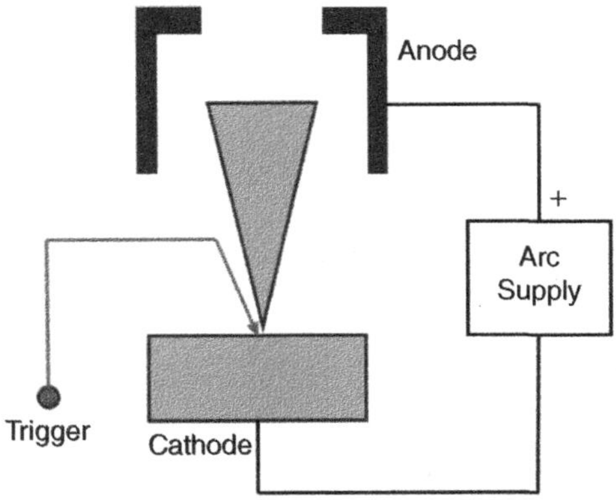

Figure 2.7 A basic cathodic arc setup.

The arc causes a rapid emission process, creating particulates along with plasma and thereby forming a cathode spot, which is small (1–10 μm) and carries a high current density (10^6–10^8 A/cm^2). The plasma is passed along a toroidal magnetic filter duct [69], in which a magnetic field of about 0.1 T is produced due to toroidal currents along the axis of the filter to clear out the undesired particulates. This is named filtered cathodic vacuum arc (FCVA), and the deposition system using this kind of ion source is termed filtered cathodic vacuum arc deposition (FCVAD). The positive ions follow the electrons of the plasma due to the electrostatic field, which was generated by the electrons of the plasma spiraling around magnetic field lines that make them follow along the filter axis. It is not possible for the particulates to follow this ambipolar transport path of the plasma around the filter, so they hit the duct wall. The particulates travel into the chamber by passing between the coils and out of the filter zone in the case of an open filter system [69, 70]. The usage of a filter in FCVA effectively increases the percent ionization of plasma from 30% to nearly 100% at the exit of the filter by having the particulates and the neutrals hit the wall. The base pressure in the FCVA process is in the order of 10^{-8} Torr, and it rises to 10^{-5} Torr when the plasma is running. A DC or RF self-bias is applied to the substrate to direct and condense the plasma onto the substrate to fabricate ta-C and to increase the mean incident ion energy of the plasma, which is about 10–30 eV at the cathode end, depending on the arc current. The single-bend FCVA setup is depicted in Figure 2.8 [69], and the Gaussian distribution of ion energies is shown in Figure 2.9 [67].

The rate of deposition in FCVAD is directly proportional to the electron current through the cathode, and the electron current to the anode is the main current in the arc, which is called arc current. The plasma current away from the cathode decreases by a factor of 10. A high arc current can generate high plasma currents as they are proportional to the arc current, and the mean energy of the ions rises to a certain extent [71]. A positive bias of about 20 V can decrease the

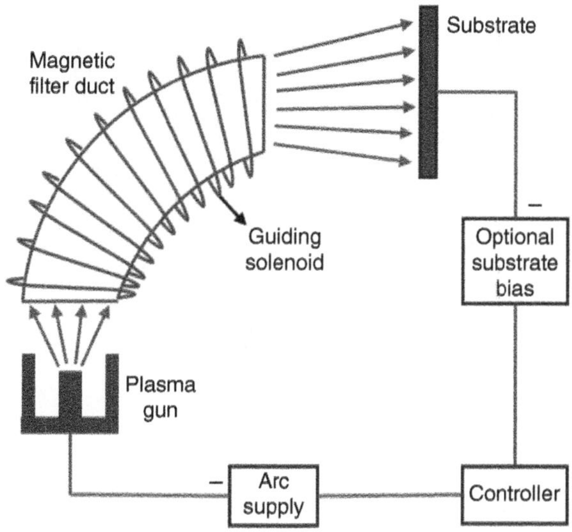

Figure 2.8 A single-bend FCVA configuration.

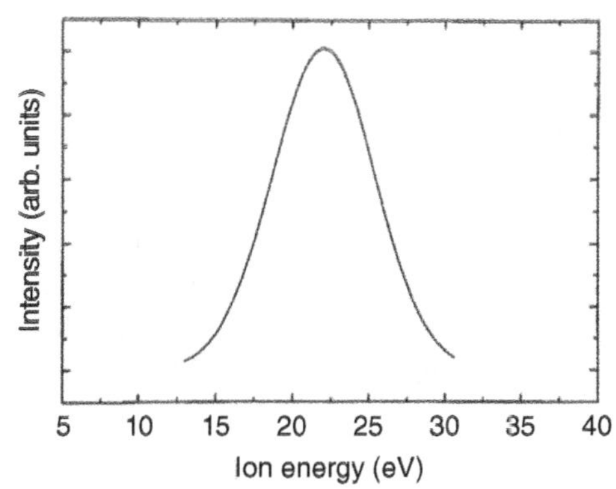

Figure 2.9 FCVA ion energy distribution. *Source:* Reproduced from [67] / with permission of AIP Publishing.

number of ions that are lost by hitting the wall while they pass through the filter duct. The uniformity of the films can be improved by making the beam diverge after it exits the filter duct through the usage of current coils to scan the beam across the substrate or by making use of the effect of having a particular kind of magnetic field configuration at the exit [72].

FCVAD is very suitable for industrial-scale depositions as the growth rates are as high as 1 nm s^{-1}, and energetic species in the highly ionized plasma with a reasonably narrow ion energy distribution, all available at fairly low capital costs. The disadvantages include an unstable cathode spot and insufficient filtering for certain applications. The submicrometer-sized particulates can still make it out by bouncing off the walls of the filter section [73]. The filtering process can be improved by running the arc in a pulsed mode rather than continuously, as the impurities tend to fall out when the beam stops. A capacitor

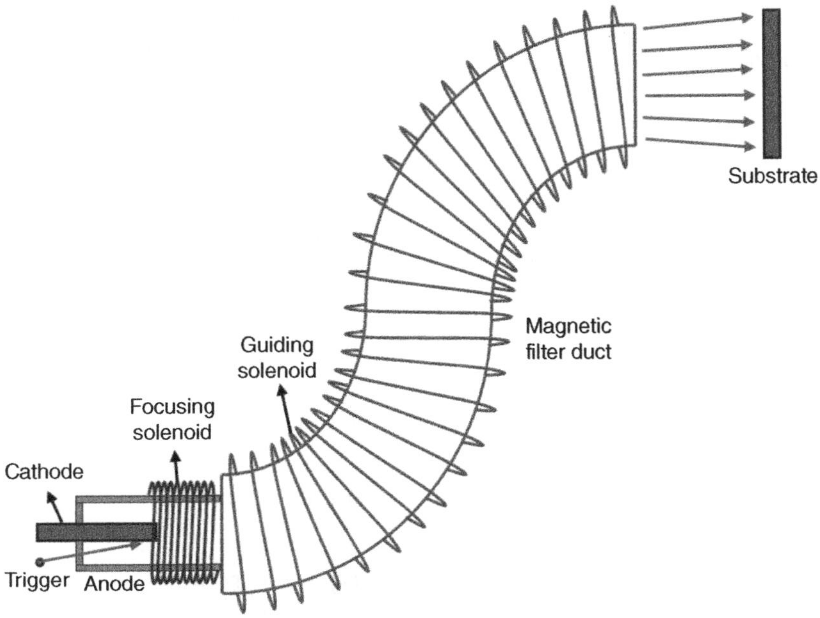

Figure 2.10 Double-bend or S-bend FCVA configuration.

bank to trigger the arc or using a pulsed laser for initiation can achieve the desired pulsed mode. A factor of 100 enhancement in the filtering can be attained by adding baffles and a second bend to the filter section. This double-bend or S-bend filter configuration is shown in Figure 2.10 [68, 74–76].

2.3.1.4 Sputtering

The other method that is especially suitable for industrial processes is sputtering; in fact, it is the most preferred industrial deposition method for fabricating DLC films because of its flexibility, ability to sputter a wide variety of materials, and scalability [77–81]. The deposition conditions can be fine-tuned by controlling the plasma power and gas pressure fairly easily. In sputtering, a graphite target is bombarded with noble gas plasma (typically Ar plasma for higher sputter yield) to get the energetic carbon species, which then condenses onto the substrate to give a-C films. A DC or RF self-bias is used for the target impingement. The kinetic energy of the carbon species helps in the formation of sp^3 bonding and improves the density of the films. Magnets are installed under the target to overcome the low sputter yield of graphite by confining the plasma mostly to the surface of the target in a variant of this method called magnetron sputtering. The electrons spiral around the magnetic field, improving the degree of ionization of the plasma due to an increase in the path length of electrons. This magnetic field can also be

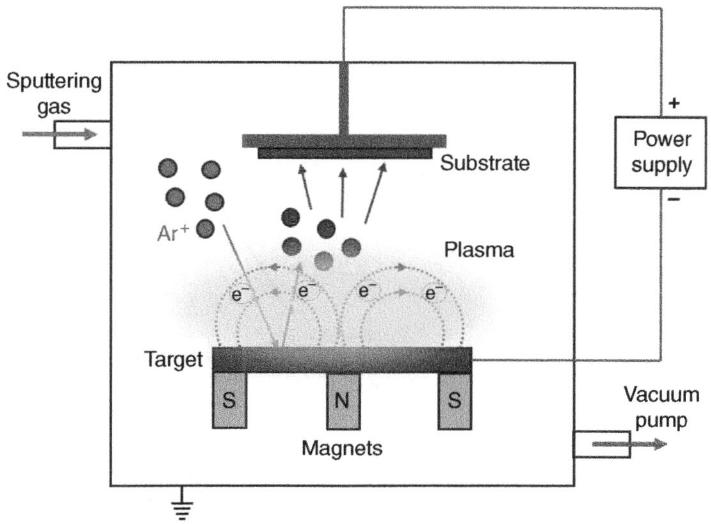

Figure 2.11 Typical magnetron sputtering system.

widened to cross the substrate in an unbalanced magnetron sputtering process, making the Ar plasma hit the substrate also, thereby further promoting the sp^3 bonding. A DC bias may be applied to the substrate to get the desired ion energy with minimal spread. Ar plasma, along with hydrogen or methane plasma, can be used in reactive sputtering to obtain hydrogenated a-C films. Similarly, doping can be achieved by assimilating different types of plasma, such as argon–nitrogen plasma, for nitrogen doping. A downside of using sputtering to grow DLC films is that the ratio of ions to neutrals within the plasma is relatively low, which results in films with a comparatively low sp^3 fraction. Several methods have been developed to overcome this shortcoming by achieving a high fraction of ions in the plasma but at the expense of growth rate [81, 82]. Figure 2.11 shows a typical magnetron sputtering system.

2.3.1.5 Ion Beam-assisted Deposition

Alternative methods involve using an ion beam as a secondary energy source for momentum transfer by impinging the growing carbon films to promote sp^3 bond formation or to improve the density of the films rather than as a primary ion source that contains the depositing carbon ion species. These types of methods are termed ion beam-assisted deposition (IBAD; or ion plating). The carbon for the deposition can come from evaporation sources such as crucibles or an e-beam evaporation source. This concept can also be incorporated into the sputtering process to get to ion beam sputtering. In this type of IBAD, an Ar ion beam sputters the graphite target to get the carbon flux, and a secondary Ar ion

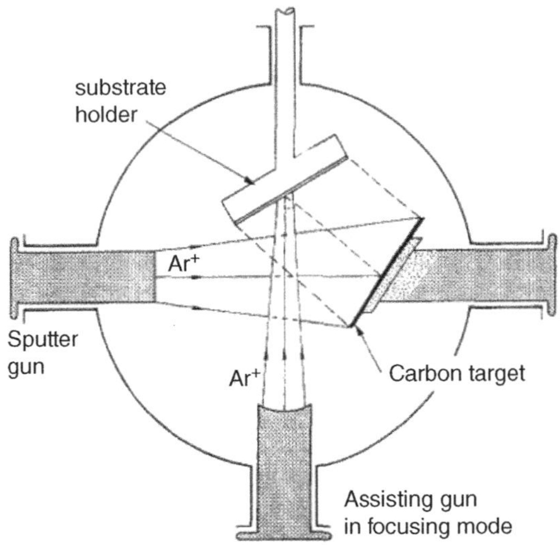

Figure 2.12 Schematic view of ion beam-assisted (sputter) deposition. *Source:* Reproduced from [83] / with permission of ELSEVIER.

beam collides with the growing carbon film to encourage the sp^3 bonding, making the films harder and denser [82, 83]. A schematic of this type of system is depicted in Figure 2.12.

2.3.1.6 Pulsed Laser Deposition

PLD is an outstanding technique for fabricating pure and doped DLC films with the availability of adequate mean ion energy at relatively low substrate temperatures. It can grow ta-C films with high sp^3 fraction even at room temperature [84–86]. PLD technique has been used to fabricate a variety of thin films, such as superconducting oxide thin films, epitaxial TiN thin films, Ge, SiO_2, and Cu, apart from DLC films [87–89]. It is one of the most preferred techniques for growing nonhydrogenated DLC thin films with over 80% sp^3 fraction. One of the earliest hydrogen-free DLC films with high uniformity and homogeneity was grown by using a hybrid PLD technique, which employs DC plasma discharge simultaneously [32]. The advantages of the PLD process include small-area uniform deposition, the ability to grow highly doped or alloyed films by co-ablation or reactive gas introduction in situ, heterostructures containing layers of other materials, or composition gradients. However, PLD is less prevalent in usage compared to other techniques described previously due to the difficulty in obtaining uniform films over large areas (as the laser-initiated plasma or plume has a strong directional character), and the presence of chunks or particulates in the plume.

In a classic PLD system, as shown in Figure 2.13, a collimated beam from the laser source is focused on the graphite target in high vacuum to ablate the top layer, having

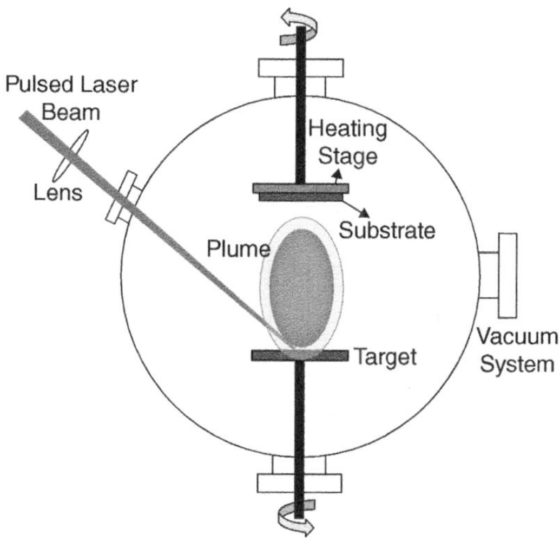

Figure 2.13 Classic PLD system schematic.

supplied sufficient kinetic energy using high laser fluence or energy density. This ablation takes place with pulses of nanosecond or femtosecond frequency and produces carbon plasma or plume containing energetic carbon ion species along with some neutrals and atomic clusters or particulates. The plume expands isothermally perpendicular to the target surface and continues adiabatically after the laser pulse finishes proceeding onto the substrate surface, which is typically placed 3–5 cm away from the target. The plume condenses onto the substrate to form ta-C films. The target is rotated to ensure the constant ablation rate and uniformity around the target surface. The substrate is equipped with a heater and negative bias when necessary. It is rotated and sometimes displaced from the rotation axis to obtain uniform films [90].

The lasers used in PLD are nanosecond or femtosecond pulsed excimer lasers, typically UV lasers like ArF (193 nm) [91–93] and KrF (248 nm) [85, 92, 94, 95]. These generate very short, intense energy pulses to produce energetic plasma, which then expands, and the kinetic energy from this expansion gives analogous mean ion energy when compared to the plasma in MSIBD and FCVAD techniques [96, 97], thus providing similar ta-C films with the resembling ion energy dependence of properties [98, 99]. Different deposition conditions during PLD determine the properties of the DLC films, with the hybridized sp^3 fraction being the important one among them. Figure 2.14 shows the chart of the influential conditions of the PLD process, with pretreatment and post-treatment steps being the generally used measures in all techniques to improve the properties of the films.

Laser wavelength and fluence are two of the most influential parameters during the PLD growth of DLC films. Laser fluence or energy density (J/cm^2) is directly related to the mean ion energy of the energetic species in the plume. Ion energy increases linearly with the increase in the laser fluence or energy density, as plotted in Figure 2.15. In turn, this laser fluence is also related to the sp^3 bond fraction

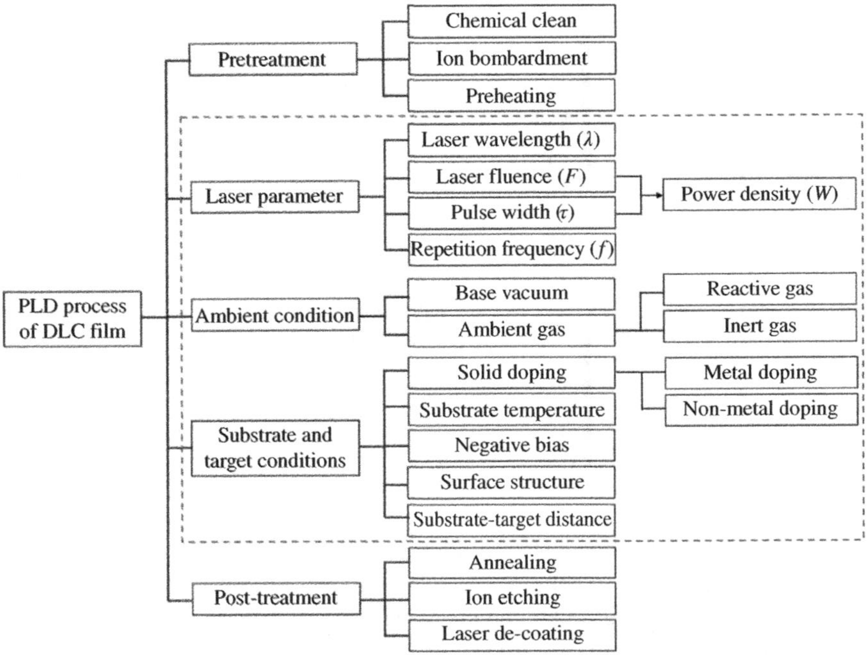

Figure 2.14 Influential conditions affecting the growth of DLC films during the PLD process. *Source:* Reproduced from [90] / with permission of ELSEVIER.

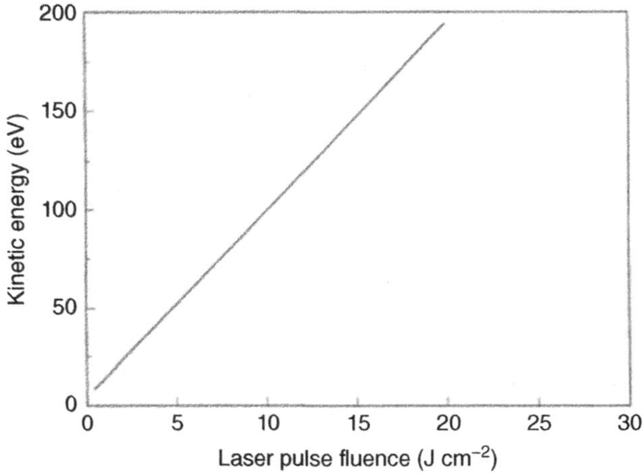

Figure 2.15 Mean ion energy versus laser fluence. *Source:* Reproduced from [3] / with permission of ELSEVIER.

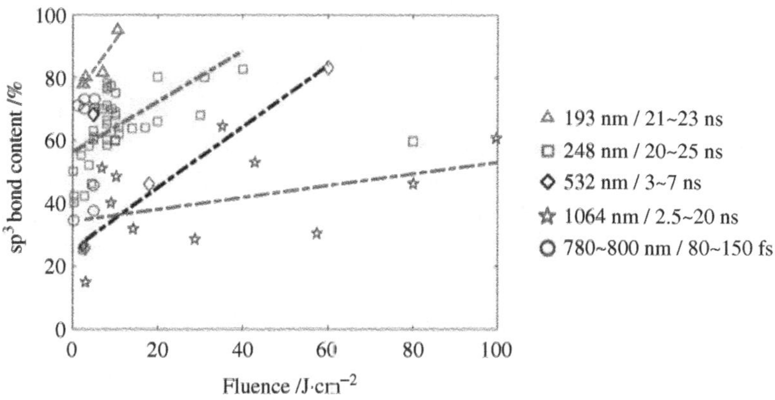

Figure 2.16 Dependence of sp^3 bond content on laser wavelength and fluence. *Source:* Reproduced from [90] / with permission of ELSEVIER.

in the films along with the laser wavelength. The sp^3 bond content is plotted against the laser fluence in Figure 2.16, with the data obtained from DLC films grown using the lasers emitting at 193 nm [91–93], 248 nm [85, 92, 94, 95], 532 nm [100–102], 1064 nm [84, 93, 103–109], and 800 nm [110–113] with varying fluences. Excluding the data from the 780–800 nm laser source, it can be safely deduced that decreasing the wavelength of the laser results in higher energy UV light, and increasing the laser fluence causes a linear rise in the sp^3 fraction. To summarize, the energy density or laser fluence is directly proportional to mean ion energy and sp^3 fraction. This helps in correlating mean ion energy to the sp^3 fraction through laser fluence. The degree of ionization, along with the ion energy, can be optimized by regulating the laser wavelength and fluence [114].

The 780–800 nm laser used a femtosecond configuration, which resulted in higher sp^3 content at relatively low fluence and was notably different from other nanosecond lasers. This is due to the difference in peak power density between the two, which depends on both laser fluence and pulse width [90].

PLD is a very powerful and versatile technique for producing hydrogen-free DLC films. However, hydrogenated DLC films can be achieved in PLD by using reactive gases such as hydrogen, methane, and other gases like oxygen and nitrogen to get doped films. Metal-doped DLC can be fabricated in multiple ways, such as using a sintered target made from graphite and doping material [115], a doping material patch made on the graphite target [116, 117], or using multiple targets ablated alternatively [118] or simultaneously by equipping two laser beams as in double-laser beam deposition [119, 120]. Multilayer films and gradient-doped multilayer films can also be fabricated by using similar strategies [121–123].

The uniformity of the DLC films can be greatly enhanced by incorporating a few tweaks into the deposition process. The simplest one of all is to slightly displace the rotation axis from the plasma axis so that the plasma axis is not at the center of the substrate [124]. Another strategy is to use a raster scan of the laser beam on the target so that the laser beam swings across the target surface, thereby shifting the plasma axis continuously, ensuring uniform deposition. This type of scanning laser is often used in industrial settings to get large-area uniform films [125, 126]. The other plan to achieve uniform large-area films involves using a rotating cylindrical graphite target. Since the plume or plasma axis is always normal to the plane tangent, the rotation of the target keeps shifting the plasma axis continuously to get uniform films [127]. The DC plasma discharge employed during the deposition in the hybrid PLD technique improved the uniformity of the DLC films significantly [32].

One other improvement to the PLD process involves the application of magnetic fields to enhance plasma quality. The magnetic field alters the plume expansion by improving the degree of ionization of the carbon ion species, thereby enhancing the ion bombardment, which promotes the sp^3 bond formation. In magnetic field-assisted PLD, confinement of the plume, increase in ionized species, and plasma instability have been observed [128]. The results include an improvement in the deposition rate (due to the rise in the number of ionized species reaching the substrate because of plasma confinement), sp^3 fraction, density, and hardness of the film with the introduction of external magnetic field into the PLD system [129]. Furthermore, it is imaginable to further optimize the PLD process by using the single-bend and S-bend magnetic filter ducts, as utilized in the FCVAD technique, to gain positive benefits.

2.3.2 Chemical Vapor Deposition

Chemical vapor deposition (CVD), in general, relies on chemical reactions among the gaseous precursors, usually hydrocarbon gases, to produce hydrogenated DLC films. The most popular of these methods involve RF power to create the plasma containing the ionized species from the decomposed hydrocarbon gas, and this plasma condenses onto the negatively self-biased substrates. Holland and Ojha [130] pioneered the technique for developing a-C:H films by using RF plasma decomposition of butane gas, which was initially named plasma deposition and later widely used as PECVD for getting hydrogenated DLC films and hydrogenated films of other materials. RF self-bias is preferred rather than DC biasing as these types of films are insulating [131]. In a classic RF-PECVD [132–136] reactor, two electrodes of different areas are equipped, with the substrate-mounted smaller electrode being the one that has RF power capacitively coupled. The earthed reactor walls act as the larger electrode. The plasma is formed between the electrodes by the RF power, and a sheath of excess ions is formed next to the electrodes due to the higher mobility of electrons compared to ions in the plasma. The plasma develops a positive bias with respect to the electrodes as the sheath has a positive space charge [137]. These phenomena make the smaller area electrode negatively biased, and the positive ions in the sheath accelerate and bombard the substrate mounted on the smaller electrode to facilitate sp^3 bond formation in the resultant hydrogenated DLC films. A low-pressure environment in the chamber is preferred to maximize the ionization and minimize the ion energy distribution [3]. However, the plasma will no longer strike the substrate if low pressure is used in a conventional PECVD. To overcome this, a plasma beam source (PBS) is incorporated into the PECVD system. In PBS, the plasma is confined by the application of a magnetic field, thereby raising ionization efficiency because of the increased electron path length [138].

The bias voltage (V_b) varies with RF power (W) and pressure (P) as [3]

$$V_b = k'WP^{1/2}. \tag{2.1}$$

At low operating pressure, the mean ion energy, E_i, is proportional to bias voltage V_b [131]

$$E_i \cong eV_b. \tag{2.2}$$

While at the typical higher operating pressures, the mean ion energy varies as

$$E_i \cong k\frac{V_b}{P^{1/2}}, \tag{2.3}$$

in which the constant k depends on factors such as electrode areas. k is ~ 0.4 at typical PECVD operating pressures of 5 Pa [3].

Plasma excitation by RF power, capacitively coupled at a frequency of 13.5 MHz, is relatively inefficient, and the produced plasma density is quite low, a maximum of 10^9 cm^{-3}. The plasma density can be increased by using higher frequencies, as in inductively coupled plasma. It can be further improved by using microwaves in a process known as electron cyclotron resonance (ECR). In an ECR-based system, the electrons oscillate at a cyclotron resonance frequency of 2.45 GHz by the application of a large static axial magnetic field to achieve high plasma densities [139, 140]. The highest plasma densities of 10^{12} cm^{-3} are realized by an electron cyclotron wave resonance source [141], which is similar to ECR.

The precursor gases used in the PECVD process have a considerable effect on the deposition rates and the properties of hydrogenated DLC films. The rate of deposition sees an exponential increase roughly with a drop in ionization energy, with few exceptions. The deposition rates of hydrogenated a-C films by PECVD were plotted [3] against the ionization potential of the precursor gas using the data from Koidi et al. [142] and Zou et al. [143], as shown in Figure 2.17.

An increase in deposition rates was also achieved independent of the precursor species by using microwave-excited plasma along with mid-frequency substrate

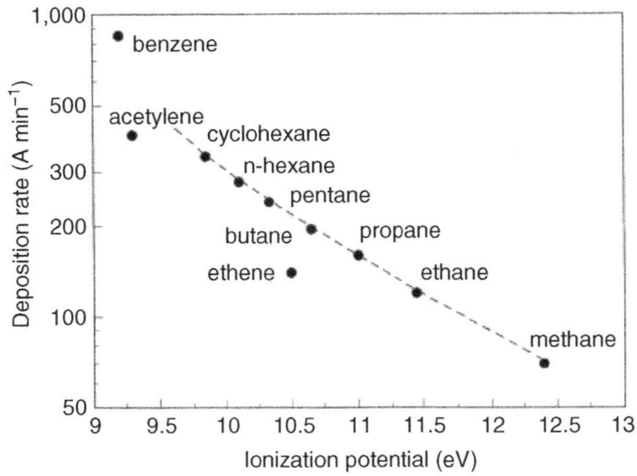

Figure 2.17 PECVD growth rate of a-C:H films versus ionization potential of the precursor gas. *Source:* Reproduced from [3] / with permission of ELSEVIER.

bias, allowing control of ion flux to the substrate surface [144]. The fabrication of doped DLC films is possible with PECVD methods. This is achieved by using suitable precursor gases containing the desired dopant elements. For example, silicon-doped DLC (Si-DLC) films were fabricated by both conventional and unconventional PECVD methods for electronic and biomedical applications. Tetramethylsilane (TMS) was utilized as a precursor gas along with argon gas in a conventional RF-PECVD process to get Si-DLC films [145]. Alternatively, TMS was added to the cyclohexane and hydrogen mixture as a second precursor for the co-deposition of hydrogenated Si-DLC films in an unconventional low-frequency PECVD process [146]. The aforementioned discussion of various deposition methods is summarized in Table 2.1 [56].

2.4 Deposition Mechanisms

The most convenient way of attempting to explain the growth mechanism of DLC films involves the interaction of ions with growing solid film. In this case, nonhydrogenated DLC films are bombarded with positive carbon ion species with energies in the range of 10–1000 eV, where most of the ordering or phase transformation takes place within the minute time scales of picoseconds and femtoseconds. Considering hydrogenated films makes the model more complicated. Although there are few relatable surface effects, most of the nucleation and growth processes occur subsurface. This process is termed subplantation [3, 51, 56, 64, 152–156], in which low-energy implantation of ions a few nanometers below the surface causes

Table 2.1 A summary of deposition methods associated with DLC films.

Deposition method	Precursor	Plasma source	Growth rate (A° min^{-1})	References
IBD	Graphite	Coaxial	80	[2]
IBD	Methane	Kaufman	120	[147]
MSIBD	Graphite	Coaxial	6–360	[148]
FCVAD	Graphite	Arc	600	[3]
Cascade arc	Ar/methane	Arc	18000	[149]
IBAD	Benzene	Kaufman	600	[150]
Ar beam sputtering	Ar/graphite	RF or DC	180	[151]
Magnetron sputtering	Ar/graphite	RF or DC	180	[79]
PLD	Graphite	Laser-induced	<180	[90, 151]
PECVD	Methane	RF	60	[142]
PECVD	Benzene	RF	900	[142]
PECVD	Acetylene	Microwave	2400	[143]

Source: Adopted from Ref. [56].

metastable transformations. The subplanting species enter the growing film or target material by either direct penetration or by knock-on collisions with the surface atoms. There may be multiple collisions subsequently, like secondary and tertiary collisions. This process also causes sputtering of target material atoms along with damage in the form of atomic displacements or the formation of vacancy defects. This damage can be estimated by calculating the number of vacancy-interstitial pairs or Frenkel defects. In the case of Ar^+ ion bombardment, 250 eV is the optimum ion energy for creating one Frenkel pair or defect in the growing film. The subplanted species undergoes nuclear stopping, which involves collisions with nuclei rather than electron cloud (electronic stopping), and finally gets trapped inside the target material. It is important to consider the quantities like the range of ion penetration (R_p), the range straggling (ΔR_p) or distribution, and the trapping efficiency of the subplanted species along with the sputtering of the target material atoms and the created damage [157].

The stopping and trapping of subplanting species, as well as the resultant transformations in the growing films, can be summarized by dividing them into multiple stages of time scales. The first is the collisional stage, where the stopping occurs through nuclear stopping or atomic collisions, causing ionization and phonon excitations in the span of about 10^{-13} seconds. The next thermalization stage happens in the time span of about 10^{-12} seconds, where the excess energy is dissipated in the target material. Diffusion, phase modifications, and chemical reactions occur between about 10^{-10} seconds and a few seconds during the long-term relaxation stage [51, 158].

The process of subplantation proceeds by the penetration of the carbon ion species to the subsurface layers. This penetration requires a minimum threshold energy called penetration threshold (E_p). The penetrated species causes local compressive stresses due to trapping and elastic expansion. It is assumed that due to the combined effect of this trapping and displacement or damage (vacancy-interstitial pairs) caused by the subplanted C ion species, a localized increase in density or stress induces transformation into metastable sp^3 fraction in the amorphous matrix surrounding the implanted species [159, 160]. Basically, the subsurface growth within a restricted volume is the driver of the transformation to get sp^3 bonding. The damage required to cause this transformation only happens when the subplanted species exceeds the damage threshold energy (E_d). The displacement or damage threshold energy is 25 and 80 eV in graphite and diamond, respectively [3]. The energy dissipation and local relaxations during the thermalization and relaxation stages, called thermal spike relaxation, promote local changes to this high-density amorphous region. These changes can either be beneficial or detrimental to the sp^3 bonding. At higher ion energy levels, the dissipation and relaxation processes provide sufficient energy for increasing the mobility and detrapping the subplanted species, thus suppressing the sp^3 fraction [51].

In contradiction to the explanation of sp^3 bonding driven by increased density, it is debated that the thermal spike induces a local high temperature and high-pressure region in which the metastable sp^3 state is the more thermodynamically stable configuration. According to this spherical thermal spike model, the sp^3 bond formation is driven by the quenching or the rapid relaxation of the thermal spike region [148, 161]. A more accurate description of this model is addressed in the cylindrical spike model by accounting for the energy losses and collision cascade effects along the track of the ions [160].

As for the experimental validation, a clear correlation exists between the sp^3 content and the surface roughness. The atomically smoother films facilitated a higher sp^3 content, while the suppression of sp^3 and promotion of the sp^2 phase caused rougher carbon films. The compressive stress in the films is another indirect measure of the sp^3 content [162]. The shallow implantation nature of the DLC film deposition led to several modeling efforts, thus furthering the field [51, 159–161]. More work is needed in order to address the several inconsistencies of various models with the experimental evidence.

Alternatively, considering the case of hydrogenated DLC films, gaseous precursors like methane, benzene, and acetylene are employed during deposition, and the carbon ion species are in the form of molecular ions in almost all similar methods. The molecular ion splits into two ions of equal energy when penetrated into the growing carbon film. Therefore, it can be treated as two carbon ion species of particular energy for modeling and calculation purposes [3].

2.5 Influence of Deposition Parameters

The mean ion energy (E) of the carbon ion species plays an important role in determining the nature of the DLC films. To accurately assess the role of ion energy, a low substrate temperature is maintained to result in insignificant thermal mobility of the trapped carbon ion species. If the ion energy E is below the critical energy E_p ($E \leq 30$ eV), there was no subsurface penetration, and the sp^2 phase is the more stable configuration under normal conditions; as such, sp^2-rich films were deposited by the surface growth process. In the regime where E is above the threshold penetration energy but below the energy where detrapping and the relaxation processes are not enough to cause suppression of sp^3 fraction (30 eV $< E \leq$ 600 eV), optimally grown ta-C films with higher than 70% sp^3 fraction were obtained. As discussed previously, the local increase in the density due to the trapping of carbon species promoted the formation of sp^3 bonds. In this regime, the growth is mostly internal or subsurface and not on the surface; as such, the films were atomically smooth. In the next dose regime (600 eV $< E \leq$ 10 keV), the energy is high enough to cause some level of suppression in the sp^3 fraction because the dissipated energy and damage increased the mobility of trapped carbon species, thereby detrapping them. The relaxation process is still internal, and there was no significant rise in roughness associated with this regime. The films showed a significant sp^3 fraction (about 40%) even at $E = 10$ keV. When the energies go even higher ($E \geq 20$ keV), the damage is high enough to cause the migration of carbon species to the surface by increasing the mobility further, thereby suppressing the densification and sp^3 bonding, growing films with almost 100% sp^2 fraction. This gave a very rough surface morphology and initiated the clustering of sp^2 bonds to give nanocrystalline graphite due to high mobilities. The measurements like hardness, density, surface, and roughness follow the same trend as the sp^3 fraction and corroborate the evidence [51].

The substrate temperature has a significant effect on the bonding dynamics of the DLC films. With increasing temperature, the mobility of the carbon interstitials increases, leading to the elimination of any densification effects due to the thermal migration of carbon interstitials to the surface. At 120 eV ion energy deposition, when the temperature exceeded the critical temperature of 150 °C, a sharp decrease in the sp^3 fraction was observed. The thermal migration of the carbon species produced much rougher and sp^2-rich films. In this case, it was surmised that the deposition rate was low enough that the subsurface thermal migration was higher than the carbon ion flux to the surface, and the deposition rate had no effect. But, in practical conditions, the nature of the deposited films will be dictated by the equilibrium that takes place between the deposition rate and the thermal migration toward the surface [51].

2.6 Properties and Analysis Methods

The properties associated with DLC films vary greatly and are dependent on a few important attributes, such as structure, crystallinity, and bonding characteristics. On one end of the spectrum, the crystalline carbon-based material with the largest sp^3 bonding fraction, diamond, exhibits a dense, cubic close-packed structure with the highest elastic modulus and hardness. The other extreme of crystalline carbon materials is graphite, with the highest sp^2 fraction, which has sheets of hexagonally packed atoms stacked on top of each other due to weak van der Waals attraction and exhibits low friction. The amorphous analogs to this are a-C with the highest sp^2 bond content and ta-C with the highest sp^3 content. Therefore, it is advantageous to compare the extreme examples to get a good understanding of how the properties of DLC films respond to changes in these key attributes. DLC films, in general, are popular choices for applications requiring high hardness, wear resistance, chemical inertness, biocompatibility, friction resistance, and electrical insulation. [163]. Table 2.2 [51] shows the properties of different carbon materials, mainly comparing the extreme examples as described earlier.

The analysis of DLC films plays a significant role in confirming if the material is suitable for the desired application. A multitude of analysis methods are available for the task, with the focus of most of these methods being the determination of metastable state or diamond-like nature of the films by determining the sp^3 fraction because the sp^3 fraction directly correlates to other material properties like hardness, surface roughness, density, chemical inertness, resistivity, and mobility gap [51]. This diamond-like quality or bonding characteristics can be analyzed by surface analysis methods such as X-ray photoelectron spectroscopy (XPS), Auger electron spectroscopy, electron diffraction, and radial distribution function using transmission electron microscope (TEM) or neutron scattering, low-energy electron energy loss spectroscopy (EELS), high-energy EELS, and Raman spectroscopy. The crystallinity of the films can be measured by using electron diffraction (typically in TEM) and X-ray diffraction [51]. Density can be evaluated via a combination of Rutherford backscattering spectroscopy and profilometry, an ultrasonic surface wave approach, as well as EELS

Table 2.2 Properties of different carbon materials.

Structure	ta-C	a-C	Diamond	Graphite
sp^3 fraction	>80%	0%	100%	0%
Nearest neighbor distance (r_{nn})	1.53 Å	1.42 Å	1.54 Å	1.42 Å
Bond angle	110°	120°	109.5°	120°
Coordination number	>3.8	3	4	3
Crystallinity	Amorphous	Amorphous	Cubic diamond ($O^7{}_h$)	Hexagonal ($D_{6h}{}^4$)
Density (g cm^{-3})		2–2.3	3.52	2.3
Rutherford backscattering spectrometry (RBS) + thickness	3.5			
Surface ultrasonic wave	3.3			
Plasmon energy	3.6			
Raman	(1) Symmetric broad line at 1550 cm^{-1}	(1) Very broad asymmetric line decreasing from 1550 to 1000 cm^{-1}	1330 cm^{-1} sharp line	(1) 1580 cm^{-1} sharp line (G line)
	(2) No 1330 cm^{-1} (visible)	(2) No distinguished D peak		(2) 1350 cm^{-1} sharp line for nanocrystalline graphite (D line)
	(3) Broad line below 1330 cm^{-1} (UV)			
	(4) No D graphitic line			
Plasmon energy	30.5–31 eV	23–24 eV	33 eV	27 eV
Resistivity	3×10^{11} Ωcm	<1 Ωcm	10^{16} Ωcm	10^{-2} Ωcm
Optical gap	2.5 eV	0	5.5 eV	0
Elastic modulus	800 GPa		1000–1200 GPa	686 GPa (‖a) low (⊥a)

(Continued)

Table 2.2 (Continued)

Structure	ta-C	a-C	Diamond	Graphite
Hardness	120 GPa	2–5	120 GPa	
Thermal stability	>700 °C	Gradually graphitizes for T > 200 °C	>700 °C	>700 °C

Source: Reproduced from [51] / with permission of ELSEVIER.

or XPS plasmon energies [164–166]. Scanning electron microscopy can provide surface morphology [167]; atomic force microscopy (AFM) measures surface roughness, features, and thickness variations [168]. Mechanical properties such as friction and wear can be evaluated using pin-on-disc measurements [169]. Optical properties can be measured from transmission and reflection in ellipsometry [29]. Electrical properties like resistivity can be determined from I–V measurements.

2.7 Applications

The unique set of properties enables DLC films to find many important practical applications in diverse fields. These applications are correlated to the relevant properties in Table 2.3 [90].

Table 2.3 Practical applications of protective DLC films.

Properties	Applications
Hardness, wear resistance, and low friction	Cutting tool coatings, wear protection of sliding and gear components, magnetic recording thin films
Optical transparency with high hardness	Protective and antireflective films for optical windows and waveguides
Biological inertness with low friction	Medical prostheses, biosensors, medical instrument coatings
Chemical inertness	Anticorrosive surfaces, fuel cell anode coatings
Wear resistance with high thermal conductivity and electric resistance	Protective and heat-dissipating surfaces for MEMS and printed circuit boards

Source: Reproduced from [90] / with permission of ELSEVIER.

2.8 Conclusions

DLC films exhibit several intriguing physical, chemical, optical, electrical, tribological, and mechanical properties due to their unique set of properties, which can be fine-tuned by adjusting critical quantities such as density and sp^3 bond fraction. This tuning can be done by tweaking important parameters during the fabrication of DLC thin films. Many deposition approaches have been employed for generating DLC films. Nonhydrogenated DLC can be formed by important techniques like PLD, cathodic arc, and IBD methods. Magnetic filters have been employed in ion beam and cathodic arc methods to achieve e/m selection for improving plasma quality. Sputtering methods are being used to get good deposition rates at relatively low costs. Magnetic fields are incorporated into sputtering systems for confining the plasma and improving ionization. A second beam of ions is also utilized to obtain IBADs. Laser-induced plasma generation is applied in PLD methods to obtain efficient small-area uniform films. Cascade arc can be used to get very high deposition rates. RF-PECVD methods are helpful in getting hydrogenated DLC films from gaseous precursors at relatively high deposition rates. Microwave frequencies can be employed for better performance. The growth mechanism was explained with reasonable correctness using models such as the subplantation model as well as spherical and cylindrical thermal spike models. The nature of the DLC films is dictated by controlling the growth parameters such as mean ion energy, substrate temperature, and deposition rate. Despite the maturation of technology in the field, further improvements in modeling and processing methods are essential to obtain optimal films.

References

1 Zia, A.W. (2020) Chapter 11 – New generation carbon particles embedded diamond-like carbon coatings for transportation industry. In: *Advances in Smart Coatings and Thin Films for Future Industrial and Biomedical Engineering Applications* (ed. A.S.H. Makhlouf and N.Y. Abu-Thabit), 307–332. Elsevier.

2 Aisenberg, S. and Chabot, R. (1971) Ion-beam deposition of thin films of diamond-like carbon. *Journal of Vacuum Science and Technology* 8 (1): 112.

3 Robertson, J. (2002) Diamond-like amorphous carbon. *Materials Science and Engineering: R: Reports* 37 (4): 129–281.

4 Angus, J.C. and Hayman, C.C. (1988) Low-pressure, metastable growth of diamond and "diamond-like" phases. *Science* 241 (4868): 913–921.

5 Dresselhaus, M.S., Dresselhaus, G., and Eklund, P.C. (1996) *Science of Fullerenes and Carbon Nanotubes: Their Properties and Applications.* Elsevier.

6 Brocklehurst, J., Kelly, B., and Gilchrist, K. (1981) The effect of substitutional boron on irradiation damage in graphite. *Chemistry & Physics of Carbon* 17: 175–231.

7 Bonelli, M., Ferrari, A.C., Fioravanti, A. et al. (2002) Structure and mechanical properties of low stress tetrahedral amorphous carbon films prepared by pulsed laser deposition. *The European Physical Journal B-Condensed Matter and Complex Systems* 25 (3): 269–280.

8 Savvides, N. and Bell, TJ (1993) Hardness and elastic modulus of diamond and diamond-like carbon films. *Thin Solid Films* 228 (1): 289–292.

9 Li, L., Song, W., Ovcharenko, A. et al. (2015) Effects of atomic structure on the frictional properties of amorphous carbon coatings. *Surface and Coatings Technology* 263: 8–14.

10 Seahra, S.S. and Duley, W.W. (1999) Extended red emission from carbon clusters in interstellar clouds. *The Astrophysical Journal* 520 (2): 719.

11 Zia, A.W., Hussain, S.A., Rasul, S. et al. (2023) Progress in diamond-like carbon coatings for lithium-based batteries. *Journal of Energy Storage* 72: 108803.

12 Hatem, A., Lin, J., Wei, R. et al. (2017) Tribocorrosion behavior of DLC-coated Ti-6Al-4V alloy deposited by PIID and PEMS + PIID techniques for biomedical applications. *Surface and Coatings Technology* 332 (A): 223–232.

13 Pagnoux, G., Fouvry, S., Peigney, M. et al. (2015) Influence of scratches on the wear behavior of DLC coatings. *Wear* 330–331: 380–389.

14 Mobarak, H.M., Masjuki, H.H., Mohamad, E.N. et al. (2014) Effect of DLC coating on tribological behavior of cylinder liner-piston ring material combination when lubricated with jatropha oil. *Procedia Engineering* 90: 733–739.

15 Benedetti, M., Fontanari, V., Torresani, E. et al. (2017) Investigation of lubricated rolling sliding behaviour of WC/C, WC/C-CrN, DLC based coatings and plasma nitriding of steel for possible use in worm gearing. *Wear* 378–379: 106–113.

16 Voevodin, A.A., Walck, S.D., and Zabinski, J.S. (1997) Architecture of multilayer nanocomposite coatings with super-hard diamond-like carbon layers for wear protection at high contact loads. *Wear* 203–204: 516–527.

17 Yu, W., Wang, J., Huang, W. et al. (2020) Improving high temperature tribological performances of Si doped diamond-like carbon by using W interlayer. *Tribology International* 146: 106241.

18 Brown, S., Lengaigne, J., Sharifi, N. et al. (2020) Durability of superhydrophobic duplex coating systems for aerospace applications. *Surface and Coatings Technology* 401: 126249.

19 Sivakumar, E.R., Senthilkumar, P., Sreenivasan, M. et al. (2020) Experimental investigation of H-DLC coated exhaust valve characteristics of a diesel engine. *Materials Today Proceedings* 33 (4): 675–681.

20 Kabir, M.S., Zhou, Z., Xie, Z. et al. (2021) Designing multilayer diamond like carbon coatings for improved mechanical properties. *Journal of Materials Science & Technology* 65: 108–117.

21 Yang, N., Yu, S., Macpherson, J.V. et al. (2019) Conductive diamond: synthesis, properties and electrochemical applications. *Chemical Society Reviews* 48 (1): 157–204.

22 Jurewicz, A.J.G., Rieck, K.D., Hervig, R. et al. (2020) Magnesium isotopes of the bulk solar wind from Genesis diamond-like carbon films. *Meteoritics & Planetary Science* 55 (2): 352–375.

23 Stock, F., Antoni, F., Aubel, D. et al. (2020) Pure carbon conductive transparent electrodes synthetized by a full laser deposition and annealing process. *Applied Surface Science* 505: 144505.

24 Ryaguzov, A.P., Nemkayeva, R.R., Yukhnovets, O.I. et al. (2019) The effect of nonequilibrium synthesis conditions on the structure and optical properties of amorphous carbon films. *Optics and Spectroscopy* 127 (2): 251–259.

25 Rao, X., Yang, J., Chen, Z. et al. (2020) Tuning C–C sp^2/sp^3 ratio of DLC films in FCVA system for biomedical application. *Bioactive Materials.* 5 (2): 192–200.

26 Zia, A.W. and Birkett, M. (2021) Deposition of diamond-like carbon coatings: conventional to non-conventional approaches for emerging markets. *Ceramics International* 47 (20): 28075–28085.

27 Wei, Q. and Narayan, J. (2000) Superhard diamond-like carbon: preparation, theory and properties. *International Materials Reviews* 45 (4): 133–164.

28 Zia, A.W., Hussain, S.A., and Baig, MMFA (2022) Optimizing diamond-like carbon coatings – from experimental era to artificial intelligence. *Ceramics International* 48 (24): 36000–36011.

29 Ohtake, N., Hiratsuka, M., Kanda, K. et al. (2021) Properties and classification of diamond-like carbon films. *Materials* 14 (2): 315.

30 Zia, A.W., Zhou, Z., and Li, L.K.-Y. (2019) Chapter 7 – Structural, mechanical and tribological characteristics of diamond-like carbon coatings. In: *Nanomaterials-based Coatings* (ed. P. Nguyen Tri, S. Rtimi and C.M. Ouellet Plamondon), 171–194. Elsevier.

31 Rajak, D.K., Kumar, A., Behera, A. et al. (2021) Diamond-like carbon (DLC) coatings: classification, properties and applications. *Applied Sciences* 11 (10): 4445.

32 Krishnaswamy, J., Rengan, A., Narayan, J. et al. (1989) Thin-film deposition by a new laser ablation and plasma hybrid technique. *Applied Physics Letters* 54 (24): 2455–2457.

33 Sánchez-López, J.C. and Fernández, A. (2008) Doping and alloying effects on DLC coatings. In: *Tribology of Diamond-like Carbon Films: Fundamentals and Applications* (ed. C. Donnet and A. Erdemir), 311–338. Boston, MA: Springer.

34 Malisz, K., Świeczko-Żurek, B., and Sionkowska, A. (2023) Preparation and characterization of diamond-like carbon coatings for biomedical applications—a review. *Materials* 16 (9): 3420.

35 Guo, Y., Guo, P., Sun, L. et al. (2019) Tribological properties of Ti-doped diamond-like carbon coatings under dry friction and PAO oil lubrication. *Surface and Interface Analysis* 51 (3): 361–370.

36 Viswanathan, S., Mohan, L., Bera, P. et al. (2017) Corrosion and wear behaviors of Cr-doped diamond-like carbon coatings. *Journal of Materials Engineering and Performance* 26 (8): 3633–3647.

37 Wang, A.-Y., Lee, K.-R., Ahn, J.-P. et al. (2006) Structure and mechanical properties of W incorporated diamond-like carbon films prepared by a hybrid ion beam deposition technique. *Carbon* 44 (9): 1826–1832.

38 Koppert, R., Goettel, D., Freitag-Weber, O. et al. (2009) Nickel containing diamond like carbon thin films. *Solid State Sciences* 11 (10): 1797–1800.

39 Ray, S.C., Pong, W.F. and Papakonstantinou, P. (2016) Iron, nitrogen and silicon doped diamond like carbon (DLC) thin films: a comparative study. *Thin Solid Films* 610: 42–47.

40 Nobili, L. and Guglielmini, A. (2013) Thermal stability and mechanical properties of fluorinated diamond-like carbon coatings. *Surface and Coatings Technology* 219: 144–150.

41 Dai, W., Gao, X., Liu, J. et al. (2016) Microstructure, mechanical property and thermal stability of diamond-like carbon coatings with Al, Cr and Si multi-doping. *Diamond and Related Materials* 70: 98–104.

42 Park, C.S., Choi, S.G., Park, H.-H. et al. (2012) Compensation effect of boron and nitrogen codoping on the hardness and electrical resistivity of diamond-like carbon films prepared by magnetron sputtering deposition. *Journal of Materials Research* 27 (23): 3027–3032.

43 de Oliveira, A., Placias, F.G., Sobrinho, A.S.S. et al. (2021) Secondary ion mass spectrometry and atomic force microscopy analysis of silver-doped diamond-like carbon films on titanium alloy (Ti6Al4V) for possible biomedical application. *Thin Solid Films* 719: 138487.

44 Orrit-Prat, J., Bonet, R., Rupérez, E. et al. (2021) Bactericidal silver-doped DLC coatings obtained by pulsed filtered cathodic arc co-deposition. *Surface and Coatings Technology* 411: 126977.

45 Wang, Q., Zhou, F., Zhou, Z. et al. (2013) Effect of titanium or chromium content on the electrochemical properties of amorphous carbon coatings in simulated body fluid. *Electrochimica Acta* 112: 603–611.

46 Wongpanya, P., Pintitraratibodee, N., Thumanu, K. et al. (2021) Improvement of corrosion resistance and biocompatibility of 316L stainless steel for joint replacement application by Ti-doped and Ti-interlayered DLC films. *Surface and Coatings Technology* 425: 127734.

47 Cao, H., Ye, X., Li, H. et al. (2021) Microstructure, mechanical and tribological properties of multilayer Ti-DLC thick films on Al alloys by filtered cathodic vacuum arc technology. *Materials & Design* 198: 109320.

48 Yang, K.-H., Riley, P., Rodenhausen, K.B. et al. (2022) Antifungal behavior of silicon-incorporated diamond-like carbon by tuning surface hydrophobicity with plasma treatment. *International Journal of Applied Ceramic Technology* 19 (5): 2545–2555.

49 Joshi, N., Shukla, S., Khosla, N. et al. (2024) Q-carbon as an emergent surface coating material for antimicrobial applications. *Thin Solid Films* 791: 140227.

50 Jacob, W. and Möller, W. (1993) On the structure of thin hydrocarbon films. *Applied Physics Letters* 63 (13): 1771–1773.

51 Lifshitz, Y. (1999) Diamond-like carbon — present status. *Diamond and Related Materials* 8 (8): 1659–1676.

52 Celii, F.G. and Butler, J.E. (1991) Diamond chemical vapor deposition. *Annual Review of Physical Chemistry* 42: 643–684.

53 Weiler, M., Sattel, S., Jung, K. et al. (1994) Highly tetrahedral, diamond-like amorphous hydrogenated carbon prepared from a plasma beam source. *Applied Physics Letters* 64 (21): 2797–2799.

54 Locher, R., Wild, C., and Koidl, P. (1991) Direct ion-beam deposition of amorphous hydrogenated carbon films. *Surface and Coatings Technology* 47 (1): 426–432.

55 Druz, B., Ostan, E., Distefano, S. et al. (1998) Diamond-like carbon films deposited using a broad, uniform ion beam from an RF inductively coupled CH_4-plasma source. *Diamond and Related Materials* 7 (7): 965–972.

56 Robertson, J. (1991) Hard amorphous (diamond-like) carbons. *Progress in Solid State Chemistry* 21 (4): 199–333.

57 Gielen, J.W.A.M., van de Sanden, M.C.M., and Schram, D.C. (1996) Plasma beam deposited amorphous hydrogenated carbon: improved film quality at higher growth rate. *Applied Physics Letters* 69 (2): 152–154.

58 Aksenov, I.I., Vakula, S.I., Padalka, V.G. et al. (1980) High-efficiency source of pure carbon plasma. *Zhurnal Tekhnicheskoi Fiz.* 50: 2000–2004.

59 Hofsäss, H., Binder, H., Klumpp, T. et al. (1994) Doping and growth of diamond-like carbon films by ion beam deposition. *Diamond and Related Materials* 3 (1): 137–142.

60 Ishikawa, J. (1992) Negative ion beam technology for materials science (invited *Review of Scientific Instruments* 63 (4): 2368–2373.

61 Hirvonen, J.-P., Koskinen, J., Lappalainen, R. et al. (1991) Direct, mass-analyzed ion-beam and arc-discharge deposition of diamond-like films. *Journal of Electronic Materials* 20 (2): 127–132.

62 McKenzie, D.R., Muller, D., Pailthorpe, B.A. et al. (1991) Properties of tetrahedral amorphous carbon prepared by vacuum arc deposition. *Diamond and Related Materials* 1 (1): 51–59.

63 Coll, B.F. and Chhowalla, M. (1994) Modelization of reaction kinetics of nitrogen and titanium during TiN arc deposition. *Surface and Coatings Technology* 68–69: 131–140.

64 McKenzie, D.R. (1996) Tetrahedral bonding in amorphous carbon. *Reports on Progress in Physics* 59 (12): 1611.

65 Falabella, S., Boercker, D.B., and Sanders, D.M. (1993) Fabrication of amorphous diamond films. *Thin Solid Films* 236 (1): 82–86.

66 Tay, B.K., You, G.F., Lau, S.P. et al. (2000) Plasma flow simulation in an off-plane double bend magnetic filter. *Surface and Coatings Technology* 133–134: 593–597.

67 Chhowalla, M., Robertson, J., Chen, C.W. et al. (1997) Influence of ion energy and substrate temperature on the optical and electronic properties of tetrahedral amorphous carbon (ta-C) films. *Journal of Applied Physics* 81 (1): 139–145.

68 Polo, M.C. andújar, J.L., Hart, A. et al. (2000) Preparation of tetrahedral amorphous carbon films by filtered cathodic vacuum arc deposition. *Diamond and Related Materials* 9 (3): 663–667.

69 Brown, I.G. (1998) Cathodic arc deposition of films. *Annual Review of Materials Research* 28: 243–269.

70 Anders, S., anders, A., and Brown, I. (1993) Macroparticle-free thin films produced by an efficient vacuum arc deposition technique. *Journal of Applied Physics* 74 (6): 4239–4241.

71 Scheibe, H.-J. and Schultrich, B. (1994) DLC film deposition by laser-arc and study of properties. *Thin Solid Films* 246 (1): 92–102.

72 Bilek, M.M.M. (1999) The effect of magnetic field configuration on plasma beam profiles in curved magnetic filters. *Journal of Applied Physics* 85 (9): 6385–6391.

73 Kitagawa, T., Miyauchi, K., Toyoda, N. et al. (2004) Optimum incident angle of Ar cluster ion beam for superhard carbon film deposition. *Japanese Journal of Applied Physics* 43 (6S): 3955.

74 Anders, A. (2002) From plasma immersion ion implantation to deposition: a historical perspective on principles and trends. *Surface and Coatings Technology* 156 (1): 3–12.

75 Xu, S., Flynn, D., Tay, B.K. et al. (1997) Mechanical properties and Raman spectra of tetrahedral amorphous carbon films with high sp^3 fraction deposited using a filtered cathodic arc. *Philosophical Magazine B* 76 (3): 351–361.

76 Anders, S. anders, A., Dickinson, M.R. et al. (1997) S-shaped magnetic macroparticle filter for cathodic arc deposition. *IEEE Transactions on Plasma Science* 25 (4): 670–674.

77 Logothetidis, S. (1996) Hydrogen-free amorphous carbon films approaching diamond prepared by magnetron sputtering. *Applied Physics Letters* 69 (2): 158–160.

78 Savvides, N. (1986) Optical constants and associated functions of metastable diamond-like amorphous carbon films in the energy range 0.5–7.3 eV. *Journal of Applied Physics* 59 (12): 4133–4145.

79 Savvides, N. (1984) Deposition parameters and film properties of hydrogenated amorphous silicon prepared by high rate dc planar magnetron reactive sputtering. *Journal of Applied Physics* 55 (12): 4232–4238.

80 Cho, N.-H., Krishnan, K.M., Veirs, D.K. et al. (1990) Chemical structure and physical properties of diamond-like amorphous carbon films prepared by magnetron sputtering. *Journal of Materials Research* 5 (11): 2543–2554.

81 Schwan, J., Ulrich, S., Roth, H. et al. (1996) Tetrahedral amorphous carbon films prepared by magnetron sputtering and DC ion plating. *Journal of Applied Physics* 79 (3): 1416–1422.

82 Cuomo, J.J., Doyle, J.P., Bruley, J. et al. (1991) Sputter deposition of dense diamond-like carbon films at low temperature. *Applied Physics Letters* 58 (5): 466–468.

83 Gissler, W., Hammer, P., and Haupt, J. (1994) Hardness and elasticity of diamond-like carbon films prepared by ion-beam assisted sputter deposition. *Diamond and Related Materials* 3 (4): 770–774.

84 Panda, M., Krishnan, R., Madapu, K.K. et al. (2019) Influence of particulate on surface energy and mechanical property of diamond-like carbon films synthesized by pulsed laser deposition. *Applied Surface Science* 484: 1176–1183.

85 Gupta, S., Bhaumik, A., Sachan, R. et al. (2018) Structural evolution of Q-carbon and nanodiamonds. *JOM* 70 (4): 450–455.

86 Zoubos, H., Koutsokeras, L.E., Anagnostopoulos, D.F. et al. (2013) Broadband optical absorption of amorphous carbon/Ag nanocomposite films and its potential for solar harvesting applications. *Solar Energy Materials and Solar Cells* 117: 350–356.

87 Narayan, J., Biunno, N., Singh, R. et al. (1987) Formation of thin superconducting films by the laser processing method. *Applied Physics Letters* 51 (22): 1845–1847.

88 Biunno, N., Narayan, J., Hofmeister, S.K. et al. (1989) Low-temperature processing of titanium nitride films by laser physical vapor deposition. *Applied Physics Letters* 54 (16): 1519–1521.

89 Sato, T., Furuno, S., Iguchi, S. et al. (1987) Deposition of diamond-like carbon films by pulsed-laser evaporation. *Japanese Journal of Applied Physics* 26 (9A): L1487.

90 Lu, Y., Huang, G., Wang, S. et al. (2021) A review on diamond-like carbon films grown by pulsed laser deposition. *Applied Surface Science* 541: 148573.

91 Xiong, F., Wang, Y.Y., Leppert, V. et al. (1993) Pulsed laser deposition of amorphous diamond-like carbon films with ArF (193 nm) excimer laser. *Journal of Materials Research* 8 (9): 2265–2272.

92 Stock, F., Antoni, F., Le Normand, F. et al. (2017) High performance diamond-like carbon layers obtained by pulsed laser deposition for conductive electrode applications. *Applied Physics A* 123 (9): 590.

93 Yamamoto, K., Koga, Y., Fujiwara, S. et al. (1998) Dependence of the sp^3 bond fraction on the laser wavelength in thin carbon films prepared by pulsed laser deposition. *Applied Physics A* 66 (1): 115–117.

94 Constantinou, M., Pervolaraki, M., Nikolaou, P. et al. (2017) Microstructure and nanomechanical properties of pulsed excimer laser deposited DLC:Ag films: enhanced nanotribological response. *Surface and Coatings Technology* 309: 320–330.

95 Jelínek, M., Zemek, J., Remsa, J. et al. (2017) Hybrid laser technology and doped biomaterials. *Applied Surface Science* 417: 73–83.

96 Siegal, M.P., Barbour, J.C., Provencio, P.N. et al. (1998) Amorphous-tetrahedral diamond-like carbon layered structures resulting from film growth energetics. *Applied Physics Letters* 73 (6): 759–761.

97 Merkulov, V.I., Lowndes, D.H., Jellison, G.E., Jr. et al. (1998) Structure and optical properties of amorphous diamond films prepared by ArF laser ablation as a function of carbon ion kinetic energy. *Applied Physics Letters* 73 (18): 2591–2593.

98 Voevodin, A.A. and Donley, M.S. (1996) Preparation of amorphous diamond-like carbon by pulsed laser deposition: a critical review. *Surface and Coatings Technology* 82 (3): 199–213.

99 Siegal, M.P., Martinez-Miranda, L.J., DiNardo, N.J. et al. (1998) Characterization of amorphous carbon films grown by pulsed laser deposition. *High-Power Laser Ablation* 3343: 885–894.

100 Kumar, I. and Khare, A. (2016) Optical nonlinearity in nanostructured carbon thin films fabricated by pulsed laser deposition technique. *Thin Solid Films* 611: 56–61.

101 Rau, J.V., Teghil, R., De Bonis, A. et al. (2010) Pulsed laser deposition of hard and superhard carbon thin films from C60 targets. *Diamond and Related Materials* 19 (1): 7–14.

102 Patsalas, P. (2011) Optical properties of amorphous carbons and their applications and perspectives in photonics. *Thin Solid Films* 519 (12): 3990–3996.

103 Panda, M., Mangamma, G., Krishnan, R. et al. (2016) Nano scale investigation of particulate contribution to diamond like carbon film by pulsed laser deposition. *RSC Advances* 6 (8): 6016–6028.

104 Grigoriev, SN, Romanov, RI and Fominskii, V.Yu. (2012) Dependence of mechanical and tribological properties of diamond-like carbon coatings on laser deposition conditions and alloying by metals. *Journal of Friction and Wear* 33 (4): 253–259.

105 Grigoriev, S.N., Fominski, V.Yu., Romanov, R.I. et al. (2014) Effect of energy fluence and Ti/W co-deposition on the structural, mechanical and tribological characteristics of diamond-like carbon coatings obtained by pulsed Nd:YAG laser deposition on a steel substrate. *Surface and Coatings Technology* 259: 415–425.

106 Gayathri, S., Kumar, N., Krishnan, R. et al. (2015) Influence of transition metal doping on the tribological properties of pulsed laser deposited DLC films. *Ceramics International* 41 (1, Part B): 1797–1805.

107 Yap, S.S., Siew, W.O., Nee, C.H. et al. (2011) Parametric studies of diamond-like carbon by pulsed Nd:YAG laser deposition. *Diamond and Related Materials* 20 (3): 294–298.

108 Hevia, S.A., Guzmán-Olivos, F., Muñoz, I. et al. (2017) Nanostructured substrate effects on diamond-like carbon films properties grown by pulsed laser deposition. *Surface and Coatings Technology* 312: 55–60.

109 Panda, M., Krishnan, R., Ravindran, T.R. et al. (2016) Spectroscopic studies on diamond like carbon films synthesized by pulsed laser ablation. *AIP Conference Proceedings* 1731 (1): 080040.

110 Sikora, A., Garrelie, F., Donnet, C. et al. (2010) Structure of diamond-like carbon films deposited by femtosecond and nanosecond pulsed laser ablation. *Journal of Applied Physics* 108 (11): 113516.

111 Katsuno, T., Godet, C., Orlianges, J.C. et al. (2005) Optical properties of high-density amorphous carbon films grown by nanosecond and femtosecond pulsed laser ablation. *Applied Physics A* 81 (3): 471–476.

112 Maddi, C., Donnet, C., Loir, A.-S. et al. (2015) High N-content a-C:N films elaborated by femtosecond PLD with plasma assistance. *Applied Surface Science* 332: 346–353.

113 Loir, A.-S., Garrelie, F., Donnet, C. et al. (2005) Mechanical and tribological characterization of tetrahedral diamond-like carbon deposited by femtosecond pulsed laser deposition on pre-treated orthopaedic biomaterials. *Applied Surface Science* 247 (1): 225–231.

114 Patsalas, P., Kaziannis, S., Kosmidis, C. et al. (2007) Optimized pulsed laser deposition by wavelength and static electric field control: the case of tetrahedral amorphous carbon films. *Journal of Applied Physics* 101 (12): 124903.

115 Foong, Y.M., Koh, A.T.T., Lim, S.R. et al. (2012) Materials properties of ZnO/diamond-like carbon (DLC) nanocomposite fabricated with different source of targets. *Diamond and Related Materials* 25: 103–110.

116 Panda, M., Krishnan, R., Krishna, N.G. et al. (2019) Tuning the tribological property of PLD deposited DLC-Au nanocomposite thin films. *Ceramics International* 45 (7, Part A): 8847–8855.

117 Gayathri, S., Kumar, N., Krishnan, R. et al. (2015) Influence of Cr content on the micro-structural and tribological properties of PLD grown nanocomposite DLC-Cr thin films. *Materials Chemistry and Physics* 167: 194–200.

118 Grigoriev, S.N., Fominski, V.Yu., Volosova, M.A. et al. (2016) Surface modification of diamond-like carbon coatings to control over run-in processes in friction pair. *Journal of Friction and Wear* 37 (1): 7–14.

119 Písařík, P., Jelínek, M., Kocourek, T. et al. (2014) Chromium-doped diamond-like carbon films deposited by dual-pulsed laser deposition. *Applied Physics A* 117 (1): 83–88.

120 Cheng, Y., Lu, Y., Huang, G. et al. (2017) Experimental research on Ge-doped diamond-like carbon films prepared by double laser beams deposition. *Chinese Journal of Lasers* 44 (3): 0303001.

121 Lu, Y., Huang, G., and Xi, L. (2019) Tribological and mechanical properties of the multilayer DLC film on the soft copper substrate. *Diamond and Related Materials* 94: 21–27.

122 Jai Won, Y. and Ki, H. (2013) Fabricating functionally graded films with designed gradient profiles using pulsed laser deposition. *Journal of Applied Physics* 113 (17): 174910.

123 Cho, H., Kim, S. and Ki, H. (2012) Pulsed laser deposition of functionally gradient diamond-like carbon (DLC) films using a 355 nm picosecond laser. *Acta Materialia* 60 (18): 6237–6246.

124 Plotnikov, V.A., Dem'yanov, B.F., Yeliseeyev, A.P. et al. (2019) Structural state of diamond-like amorphous carbon films, obtained by laser evaporation of carbon target. *Diamond and Related Materials* 91: 225–229.

125 Lu, Y., Huang, G., Cheng, Y. et al. (2020) Optical and micro-structural properties of the uniform large-area carbon-based films prepared by pulsed laser deposition. *Infrared Physics & Technology* 104: 103113.

126 Wang, W., Yang, W., Liu, Z. et al. (2014) Achieve 2-inch-diameter homogeneous GaN films on sapphire substrates by pulsed laser deposition. *Journal of Materials Science* 49 (9): 3511–3518.

127 Menzel, M., Weissbach, D., Gawlitza, P. et al. (2004) Deposition of high-resolution carbon/carbon multilayers on large areas for X-ray optical applications. *Applied Physics A* 79 (4): 1039–1042.

128 Zhang, K., Dai, J., Wu, W. et al. (2015) Development of a high magnetic field assisted pulsed laser deposition system. *The Review of Scientific Instruments* 86 (9): 095105.

129 Modabberasl, A., Kameli, P., Ranjbar, M. et al. (2015) Fabrication of DLC thin films with improved diamond-like carbon character by the application of external magnetic field. *Carbon* 94: 485–493.

130 Holland, L. and Ojha, S.M. (1976) Deposition of hard and insulating carbonaceous films on an r.f. target in a butane plasma. *Thin Solid Films* 38 (2): L17–L19.

131 Bubenzer, A., Dischler, B., Brandt, G. et al. (1983) rf-plasma deposited amorphous hydrogenated hard carbon thin films: preparation, properties and applications. *Journal of Applied Physics* 54 (8): 4590–4595.

132 Catherine, Y. and Couderc, P. (1986) Electrical characteristics and growth kinetics in discharges used for plasma deposition of amorphous carbon. *Thin Solid Films* 144 (2): 265–280.

133 Wild, Ch. and Koidl, P. (1989) Structured ion energy distribution in radio frequency glow-discharge systems. *Applied Physics Letters* 54 (6): 505–507.

134 Wild, C. and Koidl, P. (1991) Ion and electron dynamics in the sheath of radio-frequency glow discharges. *Journal of Applied Physics* 69 (5): 2909–2922.

135 Weiler, M., Lang, K., Li, E. et al. (1998) Deposition of tetrahedral hydrogenated amorphous carbon using a novel electron cyclotron wave resonance reactor. *Applied Physics Letters* 72 (11): 1314–1316.

136 Conway, N.M.J., Ferrari, A.C., Flewitt, A.J. et al. (2000) Defect and disorder reduction by annealing in hydrogenated tetrahedral amorphous carbon. *Diamond and Related Materials* 9 (3): 765–770.

137 Lieberman, M.A. and Lichtenberg, A.J. (1994) Principles of plasma discharges and materials processing. *MRS Bulletin* 30 (12): 899–901.

138 Weiler, M., Sattel, S., Giessen, T. et al. (1996) Preparation and properties of highly tetrahedral hydrogenated amorphous carbon. *Physical Review B* 53 (3): 1594–1608.

139 Zarrabian, M., Fourches-Coulon, N., Turban, G. et al. (1997) Observation of nanocrystalline diamond in diamond-like carbon films deposited at room temperature in electron cyclotron resonance plasma. *Applied Physics Letters* 70 (19): 2535–2537.

140 He, X.-M., Lee, S.-T., Bello, I. et al. (1999) Characterization and optical properties of diamond-like carbon prepared by electron cyclotron resonance plasma. *Journal of Materials Research* 14 (3): 1055–1061.

141 Weiler, M. and Yankov, R. (1998) Large-area, high-rate deposition of thin amorphous carbon films using a novel controlled plasma reactor. In: *Proceedings of 12th International Conference on Vacuum Web Coating Conference*, Reno, NV. USA (ed. R. Bakish), 60–70.

142 Koidl, P., Wild, C., Dischler, B. et al. (1990) Plasma deposition, properties and structure of amorphous hydrogenated carbon films. *Materials Science Forum* 52: 41–70.

143 Zou, J.W., Schmidt, K., Reichelt, K. et al. (1990) The properties of a-C:H films deposited by plasma decomposition of C_2H_2. *Journal of Applied Physics* 67 (1): 487–494.

144 Günther, M., Bialuch, I., Peter, S. et al. (2011) High rate deposition of hard aC: H films using microwave excited plasma enhanced CVD. *Surface and Coatings Technology* 205: S94–S98.

145 Riley, P.R., Joshi, P., Azizi Machekposhti, S. et al. (2021) Enhanced vapor transmission barrier properties via silicon-incorporated diamond-like carbon coating. *Polymers* 13 (20): 3543.

146 Sanchette, F., El Garah, M., Achache, S. et al. (2021) DLC-based coatings obtained by low-frequency plasma-enhanced chemical vapor deposition (LFPECVD) in cyclohexane, principle and examples. *Coatings* 11 (10): 1225.

147 Kaufman, H.R. (1978) Technology of ion beam sources used in sputtering. *Journal of Vacuum Science and Technology* 15 (2): 272–276.

148 McKenzie, D.R., Muller, D., and Pailthorpe, BA (1991) Compressive-stress-induced formation of thin-film tetrahedral amorphous carbon. *Physical Review Letters* 67 (6): 773–776.

149 Collins, C.B., Davanloo, F., Jander, D.R. et al. (1991) Microstructure of amorphic diamond films. *Journal of Applied Physics* 69 (11): 7862–7870.

150 Weissmantel, C., Reisse, G., Erler, H.-J. et al. (1979) Preparation of hard coatings by ion beam methods. *Thin Solid Films* 63 (2): 315–325.

151 Jansen, F., Machonkin, M., Kaplan, S. et al. (1985) The effects of hydrogenation on the properties of ion beam sputter deposited amorphous carbon. *Journal of Vacuum Science and Technology A* 3 (3): 605–609.

152 Lifshitz, Y., Kasi, S.R., Rabalais, J.W. et al. (1990) Subplantation model for film growth from hyperthermal species. *Physical Review B* 41 (15): 10468–10480.

153 Lifshitz, Y., Kasi, S., and Rabalais, J. (1990) Carbon (sp^3) film growth from mass selected ion beams: parametric investigations and subplantation model. *Materials Science Forum* 52: 237–290.

154 Robertson, J. (1994) The deposition mechanism of diamond-like a-C and a-C: H. *Diamond and Related Materials* 3 (4): 361–368.

155 Lifshitz, Y., Lempert, G.D., and Grossman, E. (1994) Substantiation of subplantation model for diamond-like film growth by atomic force microscopy. *Physical Review Letters* 72 (17): 2753–2756.

156 Aijaz, A. and Kubart, T. (2017) Ion induced stress relaxation in dense sputter-deposited DLC thin films. *Applied Physics Letters* 111 (5): 051902.

157 Narayan, J., Fathy, D., Oen, O.S. et al. (1984) Atomic structure of ion implantation damage and process of amorphization in semiconductors. *Journal of Vacuum Science and Technology A* 2 (3): 1303–1308.

158 Rehn, L.E., Greene, J., and Smidt, F.A. (1989) *Processing and Characterization of Materials Using Ion Beams*. Warrendale, PA: Materials Research Society.

159 Uhlmann, S., Frauenheim, Th., and Lifshitz, Y. (1998) Molecular-dynamics study of the fundamental processes involved in subplantation of diamond-like carbon. *Physical Review Letters* 81 (3): 641–644.

160 Hofsäss, H., Feldermann, H., Merk, R. et al. (1998) Cylindrical spike model for the formation of diamond-like thin films by ion deposition. *Applied Physics A* 66 (2): 153–181.

161 Ronning, C. (2003) Ion-beam synthesis and growth mechanism of diamond-like materials. *Applied Physics A* 77 (1): 39–50.

162 Lifshitz, Y., Lempert, G.D., Grossman, E. et al. (1995) Growth mechanisms of DLC films from C+ ions: experimental studies. *Diamond and Related Materials* 4 (4): 318–323.

163 Anne Thomson, L., Law, F.C., Rushton, N. et al. (1991) Biocompatibility of diamond-like carbon coating. *Biomaterials* 12 (1): 37–40.

164 Hu, J., Tian, X., Liu, H. et al. (2021) Enhancement of discharge and deposition rate by imposed pulse current during the cathodic arc deposition of ta-C films. *Vacuum* 193: 110515.

165 Schneider, D., Scheibe, H.-J., Schwarz, Th. et al. (1993) Characterization of thin diamond-like carbon films by ultrasonic surface waves. *Diamond and Related Materials* 2 (11): 1396–1401.

166 Pappas, D.L., Saenger, K.L., Bruley, J. et al. (1992) Pulsed laser deposition of diamond-like carbon films. *Journal of Applied Physics* 71 (11): 5675–5684.

167 Ferreira, F., Aijaz, A., Kubart, T. et al. (2018) Hard and dense diamond like carbon coatings deposited by deep oscillations magnetron sputtering. *Surface and Coatings Technology* 336: 92–98.

168 Viswanathan, S., Mohan, L., Bera, P. et al. (2018) Corrosion and wear resistance properties of multilayered diamond-like carbon nanocomposite coating. *Surface and Interface Analysis* 50 (3): 265–276.

169 Ren, Z., Qin, H., Dong, Y. et al. (2019) A boron-doped diamond like carbon coating with high hardness and low friction coefficient. *Wear* 436–437: 203031.

3

Properties of Diamond-like Carbon Coatings

Muhammad Abdul Basit, Ramsha Zafar, Nawal Haider

Department of Materials Science and Engineering, Institute of Space Technology, Islamabad, Pakistan

CONTENTS

3.1 Introduction to Diamond-like Carbon Coating, 72
3.2 General Categorization of DLC Coatings, 72
3.3 Preparation of DLC Coatings, 74
3.4 Characterization Techniques of DLC Coatings, 74
3.5 Properties of DLC Coatings, 76
3.5.1 Physical Properties of DLC Coatings, 76
3.5.1.1 Density, 77
3.5.1.2 Surface-free Energy, 77
3.5.1.3 Thermal Conductivity, 79
3.5.2 Chemical Properties of DLC Coatings, 80
3.5.2.1 Reactivity and Chemical Stability, 81
3.5.3 Anticorrosion Properties of DLC Coatings, 82
3.5.4 Biocompatibility of DLC Coatings, 83
3.5.5 Opto-electrical Properties of DLC Coatings, 84
3.5.5.1 Electrical Resistance, 84
3.5.5.2 Bandgap, 86
3.5.5.3 Refractive Index, 86
3.5.6 Tribological Properties of DLC Coatings, 87
3.5.6.1 Friction Coefficient, 88
3.5.6.2 Wear Resistance, 91
3.5.7 Adhesion of DLC Coatings to Substrates, 92
3.5.8 Mechanical Properties of DLC Coatings, 94
3.5.8.1 Hardness and Young's Modulus, 94
3.5.8.2 Fatigue Strength of DLC Coatings, 96
3.6 Improving the Properties of DLC Coatings, 96
3.6.1 Modification Through Alloy Addition, 97
3.6.2 Modification Through Multilayer and Hybrid Coatings, 98
3.7 Applications and Uses of DLC Coatings, 100
3.8 Summary and Future Outlook, 100

Applications of Diamond-like Carbon Coatings, First Edition. Abdul Wasy Zia.

3.1 Introduction to Diamond-like Carbon Coating

Carbon, classified as a metalloid, exhibits a unique amalgamation of metallic and nonmetallic properties. Within the realm of carbon's allotropic forms, graphite and diamond are prominently recognized. Graphite, characterized by a layered hexagonal structure, manifests commendable attributes, such as chemical and thermal stability, and lubricity. However, it possesses a notable drawback in its inherent softness. Conversely, diamond, acknowledged as the preeminent pinnacle of carbon hardness, showcases unparalleled resistance to wear and abrasion. Consequently, diamond coatings have garnered multifaceted applications across industries encompassing cutting tools [1], automotive [2], aerospace [3], and biomedical equipment [4]. Nonetheless, these coatings are subject to cost constraints due to their expense and limited by their reliance on single-crystal configurations, thereby imposing restrictions on their practical implementation. In addressing these concerns, researchers have redirected their focus toward amorphous carbon (a-C) coatings, commonly referred to as diamond-like carbon (DLC). Initially reported by Schmellenmeier in 1953, DLC coatings showcase remarkable hardness and wear resistance arising from their substantial diamond-like composition [5]. Diverging from natural diamonds, DLC lacks a crystalline lattice structure, thereby assuming an amorphous state. The advent of DLC coatings materialized during the 1970s when Dr. Samuel D. Roscoe and Dr. Robert H. Wentorf Jr., fortuitously generated a noncrystalline carbon form while pursuing the synthesis of diamond films via chemical vapor deposition (CVD) at the General Electric Research Laboratory [6]. Since then, DLC has undergone noteworthy advancements, prompting the development of diverse deposition and characterization methodologies. As a result, DLC coatings have emerged as a pivotal technological innovation within the materials science domain, attaining substantial commercial utility across various industries. While initially embraced within the biomedical and nuclear sectors, DLC coatings have demonstrated exceptional capabilities within the automotive, cutting tools, and optical products industries. Over the years, distinct variations of DLC coatings have been introduced, coupled with refinements in their properties accomplished through modified preparation techniques, and the incorporation of alloying elements. Providing further testament to its burgeoning prospects, revenue generated by DLC coatings has witnessed a substantial increase from 0.8 to 1.7 billion dollars over the past decade, elucidating the considerable growth trajectory for this industry in the foreseeable future [7].

3.2 General Categorization of DLC Coatings

DLC coatings, renowned for their exceptional hardness, low friction, and chemical resistance, are commonly classified into two main categories based on the presence or absence of doping materials as shown in Figure 3.1 [8]. Undoped DLC

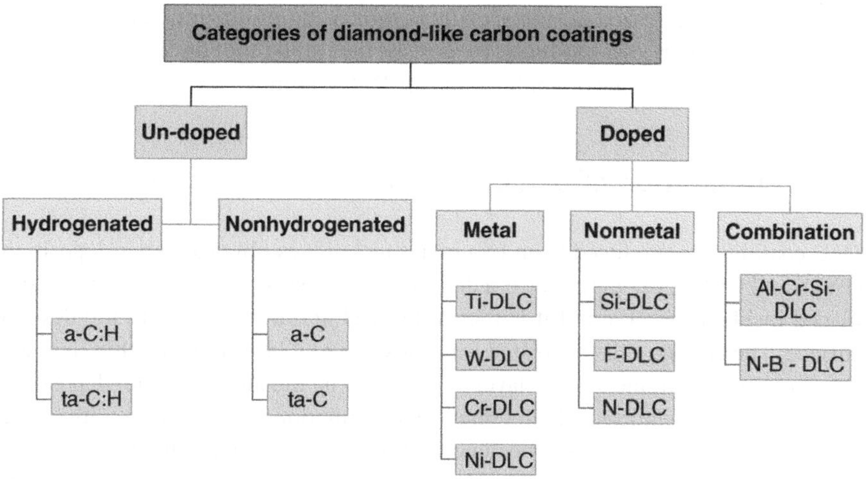

Figure 3.1 Classification of DLC coatings.

coatings, also known as pure DLC or a-C, are composed solely of carbon atoms. These coatings offer outstanding tribological properties, including high hardness, excellent adhesion, and low friction. Undoped DLC coatings find applications in various industries where a combination of hardness and low friction is desired. For example, in the medical field, DLC coatings are used on surgical instruments, implants, and medical devices to improve wear resistance, reduce friction, and enhance biocompatibility [9]. On the other hand, doped DLC coatings involve the introduction of additional elements or compounds into the DLC matrix, which can alter the coating's properties to achieve specific characteristics. Metal-doped DLC coatings, such as tungsten-doped [10], titanium-doped [11], chromium-doped [12], nickel-doped [13], and iron-doped [14] DLC coatings, exhibit enhanced hardness, improved thermal stability, and specific functionalities suitable for diverse applications. These coatings find use in cutting tools, automotive components, and industrial equipment, where exceptional wear resistance and durability are crucial. Nonmetal-doped DLC coatings, including silicon-doped [14], fluoride-doped [15], and nitrogen-doped DLC coatings [14], offer unique properties depending on the dopant material. Silicon-doped DLC coatings, for instance, exhibit increased hardness and improved adhesion, making them ideal for applications where wear resistance and coating-substrate bonding are paramount. Nitrogen-doped DLC coatings, on the other hand, demonstrate reduced friction and enhanced tribological performance, making them suitable for applications in engine components, such as piston rings and valve lifters [14]. Additionally, there is a category of doped DLC coatings that combine both metallic and nonmetallic doping elements. These hybrid coatings, such as aluminum-chromium-silicon-doped [16] and nitrogen-boron-doped DLC coatings [17], offer tailored properties based on the specific combination of dopants. They can provide a unique balance of hardness, wear resistance, and other desired characteristics, expanding the range of applications for DLC coatings.

3.3 Preparation of DLC Coatings

DLC coatings are produced from two major techniques physical vapor deposition (PVD) and CVD. PVD technique is a deposition process in which a solid material is placed inside a vacuum and it is then vaporized and deposited onto the desired surface. The coating thus deposited is not just a metal layer in fact a thin, bonded metal or metal-ceramic surface layer is produced by the atom-by-atom deposition of compound substances, which enhances the functionality and durability of the part on which the coating is to be deposited. The PVD method is further classified into sputtering, ion beam, pulse laser, and cathodic vacuum arc deposition methods [5]. Magnetron sputtering and cathodic arc evaporation are used for the synthesis of hydrogen-free DLC coatings. Sputter deposition is a method to produce carbon film on a base material with the help of a solid graphite sputter vapor source with a low coating deposition rate resulting in the formation of DLC films having no hydrogen content. The cathode arc PVD method is another technique developed in the 1970s that generates a vacuum arc discharge on a solid graphite cathode to produce a DLC layer on a base material at a negative potential. This technique is capable of producing hard, hydrogen-free DLC films due to high vaporization and ionization rates [18]. The CVD method is a process in which a solid material is deposited as a result of a chemical reaction of precursor gases taking place near the heated surface of the desired substrate. The CVD method is further classified into radio frequency method, self-discharge method, and plasma enhanced CVD method to deposit hydrogenated amorphous carbon (a-C:H) DLC films.

3.4 Characterization Techniques of DLC Coatings

DLC coatings undergo extensive characterization using a range of techniques to assess their quality and suitability for specific applications. Various characterization methods are used to gain insights into the properties and behavior of DLC coatings. Figure 3.2 illustrates the comprehensive set of characterization techniques utilized for DLC coatings. Scanning electron microscopy (SEM) enables high-resolution imaging of the DLC coating's surface morphology [19, 20], offering valuable information about its texture, grain structure, and any defects or cracks present. Atomic force microscopy provides detailed topographic information at the nanoscale, including surface roughness, thickness variations, and surface features [21]. Raman spectroscopy is used to analyze the molecular

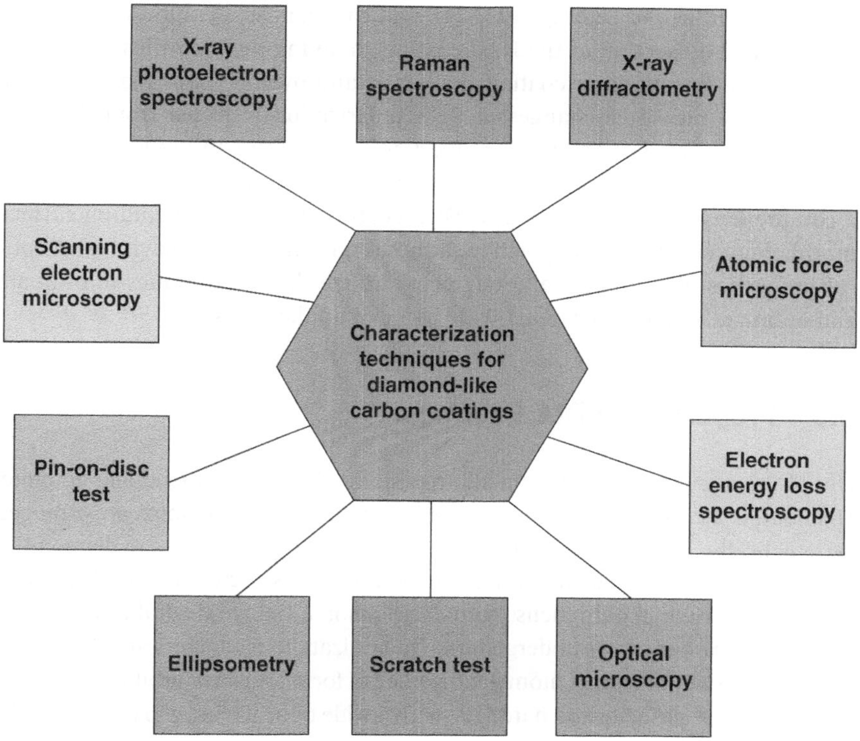

Figure 3.2 Techniques for the characterization of DLC coatings.

bonding and structure of DLC coatings [19, 20], facilitating the identification of sp^2 and sp^3 carbon bonds and shedding light on the coating's amorphous or crystalline nature. X-ray diffraction (XRD) is used to examine the crystallinity and crystal structure of the DLC coating, enabling the detection of crystalline phases and evaluating the degree of graphitization. X-ray photoelectron spectroscopy (XPS) analyzes the elemental composition and chemical state of the DLC coating's surface, providing insights into the types of carbon bonds present (sp^2 or sp^3), identifying dopants or contaminants, and offering information about the coating's chemical environment [22]. The scratch test evaluates the DLC coating's adhesion and hardness through controlled mechanical scratching, assessing its resistance to wear, delamination, and failure under load [20, 22]. The pin-on-disc test measures the coating's friction and wear properties by sliding a pin against a rotating disc, yielding information about the coefficient of friction, wear rate, and lubrication performance [22]. Electron energy loss spectroscopy is an electron microscopy technique that investigates the elemental composition and electronic structure of materials, including DLC coatings. It provides valuable insights into energy loss processes and electronic transitions occurring within the coatings [23, 24]. Optical microscopy allows for macroscopic visual inspection of the DLC coating, enabling the identification of surface defects, variations in coating thickness, and overall coating quality [22]. Ellipsometry measures changes in light polarization reflected from the DLC coating, providing data on optical properties, such as refractive index, thickness, and surface roughness [25]. Collectively, these characterization techniques offer a comprehensive understanding of DLC coating properties, including surface morphology, structure, composition, adhesion, mechanical behavior, and optical properties. Researchers and engineers utilize these techniques to evaluate and optimize DLC coatings for a wide range of applications.

3.5 Properties of DLC Coatings

Out of the many, two of the main allotropes of carbon are diamond and graphite. Diamond is a crystalline form of carbon in which the carbon atoms are arranged systematically. This cubic crystalline form exerts various properties to diamond as the arrangement of carbon atoms is systematic and the structure is densely packed, which gives rise to a hard, dense form. In diamond, the tetrahedral arrangement involves the carbon atoms undergoing sp^3 hybridization in which 1s and 3p orbitals of the same shell of carbon atom get together to form four new orbitals. Diamond has the highest elasticity and hardness, with a wide optical band gap of 5.47 eV [8]. Graphite consists of carbon atoms arranged in weakly assembled sheets or layers undergoing sp^2 hybridization resulting in a hexagonal structure. DLC coatings are a-C with a combination of both the diamond and graphite form of carbon and depending upon their content, the overall properties of DLC coatings are governed. DLC films possess high hardness, low friction, electrical insulation, chemical inertness, resistivity, optical transparency, biocompatibility, and high resistance to wear [26]. We will discuss the various advantageous properties associated with DLC, as presented in Figure 3.3, in the upcoming sections.

3.5.1 Physical Properties of DLC Coatings

Physical property is the measurable property of a material that is discerned without any change in its chemical identity. DLC is a metastable material that needs to be bombarded with energetic ions to deposit films that possess properties similar to that of the diamond. DLC coatings possess interesting physical characteristics, such as density, thermal conductivity, and surface tension, which will be discussed in the following sections. These physical properties vary with deposition parameters, film structure either amorphous or crystalline, sp^3/sp^2 ratio, and hydrogen content. Hence, it is vital to have a thorough understanding of these parameters to make use of DLC films in any of the desired applications.

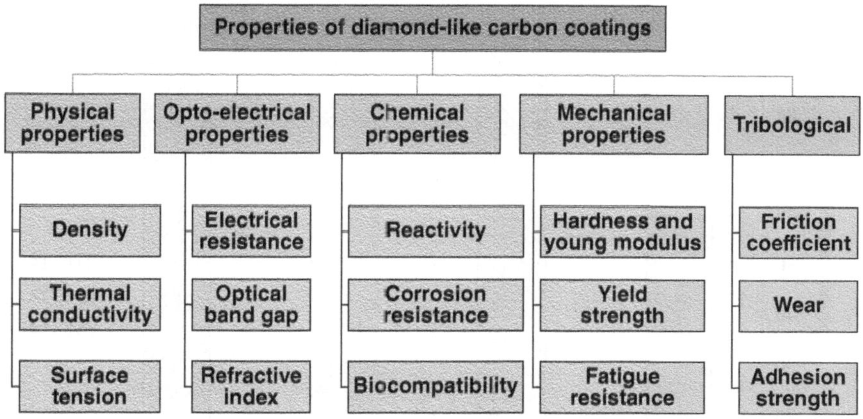

Figure 3.3 Overview of DLC properties.

3.5.1.1 Density

The density, which is mass per unit volume, of diamond is 3.5 g cm^{-3} with the atoms being densely packed, while that for graphite is 2.26 g cm^{-3}. DLC, which is a combination of both diamond and graphite, has less density than diamond depending on its various types. The density of a-C or a-C:H ranges from 1.6 to 2.2 g cm^{-3}. a-C:H films that are deposited through plasma-assisted CVD (PACVD) have a lower sp^3 content of approximately 50% along with a large amount of hydrogen, which induces a lower density as compared to other types [8]. Nonhydrogenated carbon films (a-C) exhibit higher densities approaching that of natural diamonds due to low hydrogen content and higher sp^3 content of 85%. The tetrahedral amorphous carbon films (ta-C), which are typically prepared from an energetic method of deposition-filtered cathode vacuum arc (FCVA), exhibit a density of 3.2 g cm^{-3} resulting in hard, dense films [26]. The low-density DLC coatings can provide beneficial properties overall in applications where low weight is desirable and enhance the performance of such coatings.

3.5.1.2 Surface-free Energy

Surface-free energy is the energy required to enlarge the surface area of the solid phase. DLC coatings are characterized by low surface-free energies and are known for their hydrophobic nature, which is in turn dependent on surface composition and surface modifications. Generally, it is assumed that if the contact angle for a water droplet is large enough, it becomes difficult for the liquid to penetrate into the substrate. The contact angle is the angle formed by the liquid droplet at the three-phase boundary, where three phases of solid, liquid, and vapor phases interact. Figure 3.4 depicts the contact angles for bare zinc substrate and zinc samples coated with DLC. The bare zinc sample exhibits a smaller contact angle of

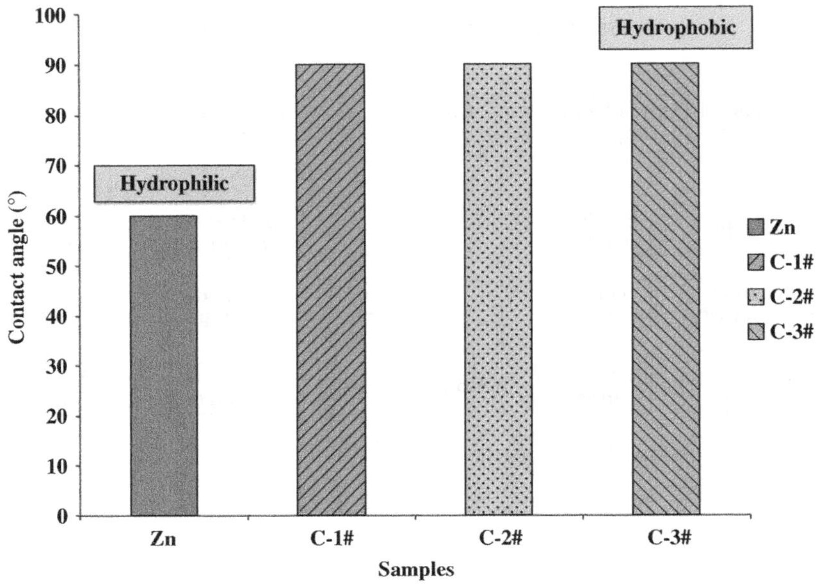

Figure 3.4 Water contact angles for coated and uncoated zinc samples.

approximately 60°, and for the DLC-coated zinc sample, the contact angle increases to approximately 90°. The larger contact angles thus implement DLC coatings in various materials to produce hydrophobic surfaces [27].

The wettability characteristics of tool steel material coated with DLC were also assessed in terms of changes in surface-free energy and contact angle measurements [28]. The contact angle on any solid surface is measured through a common technique in which a camera and ImageJ with drop analysis are used to evaluate the measurements. Steel samples with and without DLC-coated surfaces are placed on a flat surface. Then, a liquid droplet of water is utilized, followed by a high-resolution image of a liquid droplet on a solid surface taken by the camera. ImageJ software is used to analyze the results in which the area of interest is selected and the contact angle is measured using a drop analysis plug-in, which calculates the contact angle through mathematical approximations [29]. These approximations use drop height (h) and base diameter (l) to measure the contact angle through the equation $\theta = 2\ \tan^{(-1)\,(2h/l)}$. Young's equation associates the surface energies of three interfaces (solid–liquid, solid–vapor, and liquid–vapor) with the equilibrium contact angle as $\gamma SV = \gamma SL + \gamma LV \cos\theta$, as shown in Figure 3.5.

To determine the surface energy, Fowkes method is utilized, which provides estimations based on contact angle measurements:

$$\gamma s = \gamma s^{d} = \gamma 1^{2}(1 + cos\theta)^{2}/(4\gamma s^{d}), \tag{3.1}$$

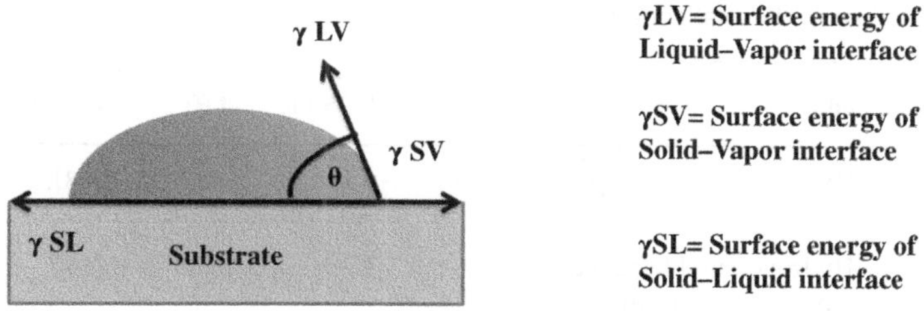

Figure 3.5 Schematic of liquid droplet deposited onto a solid substrate with the surface tension of solid, liquid, and vapor phases.

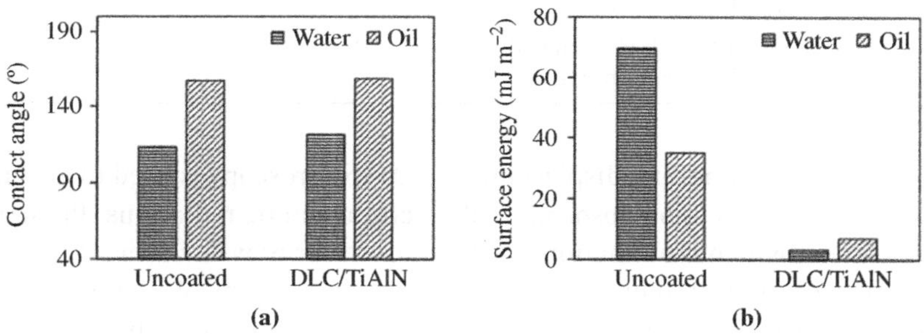

Figure 3.6 (a) Contact angles and (b) surface-free energies of DLC/TiAlN-coated and uncoated tool steel sample. *Source:* [28] / IOP Publishing / CC BY 3.0.

where θ is the contact angle, γs is surface energy, and $\gamma 1$ is the surface tension of the liquid. As depicted in Figure 3.6a,b, the uncoated tool steel surface exhibits a high surface energy of 69.73 mJ m^{-2} and a lower contact angle of 113.39° as compared to DLC/TiAlN-coated tool with low surface energy of 3.22 mJ m^{-2} and high contact angle of 121.57°. This suggests poorer wettability of DLC coating and the lubricating medium. These results could be justified as oxidation of coating could take place which in turn could affect the surface energy. Varying crystal structures of DLC coatings with different polarities also influence the surface chemistry [30].

3.5.1.3 Thermal Conductivity

Thermal conductivity refers to the potential of material to transfer heat and is denoted by K. The thermal conductivity of DLC coating varies from 400 to 1000 W mk^{-1} and depends upon the process of deposition of films [31]. By inspecting samples with various sp^3/sp^2 content and hydrogen content, their impact on the physical properties is determined [32]. The a-C:H DLC films are first deposited on a pure silicon substrate followed by measurements to determine hydrogen content, sp^3/sp^2 ratio, thermal conductivity, and thermal

Table 3.1 List of thermal conductivity of five different carbon films.

S. no.	Film	K (W mk^{-1}) at RT	References
1.	Hydrogenated amorphous carbon (a-C:H)	0.1–0.3	[18]
2.	Hydrogenated tetrahedral amorphous carbon (ta-C:H)	1.3	[18]
3.	Tetrahedral amorphous carbon (ta-C)	1.41–2.7	[33]
4.	Polymer-like hydrogenated amorphous carbon (PLCH)	0.277	[33]
5.	Graphite-like hydrogenated amorphous carbon (GLCH)	0.248	[33]

coefficient. Photothermal displacement (PD) spectroscopy is used to evaluate thermal properties upon absorption of electromagnetic radiations. PD signal ratio (A/A_{si}), which is a function of temperature varies with different samples showing variations in thermal expansion behavior of films. With the decrease in sp^3/sp^2 ratio, the thermal coefficient of the films increases, which indicates that films containing higher sp^2 content are more likely to expand or contract as temperature changes occur. It is also determined that the thermal conductivity of a-C:H film is also dependent on the sp^3/sp^2 ratio and the films richer in sp^2 possess lower thermal conductivity. Thermal conductivities for various types of carbon films are presented in Table 3.1. Graphite-like a-C:H is the type of carbon film containing hydrogen content of approximately 24 at.% and sp^3 content less than 20%. Polymer-like a-C:H (PLCH) consists of 36 at.% hydrogen content with the sp^3 content up to 70%. In ta-C:H, the sp^3 content is approximately 70% while the hydrogen content is 25–35% and the C–C sp^3 content is much greater than in PLCH carbon films [33]. The ta-C films possess the highest thermal conductivity among all types, which could be due to the presence of high sp^3 content of greater than 70%.

3.5.2 Chemical Properties of DLC Coatings

Chemical property is the property that is apparent once the material undergoes some change as a result of a chemical reaction. DLC films, mainly a-C:H films, own good corrosion resistance and are chemically inert in most hostile environments making them enticing for applications, such as mechanical, electronic, petrochemical, and food industries, etc. [34]. However, DLC coatings have several defects, such as voids, scratches, cracks, and macroparticles that severely

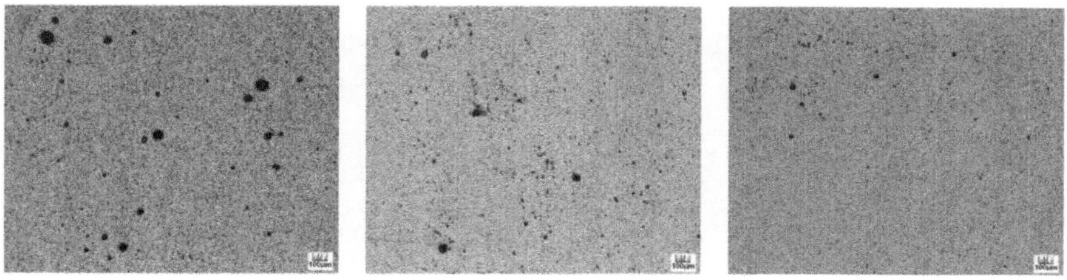

Figure 3.7 Optical micrographs of the samples with thick coatings in distinct positions placed inside the chamber, i.e. (a) lying (b) vertical, and (c) up-side-down positions. *Source:* Reprinted with permission from Ref. [35]. Copyright 2018, SciELO Brazil.

influence their characteristics, mainly corrosion resistance and chemical inertness, and it becomes easy for corrosive substances to attack the substrates and promote crevice corrosion. One such study is taken into account [35], in which three different substrates of low alloy steel, cold-worked steel, and austenitic stainless steel were used to identify superficial defects. Samples were then coated with a-C:H:Si using the PACVD technique. Micrographs were taken from optical microscopes, and image analysis software revealed the number of defects in the samples as shown in Figure 3.7. The sample that was lying consisted of a large number of defects, whereas the sample that was upside down had fewer defects. SEM was used to investigate the coated surfaces for inspecting the morphology of defects. One major type of defect is known as protuberances, which further adhere to develop larger defects or clusters. EDX analysis also confirmed that the chemical composition of these defects was similar to that of the coating composition, and impurities were also present inside these defects. These impurities act as a nucleating site for corrosion to occur; hence, the presence of defects can deteriorate the chemical properties of DLC-coated substances.

3.5.2.1 Reactivity and Chemical Stability

Reactivity refers to the ease of a substance to chemically react with another substance and likewise. Chemical stability is the likelihood of a substance to resist change caused by the interaction with the environment. DLC coatings possess high chemical stability, which renders resistance to corrosive environment, oxidation, and moisture. DLC coating is chemically inert in nature which means that it is not chemically reactive. To test the chemical reactivity, discs of precipitation-hardened Al alloy that is 6061-T and 5053 Al alloy are utilized with a diameter of 2 inches and thickness of ¼ inches as the substrate upon, which DLC layers are deposited as shown in Figure 3.8. For the determination of chemical attack, samples were tested with drops of acid. DLC coatings that were thicker provide better resistance to chemical attack than thinner coatings as they reduce the capability

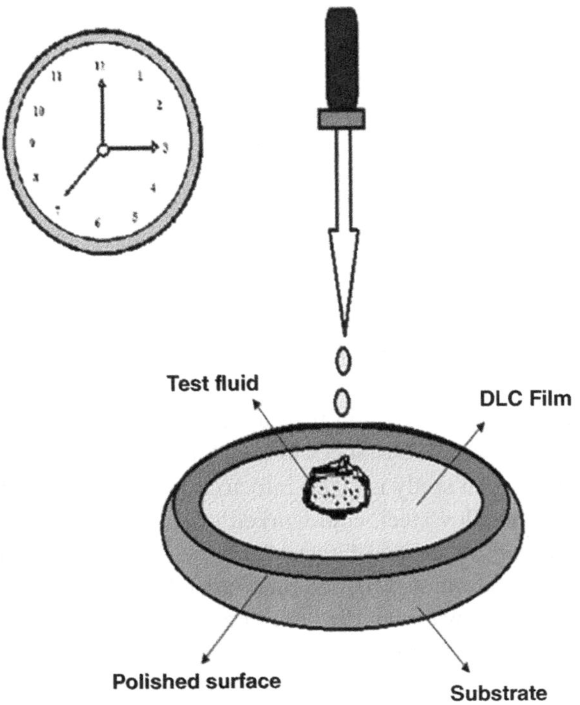

Figure 3.8 Schematic of the procedure utilized to test the chemical reactivity of DLC films.

of acid to penetrate inside. The addition of a silicon interlayer between the DLC layer and aluminum substrate increases the time for the acid to react up to 60 minutes with the sample, which in turn enhances the chemical inertness. However, it is still not possible to completely make the substrate inert as there are chances of secondary particles or debris to be present at the surfaces which might act as a nucleating site for chemical reaction to begin [26].

3.5.3 Anticorrosion Properties of DLC Coatings

Corrosion resistance is the ability of a material to withstand the damage of corrosion caused by electrochemical attack. One of the major benefits of DLC coating is its corrosion resistance, which allows us to use it in highly aggressive environments without causing the formation of rust. DLC coatings act as a barrier and prevent the direct interaction of ions that are capable to diffuse into the substrate and initiate corrosion. One such study in which the corrosion resistance property of DLC-coated zinc samples and uncoated zinc samples was determined through electrochemical and immersion tests mainly potentiodynamic polarization. Potentiodynamic polarization is a method in which current is passed between the sample and an inert electrode in an electrolyte to change the electrochemical potential of the sample. As a result, the current density for a given potential is recorded. The electrochemical evaluation showed that for a longer deposition period, the corrosion current density represented by j_{corr} decreased, showing better corrosion resistance. The corrosion resistance depicts the trend of Zn > third-coated sample > second-coated sample > first-coated sample. The immersion tests with a seven-day period also relate these findings by demonstrating the release of zinc ions in a solution of phosphate-buffered saline solution while the uncoated zinc sample had the lowest release of zinc ions. Further investigations also revealed the formation of $Zn_3(PO_4)_2$ on the DLC-coated samples as a result of the direct interaction between phosphate ions in the solution and the zinc substrate. Moreover, DLC films were thinner and by the deposition through magnetron sputtering became Nano-porous which allows water permeability resulting in the galvanic corrosion between DLC films acting as a cathode and zinc substrate acting as an anode [27].

3.5.4 Biocompatibility of DLC Coatings

Biocompatibility is the ability of a material to perform with a suitable host response without being toxic and harmful. Another fascinating characteristic of DLC coatings is their biocompatibility, which allows it to be used for various biomedical applications. The biological properties of DLC coatings can be changed by the addition of various dopants into the carbon matrix as shown in Table 3.2 [36]. An implantable medical device incorporates silicon- and fluorine-doped amorphous DLC films in a study [37]. The study came across the results that DLC films could be useful as antithrombogenic coatings for medical devices that prevent blood clot formation. Also, fluorine-doped DLC films could be efficacious for

Table 3.2 List of elements incorporated into DLC coatings and their biological advantages.

S. no.	Elements	Advantages
1.	Ag	Improves the antibacterial properties as it evinces bactericidal activity
2.	Cu	Suppress the growth of fungi species on DLC surfaces
3.	Si	Reduces the surface energy of DLC coating preventing bacterial adhesion
4.	F	Enhances hemocompatibility through nonthrombogenic Fluorinated coated DLC
5.	N	Suppress thrombus formation as its addition leads to lower platelet activation and adhesion
6.	Ti	Upgrades biocompatibility of Ti-doped DLC toward osteoblasts

preventing thrombosis in medical devices. DLC coatings can also be utilized for hip and knee joints, coronary stents, and intraocular lenses as DLC coatings are chemically inert, which makes them appropriate for medical devices. To determine the biocompatibility of DLC coatings different in vitro studies are conducted in which DLC coatings are exposed to simulated environments replicating human body conditions, for example, microphage cells [38]. DLC films were determined to be inert without causing inflammatory reactions or causing toxicity. Also using DLC coating for implant adhesion is also a requirement that could be improved by the addition of intermediate layers between the coating and metal substrate.

3.5.5 Opto-electrical Properties of DLC Coatings

Electrical property refers to the rejoinder of a material to an imposed electric field while the optical properties define the interactivity of material with the light. The electrical property of DLC coatings varies from being a semiconductor to being an insulator with a wide band gap, which promotes its insulating behavior [39]. If we talk about their optical characteristics infrared light is capable to pass through DLC films as they are transparent to it making their use possible in infrared optical windows. Due to the weak absorption in the visible spectrum, some visible light can pass through DLC films making them appear translucent in the visible range. This helps them to be used in applications demanding visible light transmission [40]. Also, the optical properties of DLC coatings, such as band gap and refractive index correlate with their electrical performance which will be discussed in the upcoming sections.

3.5.5.1 Electrical Resistance

Electrical resistance is the ability of a material to oppose the flow of current in an electrical circuit. The electrical property of DLC coatings varies from being a semiconductor to being an insulator with a wide band gap [39]. The properties are influenced by impurities, the addition of doping elements, structure, and process parameters including substrate temperature and rate of deposition. The electrical properties of DLC coatings have been examined by applying a model that supposes a specified band structure. In the following model, the band structure only contains a mobility gap that is a range of energy levels with no available states for electrons to occupy. The presence of a mobility gap leads to the semiconductor behavior of the material. Thus, the presence of mobility gap in semiconductors confined the movement of electrons making them localized particles. However, the large density of the localized gap state results in appalling carrier mobility and degrades the semiconductor characteristics of the material. The mobility of DLC at room temperature is determined to be 10^{-11}–10^{-12} $cm^2\ V^{-1}$ while the electrical resistance possessed by DLC coating is found to be 10^2–10^{16} Ω cm [40]. One such study in which the insulating properties of DLC are improved by doping with

SiO_x [41]. The study includes the implementation of amorphous hydrogenated silicon oxide containing DLC films that were deposited at room temperature on substrates of quartz glass wafer and silicon wafer through direct ion deposition method with the help of hexamethyldisiloxane vapor and H_2 mixture. Two steps were involved in the procedure with the first step of determining the optical band gap of DLC samples followed by the deposition of aluminum contacts at the surface of the sample to evaluate the electrical properties of the sample through the formation of the Al-Si-DLC-Al structure. The following sample was then exposed to high-energy X-ray photons and changes induced by radiation exposure were then analyzed.

By analyzing the transmittance and absorbance spectra, which showed that the transparency of the irradiated DLC films did not change compared to the initiated sample followed by a small increment in optical band gap as shown in Figure 3.9. The changes caused by the radiations introduced into the DLC films were mainly dependent on the conditions in which the film was deposited and also codoping which resulted in changes in sp^3/sp^2 ratio along with the hydrogen content. Another observation was the reduction in leakage current suggesting that irradiations with high-energy photons can reduce the leakage current in DLC films.

Moreover, the breakdown voltage was highly affected by the deposition conditions of the film as depicted in Table 3.3. Breakdown voltage is minimum voltage that causes some part of an insulator to undergo small breakdown and become conductor instead. The results indicate that the conductivity of irradiated sample

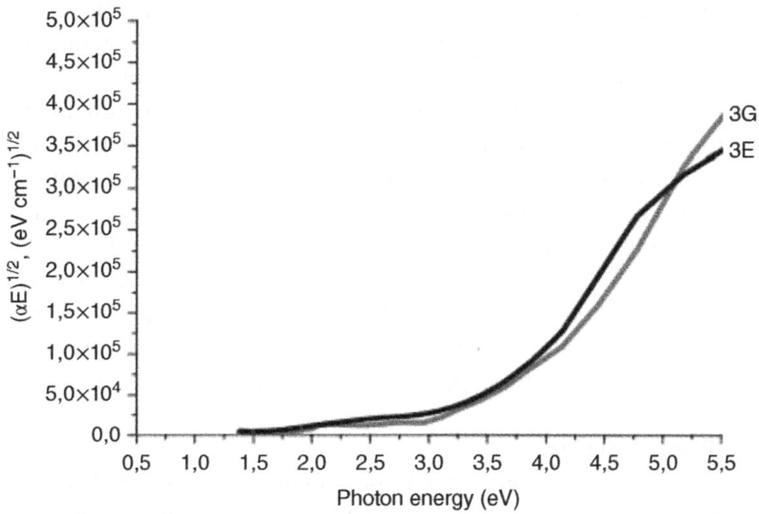

Figure 3.9 Estimated values of the optical band gap of E_τ (3E) = 3.25 eV for the initial samples and E_τ (3G) = 3.60 eV irradiated samples. *Source:* [41] / IOP Publishing / CC BY 4.0.

Table 3.3 Electrical properties of Al-Si-DLC-Al before and after irradiation of high-energy photons.

Samples	Film thickness, nm	Exposure dose (Gy)	Breakdown voltage, U_{min}, V	Breakdown strength E_{min} (MV cm^{-1})
			Φ500 μm	Φ500 μm
			Φ300 μm	Φ300 μm
Initial sample	200	2	≥46.5	≥2.33
			≥48.2	≥2.41
Irradiated sample	~200		<13.75>	0.69
			<45.1>	2.26

Source: [41] / IOP Publishing / CC BY 4.0.

was higher than the initial sample as the photon irradiation can induce additional charges promoting the movement of electrons.

3.5.5.2 Bandgap

The energy difference between the valence band and conduction band that will govern the conductivity of a material is known as the band gap. Different DLC coatings possess different band gaps. First, for hydrogenated carbon films (a-C:H) the band gap varies from 1.1 to 1.7eV. The carbon structure in carbon films is made up of both sp^3 and sp^2 hybridized orbitals. Tetrahedral coordination of carbon atoms is produced by sp^3 hybridization, while hexagonal planar coordination of carbon atoms is produced by sp^2 hybridization. The ratio of tetrahedral coordinated carbon atoms (sp^3 hybridized) to hexagonal planar coordinated carbon atoms (sp^2 hybridized) is the sp^2 ratio. The relationship between them is interrelated [25] so we can predict that the band gap for carbon films is not only determined by the sp^3 ratio but also through the hydrogen content. The correlation coefficient determined for sp^3 ratio and hydrogen content were 0.258 and 0.402, respectively, specifying a strong positive relationship among the variables [9].

3.5.5.3 Refractive Index

Refractive index is defined as ratio of velocity of light in vacuum to the velocity in a certain medium. Refractive index of DLC films varies depending upon the method of deposition and its composition. Typically, for a hydrogenated DLC film (a-C:H) the refractive index lies within 1.6–3.1. Tetrahedral a-C films (ta-C) have

Table 3.4 Properties of DLC films deposited by ion beam deposition technique.

Ion beam deposition parameters		Type of DLC films	Hardness (GPa)	Hydrogen content (at. %)	Refractive index (550 nm)
Substrate voltage (kV)	Source gas				
1	C_6H_6	ta-C:H	32.2	23.2	2.43
2	C_6H_6	ta-C:H	25.8	22.7	2.25
2	C_6H_{12}	ta-C:H	21.2	25.0	2.14
3	C_6H_6	a-C:H	23.6	20.5	2.17

Source: [42] / IOP Publishing / CC BY 4.0.

a higher refractive index of 2.0–3.0, sometimes even higher than diamond which is typically 2.41. Whereas a-C films have lower refractive index than tetrahedral carbon films ranging from 1.6 to 2.6. The optical properties of DLC films are correlated with the hardness of the films and this relationship is determined in [42]. The Ion beam deposition method is used to deposit four DLC films on a silicon substrate. Two main parameters of the process are substrate voltage and gas source that could be adjusted. Table 3.4 consists of information about the hardness, hydrogen content, and refractive index of the DLC films. From the data presented in the table, it can be declared that as the applied voltage increases the hardness of the DLC nanostructure decreases and so does the refractive index. Hydrogen content also displayed a similar trend with an increase in applied voltage. Hydrogen content was affected by the type of gas source used as with the replacement of benzene with cyclohexane the hydrogen content increased significantly. These changes observed are closely related to the energy received by carbon cations during the deposition process, which influences the arrangement of carbon bonds within the films. These sp^3 bonds lead to the hard rigid nature of films, therefore, it is vital to ensure that these bonds are sufficiently present within an appropriate energy range. We can claim that the hardness of DLC coating goes hand in hand with its refractive index. If we can control the refractive index of DLC coatings, it becomes easy to optimize their optical properties as it is used in antireflective optical coating for windows, solar cells, etc.

3.5.6 Tribological Properties of DLC Coatings

Tribology is the study of science and engineering of interacting surfaces that are in relative motion along with the study of friction, wear, and lubrication. By understanding and managing the interacting surfaces tribology is important as

Table 3.5 Friction coefficients of a-C:H and a-C type coatings prepared by pulse laser deposition.

Environment for friction tests	a-C:H nonbiased	a-C:H biased	a-C biased	a-C nonbiased
Sliding against steel ball				
1. Air with relative humidity (RH) of 50%	0.19	0.17	0.12	0.08
2. Nitrogen RH <2%	0.20	0.16	0.10	0.04
3. Vacuum 10 Pa	0.26	0.13	0.11	0.10
Sliding against sapphire ball				
1. Air with RH of 50%	0.07	0.05	0.08	0.06
2. Nitrogen RH <2%	0.10	0.06	0.07	0.03
3. Vacuum 10 Pa	0.12	0.09	0.08	0.06

Source: Reproduced from [24] / with permission of ELSEVIER.

the sliding interfaces lead to wastage of energy due to friction. Different type of DLC coatings: such as a-C and hydrogenated carbon, exhibit different tribological behavior in various environments [24]. The friction coefficients were evaluated by sliding the DLC coating against steel balls and sapphire balls. The results are presented in Table 3.5, which indicates that a-C:H type coating possesses higher friction coefficients ranging from 0.05 to 0.26 than that of a-C type with friction coefficients ranging from 0.03 to 0.12. The friction coefficients for sapphire and steel balls is comparatively less for a-C type, which shows that a-C DLC is less pretentious by counterpart material. Moreover, the two types of coating disclose higher friction coefficients in vacuum environments than in humid air and dry environments. We will further discuss in detail the main tribological characteristic of DLC coatings, which includes wear resistance and adhesion of coating.

3.5.6.1 Friction Coefficient

Another unique characteristic of DLC coating is the low friction coefficient, which is influenced by various intrinsic factors that determine the inherent properties of DLC coatings and extrinsic factors that determine the environment in which DLC coatings are to be used such as the hardness of the coating, the thickness of film, doping element, hydrogen content, sp^3/sp^2 ratio, temperature, load applied, sliding speed, etc. [43]. The following parameters need to be considered one by one:

A. *Hydrogen content*: Hydrogen plays a prominent role in determining the tribological characteristics of DLC coatings. Hydrogenated type of DLC coatings (a-C:H) exhibit lower friction coefficients under inert and dry atmospheres while for the nonhydrogenated DLC coatings, friction coefficient can be improved by increasing the humidity levels or lubricants. This is mainly due to the presence of hydrogen obtained through strain and thermal effects during sliding motion, which causes the formation of soft porous graphitic layer, which acts as a lubricant reducing friction and permitting smoother sliding motion between DLC-coated surfaces. Environmental factors including water molecules and oxygen can affect the tribological behavior of DLC films mainly their friction coefficients [44]. To determine the effect of these environmental species additional tests were carried out with hydrogen-free and hydrogenated DLC films. Sliding tests were first conducted in dry nitrogen environment. Moist lab air was then added hindering the nitrogen flow. The results depict that friction coefficient of hydrogen-free DLC films decreased to 0.25 while the hydrogenated DLC films had a 20-fold rise in friction coefficient as shown in Figure 3.10.

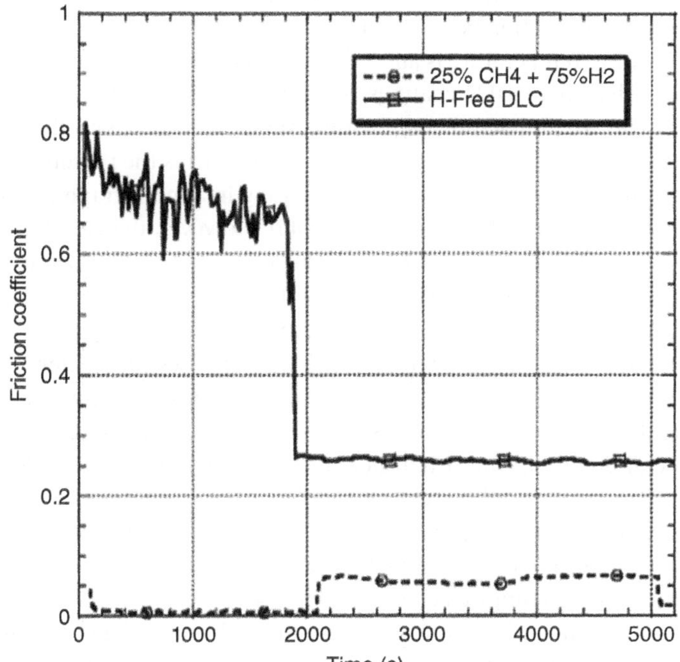

Figure 3.10 Environmental factors effect on friction coefficient of hydrogenated and hydrogen-free DLC films. *Source:* Reproduced from [44] / with permission of ELSEVIER.

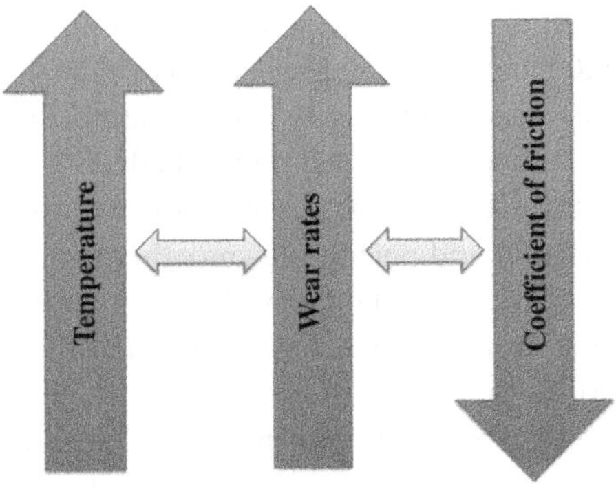

Figure 3.11 General trends in tribological properties with increase in temperature.

B. *Sp^3/Sp^2 ratio*: DLC coatings that contain a high percentage of sp^2-bonded carbon atoms possess graphitic characteristics and are relatively soft while the coatings with high percentage of sp^3 carbon-bonded atoms possess diamond-like characteristics and are relatively hard. Increasing the sp^3/sp^2 ratio also increases the internal stresses making the coatings more prone to delamination. DLC coatings with high sp^3 content have improved wear resistance with little change in friction behavior [45].

C. *Temperature*: Temperature is another critical factor that determines the tribological behavior of DLC coatings. Various studies have been conducted to understand the impact of temperature and it was indicated that at higher temperatures, the ID/IG ratio increases that confirms the presence of defects and disorder in the carbon structure resulting in a more graphitic-like structure and ameliorated friction behavior [46]. However, the elevated temperatures adversely affected the wear resistance of DLC coatings as indicated in Figure 3.11 decreasing the stability of tribofilms. In another study, a BC/a-C:H coating containing 1% boron was evaluated to have high hardness, and low friction coefficient when tested with pins on discs tribometers [47].

D. *Doping*: Metal and nonmetal doping to DLC coatings can influence their friction and wear behavior. Various studies demonstrate the effect of metal and nonmetal doping in the tribological behavior such as titanium-doped DLC (Ti-DLC) coatings exhibit excellent tribological properties, including low friction coefficients and high wear resistance. While the tungsten-doped DLC (W-DLC) coatings exhibit somehow different behavior, with lower wear resistance at high tungsten concentrations [45].

E. *Speed of Sliding and Load*: Numerous studies have depicted that when the sliding speed and applied load are increased the coefficient of friction is reduced. The wear coefficient is also reduced with an increase in applied load but increases with sliding speed. Another study conducted also determined a similar behavior with friction coefficient reducing from 0.16 to 0.12 as the load is increased from 4 to 8 N similarly with an increase in sliding speed from 0.03 to 0.27 m s^{-1} the coefficient of friction reducing from 0.1 to 0.085 [48].

F. *Thickness of DLC Coatings*: To understand the effect of thickness on tribological behavior one study [49] is taken into account. In the following study, DLC coatings were deposited on substrates of stainless steel by FCVA technique. The DLC coatings with various thicknesses of 300, 500, and 650 nm were deposited along with a 100 nm chromium interlayer that was sputtered at 200 W, 600 B bias, and 40 sccm of Argon flow. It can be seen in Figure 3.12a that the minimum friction coefficient is for 500 nm thick DLC-coated substrate and the maximum friction coefficient is for 650 nm thick DLC-coated substrate because as the thickness of the film increases so does the hardness and in respond to that internal stresses in the film also increases corresponding to the brittle failure of the film. The tribology test indicated mean friction coefficients for 100 m distance to be 0.15, 0.12, and 0.16, for 300, 500, and 650 nm DLC-coated stainless steel substrates, respectively, as depicted in Figure 3.12b.

3.5.6.2 Wear Resistance

Wear resistance refers to the capability of a material to resist the loss of material by mechanical action such as rubbing and sliding, etc. DLC coatings have good wear resistance depending upon various deposition factors. To access the wear behavior of DLC coating, we consider one research [50] in which the graphite targets were used for DC-sputtering deposition on aluminum substrate followed by deposition of DLC thin films with a total thickness of 1000 nm on the samples.

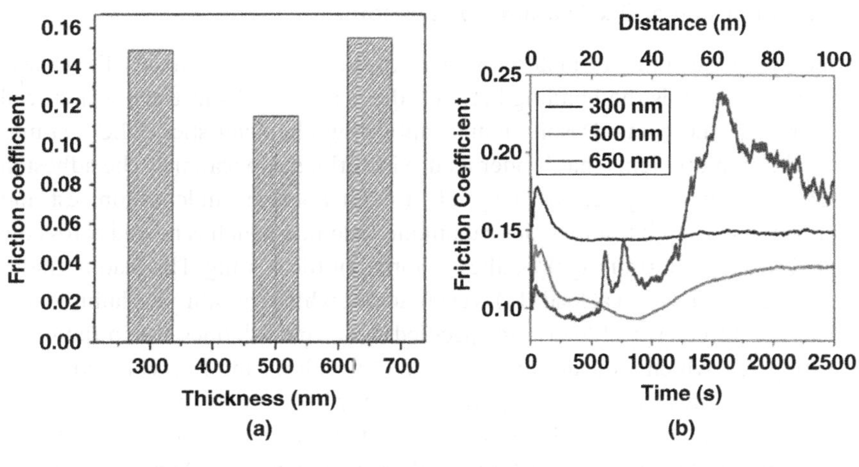

Figure 3.12 (a) Friction coefficient of DLC films with various thicknesses of 300, 500, and 650 nm. (b) Friction coefficients of DLC films with time and distance with different thicknesses. *Source:* Reproduced from [49] / with permission of Taylor & Francis Group.

Table 3.6 Results of wear tests for coated and uncoated sample.

Samples	Friction coefficients	Wear coefficients (K)
Aluminum substrate	0.95	7.52×10^{-5}
DLC-coated Aluminum substrates	0.01–0.04	3.02×10^{-6}

Friction and wear characteristics were evaluated by a pin-on-disk method in which uncoated aluminum pin and DLC-coated aluminum disk were assessed under vacuum with dry and ambient air having 0 and 40% relative humidity, respectively. XRD analysis revealed that it was an a-C with no peaks other than carbon and the highest peak also confirmed that the films were nanocrystalline. The results in Table 3.6 deduced that amorphous DLC-coated samples showed lower friction coefficients in contrast to uncoated aluminum reference samples with friction coefficients. The amorphous DLC-coated samples also revealed narrower wear tracks with a width of 230 μm while the uncoated sample exhibited wear tracks with a width of 1.5 mm. Further investigation also suggested that the wear coefficient for the uncoated sample was higher whereas DLC-coated sample had a lower wear coefficient. It was also determined that the presence of humidity reduces the friction coefficient to 0.01 compared to dry air ion which the friction coefficient was 0.03. It was due to the formation of graphitized tribofilms as a result of absorbing the humidity, which lower down the friction of DLC films.

3.5.7 Adhesion of DLC Coatings to Substrates

Adhesion is the ability of material surface to stick to another material's surface. Adhesion, which is the bonding between the coating and substrate, is a crucial property of DLC films. Despite many captivating characteristics, DLC coatings face adhesion problems that hinder its use in various applications. The adhesion of DLC coatings is generally analyzed through Rockwell indentation test and scratch tests. Scratch test includes a diamond indenter, which is moved across the coated surface determining the adhesive forces of the coating. The load is gradually increased and critical load (L_c) is evaluated at which the adhesive failure takes place. The three critical loads are specified: L_{c1}, which defines the initiation of cracking, L_{c2}, which indicates the initiation of buckling, and L_{c3}, which indicates the failure of the coating. In one particular study [51], various types of steel substrates such as die steel, high-strength steel, stainless steel, mild steel, and hard steel were selected and DLC coating was deposited using a PVD process in a pure nitrogen environment to evaluate the best adhesion among the substrates. In the

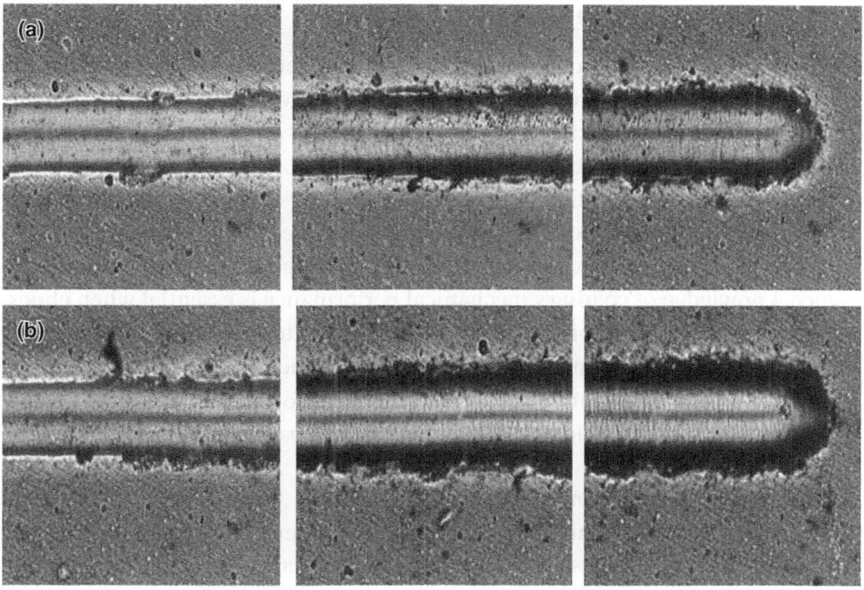

Figure 3.13 (a) Optical images of multiple forming element scratch on DLC-coated die steel at 20 N load, (b) optical images of multiscratch on DLC-coated die steel at 25N load. *Source:* Reprinted with permission from Ref. [51]. Copyright 2014, IOP Publishing.

scratch test, loads of 20 and 25N were applied to create the scratches on the coated samples followed by pre- and post-scan measurements to note the surface profile before and after the test. The scratches were inspected through the optical microscope to determine the critical loads and scan the failure mechanism. Figure 3.13a contains optical micrographs of multiscratch on DLC-coated die steel at 20N load. The first micrograph represents the region of L_{c1}, and the second and third micrograph represents the region of L_{c2} and L_{c3}, respectively. There is no detectable wear on the coated surface, which stipulates good adhesion between the coating and substrate. Figure 3.13b presents optical micrographs of die steel at 25N load with a prominent wear. The dense cracking observed within the scratch is due to the excessive load and repeated scratching in the same region. The penetration depth curves for different substrates showed that mild steel had the largest penetration depth while high-strength steel and die steel showed the least penetration depth because of their high hardness. It can be deduced that harder the substrate better the adhesion property of DLC with that substrate as compared to the soft substrate. The adhesion performance of DLC coatings also varies with operating conditions [52].

3.5.8 Mechanical Properties of DLC Coatings

Mechanical properties describe the behavior of a material when an external force is applied to it. The sp^3 bonding in DLC provides advantageous features. While considering the mechanical characteristics, it possesses such as high hardness and high elastic modulus, etc., DLC films are capable of competing with the properties of diamond. The illiterateness regarding DLC characteristics limits their broad usage. The same type of layer could have completely distinct characteristics when applied using different deposition techniques and deposition parameters. Hence, knowledge of coating's mechanical performance is essential when choosing it for certain application. Indentation and scratch tests are utilized to evaluate the hardness and elastic modulus of the coating, which will be discussed in further sections. The results of these tests will make it effortless to choose the coating and create the best technological deposition method based on its application [53].

3.5.8.1 Hardness and Young's Modulus

DLC coating possesses entrancing mechanical properties among which one of them is hardness. Hardness of a material is its resistance to plastic deformation while the Young's modulus is the measure of the stiffness of the material. DLC coatings are characterized by high hardness and high Young's modulus with hardness in the range of 10–30 GPa and Young's modulus six to seven times larger [54]. In a research [55], the use of DLC nanocoated aluminum to reduce the sliding friction of moving parts is considered. The research used radio frequency sputtering method to deposit graphite targets on aluminum substrates upon which DLC coating that was 1200-nm thick was then deposited. The hardness and elastic modulus of DLC coatings were measured by nanoindentation machine with a Berkovich indenter that applied the load. The results indicated that the highest hardness values were for the substrate deposited at higher temperature as compared to substrates deposited at room temperature as shown in Table 3.7. When the deposition power was increased from 100 to 400 W the hardness also increased till 300 W followed by reduction at 400 W. Hardness and Young's modulus is also a function of thickness of DLC films and increase as the thickness of film increases [56]. Hence, we can assume that the amalgam of high hardness and Young's modulus allows DLC coatings to resist deformation.

Multidoping is the introduction of metals to DLC is also capable to alter its mechanical performance [16]. Metal atoms, such as aluminum, silicon, and chromium are incorporated into the coatings and the residual stresses and mechanical properties are analyzed. Studies have declared that Al incorporation significantly reduces the residual stresses in the coatings. The effect on hardness and elastic modulus of AlCrSi-DLC deposited at different gas precursor fraction, which is Ar mainly in this scenario is shown in Figure 3.14. The hardness of coating remains

Table 3.7 Hardness and modulus values of samples deposited at room temperature and 500 °C, respectively.

Sample	100 W	200 W	300 W	400 W
Samples deposited at room temperature				
Hardness (GPa)	10.6	13.5	18.2	12.3
Modulus (GPa)	51.3	70.2	120	68.1
Samples deposited at 500 °C				
Hardness (GPa)	15.1	20.8	22.5	20.1
Modulus (GPa)	76.2	111.2	130.5	100.6

Source: [55] / John Wiley & Sons / CC BY 4.0.

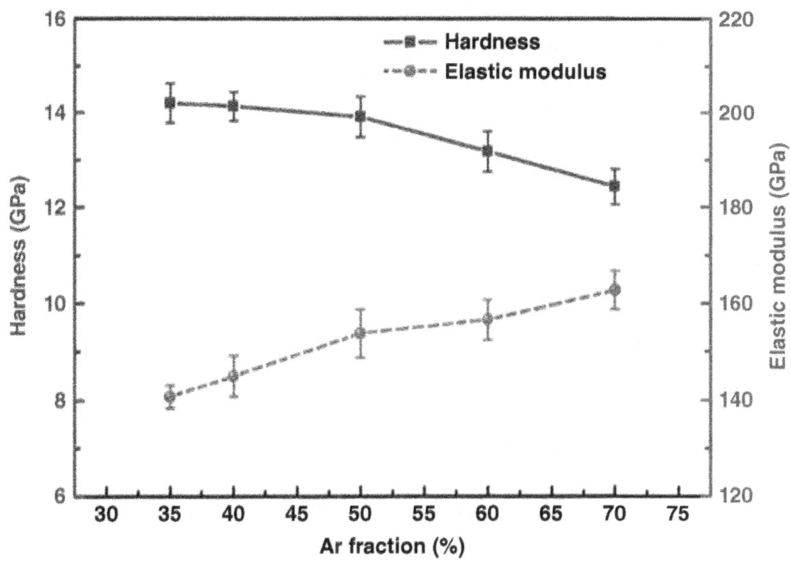

Figure 3.14 Hardness and elastic modulus of coating as a function of Ar fraction. *Source:* Reproduced from [16] / with permission of ELSEVIER.

unchanged with a decrease from 14.2 to 12.4 GPa as the Ar fraction increases from 35 to 70%. Hardness is related mainly with the sp^3 content, which remained constant as determined through XPS with increase in doping content. Therefore, this small decrease in hardness is mainly due to the reduction in residual stresses. Even with the reduction, the coating still possesses high hardness over 12 GPa which indicates that through codoping high hardness and low stress can be attained easily. Moreover, it is noticed that the elastic modulus of the coating increases with the Ar fraction, which is mainly due to the carbide formation such as Si–C or Cr–C bonds. The carbide formers increase the stiffness and rigidity of the material leading to high elastic modulus. Hence, by regulating the doping addition through varying the gas precursor, it is possible to change the hardness and elastic modulus ratio (H/E) of the coatings.

3.5.8.2 Fatigue Strength of DLC Coatings

Fatigue strength is the ability of material to combat fatigue failure, which is the failure caused by repetitive load. To determine the fatigue strength DLC-coated Ti-6Al-4V was used as a surface treatment to overcome the limitation of low mechanical properties of Ti alloys in one research [57]. The fatigue tests were performed on a rotating fatigue-testing machine at a repetition rate of 3000 rpm and a torque of 14.7 Nm. The surfaces that were fractured were observed through an optical microscope and SEM. The hardness test revealed that the hardness of the nonsurface treated specimen was HV320 while the DLC-coated specimen had the hardness approximately 1.5 times higher than the nonsurface treated sample. Also, as the coating time was increased the hardness also improved with a longer coating duration. The nonsurface treated sample had a fatigue limit of 390 MPa while the fatigue limit of DLC-coated sample increased with longer coating time. One of the coated samples exhibited the highest fatigue limit of 500 MPa crossing the nonsurface treated sample fatigue limit by 30%. This increase in hardness was mainly due to the presence of compressive stresses in the surface layer, which retarded crack propagation into the interior of the specimen hence leading to improved fatigue strength.

3.6 Improving the Properties of DLC Coatings

Apart from certain useful characteristics DLC coatings also have certain limitations that include weak adhesion, inferior thermal stability, and incapacity to bear stresses, as illustrated in Figure 3.15. DLC films have high residual stresses resulting in poor adhesion strength between the DLC films and the substrate. This causes DLC films to peel off because of difference in thermal expansion coefficients and structural mismatch between the film and substrate. Moreover, DLC films have low thermal stability also the mechanical and tribological characteristics of DLC films worsen with temperature as graphitization occurs and sp^3-C changes to sp^2-C. When the operating temperature reaches 300 °C, the immoderate graphitization causes lubrication failure [58]. All of these limitations prevent excessive use and applications of DLC coatings. Various modification techniques are used to boost the functionality of DLC coatings, including surface preparation, alloying and doping,

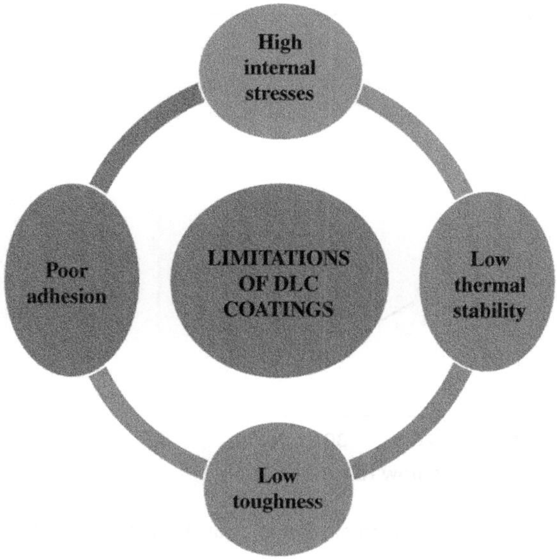

Figure 3.15 Illustration of limitations of DLC coatings.

depositing multilayer, and hybrid coatings. The researchers believe that in comparison to single-layer DLC coatings, multilayer DLC coatings with alternate hard and soft layers can provide better mechanical qualities [59]. These coatings are also capable to enhance wear performances at high contact stress [39].

3.6.1 Modification Through Alloy Addition

Doping elements, doping content, and preparation procedure impact the microstructure and properties of DLC films. Doping is considered as an important approach for adjusting the residual stresses present inside the DLC coatings as shown in Figure 3.16 and enhancing their thermal stability at high temperatures. In the oxygen-doped DLC films the internal stress reduces from 2.7 to 0.11 GPa [60]. Apart from reducing residual stresses, element doping can also reduce mechanical such as hardness. It is thus important to optimize the parameters like type of doping, doping content for improved comprehensive DLC films. Doping elements in DLC films exhibit certain solubility in the carbon matrix. When the doping content is low, the doping elements dissolve in the carbon matrix while when the doping content is high, different elements results in different states such as addition of carbide-forming elements reacts strongly with carbon causing the formation of carbides in carbon matrix whereas non carbide-forming elements prevail as nanocrystals in the carbon matrix. Excessive addition of carbide-forming elements can disturb the continuity of carbon network structure and affect the properties of DLC films. If we first take into account the residual stresses present

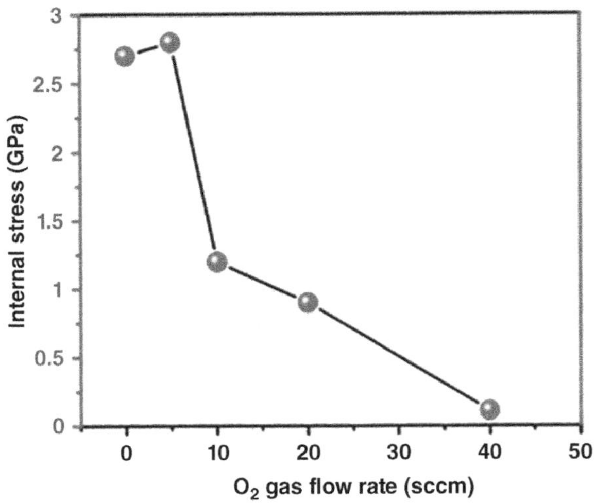

Figure 3.16 Residual stress of DLC film doped with oxygen as a function of oxygen flow rate. *Source:* [60] / IOP Publishing / CC BY 4.0.

in the DLC films, it is important to know what are the three basic kinds of residual stresses [61]. Thermal stress arises from the difference in thermal coefficient of expansion between the film and the substrate. Internal stress arises from different preparation procedures and is related to the morphology of DLC films. Finally, external stress arises from impurities from outside leading to the formation of new phases with different solubility in the carbon matrix. The creation of molecular orbitals through atomic orbital overlap where electrons are found plays an important role in atomic bonding. When carbon atoms form molecular orbitals, electrons fill into bond states resulting in high-strength covalent bonds, which causes the distortion of bond angles and bond length leading to high residual stresses in the films. The electron density increases as the atoms get close together which makes the bond angles deviate from their ideal values along with distortion of bond length. When a carbide-forming element such as W is introduced strain energy caused by bond angle distortion is reduced, hence, leading to depletion in residual stresses in the film. The carbide-forming elements alter the microstructure of films thus modifying the bonding arrangement with beneficial distribution of strain energy. Similarly, when a noncarbide-forming element such as Cu is introduced the distortion energy is smaller due to the antibonding characteristics which are also vital for reducing residual stress [62].

3.6.2 Modification Through Multilayer and Hybrid Coatings

DLC coatings exhibit excellent mechanical and chemical characteristics until they are exposed to high temperature above 300 °C at which (a-C:H) hydrogenated DLC coatings encounter graphitization at a rapid rate lowering down the thermal stability. Another type of DLC coating that is amorphous tetrahedral carbon (ta-C) possesses high thermal stability but things do become problematic for such coatings at high temperature. The addition of metals such as silicon to DLC deposited through methods such aCVD and sputtering using C_2H_2, CF_4, and tetramethylsilane as gases could enhance the thermal stability. To combat the problem of thermal degradation silicon is added to nonhydrogenated DLC coating which is applied to tungsten-carbide cobalt substrate and the Si/SiC/ta-C composite coating is developed through filtered cathode vacuum arc deposition method along with unbalanced magnetron sputtering method [63]. The samples are annealed also at different temperatures and analyzed through surface observation to understand the structural changes. The results in Figure 3.17a indicated that ta-C coating undergo carbonization and degradation at 500 °C as a result sample is partially exposed followed by complete removal of the coating whereas the Si-ta-C coating remained stable even up to 600 °C. When the silicon concentration is high up to 6.04 at.% the coating showed better oxidation resistance even at a high temperature of 700 °C exhibiting excellent thermal stability. The results are also confirmed by Raman peaks shown in Figure 3.17b, which indicate that Si-ta-C coatings preserved their structural integrity with little changes in the Raman spectra up to 500 °C with a lower thermal degradation rate. Partial weight loss of ta-C coating and Ti interlayer could be noticed at 500 °C while the complete removal could be seen at 600–700 °C. Impurities emerged in Si-ta-C coating, yet Raman spectra had negligible changes up to 500 °C, with TiO_2 peaks appearing concurrently with weight loss.

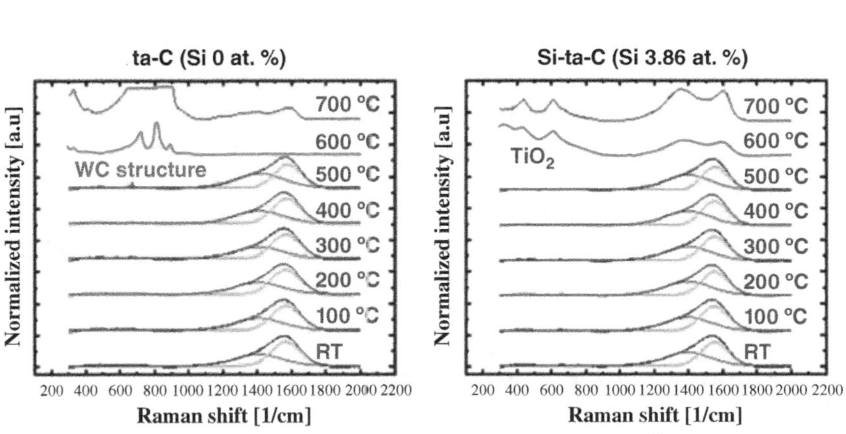

Figure 3.17 Raman spectra of (a) ta-C and (b) Si-ta-C coatings before and after annealing at different temperatures ranging from room temperature to high temperature. *Source:* [63] / Springer Nature / CC BY 4.0.

Hence by playing around with deposition parameters, incorporating various metals and nonmetals, it is possible to overcome the limitations thus, allowing a wide range of uses for DLC coatings, which will be covered next.

3.7 Applications and Uses of DLC Coatings

The increasing demand for DLC coatings in various industries has accelerated research and development efforts, leading to continuous advancements in DLC technology. In the aerospace industry, DLC coatings are used to enhance the performance and durability of various components, such as aircraft engines and turbine blades. The low friction coefficient of DLC coatings allows for reduced wear and improved fuel efficiency. In the biomedical industry, DLC coatings are utilized for various purposes, including in medical devices and implants. The biocompatibility of DLC coatings makes them suitable for applications, such as orthopedic implants, cardiovascular stents, and dental tools. The inertness of DLC coatings helps minimize the risk of adverse reactions and promotes better integration with the human body. The cutting tools industry also benefits from DLC coatings. The exceptional hardness and wear resistance of DLC coatings make them ideal for coating cutting tools like drills, taps, and end mills. These coatings increase tool life, reduce friction, and enhance cutting performance, resulting in improved efficiency and cost-effectiveness in manufacturing processes. DLC coatings also find extensive use in the automotive industry. They are applied to various engine components, such as piston rings and camshafts, to reduce friction and enhance performance. The low friction coefficient of DLC coatings improves fuel efficiency, reduces emissions, and increases the lifespan of the components, leading to improved overall vehicle performance and reliability. DLC coatings are even used for conservation purposes in the preservation of metals like brass. These coatings provide a protective barrier against corrosion and oxidation, helping to maintain the appearance and structural integrity of the metal over time. By preventing degradation, DLC coatings contribute to the long-term preservation of brass artifacts and objects. Overall, DLC coatings offer numerous benefits in various industries, including aerospace, biomedical, cutting tools, automotive, and conservation of metals.

3.8 Summary and Future Outlook

This chapter provides a synopsis of DLC coatings, which includes their general categorization, their preparation and characterization techniques, and comprehensive study of their unique characteristics, such as physical, electrical, chemical, tribological, and mechanical properties. DLC coatings are known for their advantageous properties similar to those of diamond. We present the categorization of DLC coatings based on their structure and chemical composition along with the doped and undoped coatings. To further intensify the above-mentioned properties, we focus on ways such as amendment of deposition parameters, the addition of various metals and nonmetals and producing exotic multilayer and hybrid DLC coatings. Finally, we inspect the wide range of applications of DLC coatings in fields such as biomedical, aerospace, cutting tool industries and automotive industry, etc. The future of DLC coatings is auspicious and as the development continues, we can claim that they are capable to be used in the field of engineering, science and technology contributing with the innovative functions. With the help of more research and development efforts, it is possible to invent novel deposition methods to make improvements to particular properties. Their unique properties, such as low friction coefficient and high hardness, make them valuable for enhancing performance, durability, and longevity in a wide range of applications. It is possible to engineer DLC coatings depending on the requirement that is needed in a specific application.

References

1 Martins, P.S., Almeida Magalhães Júnior, P.A., Gonçalves Carneiro, J.R. et al. (2022). Study of diamond-like carbon coating application on carbide substrate for cutting tools used in the drilling process of an Al–Si alloy at high cutting speeds. *Wear* 498–499: 204326.

2 Choleridis, A., Sao-Joao, S., Ben-Mohamed, J. et al. (2018). Experimental study of wear-induced delamination for DLC coated automotive components. *Surface and Coatings Technology* 352: 549–560.

3 Liu, L., Tang, W., Ruan, Q. et al. (2020). Robust and durable surperhydrophobic F-DLC coating for anti-icing in aircrafts engineering. *Surface and Coatings Technology* 404: 126468.

4 Neerinck, D., Persoone, P., Sercu, M. et al. (1998). Diamond-like nanocomposite coatings (a-C:H/a-Si:O) for tribological applications. *Diamond and Related Materials* 7 (2–5): 468–471.

5 Zia, A.W. (2020). New generation carbon particles embedded diamond-like carbon coatings for transportation industry. In: *Advances in Smart Coatings and Thin Films for Future Industrial and Biomedical Engineering Applications*, 307–332. Oxford: Elsevier.

6 Aisenberg, S. and Chabot, R. (1971). Ion-beam deposition of thin films of diamondlike carbon. *Journal of Applied Physics* 42 (7): 2953–2958.

7 Clausing, R.E., Horton, L.L., Angus, J.C. et al. (1991). *Diamond and Diamond-like Films and Coatings*. Boston, MA: Springer.

8 Rajak, D.K., Kumar, A., Behera, A. et al. (2021). Diamond-like carbon (DLC) coatings: classification, properties, and applications. *Applied Sciences* 11 (10): 4445.

9 Sánchez-López, J.C. and Fernández, A. (2008). Doping and alloying effects on DLC coatings. In: *Tribology of Diamond-like Carbon Films* (eds. C. Donnet and A. Erdemir), 311–338. Boston, MA: Springer.

10 Wang, A.-Y., Lee, K.-R., Ahn, J.-P. et al. (2006). Structure and mechanical properties of W incorporated diamond-like carbon films prepared by a hybrid ion beam deposition technique. *Carbon* 44 (9): 1826–1832.

11 Guo, Y., Guo, P., Sun, L. et al. (2019). Tribological properties of Ti-doped diamond-like carbon coatings under dry friction and PAO oil lubrication. *Surface Interface Analysis* 51 (3): 361–370.

12 Viswanathan, S., Mohan, L., Bera, P. et al. (2017). Corrosion and wear behaviors of Cr-doped diamond-like carbon coatings. *Journal of Materials Engineering and Performance* 26 (8): 3633–3647.

13 Koppert, R., Goettel, D., Freitag-Weber, O. et al. (2009). Nickel containing diamond like carbon thin films. *Solid State Sciences* 11 (10): 1797–1800.

14 Ray, S.C., Pong, W.F., and Papakonstantinou, P. (2016). Iron, nitrogen and silicon doped diamond like carbon (DLC) thin films: a comparative study. *Thin Solid Films* 610: 42–47.

15 Nobili, L. and Guglielmini, A. (2013). Thermal stability and mechanical properties of fluorinated diamond-like carbon coatings. *Surface and Coatings Technology* 219: 144–150.

16 Dai, W., Gao, X., Liu, J. et al. (2016) Microstructure, mechanical property and thermal stability of diamond-like carbon coatings with Al, Cr and Si multi-doping. *Diamond and Related Materials* 70: 98–104.

17 Park, C.-S., Choi, S.G., Park, H.H. (2012). Compensation effect of boron and nitrogen codoping on the hardness and electrical resistivity of diamond-like carbon films prepared by magnetron sputtering deposition. *Journal of Materials Research* 27 (23): 3027–3032.

18 Vetter, J. (2014). 60 years of DLC coatings: historical highlights and technical review of cathodic arc processes to synthesize various DLC types, and their evolution for industrial applications. *Surface and Coatings Technology* 257: 213–240.

19 Ferreira, F., Aijaz, A., Kubart, T. et al. (2018). Hard and dense diamond like carbon coatings deposited by deep oscillations magnetron sputtering. *Surface and Coatings Technology* 336: 92–98.

20 Wasy, A., Balakrishnan, G., Lee, S.H. et al. (2014). Argon plasma treatment on metal substrates and effects on diamond-like carbon (DLC) coating properties. *Crystal Research and Technology* 49 (1): 55–62.

21 Viswanathan, S., Mohan, L., Bera, P. et al. (2018). Corrosion and wear resistance properties of multilayered diamond-like carbon nanocomposite coating. *Surface Interface Analysis* 50 (3): 265–276.

22 Ren, Z., Qin, H., Dong, Y. et al. (2019). A boron-doped diamond like carbon coating with high hardness and low friction coefficient. *Wear* 436–437: 203031.

23 Sánchez-López, J.C., Erdemir, A., Donnet, C. et al. (2003). Friction-induced structural transformations of diamond like carbon coatings under various atmospheres. *Surface and Coatings Technology* 163–164: 444–450.

24 Voevodin, A.A., Donley, M.S., Zabinski, J.S. et al. (1995). Mechanical and tribological properties of diamond-like carbon coatings prepared by pulsed laser deposition. *Surface and Coatings Technology* 76–77: 534–539.

25 Ohtake, N., Hiratsuka, M., Kanda, K. et al. (2021). Properties and classification of diamond-like carbon films. *Materials* 14 (2): 315.

26 Thomson, L.A, Law, F.C., Rushton, N. et al. (1991). Biocompatibility of diamond-like carbon coating. *Biomaterials* 12 (1): 37–40.

27 Peng, F., Lin, Y., Zhang, D. et al. (2021). Corrosion behavior and biocompatibility of diamond-like carbon-coated zinc: an in vitro study. *ACS Omega* 6 (14): 9843–9851.

28 Sulaiman, M.H., Farahana, R.N., Tajul, L. et al. (2019). The wettability characteristics of DLC coating for tribological engineering applications. *IOP Conference Series: Materials Science and Engineering* 670 (1): 012053.

29 Buahom, P. (2018). Measuring the contact angle using ImageJ with contact angle plug-in. *ResearchGate*, 1. doi: 10.13140/RG.2.2.33978.85445/1.

30 Ahmed, M.H. and Byrne, J.A. (2012). Effect of surface structure and wettability of DLC and N-DLC thin films on adsorption of glycine. *Applied Surface Science* 258 (12): 5166–5174.

31 Hainsworth, S.V. and Uhure, N.J. (2007). Diamond like carbon coatings for tribology: production techniques, characterisation methods and applications. *International Materials Reviews* 52 (3): 153–174.

32 Demichelis, F., Pirri, C.F., Tagliaferro, A. et al. (1993). Mechanical and thermophysical properties of diamond-like carbon (DLC) films with different ratios. *Diamond and Related Materials* 2 (5–7): 890–892.

33 Shamsa, M., Liu, W.L., Balandin, A.A. et al. (2006). Thermal conductivity of diamond-like carbon films. *Applied Physics Letters* 89 (16): 161921.

34 Donnet, C. and Erdemir, A. (eds.) (2008). *Tribology of Diamond-Like Carbon Films: Fundamentals and Applications.* New York: Springer.

35 Delfín, F.A., Brühl, S.P., Forsich, C. et al. (2018). Carbon based DLC films: Influence of the processing parameters on the structure and properties. *Matéria (Rio J.)* 23 (2): 12059.

36 Olejnik, A., Świątek, L., and Bociąga, D. (2017). Biological evaluation of modified DLC coatings – a review. *World Scientific News* 1 (73): 61–71.

37 Toyonaga, M., Hasebe, T., Maegawa, S. et al. (2021). The property of adhesion and biocompatibility of silicon and fluorine doped diamond-like carbon films. *Diamond and Related Materials* 119: 108558.

38 Grill, A. (2003). Diamond-like carbon coatings as biocompatible materials – an overview. *Diamond and Related Materials* 12 (2): 166–170.

39 Lin, Y., Zhou, Z., and Li, K.Y. (2019). Improved wear resistance at high contact stresses of hydrogen-free diamond-like carbon coatings by carbon/carbon multilayer architecture. *Applied Surface Science* 477: 137–146.

40 Grill, A. (1999) Electrical and optical properties of diamond-like carbon. *Thin Solid Films* 355–356: 189–193.

41 Laurikaitienė, J., Meškinis, Š., Adlienė, D. et al. (2008). Electrical properties of the diamond like carbon films irradiated with high energy photons. *Journal of Physics: Conference Series* 100 (7): 072036.

42 Suasnavas, C.V., Makeev, M.O., Osipkov, A.S. et al. (2019). Relationship between hardness and optical properties of diamond-like carbon coatings. *IOP Conference Series: Materials Science and Engineering* 675 (1): 012051.

43 Zahid, R., Masjuki, H.H., Varman, M. et al. (2016). Influence of intrinsic and extrinsic conditions on the tribological characteristics of diamond-like carbon coatings: a review. *Journal of Materials Research* 31 (13): 1814–1836.

44 Erdemir, A. (2001). The role of hydrogen in tribological properties of diamond-like carbon films. *Surface and Coatings Technology* 146–147: 292–297.

45 Vengudusamy, B., Mufti, R.A., Lamb, G.D. et al. (2011). Friction properties of DLC/DLC contacts in base oil. *Tribology International* 44 (7–8): 922–932.

46 Al Mahmud, K.A.H., Varman, M., Kalam, M.A. et al. (2014). Tribological characteristics of amorphous hydrogenated (a-C:H) and tetrahedral (ta-C) diamond-like carbon coating at different test temperatures in the presence of commercial lubricating oil. *Surface and Coatings Technology* 245: 133–147.

47 Mori, H. and Tachikawa, H. (2002). Increased adhesion of diamond-like carbon–Si coatings and its tribological properties. *Surface and Coatings Technology* 149 (2–3): 224–229.

48 Kalin, M. and Polajnar, M. (2013). The correlation between the surface energy, the contact angle and the spreading parameter, and their relevance for the wetting behaviour of DLC with lubricating oils. *Tribology International* 66: 225–233.

49 Wasy, A., Balakrishnan, G., Lee, S. et al. (2015). Thickness dependent properties of diamond-like carbon coatings by filtered cathodic vacuum arc deposition. *Surface Engineering* 31 (2): 85–89.

50 Natrayan, L., Gaur, P., Merneedi, A. et al. (2022). Investigation of tribological behaviour on DLC coatings for AA5051 using DC sputtering. *Adsorption Science & Technology* 2022: 1–9.

51 Waseem, B., Alam, S., Irfan, M. et al. (2014). Optimization and characterization of adhesion properties of DLC coatings on different substrates. *IOP Conference Series: Materials Science and Engineering* 60: 012054.

52 Horiuchi, T., Yoshida, K., Kano, M. et al. (2009). Evaluation of Adhesion and wear resistance of DLC films deposited by various methods: evaluation of adhesion and wear resistance of DLC films. *Plasma Processes Polymers* 6 (6–7): 410–416.

53 Piotrowska, K., Granek, A., and Madej, M. (2020). Assessment of mechanical and tribological properties of diamond-like carbon coatings on the Ti13Nb13Zr alloy. *Open Engineering* 10 (1): 536–545.

54 Wang, C., Shan, D., and Guo, B. (2015). DLC-coated tools for micro-forming. In: *Micromanufacturing Engineering and Technology*, 487–512. Oxford: Elsevier.

55 Natrayan, L., Merneedi, A., Veeman, D. et al. (2021). Evaluating the mechanical and tribological properties of DLC nanocoated aluminum 5051 using RF sputtering. *Journal of Nanomaterials* 2021: 1–7.

56 Kodali, P., Walter, K.C., and Nastasi, M. (1997) Investigation of mechanical and tribological properties of amorphous diamond-like carbon coatings. *Tribology International* 30 (8): 591–598.

57 Nishida, S.I., Hattori, N., Nakabaru, Y. et al. (2007). Fatigue strength improvement of Ti alloy with DLC coating. *Computer Methods and Experimental Measurements for Surface Effects and Contact Mechanics* VIII I: 3–11.

58 Sun, H., Yang, L., Wu, H. et al. (2023). Effects of element doping on the structure and properties of diamond-like carbon films: a review. *Lubricants* 11 (4): 186.

59 Kabir, M.S., Zhou, Z., Xie, Z. et al. (2021). Designing multilayer diamond like carbon coatings for improved mechanical properties. *Journal of Materials Science & Technology* 65: 108–117.

60 Zarei Moghadam, R., Ehsani, M.H., Rezagholipour Dizaji, H. et al. (2021). Oxygen doping effect on wettability of diamond-like carbon films. *Materials Research Express* 8 (3): 035601.

61 Pauleau, Y. (2008). Residual stresses in dlc films and adhesion to various substrates. In: *Tribology of Diamond-Like Carbon Films* (eds. C. Donnet and A. Erdemir), 102–136. Boston, MA: Springer.

62 Guo, P., Li, X., Sun, L. et al. (2017). Stress reduction mechanism of diamond-like carbon films incorporated with different Cu contents. *Thin Solid Films* 640: 45–51.

63 Jang, Y.J., Kim, J.I., Kim, W. et al. (2022) Thermal stability of Si/SiC/ta-C composite coatings and improvement of tribological properties through high-temperature annealing. *Scientific Reports* 12 (1): 3536.

Section II

Diamond-like Carbon Coatings for Mechanical and Transportation Applications

4

Diamond-like Carbon Coatings for Tools and Molds

Farooq Ahmad, Raies Abdullah, Jamile Mohammadi Moradian

School of Materials Science and Engineering, Jiangsu University, Zhenjiang, China

CONTENTS

4.1 Introduction, 109
4.2 Hot Forging Die Tool Failure Mechanisms, 111
4.3 Surface Treatments in Die Tools and Molds, 111
4.4 Deposition of DLC Coatings, 113
  4.4.1 Physical Vapor Deposition, 113
  4.4.2 Chemical Vapor Deposition, 114
  4.4.3 Hybrid Coating Techniques, 115
4.5 Structural Properties of DLC Coatings, 115
4.6 Mechanical and Tribological Properties of DLC Coatings, 117
4.7 Applications of DLC Coatings in Die Tools and Molds, 118
4.8 Challenges and Future Perspective of DLC Coatings, 121

4.1 Introduction

Manufacturing components such as automobile parts, turbines, propellers, plastic components, and others require a variety of equipment, including molds and dies, forging, gauging, jigs and fixtures, and cutting tools. According to a survey by Allied Market Research, the size of the tooling market worldwide was estimated at \$212 500.00 million in 2020 and is expected to reach \$439 994.9 million by 2030, showing a compound annual growth rate of 7.5% from 2021 to 2030 [1]. Die tools and molds play an important role in modern manufacturing industries, enabling mass production, enhancing precision and efficiency, and being versatile and continually evolving to suit the changing needs of the industry. In recent years, extending the lifespan of die tools has been one of the major issues in forming industry.

Applications of Diamond-like Carbon Coatings, First Edition. Abdul Wasy Zia.

One of the key factors shortening the lifespan of components in the business is surface-to-surface contact between the parts [2]. An essential difficulty in modern materials research is exploring materials with significant strength, enabling them to retain their structural integrity when subjected to external load. The majority of materials exhibit performance characteristics in the presence of cyclic load, wear cavitation, corrosion, and extreme temperature conditions, such as cryogenic or high-temperature settings. These conditions typically result in extreme stress conditions within the surface layer of materials. The overall operational properties of the components are significantly influenced by the quality and features of the surface layer. Generally, the failure modes of molding and die tools are classified into four groups: wear, galling, plastic deformation, and chipping and cracking. To enhance the physical and mechanical properties, performance, and service life of dies and molding tools, protective, hard, and wear-resistant coatings can be applied. Protective coatings with thicknesses ranging from a few to tens of microns are commonly utilized on various molding tools. Ceramics based on nitrides, carbides, borides, oxides, transition metal silicide, and DLC are the most frequent hard materials for them.

In recent years, there has been an increasing interest among physicists and engineers in the study of diamond-like carbon (DLC) coatings. DLC coatings are thin films made of amorphous carbon with a diamond-like structure, hence the name. These coatings are well-suited for die tool and mold applications because of their unique mix of hardness, low friction, wear resistance, and chemical inertness [3]. The issue of insufficient adhesion of thickness ($\geq$0.5 μm) coating is one of the major limiting factors on the path to expanding applications of DLC. The coatings with poor adhesion have significant levels of internal mechanical stresses, which prevents them from being fabricated thicker up to submicrometers. This is insufficient for many practical applications. Furthermore, hard surfaces may not be strong enough to provide adequate strength. The thicker DLC-based coatings with the best possible balance hardness, strength, wear resistance, and tribological characteristics can now be obtained through several different approaches. These include the use of transitional, gradient, and multilayer layers, coating alloying, and nanoscale layered and composite architectures.

Overall, DLC coatings are very desirable for die tool and mold applications due to their distinctive attributes, including elevated hardness, reduced friction, enhanced wear resistance, and chemical inertness. These coatings improve die tool performance, durability, and efficiency, resulting in longer tool life, reduced wear, and enhanced dimensional accuracy. DLC coatings help to reduce friction, increase release qualities, and improve product quality in mold applications. As a result, DLC coatings hold enormous promise for transformative advances in the die tool and mold sectors [4–6].

4.2 Hot Forging Die Tool Failure Mechanisms

Die failure can take one of three different types during the hot forging process: Manufacturing failure, which results from inadequate raw materials and production methods; hot forging die materials are assessed based on characteristics like wear and plastic resistance, uniformity of hardening, resistance to heat checking, thermal fatigue, and mechanical fatigue. Steel, copper, aluminum, and brass are the materials that are hot forged the most frequently [7]. Low-carbon steels with high and medium alloying elements maintain hardness at high temperatures and have good strength and toughness, making them ideal for hot work die steels [8]. Hardness and toughness can be enhanced by adding tungsten; heat-checking resistance can be increased by adding vanadium; and abrasion resistance can be enhanced by adding molybdenum in high enough concentrations.

Die surface temperature rises during firm contact due to heat passage from the forming component to the die. High-temperature effects are reduced by cooling the die. The cyclic heating and cooling cause thermal strains and catastrophic breakdowns (Figure 4.1) [9]. After 2400 forging cycles, Figure 4.1a shows the beginning damage area creation. The creation of a V crack causes tool surface failure. The creation of microfractured species can also operate as third bodies. As the number of cycles increased to 4200 (Figure 4.1b, practically the entire surface was destroyed. Ductile fracture and micro abnormalities were observed. Cracks on the die surface indicate this failure, which can impair tool surface quality and end product quality.

The main causes of thermal fatigue are cracks. The failure mechanism during hot forging is surface, subsurface, or simultaneous. This failure is typically caused by inappropriate loading and heating. The major failure mechanism of molding tools is demonstrated in Figure 4.2.

4.3 Surface Treatments in Die Tools and Molds

AISI H13 tool steel is widely used for die tools due to its high-temperature thermal shock and softening resistance. Medium carbon chromium–molybdenum–vanadium alloy AISI H13 has a moderate thermal expansion, high thermal conductivity, high hot strength, tempering resistance, creep strength, and ductility. Due to its larger fraction coefficient, AISI H13 tool steel has poor sliding wear resistance and lower mechanical properties during heat cycling. Several studies have used plasma nitriding to increase AISI H13's tribological characteristics and die tool life [10, 11]. AISI H13 steel's alloying elements chromium, vanadium,

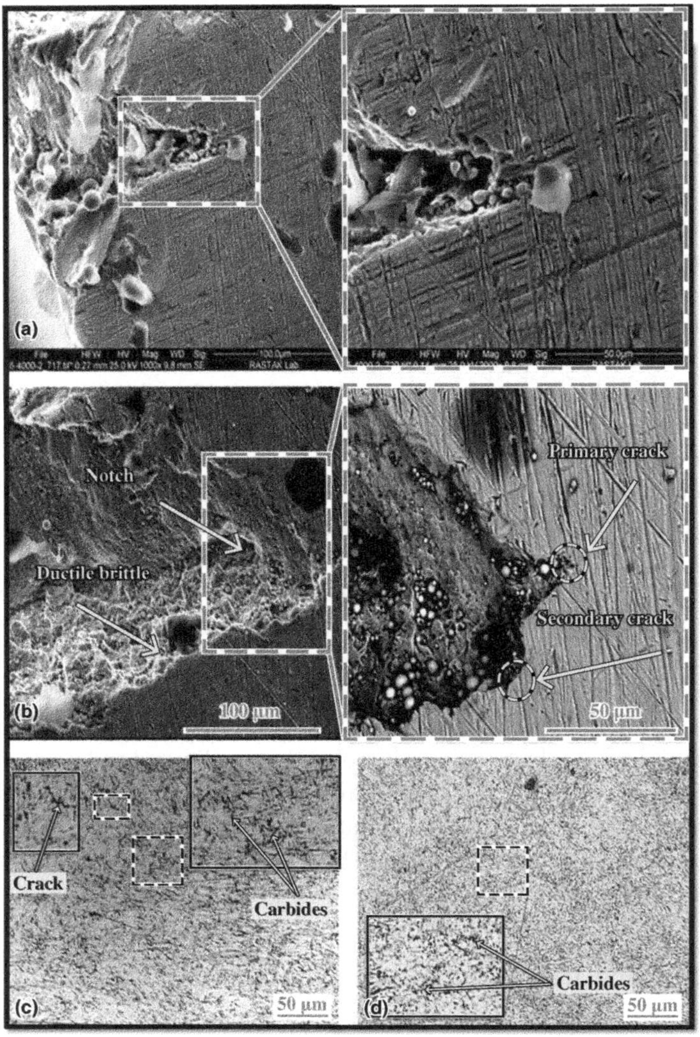

Figure 4.1 H21 steel hot forging dies exterior radius thermal fracture patterns after forging (a) 2400 cycles, (b) 4200 cycles, (c, d) optical microscopy images after 5400 cycles [9].

and molybdenum precipitate to nitrides at plasma nitriding temperatures above 450 °C, unlike other carbon steels, thereby increasing tensile stress, hardness, and wear resistance. In addition, coatings with specific surface morphologies that enhance wear resistance can be applied to die tools for improved performance. Different surface treatment processes to improve the performance of die tools are presented in Figure 4.3.

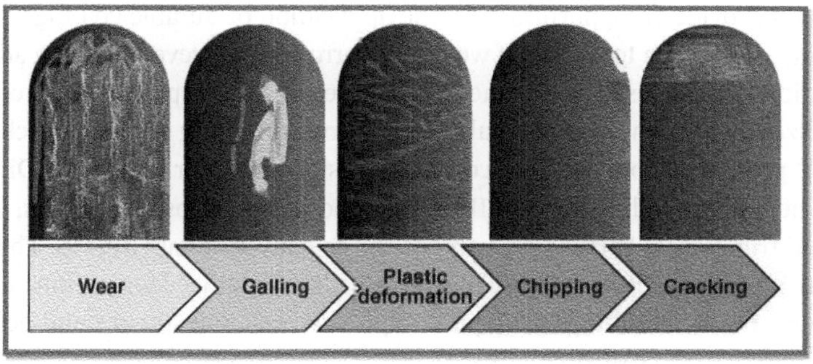

Figure 4.2 Graphical illustration of failure mode of molding tools surface.

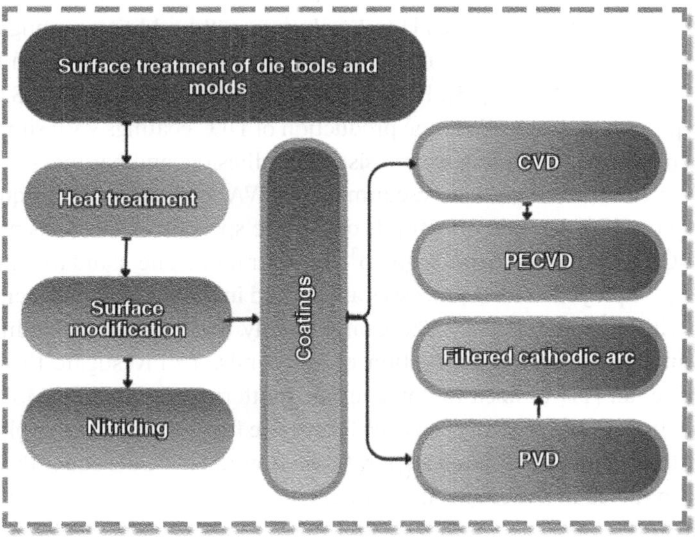

Figure 4.3 The surface treatment process of die tools and molds.

4.4 Deposition of DLC Coatings

There are various processes for depositing DLC coatings onto die tools, each with its own set of benefits and drawbacks. The most frequent DLC coating processes are shown in Figure 4.3.

4.4.1 Physical Vapor Deposition

DLC coatings are deposited onto die tool surfaces using physical vapor deposition (PVD) technologies such as magnetron sputtering or arc evaporation. These coating techniques provide an opportunity for the application of durable coatings that can effectively protect the tool against wear and corrosion; however, limiting adhesion with various substrates is one of the major issues in PVD deposited DLC coatings. The utilization of high-power impulse magnetron sputtering (HiPIMS) technology as a PVD process can optimize the coating adhesion and wear resistance. DLC coatings, when deposited using HiPIMS technology with positive pulses, exhibit improved tribological behavior in comparison to conventional HiPIMS. The incorporation of positive pulses during deposition leads to enhanced ionization of carbon and argon species, resulting in higher compressive residual stresses and densification of coating with positive pulses [12].

The coatings display higher sp^3 content, and the lower D-band and G-band intensity ratio leads to an increase in hardness [13]. The utilization of HiPIMS with a positive pulse also leads to improved adhesion of the coatings, wherein critical load values of 48.5 N are achieved [14]. Furthermore, the addition of tungsten (W) as a dopant in DLC coatings results in further enhancement of adhesion for high critical load in scratch tests [15]. In conclusion, the utilization of HiPIMS technology with positive pulses leads to the production of DLC coatings with superior tribological properties, including wear resistance, adhesion, and hardness.

However, the filtered cathode vacuum arc (FCVA) deposition technique is not frequently used by the industry [16]. Is one of the specialized methods needed to create ta-C DLC coatings with high sp^3 concentration. The standard sputtering techniques employed in PVD processes are limited in their ability to generate DLC coatings with sp^3 content above 40%. Consequently, these approaches exhibit deficiencies in facilitating the production of sp^3 bonds. To investigate this matter, Bernhardt et al. [17] utilized the magnetron sputtering ion plating (MSIP) technique to fabricate ta-C coatings on a WC substrate for glass molding applications. The resulting coatings exhibited a sp^3 ratio of around 60%. Subsequently, annealing was conducted in a vacuum environment at two distinct temperatures, namely 590 and 630 °C.

4.4.2 Chemical Vapor Deposition

DLC coatings deposited using chemical vapor deposition (CVD) technology exhibit both advantages and disadvantages. One notable benefit is their exceptional mechanical and tribological characteristics, which encompass elevated hardness and reduced friction coefficient. DLC coating confronts challenges associated with thermal instability at elevated temperatures. To enhance their thermal stability, various approaches have been employed, such as integrating nonmetallic or metallic elements into the DLC coatings [18]. Plasma-enhanced chemical vapor deposition (PECVD) is a widely used technology to deposit DLC coatings. Hydrocarbon gases are introduced into a reactor chamber and break down to generate a carbonaceous plasma in CVD. The carbon atoms are subsequently deposited onto the die tool substrate, resulting in the formation of the DLC coating. CVD provides for fine control of coating characteristics and the application of conformal coatings on complex die tool geometries [19].

4.4.3 Hybrid Coating Techniques

The effect of DLC coatings' properties varies depending on the deposition technique utilized. The highest adhesion between the DLC coatings and the substrate is achieved through the combination of laser texturing and nitriding pretreatments [20]. The magnetron sputtering technique yields DLC coatings with a significant reduction in friction coefficient, particularly in severe working conditions. The vacuum ion plasma technology produces DLC coatings with a gradient distribution of sp^3/sp^2 configuration of carbon, which decreases with the depth of the coating. PVD-coated DLC coatings on stainless steel substrates enhance the tribological properties and are devoid of pores and microcracks. Nitrogen-doped DLC coatings exhibit lower corrosion resistance than DLC coatings, with a greater reduction of sp^2 bonds observed after immersion in a NaCl solution [21]. Hybrid coating methods proffer numerous advantages over single-deposition techniques. Firstly, the integration of supplementary materials in the hybrid coatings can amplify the mechanical properties and sturdiness of the coatings, as evidenced by the improved drying time and surface hardness of the hybrid lacquer coating [22]. Secondly, hybrid coatings can showcase exceptional performance concerning liquid repellence, anti-smudge properties, and resistance to ink, dust, and bacteria adhesion, as demonstrated by the liquid-repellent, anti-smudge hybrid coating [23].

Thirdly, hybrid coatings can furnish better corrosion resistance and oxidation resistance compared to single-deposition coatings, as observed in the case of chromium nitride coatings deposited by hybrid HiPIMS [24]. On the whole, resorting to hybrid coating techniques facilitates the creation of coatings with enhanced properties and performance, rendering them suitable for a wide range of applications. To obtain optimized DLC coatings, hybrid coating techniques combine the benefits of different deposition procedures. Hybrid processes, for example, may use a combination of PVD and CVD, or the addition of other components such as metal dopants to improve specific DLC coating qualities. These hybrid techniques seek to customize the coating's properties to the unique needs of die tools [25].

4.5 Structural Properties of DLC Coatings

DLC is a microcrystalline structure composed of sp^2 and sp^3 carbon–hydrogen bonds, whose properties vary depending on the proportion of each bond. Table 4.1 presents the overview of various DLC coating techniques and the resultant structure of the coatings. Carbon hybridization gives rise to sp^1, sp^2, and sp^3 orbitals,

Table 4.1 Overview of various DLC coating techniques, benefits, and limitations.

Types of DLC coatings	Deposition technique	Substrate	Deposition conditions	Benefits	Limitations	Further reading
a-C-DLC	PVD	Tool steel, titanium, ceramics, polymers	Vacuum chamber, 200–400 °C	High hardness, low CoF, good wear resistance	Limited adhesion to certain substrates, limited thermal stability	[26]
a-C: H-DLC	CVD	Aluminum, copper, glass, polymers	Hydrocarbon gas, 350–550 °C	Excellent wear resistance, low CoF, good adhesion to various substrates	High deposition temperature, hydrogen content may affect properties	[27]
ta-C-DLC	FCVA	Steel, iron ceramics, diamond	High vacuum, 300–500 °C	Very low CoF, extremely high hardness, exceptional wear resistance	Deposition technique complexity, high equipment cost, limited availability	[28]
Nanocomposite DLC	Hybrid (PVD and CVD techniques)	Al, Ti-coated substrates (TiN, CrN, TiCN)	PVD and CVD techniques, 300–450 °C	Tailored properties for specific needs, improved adhesion to diverse substrates, enhanced wear resistance	Complex deposition process, cost and equipment requirements	[29]

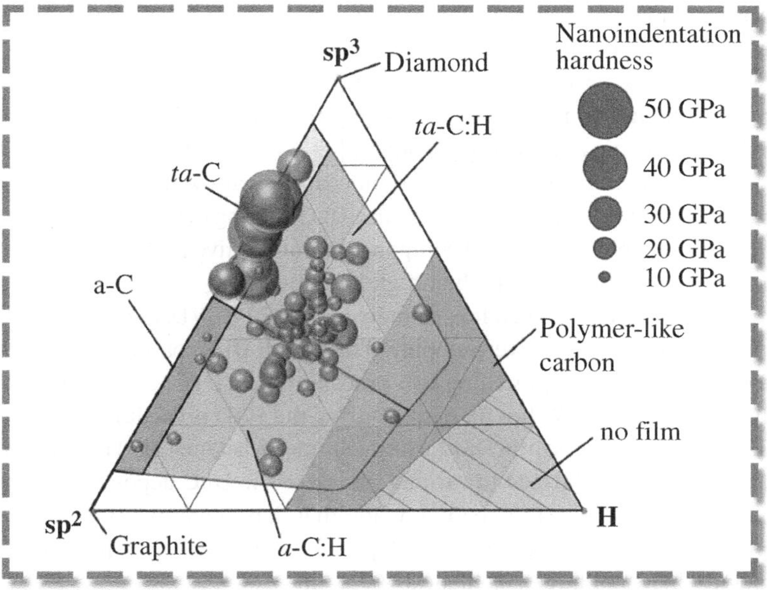

Figure 4.4 On the ternary diagram, there are 74 different forms of carbon films distributed. Each carbon film's nanoindentation hardness is represented by the diameter of the circle [32] / MDPI / CC BY 4.0.

which results from hybridization of s and p sublevels. The sp^3 tetrahedral bonding in DLC is identical to that of natural diamond and is responsible for its high strength, as reported by Liu et al. [30]. The carbon ternary diagram, as shown in Figure 4.4, illustrates various possibilities for structural organization, including diamond (located at the top), graphite (positioned on the left), and hydrogenated carbon (situated on the right) [31, 32]. The term "a-C" denotes amorphous carbon in the absence of hydrogen, whereas "ta-C" represents tetrahedral amorphous carbon. On the other hand, "a-C: H" signifies hydrogenated amorphous carbon, while "ta-C:H" denotes hydrogenated amorphous carbon.

4.6 Mechanical and Tribological Properties of DLC Coatings

The enhancement of surface performance in tools and molds relies heavily on the mechanical qualities exhibited by DLC coatings. The mechanical properties of DLC coatings are influenced by a multitude of factors: the deposition parameters, including the biased voltage, gas pressure, and substrate-to-target distance [33]. The literature survey of DLC coatings and their properties are provided, and the nanoindentation hardness is presented in a ternary diagram [34]. Some of the key points mentioned in this chapter are the following: The sp^3 (C) ratio strongly positively correlates with nanoindentation hardness and somewhat negatively with hydrogen content. sp^2 particle size positively correlates with hardness but is less than the sp^3 (c) ratio and hydrogen content [33, 34]. The mechanical characteristics of DLC-containing amorphous carbon coatings vary with substantial sp^3 (C) ratio changes. Authors also suggested that DLC coatings with ta-C structure are favorable for various mechanical components, automotive parts machining tools, cutting tools, and molds due to their high hardness [12].

Three plasma-based procedures, namely HiPIMS, PECVD, and plasma immersion ion deposition (PIID), were employed to deposit DLC coatings onto steel substrates [35]. During the preceding two procedures, a Ti/TiN adhesion layer was produced employing a W hot filament. During the PIID procedure, an interlayer composed of SiCx was synthesized using tetramethylsilane gas. The DLC-coated systems were then subjected to various analyses. The DLC deposited by PECVD was found to be the hardest coating, while that deposited by HiPIMS and PIID showed wedging spallation. The DLC deposited by PECVD also showed the best tribological behavior due to its highest H/E ratio.

The wear test using a ball-on-disc method was conducted to study the tribological properties of different coated samples including DLC, N-DLC, and N-DLC/DLC, deposited on the H13 tool steel substrate depicted in Figure 4.5 [36]. The tool steel sample had the highest friction coefficient, while the DLC coatings had the lowest coefficient of friction up to 230 m of sliding distance. The frictional coefficient found in the N-DLC coatings exhibits a gradual increase, ultimately reaching a value of 0.24. In contrast, the DLC coatings reach a friction coefficient of 0.3 at the end of the test. The wear characteristics shown by N-DLC/DLC coatings differ from those of other specimens, as evidenced by a coefficient of friction of about 0.15 observed after the test. The N-DLC/DLC specimen has the smallest wear rate, measuring $6.52 \times 10^{-5}\ mm^3\ N\ m^{-1}$. In contrast, the primary metal specimen displays the highest wear rate, which can be attributed to its lower hardness and greater friction coefficient. The cross-sectional profile of the wear track is shown in Figure 4.5, with the DLC coatings having poor adhesion to the substrate, the N-DLC/DLC specimen having the lowest wear rate, and the base metal (H13) having the highest wear volume.

4.7 Applications of DLC Coatings in Die Tools and Molds

In recent years, surface treatment has been significantly enhancing the performance and durability of die tools and molds to cater for the constantly evolving industrial requirements. Among the latest surface treatments, DLC coatings have emerged as a leading contender due to their unique characteristics, including

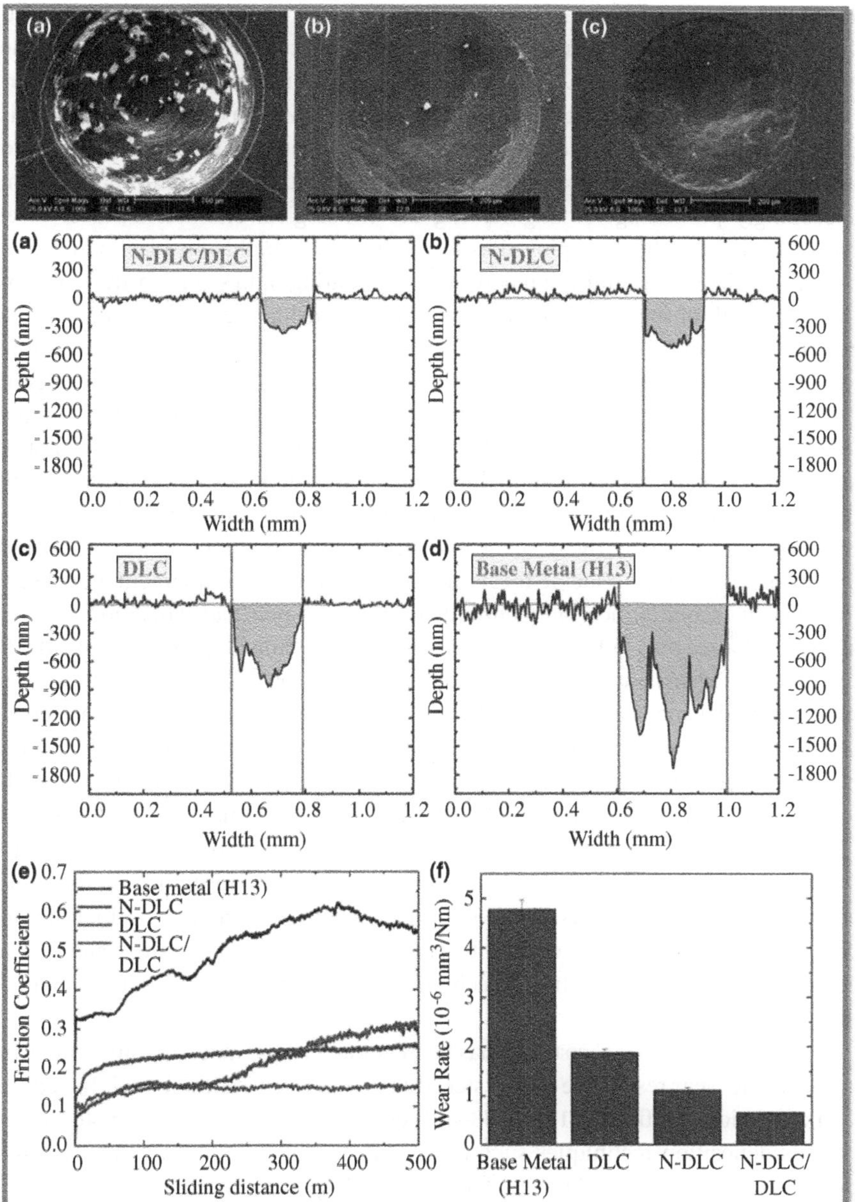

Figure 4.5 The micrograph of Rockwell indentation of DLC coatings: (a) N-DLC/DLC, (b) N-DLC, (c) DLC). (Top) The cross-sectional profile of different coatings and H13 tool steel. (a–d) 2D cross-section of wear track, (e) friction coefficient curve, (f) wear rate (bottom) Reproduced from [36] / with permission of ELSEVIER.

excellent surface smoothness, low friction coefficient, high hardness, wear resistance optical properties, antibacterial properties, and corrosion resistance [3]. The atomic composition of DLC coatings exhibits an amorphous arrangement of carbon atoms, forming covalent bonds with a hybridization of sp^2 and sp^3. This unique structure gives rise to characteristics that lie between those of graphite and diamond [33, 37, 38]. DLC coatings exhibit a notable quality that renders them exceedingly promising for augmenting the tribological characteristics of steel tools, specifically those employed as forming tools or molds for plastic injection [35]. Nevertheless, the adhesion between DLC coatings and steel substrates is limited as a result of the significant internal tensions inside the coatings and the disparities in chemical bonding at the interface between the coating and substrate [36]. Frequently, the occurrence of adhesive failures and delamination results in the premature failure of the coated steel tool [39]. To mitigate this issue, suitable interlayers are used to establish chemical compatibility between the steel substrate and the carbon layer. While the enhancement of adhesive behavior is seen, it is important to note that coating detachments may arise due to the plastic deformation of the comparatively malleable steel substrate when subjected to high applied stresses [40]. To tackle this issue, it is imperative to implement a precisely targeted surface modification technique on the steel material. This modification aims to enhance the load-bearing capability of the coating [39]. The plasma nitriding process, for instance, generates a graded hardness profile in the steel substrate, which provides mechanical support and improves the adhesion of the hard coating, thus enhancing overall tribological properties [40]. Several research studies have employed the duplex treatment approach, which incorporates plasma nitriding and DLC deposition, on various tool steel types including AISI H13 [41], H11 [39], 316L [42], WC [12], and 1040 [40]. The results of these investigations demonstrated a significant improvement in the tribological performance of the DLC coating as a result of increased load-carrying capacity and enhanced adherence to the substrate. Hence, it is crucial to regard the coating/substrate system as a composite to enhance the tribo-mechanical characteristics of steel tools coated with DLC.

The utilization of DLC coatings has been demonstrated in numerous studies to enhance the wear resistance of tool steel due to improved adhesion and the reduction of friction coefficient [16, 36, 43–45] regulated by interfacial layer [46] using various deposition techniques such as plasma-activated chemical vapor deposition [44], PECVD [47], FCVA [16], and HiPIMS [24]. Bernhardt et al. [17] employed MSIP to deposit ta-C coatings on precision glass molding tools. Ion plating induces carbon bonds to shift toward the sp^3 form, increasing internal coating stress. Enhance the sp^3 ratio to approximately 60%, a value that is frequently lower than 40% in standard sputtering methods.

Zhao et al. [48] looked into the mechanics of adhesive wear between optical glass D-FK95 mold surfaces in precision glass molding. This study investigates the suitability of WC molds coated with ta-C, AlCrN, and AlTiN in terms of their thermal properties and surface energy characteristics. The study delineates three distinct mechanisms of adhesion wear in open-air atmospheres, as well as an additional four-step process in an inert atmosphere. The experimental setup is depicted in Figure 4.6. The adhesion stress exhibits a positive correlation with temperature and pressure while demonstrating a negative association with coating. The research findings indicate that ta-C-coated WC molds exhibit superior anti-adhesive properties in comparison to coatings of AlCrN and AlTiN.

DLC coatings have a low coefficient of friction, which decreases frictional forces during die forming [5]. This reduces the amount of power required to distort materials, which can boost productivity, lessen the amount of wasteful heat produced, and improves the quality of the tool surface. Lower friction minimizes the likelihood of galling, adhesion, and material transfer, resulting in smoother operation and higher output.

4.8 Challenges and Future Perspective of DLC Coatings

DLC coatings provide numerous advantages, including high hardness, low friction coefficient, and excellent wear resistance. These characteristics make DLC coatings find a wide range of applications in automotive, aerospace, medical, and electronic industries. Despite their numerous advantages, there are still challenges to overcome and future opportunities to explore for wider applications and improved performance.

One of the basic issues with DLC coatings is their relatively high cost compared to other hard coating materials. Complex deposition processes and specialized equipment required for DLC coatings contribute to the higher cost. To overcome this problem, researchers focus on developing more cost-effective deposition techniques along with quality concerns of coatings. Finding alternative doping materials and optimizing process perimeters of coating deposition could help reduce the overall cost of DLC coating, opening them to a wider range of industries and applications [4].

Another problem is achieving strong adhesion to substrate materials. While DLC coatings generally have acceptable adhesion to many substrates, it can be difficult to achieve strong adhesion to certain materials, such as polymers or ceramics. This restriction could limit the usefulness of DLC coatings in various applications. Exploring surface pretreatment methods, exploring surface pretreatment techniques, and gaining insight into new adhesion interlayers are all steps in addressing the problem of low adhesion. The potential uses of DLC coatings across numerous industries would increase with improved adhesion to different substrates.

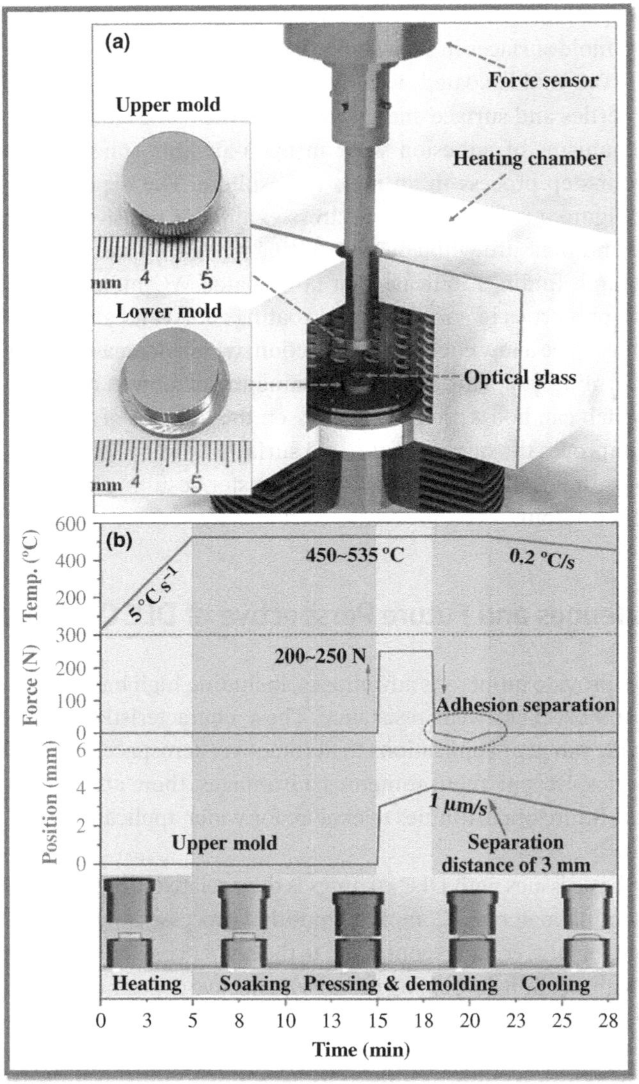

Figure 4.6 A schematic diagram presenting the molding time in terms of temperature, force, and motion of the top mold serves to demonstrate the experimental setting and testing method Reproduced from [48] / with permission of ELSEVIER.

To address adhesion challenges, Li et al. [49] reported a one-step duplex process involving plasma nitriding and graphite-like carbon (GLC) film deposition on M50NiL steel, enhanced by an argon ion sputtering pretreatment. This pretreatment resulted in thicker GLC films due to reduced surface roughness and higher polar surface energy, which facilitated nucleation. It also improved adhesion strength, evidenced by higher critical loads (Lc2). Furthermore, the process yielded the lowest coefficient of friction and wear rate, attributed to the GLC film's self-lubricating properties, the gradient hardness of the nitrided layers, and robust adhesion. Additionally, the pretreatment increased substrate hardness, mitigating mechanical mismatch and enhancing coating resistance to failure. These findings highlight the advantages of argon ion sputtering pretreatment in enhancing the performance of DLC coatings for die tools and molds.

DLC coatings exhibit excellent long-term stability up to 300 °C, when exposed to environmental conditions, maintaining their structural stability, hardness, and smoothness over time [27]. However, the thermal stability of DLC coatings may be insufficient for high-temperature applications, such as those encountered in some industrial processes. The thermal stability of DLC coatings is influenced by the hydrogen content, with higher hydrogen content leading to a decrease in hardness and structural stability at higher temperatures [50]. The thermal stability of hard coatings can be improved by introducing other elements in DLC coatings [18]. Future research will aim to improve DLC coatings' thermal stability, allowing them to withstand higher temperatures without degradation. Developing new deposition techniques and experimenting with various DLC formulations could improve thermal stability and broaden the range of high-temperature applications.

In conclusion, DLC coatings have impressive features and benefits, but there are also challenges to be overcome and prospects to explore. Addressing the high capital cost, limited adhesion to particular substrates, and enhancing thermo-mechanical stability, will extend DLC applications in tools and molds industries.

References

1 Amar, C. and Suman, O. (2021). *Tooling Market Size, Share, Competitive Landscape and Trend Analysis Report, by Product Type and End-user Industry: Global Opportunity Analysis and Industry Forecast, 2021–2030*. Allied Market Research, United Kingdom.

2 Zhang, D., Cheng, Y., Jiang, R. et al. (2018). *Turbine Blade Investment Casting Die Technology*. Berlin, Heidelberg: Springer Berlin Heidelberg. doi: 10.1007/978-3-662-54188-3.

3 Rajak, D.K., Kumar, A., Behera, A. et al. (2021). Diamond-like carbon (DLC) coatings: classification, properties, and applications. *Applied Sciences* 11: 4445. doi: 10.3390/app11104445.

4 Zia, A.W., Zhou, Z., and Li, L.K.-Y. (2019). Structural, mechanical, and tribological characteristics of diamond-like carbon coatings. In: *Nanomaterials-based Coatings*, 171–194. Oxford: Elsevier. doi: 10.1016/B978-0-12-815884-5.00007-7.

5 Wang, L., Liu, Y., Chen, H. et al. (2022). Modification methods of diamond like carbon coating and the performance in machining applications: a review. *Coatings* 12: 224. doi: 10.3390/coatings12020224.

6 Akhtar, A. and Ruan, H. (2022). Review on thin film coatings for precision glass molding. *Surfaces and Interfaces* 30: 101903. doi: 10.1016/j.surfin.2022.101903.

7 Marashi, J., Yakushina, E., Xirouchakis, P. et al. (2017). An evaluation of H13 tool steel deformation in hot forging conditions. *Journal of Materials Processing Technology* 246: 276–284. doi: 10.1016/J.JMATPROTEC.2017.03.026.

8 Krylova, S.E., Romashkov, E.V., Gladkovsky, S.V. et al. (2019). Special aspects of thermal treatment of steel for hot forming dies production. *Materials Today: Proceedings* 19: 2540–2544. doi: 10.1016/J.MATPR.2019.08.197.

9 Emamverdian, A.A., Sun, Y., Cao, C. et al. (2021). Current failure mechanisms and treatment methods of hot forging tools (dies)—a review. *Engineering Failure Analysis* 129: 105678. doi: 10.1016/j.engfailanal.2021.105678.

10 Peng, J., Zhu, Z., and Su, D. (2019). Sliding wear of nitrided and duplex coated H13 steel against aluminium alloy. *Tribology International* 129: 232–238. doi: 10. 1016/j.triboint.2018.08.017.

11 Fernandes, F.A.P., Heck, S.C., Picone, C.A. et al. (2020). On the wear and corrosion of plasma nitrided AISI H13. *Surface and Coating Technology* 381: 125216. doi: 10.1016/j.surfcoat.2019.125216.

12 García, J.A., Rivero, P.J., Barba, E. et al. (2020). A comparative study in the tribological behavior of DLC coatings deposited by HiPIMS technology with positive pulses. *Metals (Basel)* 10: 174. doi: 10.3390/met10020174.

13 Santiago, J.A., Fernández-Martínez, I., Kozák, T. et al. (2019). The influence of positive pulses on HiPIMS deposition of hard DLC coatings. *Surface and Coatings Technology* 358: 43–49. doi: 10.1016/j.surfcoat.2018.11.001.

14 Gómez, I., Claver, A., Santiago, J.A. et al. (2021). Improved adhesion of the DLC coating using HiPIMS with positive pulses and plasma immersion pretreatment. *Coatings* 11: 1070. doi: 10.3390/coatings11091070.

15 Hviščová, P., Kvetková, L., Lofaj, F. et al. (2015). Mechanical and tribological properties of HiPIMS and HiTUS W-C based coatings. In: *Key Engineering Materials*, 99–102. Wollerau: Trans Tech Publications Ltd. doi: 10.4028/KEM.6 62.99.

16 Konkhunthot, N., Euaruksakul, C., Photongkam, P. et al. (2013). Characterization of diamond-like carbon (DLC) films deposited by filtered cathodic vacuum arc technique. *Journal of Metals, Materials and Minerals* 23: 35–40.

17 Bernhardt, F., Georgiadis, K., Dolle, L. et al. (2013). Development of a ta-C diamond-like carbon (DLC) coating by magnetron sputtering for use in precision glass molding. *Materialwissenschaft und Werkstofftechnik* 44: 661–666. doi: 10.1002/mawe.201300171.

18 Zeng, Q. and Ning, Z. (2021). High-temperature tribological properties of diamond-like carbon films: a review. *Reviews on Advanced Materials Science* 60: 276–292. doi: 10.1515/rams-2021-0028.

19 Emamverdian, A.A., Sun, Y., Cao, C. et al. (2021). Current failure mechanisms and treatment methods of hot forging tools (dies)—a review. *Engineering Failure Analysis* 129: 105678. doi: 10.1016/j.engfailanal.2021.105678.

20 Zhang, Y., Fu, H., Wang, X. et al. (2020). Additional tribological effect of laser surface texturing and diamond-like carbon coating for medium carbon steel at near room temperature. *Coatings* 10: 929. doi: 10.3390/coatings10100929.

21 Sharifahmadian, O., Mahboubi, F., and Yazdani, S. (2019). Comparison between corrosion behaviour of DLC and N-DLC coatings deposited by DC-pulsed PACVD technique. *Diamond and Related Materials* 95: 60–70. doi: 10.1016/j.diamond.2019.04.007.

22 Wang, Y. and He, J. (2021). Fabrication of ultra-smooth hybrid thin coatings towards robust, highly transparent, liquid-repellent and antismudge coatings. *Journal of Colloid and Interface Science* 594: 781–790. doi: 10.1016/J.JCIS.2021.03.077.

23 Chen, Y., Zhang, G., Zhang, G. et al. (2021). Rapid curing and self-stratifying lacquer coating with antifouling and anticorrosive properties. *Chemical Engineering Journal* 421: 129755. doi: 10.1016/J.CEJ.2021.129755.

24 Ailsworth, Z., Chen, W.B., Lu, Y. et al. (2021). A hybrid image segmentation approach for thermal barrier coating quality assessments. In: *Proceedings-4th International Conference on Multimedia Information Processing and Retrieval, MIPR 2021*, 172–178. Tokyo: IEEE. doi: 10.1109/MIPR51284.2021.00033.

25 Kasiorowski, T., Lin, J., Soares, P. et al. (2020). Microstructural and tribological characterization of DLC coatings deposited by plasma enhanced techniques on steel substrates. *Surface and Coatings Technology* 389: 125615. doi: 10.1016/j.surfcoat.2020.125615.

26 Waseem, B., Alam, S., Irfan, M. et al. (2015). Optimization and characterization of adhesion properties of DLC coatings on different substrates. *Materials Today: Proceedings* 2: 5308–5312. doi: 10.1016/j.matpr.2015.11.041.

27 Cloutier, M., Harnagea, C., Hale, P. et al. (2014). Long-term stability of hydrogenated DLC coatings: effects of aging on the structural, chemical and mechanical properties. *Diamond and Related Materials* 48: 65–72. doi: 10.1016/j.diamond.2014.07.002.

28 Zia, A.W. and Birkett, M. (2021). Deposition of diamond-like carbon coatings: conventional to non-conventional approaches for emerging markets. *Ceramics International* 47: 28075–28085. doi: 10.1016/j.ceramint.2021.07.005.

29 Grigoriev, S., Volosova, M., Fyodorov, S. et al. (2019). DLC-coating application to improve the durability of ceramic tools. *Journal of Materials Engineering and Performance* 28: 4415–4426. doi: 10.1007/s11665-019-04149-1.

30 Patnaik, L., Maity, S.R., and Kumar, S. (2021). Effect of lubricated sliding wear against CFRPEEK on the nanomechanical properties of Ag alloyed Cr/DLC thin film. *Journal of the Mechanical Behavior of Biomedical Materials* 118: 104478. doi: 10.1016/j.jmbbm.2021.104478.

31 Al-Asadi, M.M. and Al-Tameemi, H.A. (2022). A review of tribological properties and deposition methods for selected hard protective coatings. *Tribology International* 176: 107919. doi: 10.1016/j.triboint.2022.107919.

32 Ohtake, N., Hiratsuka, M., Kanda, K. et al. (2021). Properties and classification of diamond-like carbon films. *Materials* 14: 1–26. doi: 10.3390/ma14020315.

33 Reshetnyak, E.N., Luchaninov, A.A., and Strel'nitskij, V.E. (2023). Optimization of diamond-like carbon coatings for mechanical and tribological applications. Review. *Problems of Atomic Science and Technology* 2023: 74–93. doi: 10.46813/2023-144-074.

34 Ohtake, N., Hiratsuka, M., Kanda, K. et al. (2021). Properties and classification of diamond-like carbon films. *Materials* 14: 1–26. doi: 10.3390/ma14020315.

35 Kasiorowski, T., Lin, J., Soares, P. et al. (2020). Microstructural and tribological characterization of DLC coatings deposited by plasma enhanced techniques on steel substrates. *Surface and Coatings Technology* 389: 125615. doi: 10.1016/j.surfcoat.2020.125615.

36 Sharifahmadian, O. and Mahboubi, F. (2019). A comparative study of microstructural and tribological properties of N-DLC/DLC double layer and single layer coatings deposited by DC-pulsed PACVD process. *Ceramics International* 45: 7736–7742. doi: 10.1016/j.ceramint.2019.01.076.

37 Bewilogua, K. and Hofmann, D. (2014). History of diamond-like carbon films—from first experiments to worldwide applications. *Surface and Coatings Technology* 242: 214–225. doi: 10.1016/J.SURFCOAT.2014.01.031.

38 Sagar, K.G. and Fernandes, F. (2023). An overview of nano-structural hard coating synthesis, design, and applications. *Preprints*. doi: 10.20944/preprints202304.0743.v1.

39 Tillmann, W., Dias, N.F.L., and Stangier, D. (2019). Influence of plasma nitriding pretreatments on the tribo-mechanical properties of DLC coatings sputtered on AISI H11. *Surface and Coatings Technology* 357: 1027–1036. doi: 10.1016/j.surfcoat.2018.11.002.

40 Teshima, P.H., Binder, C., Hammes, G. et al. (2016). Effects of different plasma nitrided layers on the tribological performance of DLC coatings. *Materials Research* 19: 1180–1188. doi: 10.1590/1980-5373-MR-2015-0455.

41 Conde, F.F., Diaz, J.A.Á., Silva, G.F.D. et al. (2019). Dependence of wear and mechanical behavior of nitrocarburized/CrN/DLC layer on film thickness. *Materials Research* 22: 1–10. doi: 10.1590/1980-5373-MR-2018-0499.

42 Zhang, X., Chang, W., Zhang, H. et al. (2019). Al content effects on mechanical and tribological properties of Cr/CrN/CrAlN multilayer nanocomposite coatings. In: *IOP Conference Series: Materials Science and Engineering*. 678: 012163. doi: 10.1088/1757-899X/678/1/012163.

43 Beake, B.D. (2022). Nano-and micro-scale impact testing of hard coatings: a review. *Coatings* 12: 793. doi: 10.3390/coatings12060793.

44 Sharifahmadian, O., Mahboubi, F., and Oskouie, A. (2019). Structural evolution and tribological behavior of nitrogen-doped DLC coatings deposited by pulsed DC PACVD method. *Diamond and Related Materials* 91: 74–83. doi: 10.1016/j.diamond.2018.11.004.

45 dos Santos de Almeida, E.A., Milan, J.C.G., da Costa, C.E. et al. (2021). Combined use of surface texturing, plasma nitriding and DLC coating on tool steel. *Coatings* 11: 201. doi: 10.3390/coatings11020201.

46 Humphrey, E., Morris, N.J., Rahmani, R. et al. (2020). Multiscale boundary frictional performance of diamond like carbon coatings. *Tribology International* 149: 105539. doi: 10.1016/j.triboint.2018.12.039.

47 Kumar, S. and Kumar, M. (2022). Tribological and mechanical performance of coatings on piston to avoid failure—a review. *Journal of Failure Analysis and Prevention* 22: 1346–1369. doi: 10.1007/s11668-022-01436-3.

48 Zhao, H., Gain, A.K., Li, Z. et al. (2023). Wear of mold surfaces: interfacial adhesion in precision glass molding. *Wear* 524–525: 204847. doi: 10.1016/j.wear.2023.204847.

49 Li, J., Hong, H., Sun, L. et al. (2024). Argon ion sputtering bridging plasma nitriding and GLC film deposition: effects on the mechanical and tribological properties. *Surface and Coatings Technology* 479: 130559. doi: 10.1016/j.surfcoat.2024.130559.

50 Ito, H., Yamamoto, K., Masuko, M. (2008). Thermal stability of UBM sputtered DLC coatings with various hydrogen contents. *Thin Solid Films* 517: 1115–1119. doi: 10.1016/j.tsf.2008.08.163.

5

Diamond-like Carbon Coatings for Solid Lubrication: Production Techniques, Properties, and Applications

Deepak Kumar[1,2], Virendra Pratap Singh[3], Himanshu Rai[4]

[1] *Department of Mechanical Engineering, Carnegie Mellon University, Pittsburgh, PA, USA*
[2] *Department of Materials Design and Innovation, University at Buffalo, New York, NY, USA*
[3] *Department of Mechanical Engineering IES College of Technology, Bhopal, India*
[4] *Department of Control Engineering, Faculty of Electrical Engineering, Czech Technical University in Prague, Prague, Czech Republic*

CONTENTS

5.1 Introduction, 130
5.2 Classifications of DLC Coatings, 131
  5.2.1 Hydrogenated Dip-layer Capacitors (a-DLC), 132
  5.2.2 DLC Coatings Devoid of Hydrogen (ta-C), 132
  5.2.3 DLC Coatings Doped With Nitrogen (a-C:N), 132
  5.2.4 Fluorinated DLC Coatings (a-C:F), 133
  5.2.5 Silicon-doped DLC Coatings (a-C:Si), 133
  5.2.6 Anhydrous Carbon Nitride Amorphous (a-CN:H), 133
5.3 Deposition Methods of DLC Coatings, 133
  5.3.1 Sputtering, 133
    5.3.1.1 Direct Current Magnetron Sputtering, 133
    5.3.1.2 Sputtering at Radio Frequency, 134
  5.3.2 Ion Beam Deposition, 135
  5.3.3 Cathodic Arc, 135
  5.3.4 Pulsed Laser Deposition, 136
  5.3.5 Chemical Vapor Deposition, 137
  5.3.6 Physical Vapor Deposition, 138
5.4 Tribology of DLC Coatings, 140
  5.4.1 Effect of Surface Roughness on Wear and Friction, 140
  5.4.2 Effect of Environment on Wear and Friction, 141
  5.4.3 Effect of Substrate on Wear and Friction, 142
5.5 Modification of DLC Coatings for Reduced Wear and Friction, 143
  5.5.1 DLC Coatings With Interlayers, 144
  5.5.2 Metal-doped DLC, 144
  5.5.3 Cr-based Interlayers DLC Coatings, 145
5.6 Applications of DLC Coatings, 145
5.7 Summary and Future Directions, 146

Applications of Diamond-like Carbon Coatings, First Edition. Abdul Wasy Zia.

5.1 Introduction

The investigation and utilization of diamond-like carbon (DLC) coatings has become a noteworthy subject in the field of solid lubrication [1]. Due to their exceptional characteristics and wide-ranging applications, DLC coatings have garnered significant attention from engineers and scientists in a variety of fields [2]. This chapter explores the various dimensions of DLC coatings utilized in solid lubrication [3]. This includes an exploration of their methods of synthesis, mechanisms of lubrication, tribological characteristics, and diverse range of applications [4]. The quest for effective lubrication solutions has grown in significance across various industries in contemporary engineering, including automotive, aerospace, manufacturing, and biomedical. Traditional lubricants frequently encounter drawbacks including instability in viscosity, vulnerability to contamination, and restricted operational parameters. These concerns have spurred the investigation of alternative methods such as DLC coatings [5–8].

DLC compounds possess a unique microstructure that resembles that of diamond. This structure is distinguished by a hybridized carbon network consisting of sp^2 and sp^3 bonds, with hydrogen incorporation differing in degree [9]. The distinctive configuration of DLC coatings bestows upon them a multitude of remarkable characteristics, such as elevated hardness, diminished coefficient of friction, exceptional resistance to abrasion, chemical inertness, and biocompatibility. Due to these characteristics, DLC coatings are exceptionally appealing in situations where traditional lubricants are inadequate [10, 11]. The ternary phase diagram of DLC is shown in Figure 5.1.

The purpose of this review chapter is to provide a thorough explanation of DLC coatings utilized in solid lubrication. It will commence by examining the synthesis methods that are utilized in their fabrication. A comprehensive examination will be given to various techniques, including physical vapor deposition (PVD), chemical vapor deposition (CVD), and plasma-enhanced CVD, with an emphasis on their respective functions in customizing DLC coatings to suit particular applications demand [12–17]. Furthermore, elucidating the fundamental lubrication mechanisms that regulate the functionality of DLC coatings is discussed. Understanding the mechanisms by which protective tribofilms are formed, interfacial adhesion is reduced, or surfaces passivate is crucial for optimizing the tribological performance of DLC coatings in a variety of operating environments [18].

By conducting an extensive review of the relevant literature, this chapter seeks to outline the wide range of solid lubrication applications in which DLC coatings have proven to be effective. Diverse industrial sectors, including micro-electromechanical systems (MEMS), biomedical implants, automotive engine components, and cutting instruments, have all demonstrated the

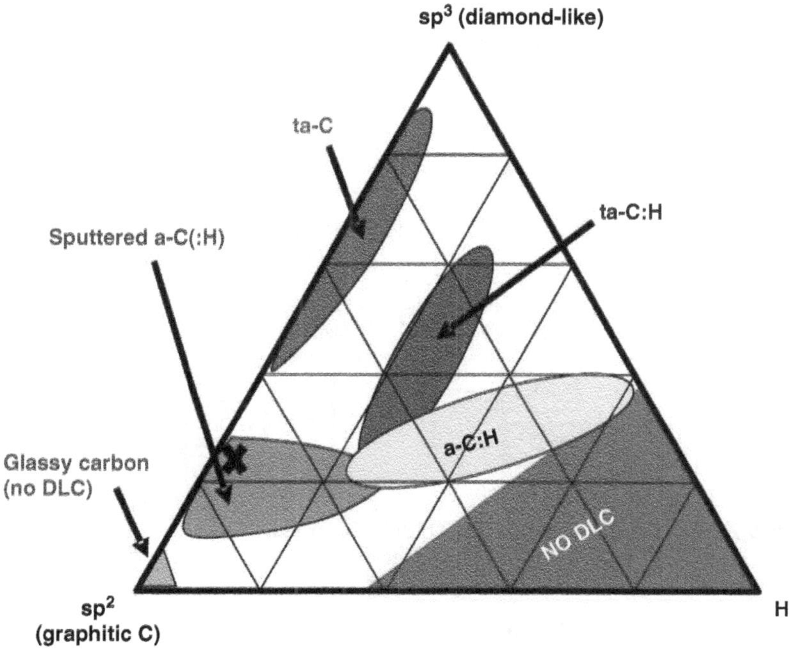

Figure 5.1 The phase diagram of carbon: regions of tetrahedral amorphous carbon (ta-C), amorphous carbon (a-C), hydrogenated ta-C (ta-C:H), and hydrogenated a-C (a-C:H). Reproduced from [10] / with permission of ELSEVIER.

adaptability and effectiveness of DLC coatings [19–23]. DLC coatings are, at their core, an auspicious new frontier in the field of solid lubrication, as they combine a compelling combination of exceptional properties and adaptable applications [24]. The purpose of this review is to consolidate recent research discoveries and developments in DLC coatings so as to offer significant perspectives on their contribution to the progression of solid lubrication technologies and their ability to resolve the lubrication obstacles faced in modern engineering applications.

5.2 Classifications of DLC Coatings

A multitude of varieties of DLC coatings exist, each possessing distinct properties and attributes. Common DLC coating varieties are shown in Figure 5.2.

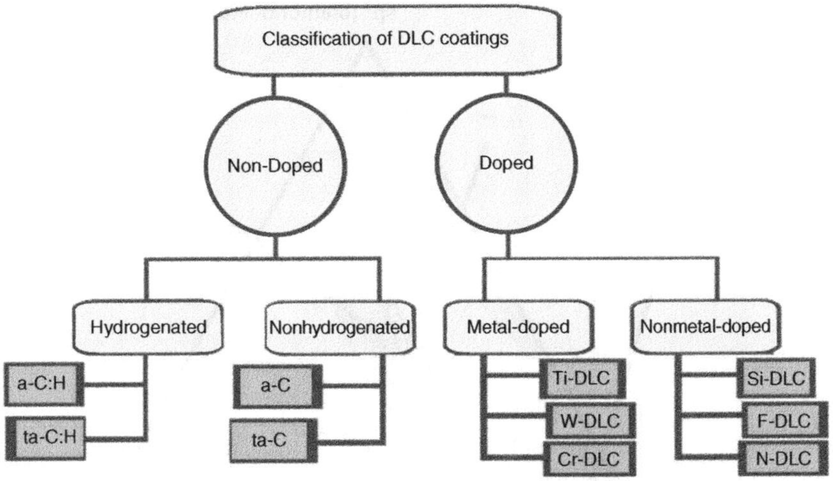

Figure 5.2 Types of DLC coatings [25] / MDPI / CC BY 4.0.

5.2.1 Hydrogenated Dip-layer Capacitors (a-DLC)

They are characterized by the incorporation of hydrogen ions into the DLC matrix. In addition to improving tribological properties such as adhesion and abrasion resistance, the inclusion of hydrogen in the coating enhances its flexibility and ameliorates internal stresses.

5.2.2 DLC Coatings Devoid of Hydrogen (ta-C)

These DLC coatings are distinguished by a greater proportion of sp^3 carbon bonds and the absence of hydrogen. These coatings are alternatively referred to as tetrahedral amorphous carbon. These coatings are suitable for use in severe environments due to their exceptional hardness, low friction, high abrasion resistance, and chemical inertness.

5.2.3 DLC Coatings Doped With Nitrogen (a-C:N)

The properties of nitrogen-doped DLC are modified by introducing nitrogen atoms into the DLC matrix. In comparison to traditional DLC coatings, these coatings frequently demonstrate improved adhesion, decreased friction, and inclines.

5.2.4 Fluorinated DLC coatings (a-C:F)

The utilization of fluorine atoms within the DLC matrix to produce results in the incorporation of distinctive characteristics, including diminished friction, enhanced anti-sticking capabilities, and decreased surface energy. Applications that require nonstick characteristics frequently employ these coatings.

5.2.5 Silicon-doped DLC coatings (a-C:Si)

They modify the DLC matrix's characteristics, including adhesion, hardness, and chemical resistance, by introducing silicon atoms into the DLC matrix. Given their ability to adhere to silicon-based substrates, these coatings are utilized in the semiconductor, MEMS, and optical sectors.

5.2.6 Anhydrous Carbon Nitride Amorphous (a-CN:H)

These coatings are composed of carbon and nitrogen in combination with hydrogen in varying proportions. In addition to enhanced adhesion, minimal friction, and superior wear resistance, they provide a balance of characteristics between DLC and CN coatings.

In order to customize the characteristics of DLC coatings for particular applications, they may also be doped with metals including titanium, chromium, or tungsten. The hardness, abrasion resistance, and thermal stability of metal-doped DLC coatings are frequently improved.

5.3 Deposition Methods of DLC Coatings

There are several methods that can be used for the deposition of DLC coatings, which discussed in detail in the following sections.

5.3.1 Sputtering

Sputtering is a PVD technique that is extensively employed in the application of DLC coatings onto substrates. For DLC deposition, primarily two sputtering techniques are utilized. Figure 5.3 shows the sputtering process.

5.3.1.1 Direct Current Magnetron Sputtering

It involves the application of a negative direct current (DC) voltage to a target material, which is typically a carbon-containing target, which is positioned within a vacuum chamber. Ions ejected from the target surface upon collision with a

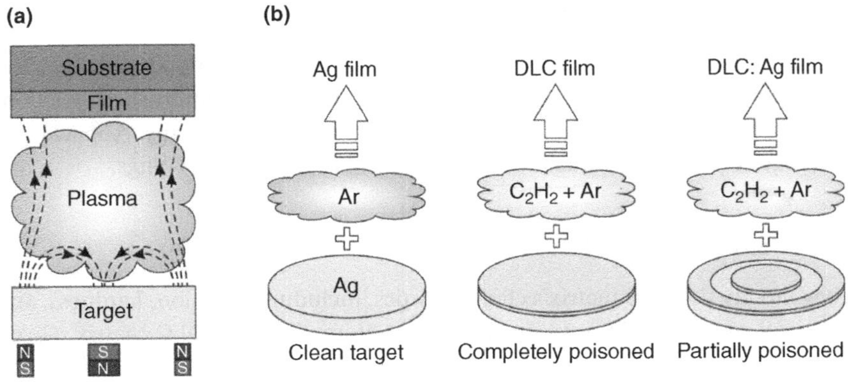

Figure 5.3 Sputtering process for DLC deposition. Reproduced from [26] / with permission of ELSEVIER.

high-energy discharge (plasma) dislodge atoms or molecules [27, 28]. The DLC coating is formed when these discharged species deposit onto the substrate positioned on the opposite side of the chamber in a straight line.

High levels of ion bombardment may result from DC sputtering, which is a relatively straightforward and economical process; nevertheless, this can cause increased film stress and decreased adhesion.

5.3.1.2 Sputtering at Radio Frequency

To produce plasma in the chamber, radio frequency (RF) sputtering requires the application of a high-frequency (typically RF) electrical field to the target material. The target surface is bombarded by the ions in the plasma, which results in the sputtering and deposition of atoms or molecules onto the substrate [29]. Compared to DC sputtering, RF sputtering provides enhanced control over ion energy and film properties. By manipulating parameters including gas pressure, target power, and substrate bias, DLC coatings with customized adhesion, friction, and hardness can be produced via RF sputtering [30].

DC and RF sputtering techniques each have their own benefits and drawbacks with regard to DLC deposition. Despite being more straightforward and economical, DC sputtering may generate greater ion bombardment, which can subsequently cause an escalation in film stress. While RF sputtering offers enhanced control over film properties, it may necessitate the use of more advanced apparatus and specialized knowledge. Moreover, the aforementioned approaches may be integrated with supplementary methodologies, including reactive sputtering or ion beam assistance, in order to augment the characteristics of DLC films or customize their composition. Sputtering techniques provide a flexible and efficient approach to applying DLC coatings that possess advantageous characteristics, catering to a wide range of industrial uses.

5.3.2 Ion Beam Deposition

By utilizing Ar ions to dislodge carbon atoms from a graphite target, ion beam sputtering or deposition (IBD) permits the production of high-quality coatings at comparatively low temperatures [13]. However, the rate of deposition is lethargic, and as a result of limited atomic motion during deposition, films frequently have a low density. In order to achieve consistent deposition, substrate manipulation is required. In order to tackle these obstacles, favored methodologies include ion plating and ion beam-assisted deposition (IBAD). Components are negatively charged during ion plating, and the coating flux is ionized via a variety of processes. In contrast, IBAD employs a secondary Ar ion bombardment for the purpose of compacting the coating and promoting sp^3 bonding [31]. However, it is crucial to exercise caution in order to avoid substrate overheating. Commercial viability is attributed to IBAD as a deposition method. Figure 5.4 shows the schematic of IBD unit.

5.3.3 Cathodic Arc

Cathodic arc deposition for DLC coatings includes vaporizing a carbon target by use of a high-voltage electrical arc inside a vacuum chamber [33]. During this technique, a plasma plume of carbon atoms and ions is produced. These particles are subsequently deposited onto a substrate, resulting in the formation of a DLC coating that is both hard and long-lasting [34].

High deposition rates, great adherence to a wide variety of substrates, and the flexibility to tune coating qualities by altering process parameters are some of the benefits that may be gained by using this approach [35]. The fact that cathodic arc DLC coatings have desired properties such as low friction, high hardness, and chemical inertness makes them appropriate for a broad variety of applications.

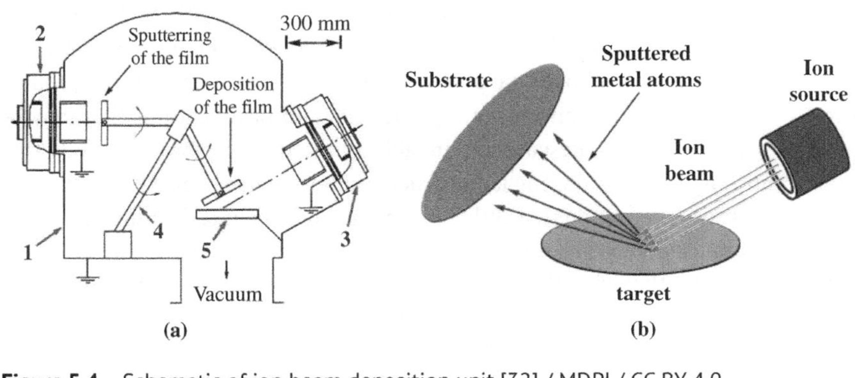

Figure 5.4 Schematic of ion beam deposition unit [32] / MDPI / CC BY 4.0.

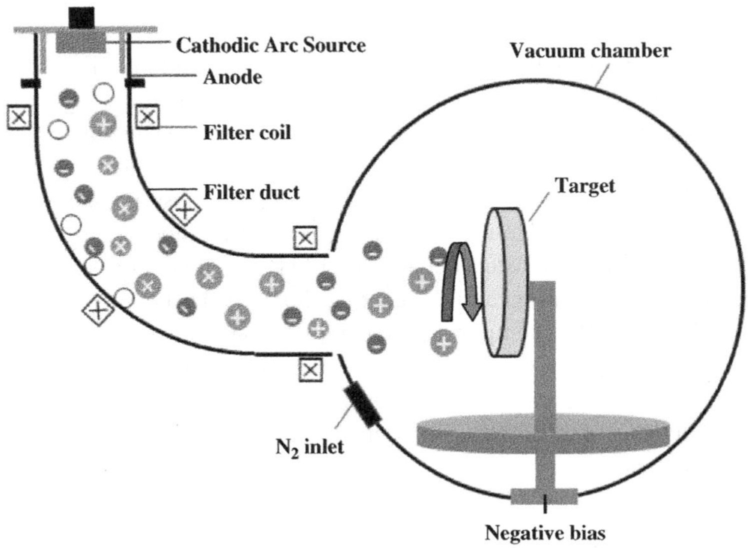

Figure 5.5 Schematic of cathodic vacuum arc deposition [34] / MDPI / CC BY 4.0.

Some examples of these applications include cutting tools, biomedical implants, and components for automobiles. Figure 5.5 shows the schematic of cathodic vacuum arc deposition.

5.3.4 Pulsed Laser Deposition

When it comes to DLC coatings, the process of pulsed laser deposition (PLD) includes the use of a high-energy laser to ablate a carbon target inside a vacuum chamber. This results in the formation of a plasma plume that is abundant in carbon species. Following this, the plasma plume is directed onto a substrate, where it continues to condense and eventually creates a thin coating of DLC [36].

A schematic diagram of pulsed laser deposition is shown in Figure 5.6. The use of PLD provides fine control over the deposition process, which enables the production of coatings that are uniform, of high quality, and have features that may be adapted to specific needs. Researchers are able to fine-tune the structure and properties of the DLC film by altering factors such as laser fluence, deposition temperature, and substrate conditions [38, 39]. These parameters include the film's hardness, adhesion, and friction coefficient. PLD is especially useful for depositing DLC coatings on substrates that are either complicated or heat-sensitive. Additionally, it has applications in a variety of disciplines, including microelectronics, optics, and biomedical devices, all of which need high-performance surface coatings [40, 41].

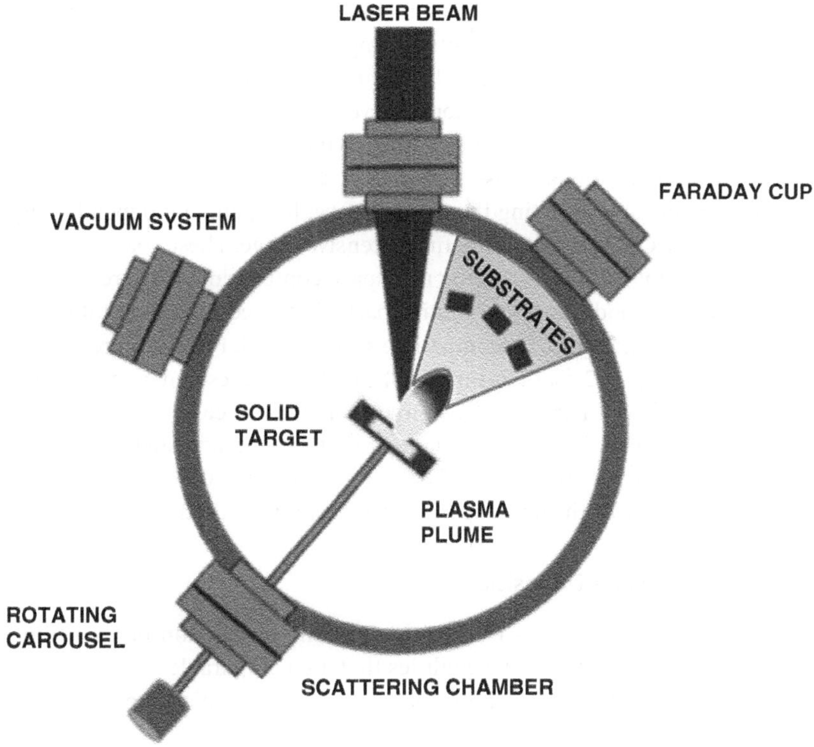

Figure 5.6 Pulsed laser deposition schematic [37] / MDPI / CC BY 4.0.

5.3.5 Chemical Vapor Deposition

CVD is a flexible process that is used to produce thin layers of carbon-based films that have characteristics that are similar to those of genuine diamonds [42]. This technique is used for DLC coatings, which stand for DLC. A reaction chamber that is kept at high temperatures, normally ranging from several hundred to a few thousand degrees Celsius, is used in this procedure. The precursor gases that contain carbon atoms, such as hydrocarbons or gasses like methane or acetylene, are delivered into the chamber. During the course of these chemical interactions, these gasses are broken down into their component carbon species [43]. These carbon species are subsequently deposited into a substrate surface, which results in the formation of the DLC coating. The controlled deposition environment makes it possible to precisely manipulate the attributes of the film, such as its thickness, homogeneity, and composition. Through the manipulation of factors such as gas flow rates, deposition temperature, pressure, and substrate conditions, researchers are able to customize the properties of the DLC coating in order to meet certain needs. These criteria may include a wide variety of characteristics, including extraordinary hardness, outstanding wear resistance, low friction coefficient, biocompatibility, and optical clarity, among others. Because the DLC coatings that are produced as a consequence have good adhesion to a wide range of substrates, such as metals, ceramics, and polymers, they are appropriate for a wide variety of applications across a number of different categories of businesses [44]. When it comes to boosting the performance, lifespan, and dependability of key components, CVD DLC coatings find extensive usage. These coatings are used in a variety of applications, including aerospace components that are exposed to intense operating conditions, automotive parts that need greater durability, and medical devices that require biocompatible surfaces [45]. Furthermore, the capability of depositing DLC coatings on complicated structures and complex geometries further widens their use. This process makes it easier to manufacture innovative materials with surface qualities that can be adjusted to match the requirements of current technical and commercial applications [46]. Figure 5.7 shows the schematic of the chemical vapor deposition method.

5.3.6 Physical Vapor Deposition

A complex method known as PVD is used for the production of thin films of carbon-based materials that have qualities that are comparable to those of genuine diamonds [48]. The first step in the process involves the placement of a carbon

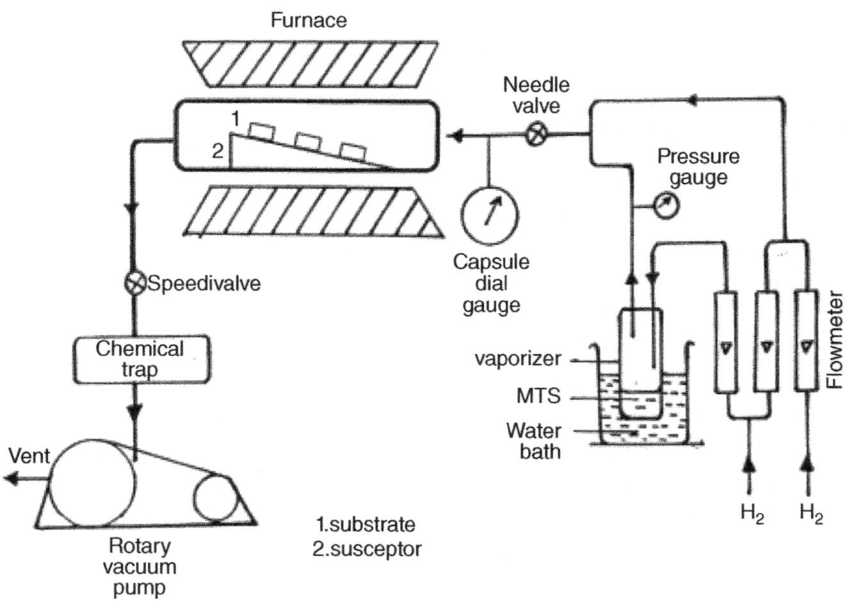

Figure 5.7 Schematic diagram of chemical vapor deposition method. Reproduced from [47] / with permission of ELSEVIER.

source material, which is commonly in the form of a graphite target, inside of a vacuum chamber [49]. Atoms are expelled off the surface of the carbon target as a result of the bombardment of high-energy ions or electrons, which results in the formation of a vapor or plasma phase [50]. Afterward, the vaporized carbon material condenses onto a substrate that is located in close proximity, resulting in the formation of a thin layer of DLC coating. Figure 5.8 shows the schematic of the PVD method.

PVD offers a number of major benefits, one of which is its remarkable control over the deposition process [52]. This control enables exact modification of factors such as the rate of deposition, the temperature of the substrate, and the ion energy inside the material. With the help of this control, engineers and researchers are able to carefully adjust the characteristics of the DLC coating in order to fulfill certain performance criteria [53]. By way of illustration, by modifying the parameters of the deposition process, it is possible to produce DLC coatings that demonstrate varied degrees of hardness, adhesion strength, wear resistance, and even optical qualities [54].

In addition, PVD DLC coatings may be deposited at temperatures that are quite low, which reduces the likelihood of substrate deterioration or distortion [55]. Having this property is very advantageous when it comes to covering heat-sensitive objects or intricate geometries. In addition, the high vacuum environment in which PVD is carried out guarantees the purity of the deposited DLC films, which are devoid of any impurities that may potentially affect their qualities [56].

As a result of their remarkable hardness and low friction coefficient, PVD DLC coatings are extremely sought for a broad variety of applications across a variety of industries. When it comes to the automobile industry, PVD DLC coatings are applied to improve the wear resistance of engine components, hence enhancing fuel economy and increasing the lifetime of essential components [57]. When used in aerospace applications, these coatings provide protection against abrasion and

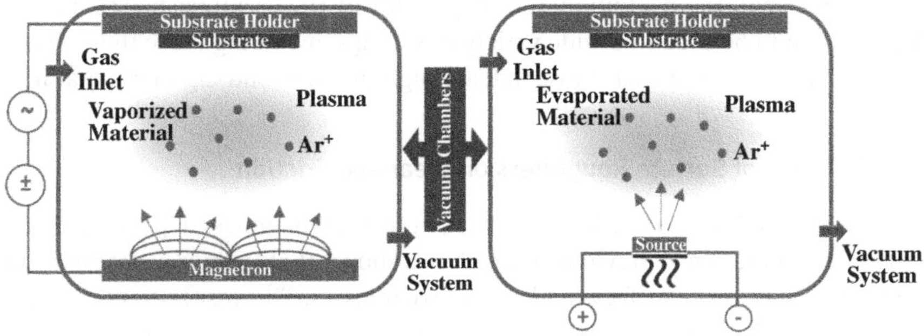

Figure 5.8 Physical vapor deposition methods: (a) sputtering and (b) evaporating using Ar gas [51] / MDPI / CC BY 4.0.

corrosion for aircraft components that are exposed to harsh operating conditions. Additionally, PVD DLC coatings are used extensively in cutting tools, medical implants, and consumer electronics [58]. This is due to the fact that their higher performance features lead to greater durability, dependability, and usefulness. When it comes to improving the functionality and durability of important components in a wide variety of technical and industrial applications, PVD DLC coatings are a vital tool due to their flexibility, accuracy, and excellent qualities [59, 60].

5.4 Tribology of DLC Coatings

Due to their outstanding hardness, low friction coefficient, and wear resistance, DLC coatings have received a large amount of interest in the field of tribology. This is because these features make them very desired for a variety of technical applications. High hardness, chemical inertness, and low surface energy are some of the characteristics that are shown by DLC coatings, which make them a unique mix of attributes that are similar to those of genuine diamonds. These characteristics contribute to a reduction in friction and wear, which ultimately results in an improvement in the durability and performance of components that are exposed to contact that is abrasive, sliding, or rolling. DLC coatings have a smooth and uniform surface morphology, which reduces surface roughness and contact stresses. This is the source of the inherent lubricity that these coatings possess. Additionally, the sp^3-hybridized carbon bonds that are present in DLC films are some of the factors that contribute to the exceptional mechanical capabilities of these films. These features include high compressive strength and resistance to deformation when subjected to stress. Because of these properties, DLC coatings are especially well suited for applications in which high levels of friction, wear, and dependability are of utmost importance. Some examples of such applications include cutting tools, aircraft bearings, medical implants, and components of automobile engines. In addition, the adaptability of DLC coatings makes it possible for them to be used in a wide variety of settings, including those that are corrosive or abrasive, as well as those that are high-pressure and high-temperature.

5.4.1 Effect of Surface Roughness on Wear and Friction

It is possible for there to be a large variance in surface roughness depending on the kind of diamond films. Previous research has shown that there is a clear association between rougher surfaces and greater wear losses. This association stays true regardless of the testing environment that is being used [61]. According to findings from recent studies, ultrasmooth nanocrystalline or polished diamond films have friction values that are very low, reaching as low as 0.01 in humid air. On the

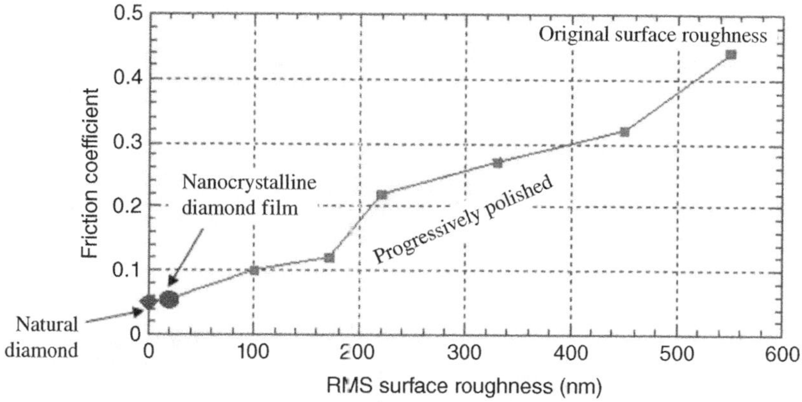

Figure 5.9 Friction coefficient versus surface roughness for DLC coatings (in general). Reproduced from [66] / with permission of ELSEVIER.

other hand, it has been shown that rough microcrystalline films have substantially greater friction values, which may reach as high as 0.7 [62]. For example, microcrystalline diamond films are a good example of such rough surfaces. These surfaces have the ability to cause harm to mating surfaces, especially those made of softer materials like steel, by activities such as cutting or scratching [63].This has the potential to cause significant abrasion, particularly on counterfaces that are softer. Furthermore, in the case of fragile materials like ceramics, the presence of sharp edges on rough diamond surfaces may cause microcracking as a consequence of concentrated pressures, which ultimately leads to the loss of material. Additionally, repeated sliding passes over the same contact region may have the effect of inducing microfatigue, which can result in structural weakness and further exacerbate the wear process [64]. Figure 5.9 shows the friction coefficient with surface roughness.

5.4.2 Effect of Environment on Wear and Friction

The surrounding environment has a significant impact on the tribological properties of a DLC film. This effect is further complicated by the complex interactions between the deposition parameters, atmospheric conditions, lubricious, mechanical, and chemical properties of the coating, and the counterface characteristics. Often, the formation of transfer layers (or tribolayers) controls wear and friction on interacting surfaces. Figure 5.10 shows the effect of environment on friction coefficient.

Experiments examining different combinations of coatings and oils with different chemical compositions under extreme boundary lubrication conditions

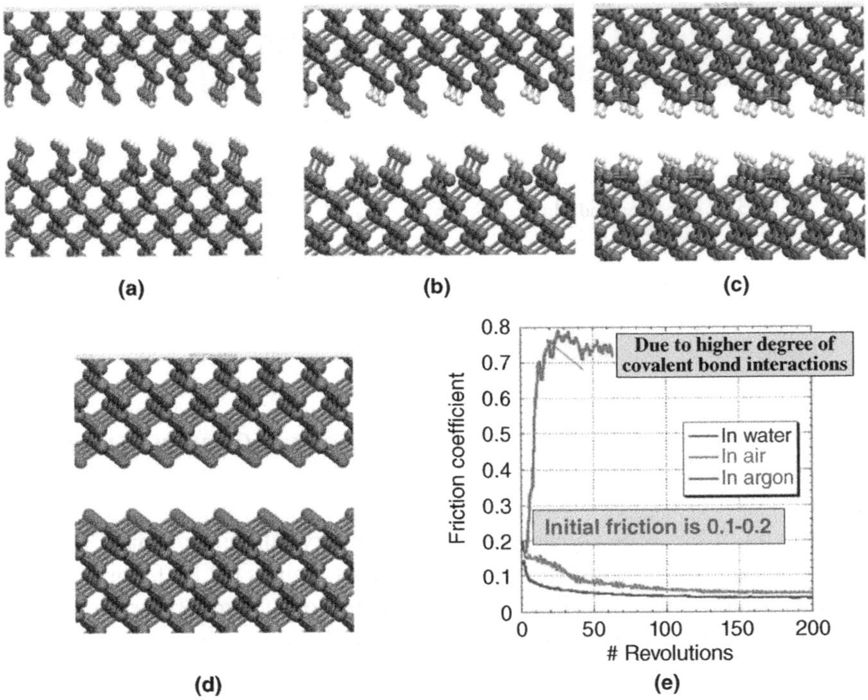

Figure 5.10 Effect of environment on friction coefficient. Reproduced from [66] / with permission of ELSEVIER.

showed that irrespective of the oil type, uncoated steel/steel contacts had less wear than steel/DLC and DLC/DLC configurations [67]. With reductions of up to 80% for tungsten-doped DLC and up to 60% for a-C:H, the addition of additives greatly reduced the wear rate for both DLC/DLC contacts, demonstrating the efficient boundary lubrication offered by these inert coatings. Elevating the concentration of extreme pressure (EP) additives has been shown to reduce wear and friction in W DLC/steel contacts, but has no effect on pure DLC coatings [68].

5.4.3 Effect of Substrate on Wear and Friction

Generally, DLC films have an overall thickness of only 1–4 -μm, occasionally even less. This thinness emphasizes how important the substrate is to supporting the imposed load in a variety of applications. If the substrate is not sufficiently rigid and strong, it might be subjected to plastic deformation or material flexure, which can accelerate the coating's breakdown. An array of techniques has been developed to increase the substrate's resistance to such risks. One such strategy uses a duplex treatment plan, as demonstrated by Podgornik and Vizintin [69]. Here, plasma nitriding is applied to AISI 4140 steel prior to coating deposition. Adhesion studies show that this approach improves a ta-C coating's binding strength in addition to its wear resistance [70].

The material selection for the counterface has a significant impact on how DLC coatings wear and frictionally behave. A number of variables are taken into account when analyzing how counterface materials affect DLC coatings. First, frictional behavior is greatly influenced by the adhesion between the counterface material and the DLC coating [71]. While certain materials may have lesser adhesion and so reduced friction, others may stick to DLC coatings more firmly, producing larger frictional forces. Frictional behavior is also influenced by surface roughness and the chemical makeup of the counterface material; smoother surfaces often result in less interfacial friction.

Furthermore, different counterface materials might cause distinct DLC coating wear processes. During tribological testing, surface fatigue, adhesive wear, and abrasive wear are often seen in wear processes [69]. The predominant wear process is determined in part by the counterface material's chemical reactivity, hardness, and surface roughness. For example, stiffer counterface materials may result in adhesive wear, whereas softer counterface materials may cause abrasive wear on the DLC covering.

Long-term performance also depends on the DLC coating and counterface material's chemical compatibility. Over time, tribological characteristics may be lost and the DLC coating may undergo chemical reactions with reactive counterface materials [72]. On the other hand, materials used as counterfaces that are chemically inert retain their integrity and do not negatively impact the performance of the DLC coating. Lastly, another important consideration is the counterface material's surface polish. The criteria for an ideal surface finish differ based on the particular application and the required level of wear and friction. In certain situations, highly polished surfaces might be ideal for reducing wear and interfacial friction, yet regulated surface roughness might be helpful to encourage the development of transfer layers [73].

5.5 Modification of DLC Coatings for Reduced Wear and Friction

In many engineering applications, developed DLC coatings present a promising way to cut wear and friction. These coatings offer a strong defense against wear and surface deterioration because of their remarkable hardness, low friction coefficient, and chemical inertness. DLC coatings can be precisely controlled to display particular tribological qualities, such as increased wear resistance and decreased friction, by using sophisticated deposition processes. Researchers have significantly improved the tribological performance of DLC coatings by adding additives or changing the coating structure; as a result, DLC coatings are now the go-to option for increasing the longevity and effectiveness of mechanical components in industrial, automotive, and aerospace systems.

5.5.1 DLC Coatings With Interlayers

Si, SiH, Cr, CrN, Ti, and TiN are among the interlayers that have been used to improve the adherence of DLC coatings on metallic substrates [74–77]. These interlayers function to form a strong adhesion layer between the interlayer and the metallic substrate and a transition layer between the interlayer and the DLC coating, resulting in a sandwich structure. For tribological applications, this interlayer architecture guarantees excellent DLC coatings on metallic substrates. One prominent example is the CrN interlayer, which consists of an amorphous interfacial zone between the columnar chromium nitride and the crystalline steel substrate that is about 5 nm thick. This special mix of layer structures improves the top layer DLC coating's adherence, leading to great wear resistance and strong corrosion resistance.

5.5.2 Metal-doped DLC

One potential method to improve the properties of DLC coatings is to employ metal-doped (Me-DLC) or carbide-doped (MeC-DLC) coatings. As hydrogen atoms are linked to carbon and integrated into the random network, they stabilize sp^3 hybridization, which adds to the network's noteworthy stiffness and hardness [78]. But as a result of this stabilization process, the interconnectedness of the sp^3-hybridized carbon sites decreases, which has an impact on the film's mechanical characteristics. Adhesion and tribological qualities are significantly improved when metals like W, WC, Mo, Ti, and Cr are infused into DLC coatings. This is mainly because the internal residual stress within the DLC coating is reduced. The reduction of stress improves the DLC coating's adhesion to metallic substrates [79]. The doped metals in DLC coatings frequently form carbide nanoclusters inside the amorphous carbon matrix. Furthermore, multilayer structures made of different DLC coatings have been successfully used to improve adhesion in tribological applications. Typically, these structures are created by utilizing a magnetron sputter reactor to develop metal-doped DLC in a rotating fashion, producing coatings with a multilayer architecture and columnar morphology [80]. The increased qualities are a result of these alternating layers that are enriched with carbon and metal. Notably, the carbides included in these coatings – which are often tougher than the amorphous carbon matrix – help to reduce stress in the DLC coating, strengthening its adhesion qualities [81].

5.5.3 Cr-based Interlayers DLC Coatings

Using chromium-based interlayers such as Cr, CrN, and CrC has greatly improved the tribological properties of the DLC top layer [82]. These interlayers have a columnar chromium nitride and a crystalline steel substrate sandwiched between amorphous interfacial zones that is about 5 nm thick. Because of the diffusion of chromium nitride into the steel substrate, the contact between this interphase and the underlying steel has a somewhat rougher surface on the upper side [83]. The remarkable adherence of the chromium layer to the steel substrate is facilitated by this interfacial area, which is typified by a high atomic density phase. The interphase first shows more contrast, which suggests a thicker mass. This is probably because chromium first diffuses into the steel substrate, and then an amorphous chromium region forms. This is thought to be the point at which the chromium interlayer forms as a result of the nucleation of the amorphous layer with a columnar structure [54, 84]. The crystalline steel substrate and the columnar chromium are usually separated by an amorphous high-atomic-density phase in this interlayer design. The coating structure then has a columnar interlayer of cubic chromium centered around the body.

5.6 Applications of DLC Coatings

DLC coatings are widely used in tribology because of their unique blend of characteristics, which includes high hardness, low friction, and wear resistance. DLC coatings are frequently used in automobile engineering on a variety of parts, including camshafts, valve lifters, and piston rings, to lessen wear and friction and extend engine life. Furthermore, DLC coatings are widely used in the aerospace sector for components – such as gears, bearings, and hydraulic systems – that are exposed to high loads and harsh environments. These coatings offer superior wear protection and minimize maintenance needs. Moreover, DLC coatings have made their way into the medical industry, where they are utilized in surgical tools and orthopedic implants to reduce wear and friction, extending the life and functionality of these devices. DLC coatings are also used in cutting tools and machining components, where they improve surface polish, prolong tool life, and improve cutting performance in metal cutting processes. Figure 5.11 shows the machining applications of DLC coatings.

Furthermore, to lower friction and increase endurance, DLC coatings are being used more and more in consumer electronics, such as hard disk drives and mobile device components. DLC coatings have a wide range of applications in tribology, which demonstrates their adaptability and efficacy in decreasing wear, reducing friction, and improving the lifespan and performance of numerous engineering components in a variety of industries.

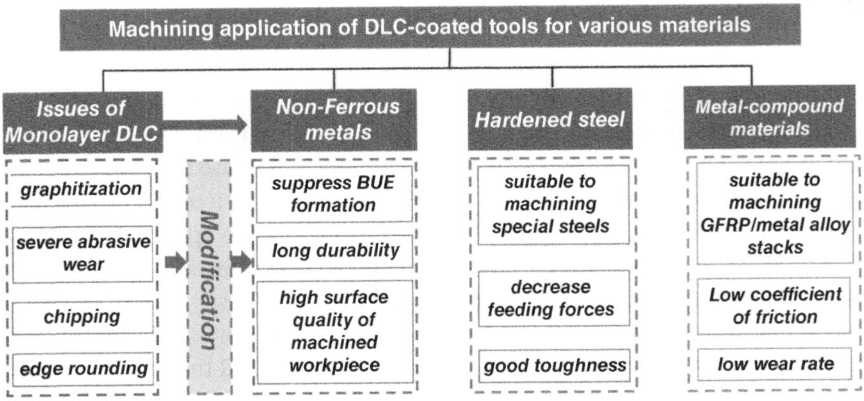

Figure 5.11 Machining application of DLC coated tool [85] / MDPI / CC BY 4.0.

5.7 Summary and Future Directions

The chapter on tribology's DLC coatings examines the many uses and underlying processes that make DLC coatings so appealing to a range of sectors. It explores the remarkable qualities of DLC coatings, including their high hardness, low friction, and superior wear resistance, which make them indispensable for lowering wear, decreasing friction, and enhancing the functionality and lifespan of engineering components. The chapter highlights the adaptability and efficiency of DLC coatings in resolving tribological difficulties by discussing the wide range of uses of these materials across industries, including automotive, aerospace, medical, and consumer electronics. It also looks at the developments in DLC coating technology, such as the creation of DLC coatings doped with metals and carbides and the use of interlayers to improve adhesion and tribological characteristics. All things considered, the chapter offers a thorough review of DLC coatings in tribology, highlighting their importance in contemporary technical applications and their potential for further development.

In the future, there are numerous opportunities for more study and advancement in DLC coatings for tribological applications. First and foremost, in order to provide coatings with improved performance and attributes that are suited to particular application needs, ongoing efforts are required to optimize DLC coating deposition processes and parameters. Furthermore, investigating cutting-edge coating topologies and new doping materials and techniques may help to further enhance the tribological characteristics of DLC coatings. In addition, to comply with changing industry requirements and laws, research on the creation of sustainable and eco-friendly DLC coating methods will be crucial. Additionally, examining the possible synergistic effects of mixing DLC coatings with other materials and surface treatments may open up new avenues for improving tribological performance. Last but not least, research institutes, industry, and academia must continue to work together to foster innovation and advance the use of DLC coatings in tribological applications. All things considered, these new avenues for research and development show promise for extending the usefulness and scope of DLC coatings in tribology and opening doors for improved longevity and performance in a range of engineering components and systems.

References

1 Robertson, J. (2002). Diamond-like amorphous carbon. *Materials Science and Engineering: R: Reports* 37 (4–6): 129–281.

2 Evtukh, A.A., Litovchenko, V.G., Litvin, Y.M. et al. (2003). Silicon doped diamond-like carbon films as a coating for improvement of electron field emission. *Journal of Vacuum Science & Technology B: Microelectronics and Nanometer Structures Processing, Measurement, and Phenomena* 21 (1): 627–630.

3 Treutler, C.P. (2005). Industrial use of plasma-deposited coatings for components of automotive fuel injection systems. *Surface and Coatings Technology* 200 (5–6): 1969–1975.

4 Donnet, C. and Erdemir, A. (eds.) (2007). *Tribology of Diamond-like Carbon Films: Fundamentals and Applications*. New York, NY: Springer Science & Business Media.

5 Goglia, P.R., Berkowitz, J., Hoehn, J. et al. (2001). Diamond-like carbon applications in high density hard disc recording heads. *Diamond and Related Materials* 10 (2): 271–277.

6 Kumar, D., Jain, J., and Gosvami, N.N. (2020). In situ study of role of microstructure on antiwear tribofilm formation on AZ91 magnesium alloy under zinc dialkyldithiophosphate containing lubricant. *Advanced Engineering Materials* 22 (8): 2000335.

7 Singh, V.P., Kumar, D., Mahesh, V. et al. (2023). Role of interfacial shear strength in nanotribological behaviour of PTA deposited Colmonoy 6 coating on SS304 steel. *Materials Letters* 353: 135287.

8 Kumar, P., Philip, J.T., Wani, M.F. et al. (2023). Study of tribological properties of EN8 steel against Inconel X-750 alloy under dry and lubricated conditions. *Transactions of the Indian Institute of Metals* 76 (9): 2353–2361.

9 Ferrari, A.C. (2004). Diamond-like carbon for magnetic storage disks. *Surface and Coatings Technology* 180: 190–206.

10 Vallinot, I.B., de la Guerra Ochoa, E., Otero, J.E. et al. (2020). Individual and combined effects of introducing DLC coating and textured surfaces in lubricated contacts. *Tribology International* 151: 106440.

11 Kumar, D. (2023). Recent advances in tribology of high entropy alloys: a critical review. *Progress in Materials Science* 136: 101106.

12 König, H. and Helwig, G.Z. (1951). Thin layers formed from hydrocarbons by electron or ion bombardment. *Physik* 129: 491–503.

13 Aisenberg, S. and Chabot, R. (1971). Ion-beam deposition of thin films of diamond-like carbon. *Journal of Applied Physics* 42 (7): 2953–2958.

14 Kumar, D., Jain, J., and Gosvami, N.N. (2022). Macroscale to nanoscale tribology of magnesium-based alloys: a review. *Tribology Letters* 70 (1): 27.

15 Sreesha, R.B., Kumar, D., Chandraker, S. et al. (2021). Room temperature sliding wear behavior of Ti6Al4V: a review. In: *AIP Conference Proceedings* (Vol. 2341, No. 1: 040041.). Jamshedpur: AIP Publishing.

16 Kumar, D., Gosvami, N.N., and Jain, J. (2020). Influence of crystallographic orientation on nanoscale friction and wear mechanisms of the AZ91 alloy. *Tribology Letters* 68: 1–10.

17 Kumar, D., Deepak, K.B., Muzakkir, S.M. et al. (2018). Enhancing tribological performance of Ti-6Al-4V by sliding process. *Tribology-Materials, Surfaces & Interfaces* 12 (3): 137–143.

18 Whitmell, D.S. and Williamson, R. (1976). The deposition of hard surface layers by hydrocarbon cracking in a glow discharge. *Thin Solid Films* 35 (2): 255–261.

19 Bewilogua, K. and Wagner, D. (1991). The effect of secondary electrons in the ion plating deposition of amorphous hydrogenated carbon (aC: H) films. *Vacuum* 42 (7): 473–476.

20 Ohgoe, Y., Hirakuri, K.K., Saitoh, H. et al. (2012). Classification of DLC films in terms of biological response. *Surface and Coatings Technology* 207: 350–354.

21 KP, L., Kumar, D., and Hirani, H. (2017). Synthesis and field dependent shear stress evaluation of stable MR fluid for brake application. *Industrial Lubrication and Tribology* 69 (5): 655–665.

22 Lijesh, K.P., Kumar, D., and Hirani, H. (2017). Effect of disc hardness on MR brake performance. *Engineering Failure Analysis* 74: 228–238.

23 Hiratsuka, M., Nakamori, H., Kogo, Y. et al. (2013). Correlation between optical properties and hardness of diamond-like carbon films. *Journal of Solid Mechanics and Materials Engineering* 7 (2): 187–198.

24 Lifshitz, Y., Lempert, G.D., Grossman, E. et al. (1995). Growth mechanisms of DLC films from C+ ions: experimental studies. *Diamond and Related Materials* 4 (4): 318–323.

25 Rajak, D.K., Kumar, A., Behera, A. et al. (2021). Diamond-like carbon (DLC) coatings: classification, properties, and applications. *Applied Sciences* 11 (10): 4445.

26 Jurkevičiūtė, A., Dolmantas, P., Vasiliauskas, A. et al. (2023). Magnetron sputtering process for deposition of multilayered thin diamond-like carbon films with silver nanoparticles for anti-reflective coatings and refractometric sensing. *Materials Chemistry and Physics* 309: 128425.

27 Natrayan, L., Gaur, P., Merneedi, A. et al. (2022). Investigation of tribological behaviour on DLC coatings for AA5051 using DC sputtering. *Adsorption Science & Technology* 2022: 4574218.

28 Bewilogua, K., Wittorf, R., Thomsen, H. et al. (2004). DLC based coatings prepared by reactive DC magnetron sputtering. *Thin Solid Films* 447: 142–147.

29 Chowdhury, S., Laugier, M.T., and Rahman, I. Z. (2004). Characterization of DLC coatings deposited by RF magnetron sputtering. *Journal of Materials Processing Technology* 153: 804–810.

30 Sánchez, N.A., Rincon, C., Zambrano, G. et al. (2000). Characterization of diamond-like carbon (DLC) thin films prepared by RF magnetron sputtering. *Thin Solid Films* 373 (1–2): 247–250.

31 Harper, J.M. Ion beam deposition.In: Vossen J. L., Kern, W., (Eds.) (1978). *Thin Film Processes* 175–206, New York, NY: Acdemic Press.

32 Sharko, S.A., Serokurova, A.I., Novitskii, N.N. et al. (2022). A new approach to the formation of nanosized gold and beryllium films by ion-beam sputtering deposition. *Nanomaterials* 12 (3): 470.

33 Leu, M.S., Chen, S.Y., Chang, J.J. et al. (2004). Diamond-like coatings prepared by the filtered cathodic arc technique for minting application. *Surface and Coatings Technology* 177: 566–572.

34 Zhang, Z., Zhang, L., Yuan, H. et al. (2022). Tribological behaviors of super-hard TiAlN coatings deposited by filtered cathode vacuum arc deposition. *Materials* 15 (6): 2236.

35 Lukaszkowicz, K., Sondor, J., Balin, K. et al. (2014). Characteristics of CrAlSiN+ DLC coating deposited by lateral rotating cathode arc PVD and PACVD process. *Applied Surface Science* 312: 126–133.

36 Voevodin, A.A., Donley, M.S., and Zabinski, J.S. (1997). Pulsed laser deposition of diamond-like carbon wear protective coatings: A review. *Surface and Coatings Technology* 92 (1–2): 42–49.

37 Cutroneo, M., Havranek, V., Flaks, J. et al. (2021). Pulsed laser deposition and laser-induced backward transfer to modify polydimethylsiloxane. *Coatings* 11 (12): 1521.

38 Lu, Y., Huang, G., Wang, S. et al. (2021). A review on diamond-like carbon films grown by pulsed laser deposition. *Applied Surface Science* 541: 148573.

39 Konov, V.I., Kononenko, T.V., Loubnin, E.N. et al. (2004). Pulsed laser deposition of hard coatings in atmospheric air. *Applied Physics A* 79: 931–936.

40 Voevodin, A.A., Donley, M.S., Zabinski, J.S. et al. (1995). Mechanical and tribological properties of diamond-like carbon coatings prepared by pulsed laser deposition. *Surface and Coatings Technology* 76: 534–539.

41 Dumitru, G., Romano, V., Weber, H.P. et al. (2003). Laser treatment of tribological DLC films. *Diamond and Related Materials* 12 (3–7): 1034–1040.

42 Kleps, I., Nicolaescu, D., Stamatin, I. et al. (1999). Field emission properties of silicon carbide and diamond-like carbon (DLC) films made by chemical vapour deposition techniques. *Applied Surface Science* 146 (1–4): 152–157.

43 Choi, S.S., Kim, D.W., Joe, J.W. et al. (1997). Deposition of diamond-like carbon films by plasma enhanced chemical vapour deposition. *Materials Science and Engineering: B* 46 (1–3): 133–136.

44 Baba, K., Hatada, R., Flege, S. et al. (2009). Deposition of silicon-containing diamond-like carbon films by plasma-enhanced chemical vapour deposition. *Surface and Coatings Technology* 203 (17–18): 2747–2750.

45 Pan, Y.Q. and Yin, Y. (2007). Diamond-like carbon films with End-Hall ion source enhanced chemical vapour deposition. *Diamond and Related Materials* 16 (2): 220–224.

46 Bendavid, A., Martin, P.J., Randeniya, L. et al. (2009). The properties of fluorine containing diamond-like carbon films prepared by plasma-enhanced chemical vapour deposition. *Diamond and Related Materials* 18 (1): 66–71.

47 Choy, K.L. (2003). Chemical vapour deposition of coatings. *Progress in Materials Science* 48 (2): 57–170.

48 Nabot, J.P., Andre, B., and Chabert, J.P. (1991). Hard diamond-like carbon films prepared by various physical vapour deposition techniques. *Surface and Coatings Technology* 45 (1–3): 359–367.

49 Mehran, Q.M., Fazal, M.A., Bushroa, A.R. et al. (2018). A critical review on physical vapor deposition coatings applied on different engine components. *Critical Reviews in Solid State and Materials Sciences* 43 (2): 158–175.

50 Ma, W.J., Ruys, A.J., Mason, R.S. et al. (2007). DLC coatings: effects of physical and chemical properties on biological response. *Biomaterials* 28 (9): 1620–1628.

51 Baptista, A., Silva, F., Porteiro, J. et al. (2018). Sputtering physical vapour deposition (PVD) coatings: a critical review on process improvement and market trend demands. *Coatings* 8 (11): 402.

52 Song, R., Chen, S., Liu, Z. et al. (2023). Effect of W-doping on the structure and properties of DLC films prepared by combining physical and chemical vapor deposition. *Diamond and Related Materials* 132: 109687.

53 Baragetti, S., Gerosa, R., and Villa, F. (2017). Physical vapour deposition of diamond-like carbon coatings on a 7075-T6 substrate for corrosion protection at long and short fatigue lives. *International Journal of Structural Integrity* 8 (5): 576–584.

54 Muthuraja, A., Naik, S., Rajak, D.K. et al. (2019). Experimental investigation on chromium-diamond like carbon (Cr-DLC) coating through plasma enhanced chemical vapour deposition (PECVD) on the nozzle needle surface. *Diamond and Related Materials* 100: 107588.

55 Oguri, K. and Arai, T. (1992). Two different low friction mechanisms of diamond-like carbon with silicon coatings formed by plasma-assisted chemical vapor deposition. *Journal of Materials Research* 7 (6): 1313–1316.

56 Khrushchov, M.M., Marchenko, E.A., Petrzhik, M.I. et al. (November, 2018). The effect of structure on tribological behavior of chromium–carbon coatings obtained by plasma-assisted physical vapor deposition. *Journal of Physics: Conference Series* 1121 (1): 012017.

57 Ichou, H., Arrousse, N., Berdimurodov, E. et al. (2024). Exploring the advancements in physical vapor deposition coating: a review. *Journal of Bio- and Tribo-Corrosion* 10 (1): 3.

58 Liskiewicz, T.W. and Al-Borno, A. (2014). DLC coatings in oil and gas production. *Journal of Coating Science and Technology* 1 (1): 59–68.

59 Deng, Y., Chen, W., Li, B. et al. (2020). Physical vapor deposition technology for coated cutting tools: a review. *Ceramics International* 46 (11): 18373–18390.

60 Zia, A.W. and Birkett, M. (2021). Deposition of diamond-like carbon coatings: conventional to non-conventional approaches for emerging markets. *Ceramics International* 47 (20): 28075–28085.

61 Hayward, I.P. (1991). Friction and wear properties of diamonds and diamond coatings. *Surface and Coatings Technology* 49 (1–3): 554–559.

62 Erdemir, A.G.R.A.R.D.M.T.R.T., Fenske, G.R., Krauss, A.R. et al. (1999). Tribological properties of nanocrystalline diamond films. *Surface and Coatings Technology* 120: 565–572.

63 Erdemir, A.M.G.R.A., Halter, M., Fenske, G.R. et al. (1997). Durability and tribological performance of smooth diamond films produced by Ar-C60 microwave plasmas and by laser polishing. *Surface and Coatings Technology* 94: 537–542.

64 Bowden, F.P. and Young, J.E. (1951). Friction of diamond, graphite, and carbon and the influence of surface films. *Proceedings of the Royal Society of London. Series A. Mathematical and Physical Sciences* 208 (1095): 444–455.

65 Bhushan, B., Subramaniam, V.V., Malshe, A. et al. (1993). Tribological properties of polished diamond films. *Journal of Applied Physics* 74 (6): 4174–4180.

66 Erdemir, A. and Martin, J.M. (2018). Superior wear resistance of diamond and DLC coatings. *Current Opinion in Solid State and Materials Science* 22 (6): 243–254.

67 Kalin, M., Vižintin, J., Vercammen, K. et al. (2006). The lubrication of DLC coatings with mineral and biodegradable oils having different polar and saturation characteristics. *Surface and Coatings Technology* 200 (14–15): 4515–4522.

68 Podgornik, B., Hren, D., and Vižintin, J. (2005). Low-friction behaviour of boundary-lubricated diamond-like carbon coatings containing tungsten. *Thin Solid Films* 476 (1): 92–100.

69 Podgornik, B. and Vižintin, J. (2001). Influence of substrate treatment on the tribological properties of DLC coatings. *Diamond and Related Materials* 10 (12): 2232–2237.

70 Jiang, J., Arnell, R.D., and Tong, J. (1997). The effect of substrate properties on tribological behaviour of composite DLC coatings. *Tribology International* 30 (8): 613–625.

71 Łępicka, M., Grądzka-Dahlke, M., Pieniak, D. et al. (2017). Effect of mechanical properties of substrate and coating on wear performance of TiN-or DLC-coated 316LVM stainless steel. *Wear* 382: 62–70.

72 Ching, H.A., Choudhury, D., Nine, M.J. et al. (2014). Effects of surface coating on reducing friction and wear of orthopaedic implants. *Science and Technology of Advanced Materials* 15 (1): 014402.

73 Arslan, A., Masjuki, H.H., Varman, M. et al. (2015). Effects of texture diameter and depth on the tribological performance of DLC coating under lubricated sliding condition. *Applied Surface Science* 356: 1135–1149.

74 Cemin, F., Boeira, C.D., and Figueroa, C.A. (2016). On the understanding of the silicon-containing adhesion interlayer in DLC deposited on steel. *Tribology International* 94: 464–469.

75 Madej, M. (2014). The effect of TiN and CrN interlayers on the tribological behavior of DLC coatings. *Wear* 317 (1–2): 179–187.

76 Wu, G., Sun, L., Dai, W. et al. (2010). Influence of interlayers on corrosion resistance of diamond-like carbon coating on magnesium alloy. *Surface and Coatings Technology* 204 (14): 2193–2196.

77 Kasiorowski, T., Lin, J., Soares, P. et al. (2020). Microstructural and tribological characterization of DLC coatings deposited by plasma enhanced techniques on steel substrates. *Surface and Coatings Technology* 389: 125615.

78 Bewilogua, K. and Hofmann, D. (2014). History of diamond-like carbon films—from first experiments to worldwide applications. *Surface and Coatings Technology* 242: 214–225.

79 Yetim, A.F., Kovacı, H., Kasapoğlu, A.E. et al. (2021). Influences of Ti, Al and V metal doping on the structural, mechanical and tribological properties of DLC films. *Diamond and Related Materials* 120: 108639.

80 Wang, Y., Wang, Y., Li, X. et al. (2019). The friction and wear properties of metal-doped DLC films under current-carrying condition. *Tribology Transactions* 62 (6): 1119–1128.

81 Gayathri, S., Kumar, N., Krishnan, R. et al. (2015). Influence of transition metal doping on the tribological properties of pulsed laser deposited DLC films. *Ceramics International* 41 (1): 1797–1805.

82 Santiago, J.A., Fernández-Martínez, I., Sánchez-López, J.C. et al. (2020). Tribomechanical properties of hard Cr-doped DLC coatings deposited by low-frequency HiPIMS. *Surface and Coatings Technology* 382: 124899.

83 Weicheng, K., Zhou, Y., and Jun, H. (2021). Effect of carburizing treatment on microstructural, mechanical and tribological performances of Cr doped DLC coating deposited on Ti6Al4V alloy. *Ceramics International* 47 (24): 34425–34436.

84 Viswanathan, S., Mohan, L., Bera, P. et al. (2017). Corrosion and wear behaviors of Cr-doped diamond-like carbon coatings. *Journal of Materials Engineering and Performance* 26: 3633–3647.

85 Wang, L., Liu, Y., Chen, H. et al. (2022). Modification methods of diamond like carbon coating and the performance in machining applications: a review. *Coatings* 12 (2): 224.

6

Diamond-like Carbon Coatings and Green Lubricants

Irfan Nadeem[1], *Muhammad Shahid Arshad*[1,2], *Marko Polajnar*[1]

[1] *Laboratory for Tribology and Interface Nanotechnology, University of Ljubljana, Ljubljana, Slovenia*
[2] *Department for Nanostructured Materials, K7, Jožef Stefan Institute, Ljubljana, Slovenia*

CONTENTS

6.1 Introduction, 155
6.2 DLC Coatings in Sustainable Tribology: Navigating Green Engineering Concepts, 158
6.2.1 Biodegradable Oils, 159
6.2.2 Low-SAPS Lubricants, 160
6.2.3 Fatty Acids, 161
6.2.4 Nanoparticles as Lubricant Additives, 162
6.2.5 Ionic Liquids, 167
6.2.6 Water-based Lubricants and Polyhydric Alcohols, 169
6.3 Role and Importance of Doping Elements on DLC Lubricating Properties, 171
6.3.1 Influence of Dopants on the Surface Energy of DLC Coatings, 171
6.3.2 Influence of Dopants on the Wetting Properties of DLC Coatings, 174
6.4 Interactions of DLC Coatings with Green Additives, 176
6.5 Lubrication Behavior of DLC Coatings with Base Lubricants, 179
6.6 Conclusion, 182

6.1 Introduction

In the context of rising global challenges, understanding and managing energy consumption has become crucial. Over the past five decades, global energy consumption has increased exponentially and is projected to continue growing. A significant yet often overlooked factor in this increased energy use is the role of friction and wear, which substantially contribute to energy loss and inefficiency.

Applications of Diamond-like Carbon Coatings, First Edition. Abdul Wasy Zia.

It is well-documented that nearly 23% of the world's total energy is lost to frictional heat, leading to environmental dilemmas [1].

The growing demand for transportation is the biggest obstacle to reducing greenhouse gas emissions. The International Energy Agency reported in 2012 that the transportation sector consumes 64% of global petroleum [2]. Another scientific study found that 33% of fuel energy consumed in passenger vehicles is used to overcome friction [3]. In simpler terms, the annual expenditure of approximately 208 billion liters of fuel is attributed to the mitigation of frictional forces, with a significant portion, about 70 billion liters, used explicitly to mitigate friction in the engine compartment.

Given these energy and efficiency issues, electric vehicles (EVs) are taking over the global market. According to Bloomberg New Energy Finance, it is predicted that by 2040, one-third (33%) of all vehicles worldwide will be electric, while about one billion internal combustion engine vehicles will remain on the market (Figure 6.1) [4]. This paradigm shift presents a substantive challenge concerning energy requisites and environmental implications. This transition to EVs not only represents a change in vehicle technology but also brings forth new challenges in energy management, particularly in how we address friction and wear in these emerging vehicle types.

Recognizing these challenges, it is imperative that we take immediate action to reduce energy consumption and emissions. Our research aims to contribute to this effort by focusing on innovative solutions to mitigate friction and wear, thereby enhancing the efficiency of both traditional and EVs. The goal is to limit carbon dioxide (CO_2) and particulate emissions from internal combustion engines (ICE), paving the way for a more sustainable future in transportation.

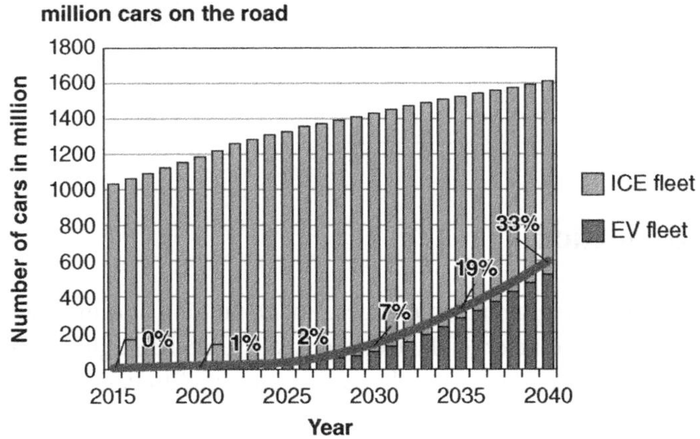

Figure 6.1 Estimated global comparison of light-duty vehicles having ICE versus EVs. *Source:* Reproduced from [4] / Bloomberg Finance L.P.

Considering that, one of the best-known and most studied ways of reducing friction is using solid and liquid lubricants. Liquid lubricants, consisting of base oil and additives, are the most obvious and common means of reducing friction and wear by interacting with materials and surface coatings to form wear-protective and low-friction boundary layers. These fully formulated oils consist of base oil and chemical additives. Among all the additives, antiwear (AW) and friction modifiers play an essential role in protecting surfaces and reducing friction in the boundary lubrication regime. Zinc dialkyl dithiophosphate (ZDDP) is the most commonly used AW additive that protects the surface by forming a polyphosphate tribofilm by tribochemical reaction. In contrast, molybdenum dithiocarbamate is the most extensively used friction modifier that improves the friction in boundary and mixed lubrication regime by forming a low shearing tribofilm of MoS_2. Despite their excellent lubricating properties, they are made of phosphorus, zinc, sulfur, and other environmental polluting elements that pollute the environment and cause harmful emissions [5]. Until now, conventional lubricants used to overcome friction and wear are made from phosphorus, zinc, sulfur, and other hazardous elements [6–8]. Additionally, most lubricant additives have a narrow range of operating temperatures. Also, they are optimized for steel surfaces and do not provide adequate lubrication for advanced and newly developed engineering materials and coatings. Nowadays, international legislation has placed severe constraints on the use of these lubricant additives [9].

Thus, tribologists are now focusing on alternative solutions to overcome friction and wear with improved lubrication in an environment-friendly manner to reduce greenhouse gas emissions and improve fuel economy. At the Fourth World Tribology Congress 2009, H. P. Jost introduced the term green tribology as a scientific term. Green tribology is the science and technology that involves using natural materials, reducing natural petroleum resources, increasing energy efficiency by reducing pollution and emissions, shortening the service cycle, and improving equipment performance for sustainable and environmentally friendly design [5]. The application of tribology in sustainable energy consumption depends on many factors, such as effective manufacturing and usage, system efficiency, greenhouse gas emissions, and reducing dependence on nonrenewable sources while minimizing environmental pollution [10]. Therefore, optimizing technological advances while considering their environmental impact can be not only challenging but also beneficial, as the future of the planet and society may depend on the application of green tribology for technological development.

As mentioned earlier, most lubricant additives are tailored for optimum performance on conventional engineering steel surfaces and often do not provide adequate lubrication for novel materials and coatings. Diamond-like carbon (DLC) coatings are becoming increasingly attractive in the search for the best tribological solutions under lubricated conditions due to their inherent low-wear

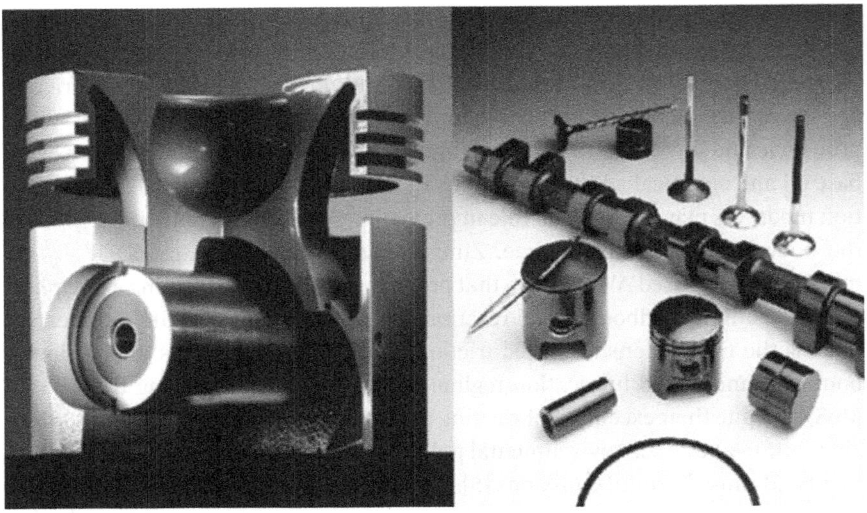

Figure 6.2 Common components found in automotive applications that frequently undergo DLC coatings. *Source:* Adapted with permission from Ref. [5].

and low-friction properties. However, the main advantage and, at the same time, a complication associated with DLC coatings and their lubrication stem from their limited reactivity. Their exceptionally low adhesion gives them favorable frictional properties; conversely, this leads to weak and unpredictable interactions with lubricants [5]. This paradox makes DLC coatings both a challenge and an opportunity, as they offer significant potential for innovative solutions in the field of lubrication, including green tribology. To date, these coatings have been successfully integrated into numerous automotive components, including camshafts, gears, and piston rings (see Figure 6.2).

This chapter aims to explore the interaction between DLC coatings and green lubricants. Special emphasis is placed on understanding the fundamentals of how these lubricants interact with different types of DLC coatings, including hydrogenated, hydrogen-free, and doped variants.

6.2 DLC Coatings in Sustainable Tribology: Navigating Green Engineering Concepts

Over the past two decades, our research group has investigated several concepts of green tribology involving DLC coatings. Several studies have been conducted to explore green tribological solutions with DLC coatings under lubricated conditions, such as the use of biodegradable oils [11–14], low-sulfated ash, phosphorus, and sulfur (SAPS) oils [15–18], simple mild organic additives [19–21], self-lubricating nanoparticles [22–25], water-based lubricants [26–30], and ionic liquids (ILs) [31–33].

6.2.1 Biodegradable Oils

Biodegradable and natural lubricants are considered one of the pillars of green tribology [5, 10]. The use of biodegradable oils was the first and most obvious way to reduce the environmental hazards and damage caused by lubricating oils. Biodegradable oils, engineered for natural degradation through microbial and ecological processes, are recognized for their environmental friendliness and sustainability. Bio-based lubricants offer several advantages, including biodegradability, superior lubricity, higher flash points, high viscosity index, and base stock renewability [11]. Vegetable oils such as rapeseed, sunflower, and castor oil are the most commonly studied biodegradable naturally occurring oils. Findings from our research group demonstrated that the biodegradable lubricant, in conjunction with AW and extreme pressure (EP) additives, performs very well and considerably improves the tribological performance of DLC/DLC contacts under boundary lubrication conditions (see Figure 6.3) [11]. In particular, they are considered highly polar owing to the presence of a large number of polar functional groups that make them more reactive on the DLC surface than conventional mineral and synthetic oils.

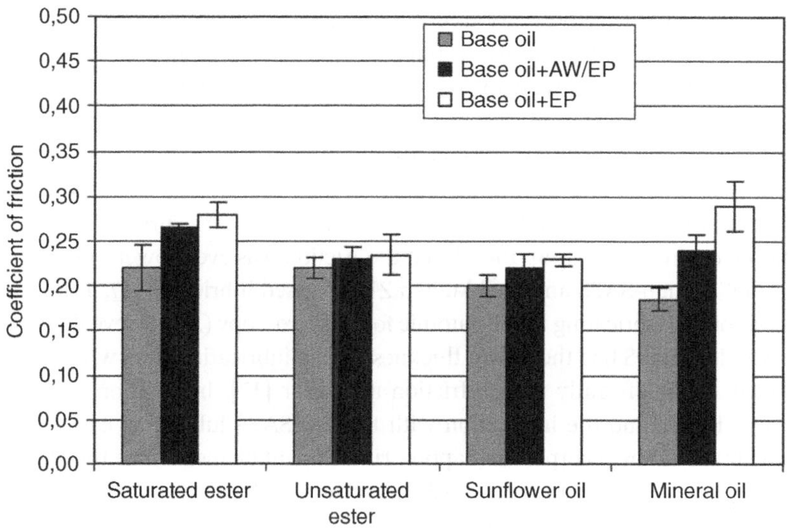

Figure 6.3 Variation in the coefficient of friction in self-mated DLC contacts using biodegradable and mineral oils. *Source:* Reproduced from [11] / with permission of ELSEVIER.

Tribo-induced phase transformation from sp^3- to sp^2-hybridized carbon on the surface of amorphous carbon coatings is considered one of the most widely accepted lubrication mechanisms. Recently, Long et al. [14] demonstrated the effectiveness of castor oil between steel and a-C contact. Superlubricity with a coefficient of friction (COF) of 0.004 was observed due to the increased proportion of sp^2 carbon on the top-most surface and the passivation effect by $-(CH_2-CH_2)n-$ oligomers formed on the surface due to the friction-induced decomposition of castor oil.

However, the downside of biodegradable lubricants is that most of them have poor oxidation and thermal stability, poor cold flow behavior, and narrow viscosity range, which hampered their use in real-time industrial applications. Therefore, several chemical modifications need to be employed to improve their oxidation stability and cold flow behavior [34, 35].

6.2.2 Low-SAPS Lubricants

Phosphorus and sulfur are the most commonly used additives in conventional lubricants due to their impressive AW, antifriction, and anti-oxidation properties. However, the use of these additives has led to the appearance of SAPS, which not only affects engine performance and fuel efficiency but is also considered a serious threat to the environment. SAPS is an abbreviation for sulfated ash phosphorus, and sulfur. In the past two decades, several studies have investigated the chemical interactions of EP and AW additives with DLC coatings. Indeed, due to the low reactivity of the DLC, these interactions were weaker compared to the steel surface [5]. Therefore, various doping strategies can be employed to tailor the properties of the DLC coatings to improve their interaction with the lubricant additives. These strategies are explained in detail in Section 6.3.

Accordingly, considering the international legislation and the Sustainable Development Goals (SDGs) requirements, many studies have investigated the interaction of DLC with low-SAPS lubricants [15–18]. The findings from these studies reveal that they can effectively reduce friction and protect the surface from wear. In addition, our research group found that the friction was even lower in many cases than that of high-SAPS and well-known ZDDP-based lubricants [17, 18]. Detailed analysis of the lubricating film by atomic force microscopy (AFM) revealed that the oils with low-SAPS had the lowest thickness of the lubricating film, which was still sufficient to significantly reduce friction and wear [17]. In addition, the surface activity of DLC and the interaction with the low-SAPS lubricants can be significantly improved by incorporating appropriate dopant elements into the DLC coatings, which is discussed in Section 6.3. However, very little research has been done to investigate the influence of doping elements on the interaction of DLC with the low-SAPS lubricants, and there is an extreme scarcity on the lubrication mechanisms responsible for their tribological performance.

6.2.3 Fatty Acids

The use of simple mild organic additives such as fatty acids is extensive in the petroleum industry to improve the lubrication performance in boundary lubrication. Fatty acids are carboxylic acids with aliphatic chains of an even number of carbon atoms [36]. They existed in both saturated and unsaturated form. Fatty acids provide excellent lubricating performance owing to their ability to form self-assembled monolayers on polar surfaces, preventing direct contact between sliding surfaces [19, 20]. The lubrication behavior of fatty acids associated with DLC highly depends on the type of DLC. For example, a COF greater than 0.04 is observed for self-mated hydrogenated DLC (a-C:H) lubricated with oleic acid, while self-mated tetrahedral amorphous carbon (ta-C) achieves superlubricity under similar conditions [37]. The phenomenon of superlubricity between self-mated ta-C arises from tribo-induced graphitization and sequential decomposition of fatty acids into smaller fragments that passivate the surface [38]. The decomposition of fatty acid occurs due to the exceptionally high local contact pressure arising from the high elastic modulus of ta-C [39]. In contrast, this decomposition of fatty acids was not observed in self-mated a-C:H contacts. Besides this, the adsorption ability of carboxyl groups with surface carbon atoms also facilitates the reduction in friction [21].

Our research group conducted several studies to investigate the adsorption mechanism of fatty acids to DLC surfaces using neutron reflectometry and AFM [40–43]. The former technique can measure the boundary layer of less than 1 nm. Indeed, it was found that the fatty acid adsorbed onto the surface of DLC coatings and partially covered the surface with a surface coverage of 50% [40]. Moreover, the boundary layer thickness was sub-nano size (i.e. less than 1 nm), which agreed with the quantum-chemistry and molecular dynamic studies conducted by Loehlé et al. [44].

Sun et al. [21] recently explored that an aqueous 3-hydroxypropionic acid (3HPA) solution between steel and a-C contact provides superlubricity. This remarkable lubrication efficiency is attributed to a synergistic interaction of solid-liquid lubrication between the a-C film and the 3HPA solution. In particular, the tribochemical reaction at the friction interface led to the chemisorption of 3HPA molecules onto both a-C and steel surfaces, which could form a low-shear interface by forming a hydrated film due to the hydrogen bond interaction with the water molecules. In addition, the potential graphitization of the a-C coating plays a pivotal role in the realization of superlubricity (see Figure 6.4).

However, despite the remarkable advancement in the fatty acids for achieving superior lubrication performance in DLC contacts, intensifying efforts are required to improve the synergism between fatty acids and DLC films. In particular, adopting fatty acids derived from renewable sources, tailoring the DLC with

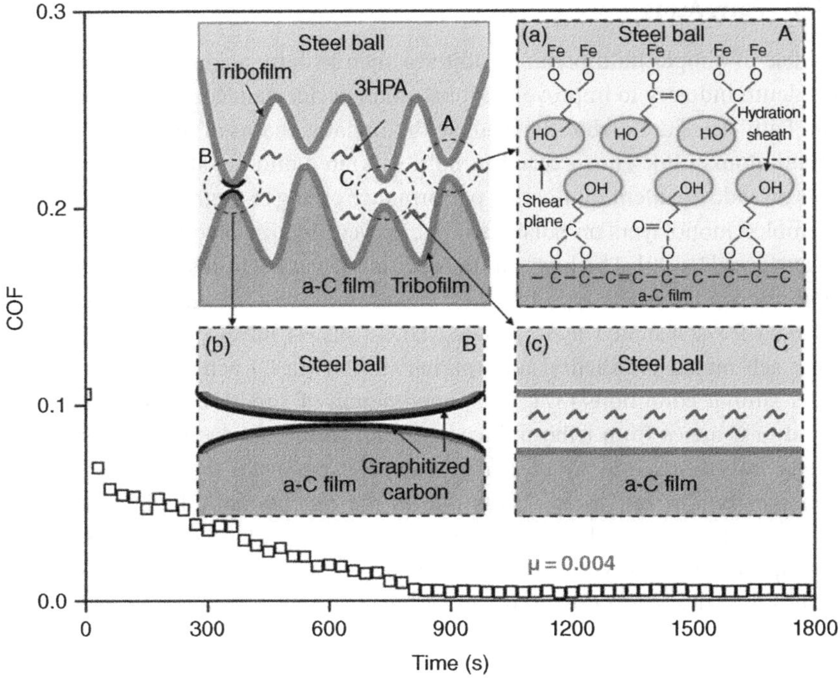

Figure 6.4 The excellent frictional performance of 3HPA showing superlubricity with a COF as low as 0.004; inset showing the schematic of the lubrication mechanism between steel and a-C contact, (a) shear-plane in the middle of the hydration shells formed above the chemisorption layers of 3HPA, (b) graphitized carbon and the tribochemically derived tribofilm preventing the direct asperity contact, and (c) showing the schematic of the elastohydrodynamic effect between the tribo-pairs. *Source:* Reproduced from [21] / with permission of ELSEVIER.

different doping elements, and varying the sp^2 and sp^3 bonding fractions could open new avenues for achieving green superlubricity.

6.2.4 Nanoparticles as Lubricant Additives

Since it has become evident that DLC coatings are insufficiently active against chemical-based additives, another possible avenue primarily involving physical-based mechanisms appears to be highly favorable as a prospective lubrication option for green DLC technology. The idea of lubrication assisted with self-lubricating nanoparticles has been considered a promising concept for reducing friction in DLC contacts. Nanoparticles are becoming more important as lubricant additives due to their potential to reduce emissions and improve fuel economy. Their characteristics size (<100 nm) suitable to enter contact regions, better thermal and chemical stability in comparison with organic additives, considerably lesser toxicity and harmful emissions, and diversity of particle chemistries make them suitable as additives for lubricants [45, 46]. A stable suspension of nanoparticles in liquid lubricants is commonly known as nanolubricants [47]. High friction and wear are considered a major concern in boundary lubrication regime, especially in automotive lubrication. Figure 6.5 shows that the nanolubricants play an important role in the boundary lubrication regime and greatly improve the friction and wear performance by effectively filling the nanovalleys between the micro-asperities on the surface.

Following the advances in the field of conventional solid lubrication, Figure 6.6 presents the physical-based lubrication mechanisms of nanoparticles in self-mated DLC contacts. In recent years, several studies have explored the tribological performance of nanolubricants between DLC contacts. Metal sulfides (such as MoS_2 and WS_2) and carbon-based nanomaterials are considered as the promising lubricant additives. Inorganic MoS_2 and WS_2 are regarded as one of the best-known solid lubricants owing to their lamellar structure and inherent shearing property of basal planes [24, 48, 49]. They have been shown to effectively improve the tribological performance of poorly reactive surfaces such as DLC coatings through physically based lubrication mechanisms, namely (i) exfoliation and formation of low-shear film, (ii) reducing friction by separating the sliding contacts,

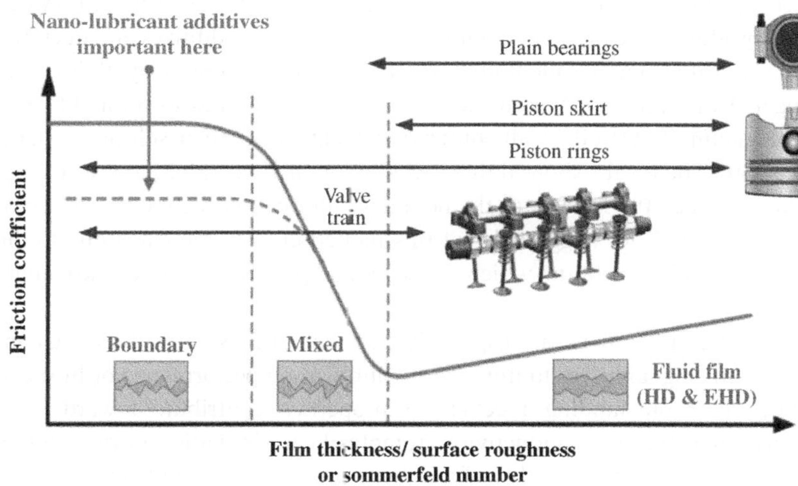

Figure 6.5 Schematic illustration of Stribeck curve representing the role of nanolubricants in different lubrication regimes. *Source:* Reproduced from [47] / with permission of ELSEVIER.

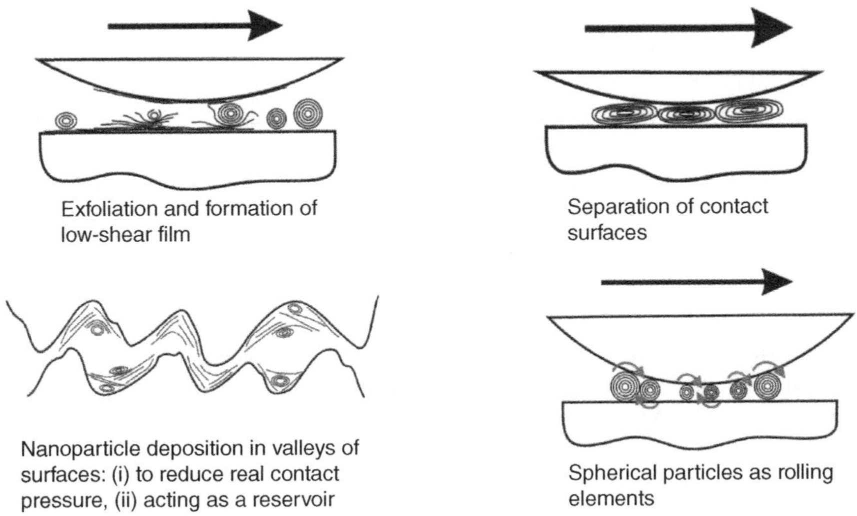

Figure 6.6 Possible physical-based lubrication mechanisms of nanoparticles in self-mated DLC contacts. *Source:* Adapted with permission from Ref. [5] / Fakulteta za strojništv / CC BY 4.0.

(iii) nano-valleys filling, and (iv) rolling effect (https://www.ukri.org/events/epsrc-health-data-for-research-workshop/6) [5]. Secondly, they are chemically stable and non-reactive compounds with highly inert basal planes and thus do not produce harmful emissions [50]. Given their ability to reduce emissions, they can potentially address the problem of eliminating high-SAPS additives and open the door to innovative, environmentally friendly lubrication technology. Kalin et al. investigated the lubrication performance of MoS_2 nanotubes as an additive in Polyalphaolefin (PAO) oil in all lubrication regimes between self-mated DLC contacts [50]. The results showed that the addition of 2 wt% MoS_2 to PAO oil significantly reduced the friction in the boundary lubrication regime by 50% compared to PAO oil. They suggested that this reduction in friction was due to the formation of a thick tribofilm, which exfoliates and partially separates the contact asperities.

Carbon-based nanomaterials, especially graphene and its derivatives, are considered promising additives to improve the lubricating performance of lubricating oils. Their outstanding mechanical properties contribute toward wear resistance, high thermal conductivity promotes heat dissipation during sliding motion, while the impermeability of graphene protects tribofilm from oxidation and corrosion [51, 52]. Moreover, graphene is much greener than carbon nanotubes [53] and has better frictional properties than other carbon nanomaterials [54]. In 2019, kogovšek and Kalin [55] investigated the macroscale tribological behavior of self-mated DLC contacts using graphene nanoplatelets (GNPs) as an additive in PAO oil in all lubrication regimes. It can be seen from the Stribeck curve, as shown in Figure 6.7a, that the COF reduces up to 50% in the boundary lubrication regime. It was observed that the COF decreases with an increase in the concentration of GNPs (up to 5 wt%) in PAO oil. Furthermore, GNPs do not affect the elastohydrodynamic regime. The decrease in COF is attributed to the formation of tribofilm. It is inferred from the results that with a decrease in the contact velocity, more and more GNPs entrapped between contacting surfaces, forming a large quantity of tribofilm, as shown in Figure 6.7b.

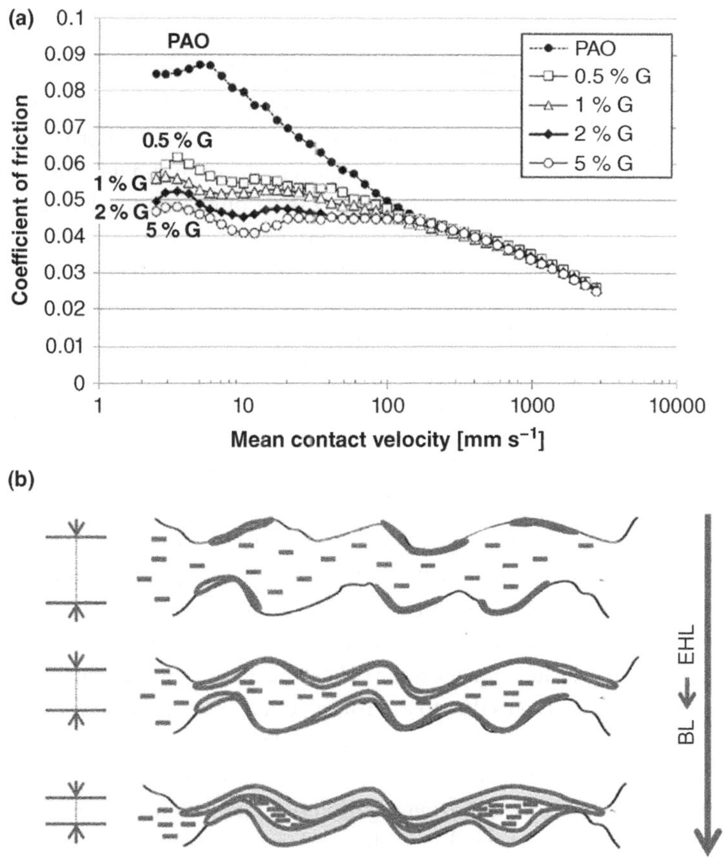

Figure 6.7 Effect of PAO oil containing GNPs on COF of self-mated DLC contacts: (a) Effect of different concentrations of graphene nanoplatelets, and (b) schematic illustration of the effect of different lubrication regimes on tribofilm formation. *Source:* Reproduced from [55] / with permission of ELSEVIER.

Despite the great potential of graphene, several following aspects stand in the way of its large-scale use as an additive in oil-based lubricants and that need to be addressed to make it a more effective lubricant additive [51, 56, 57]:

i. Strong $\pi - \pi$ interaction between interlayers of graphene leads to aggregation and poor dispersion stability of graphene.
ii. The impermeability of graphene restricts oxidation and corrosion, but the chemical inertness restricts the formation of tribofilm.
iii. Restricted interlayer slippage due to strong $\pi - \pi$ interaction between interlayers of graphene.
iv. Inferior crack resistance ability, crack even lower strain less than 5% because it has large in-plane stiffness and rigid C–C structure provides little energy dissipation.

With the advancement in nanotechnology and optoelectronics, in 2006, Sun et al. developed stable photoluminescent carbon dots (CDs) [58]. Carbon-based quantum dots, especially graphene quantum dots (GQDs) derived from graphene oxide (GO), are considered a potential replacement of graphene and GO as an additive in liquid base lubricants, thanks to their outstanding physicochemical properties and zero-dimensional spherical nanostructure [59, 60]. More interestingly, GQDs are even less toxic than GO [59]. Unlike CDs, GQDs are considered a smaller version of GO containing graphene structure inside the dots, which is responsible for their excellent physical properties [61]. Recently, Tang et al. [23] compared the tribological performance of CDs and GO nanosheets in deionized water for self-mated amorphous carbon (a-C) contacts in reciprocating mode. With the same concentration, CDs provide superior lubrication. However, because of the large size of GO nanosheets in the lateral direction, they tend to squeeze out from the rubbing surfaces instead of penetrating the nanovalleys, which gives rise to the COF (Figure 6.8a). In contrast, 0.1 wt% of CDs improve the frictional performance by 33% owing to their nanobearing and nanofilling effect (Figure 6.8b). They further explain that the surface of CDs is activated due to broken oxygen functional groups on the surface of CDs due to shear stress. These activated CDs aggregate, act as a nanobearing, and reduce the shear stress between contacting surfaces due to the rolling effect, see Figure 6.8c. The contact stresses are reduced due to the smoothening effect provided by the nanofilling. They also found that when the concentration of CDs is increased from the optimal concentration (i.e. 0.1 wt%), aggregation of nanoparticles occurs, which can lead to a plowing effect instead of filling the nanovalleys. Despite the superior lubrication performance of carbon-based quantum dots there is extreme scarcity on how the relevant parameters such as sliding motion, sliding speed, contact pressure, and nature of tribopair effects the macroscale tribological performance. Also, the lubrication mechanism responsible for their tribological performance is

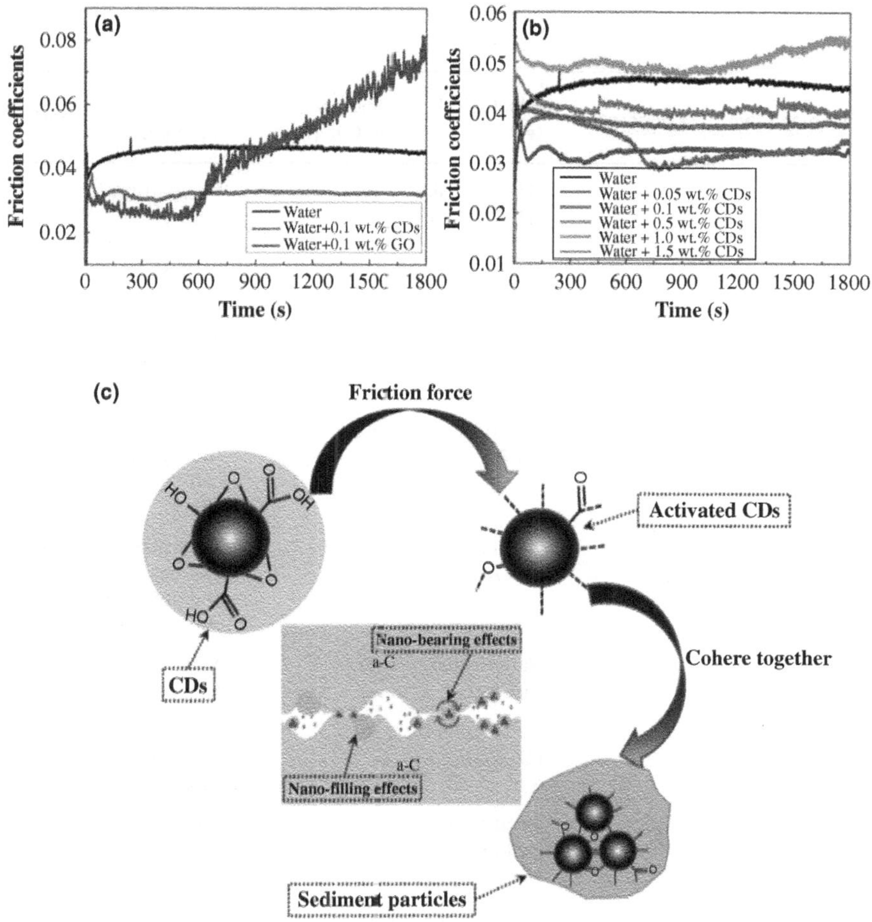

Figure 6.8 Tribological performance of CDs as an additive in deionized water for self-mated amorphous carbon contact: (a) Variation in COF with time using deionized water, 0.1 wt% CDs in water, and 0.1 wt% GO in water. (b) Variation in COF with different concentration of CDs in deionized water. (c) Schematic illustration of the formation of activated CDs and sedimentation of CDs during tribological tests. *Source:* Reproduced from [23] / with permission of ELSEVIER.

not well understood. Table 6.1 summarizes the lubrication mechanism and percentage reduction in COF by various nanoparticles.

6.2.5 Ionic Liquids

ILs are usually defined as compounds that consists of cations and anions and have a melting point below 100 °C. Their nonflammable nature, high thermal stability, wide electrochemical range, and low vapor pressure qualify them as

Table 6.1 Effect of different nanoparticles on the coefficient of friction reduction and their relevant mechanisms.

Nano-additive	Lubricant	Tribological test	Improvement in COF (%)	Lubrication mechanism	Ref.
MoS_2 nanotubes	PAO	Ball-on-disk (DLC/DLC)	50	Exfoliation and deposition of MoS_2 sheets	[50]
Graphene oxide (GO)	Aqueous ethanediol	Ball-on-disk (Si_3N_4/SiDLC)	Superlubricity	Synergistic effect of silica colloidal layer and GO	[62]
Nano boron nitride	PAO	Ball-on-disk (Si_3N_4/DLC)	Superlubricity	Internal sliding and ball-bearing effect	[63]
Zirconia nanoparticles	PAO	Ball-on-disk (a-C/a-C)	13	Synergetic ball-bearing and polishing effect	[64]
Nanodiamonds	water	Ball-on-disk (a-C/a-C)	30	Nano-bearing effect	[65]
GNPs	PAO	Ball-on-disk (DLC/DLC)	50	Formation of protective tribofilm	[55]
CDs	Water	Reciprocating ball-on-disk (a-C/a-C)	33	Nano-filling and nano-bearing effect	[23]

excellent candidates for lubrication additives [31]. The highly reactive nature of ILs makes them particularly suitable for interacting with DLC coatings, including doped variants. Research in this area, especially on IL-based lubricants for DLC contacts, has been limited. However, our research group's studies on the interaction between ILs and both standard and doped DLC have shown promising results. These studies indicate that ILs can achieve exceptionally low-friction levels, nearing superlubricity, in the boundary lubrication regime. This opens up the potential for specialized applications of green lubrication technologies, utilizing the unique properties of ILs.

The research of Arshad et al. [31, 32] delves into the interaction of ILs with tungsten-doped DLC (WDLC) coatings. In their studies, they found that certain IL additives, particularly dimethylphosphate-containing ones, can significantly reduce the COF in boundary lubrication conditions. These ILs form a phosphate-based tribofilm on the WDLC surface, contributing to friction reduction. Their research, conducted under specific conditions like a normal load of 10 N and a temperature of 100 °C, demonstrated a remarkably low COF of 0.024. This discovery holds promise for the application of ILs as green lubrication technology, potentially achieving superlubricity in certain scenarios.

6.2.6 Water-based Lubricants and Polyhydric Alcohols

Water-based lubricants have attracted considerable attention in recent years due to their environmental friendliness, low cost, and excellent heat dissipation properties. However, due to their low viscosity, poor corrosion resistance, and lubricity, they are not considered suitable candidates for conventional engineering steels. Therefore, surface engineering approaches, such as the development of DLC coatings on steel surfaces, have been adapted to overcome the prevalent issues associated with water lubrication. Water-lubricated DLC has proven to be a successful tribological combination, especially for hydrogen-free DLC [66]. In several cases, water-lubricated DLC resulted in very low wear, and COF of less than 0.1 can be observed when sliding DLC against steel [67].

Our research group has previously shown that surface engineering with DLC coatings in water can significantly improve the performance of a proportional 4/3 hydraulic water valve [27] (see Figure 6.9). In this work, a tribological modeling

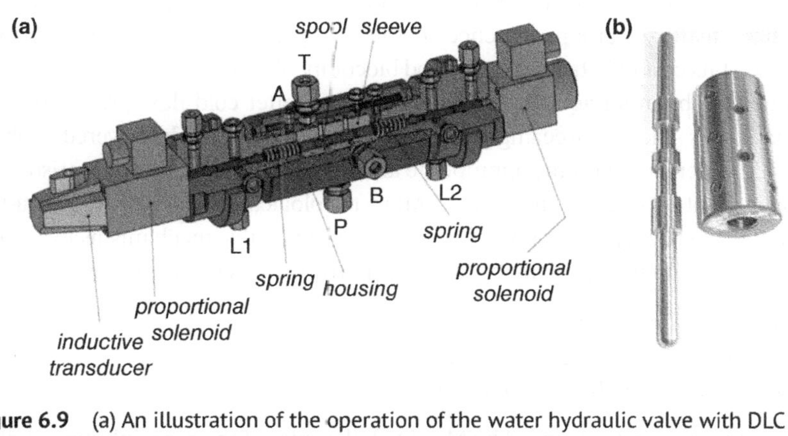

Figure 6.9 (a) An illustration of the operation of the water hydraulic valve with DLC coating operating within a water environment, (b) Samples of the spool and sleeve used for testing in the 4/3 water hydraulic valve configuration. *Source:* Reproduced from [27] / with permission of ELSEVIER.

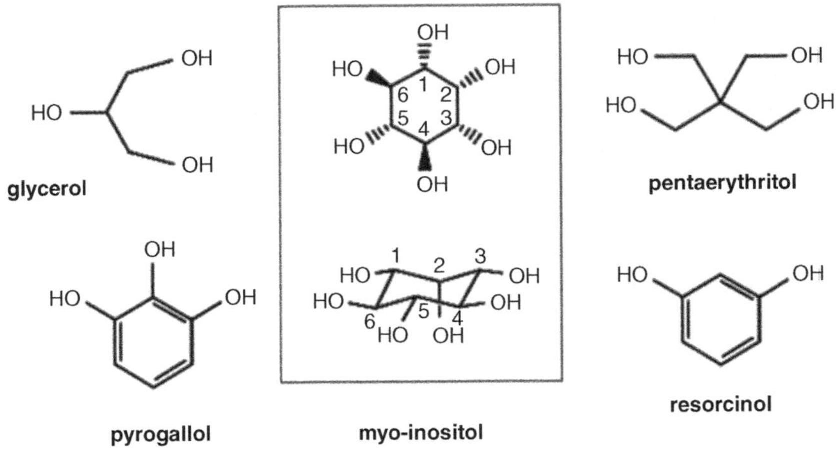

Figure 6.10 Chemical formula of some most commonly used polyhydric alcohols. *Source:* Reproduced from [70] / with permission of American Physical Society.

study and a full-scale test with hydraulic valve's life testing were carried out. The DLC-coated components remained within an acceptable 1% leakage during the life test, compared to 20% leakage for steel contacts, and tribological model tests revealed up to six times lower friction.

Polyhydric and polyhydroxy alcohols such as glycerol, myo-inositol, ethylene glycol, 1,3-propylene glycol, and 1,4-butylene glycol are considered as green lubricants [59]. Figure 6.10 shows the chemical formula of some conventional polyhydric alcohols. It should be noted that all these molecules do not contain aliphatic chains and have hydroxyl groups attached to them. Glycerol is most commonly used because of its cost-effectiveness and good biocompatibility. It is also easy to dissolve in water to obtain an aqueous solution. It offers better cold flow properties than vegetable oils since the freezing point of aqueous glycerol can be lowered to about −50 °C by blending with the appropriate amount of water [68]. In comparison, the freezing point of pure glycerol is 18 °C. Many tribological studies reported ultra-low friction using polyhydric alcohols [26, 68–71]. The possible mechanisms to achieve ultra-low friction using polyhydric alcohols in the various tribo-systems are [72]:

1. Tribofilm formation
2. Formation of the hydration layer
3. Formation of hydrogen bond network

A pioneer research group led by Martin et al. explored the superlubricity of glycerol and other polyhydric alcohols. In one of their works [70], they observed the macroscale superlubricity behavior of ta-C lubricated with glycerol in a mixed lubrication regime at 80 °C. This superlubricity behavior was observed due to the grafting of hydroxyl-terminated groups due to the tribochemical reaction between ta-C surface and the alcohol groups, which promotes easy sliding. Secondly, due to hydrogen bond interaction, glycerol adsorbed on the hydroxyl-terminated contacting surfaces, intercalating between them and promoting sliding due to rapid change in hydrogen bond location. Thirdly, experimental and numerical evidence shows that glycerol degrades under high pressure and shear, forming the tribo-induced film consisting of water molecules and organic acid, which is responsible for superlubricity. In another famous work [26], they achieved superlubricity, i.e. COF of 0.004, by sliding a steel ball against ta-C coating in glycerol maintained at 50 °C (see Figure 6.11a). X-ray photoelectron spectroscopy (XPS) analysis reveals that the iron oxyhydroxide (FeOOH) formed at the terminal ends of tribopairs due to the transfer of iron atoms from the steel ball to the ta-C layer. The super-low friction is due to the formation of the self-sustained thin film consisting of glycerol and its degradation products along with hydroxyl-terminated groups, as shown in Figure 6.11b.

Polyhydric alcohols provide ultra-low friction, but a long running-in period is required to achieve stable super-low friction, causing mechanical deterioration of the tribo-pairs and severe wear. Therefore, several solutions have been proposed to shorten the running time, such as the incorporation of graphene-based nanoparticles, modification using ILs, and mixing polyhydroxy alcohols with acidic solutions [32, 62, 71]. Recently, Yi et al. [62] reported superlubricity by introducing GO into an aqueous ethanediol solution between silicon nitride and silicon-doped DLC coating. They proposed that the tribochemical reaction on both tribo-pairs leads to the formation of a colloidal silica layer; meanwhile, the physical adsorption of GO on the tribo-pairs further reduces the shear stress. Therefore, there is a dire need for extensive research on the development of water-based lubricants that can offer superior lubrication performance with extremely low wear and friction.

6.3 Role and Importance of Doping Elements on DLC Lubricating Properties

6.3.1 Influence of Dopants on the Surface Energy of DLC Coatings

DLC coatings have been for a longtime estimated as chemically inert surfaces [73], which raised doubts about their ability to interact with lubricant additives. However, DLC coatings are not a uniform group possessing the same mechanical and chemical properties, but their properties can be tailored, e.g. based on the deposition technique, sp^2 to sp^3 ratio, hydrogenation, and the presence of different doping elements [74]. Due to this fact, their ability to interact with lubricants and additives can differ greatly. To evaluate the surface ability to interact with other

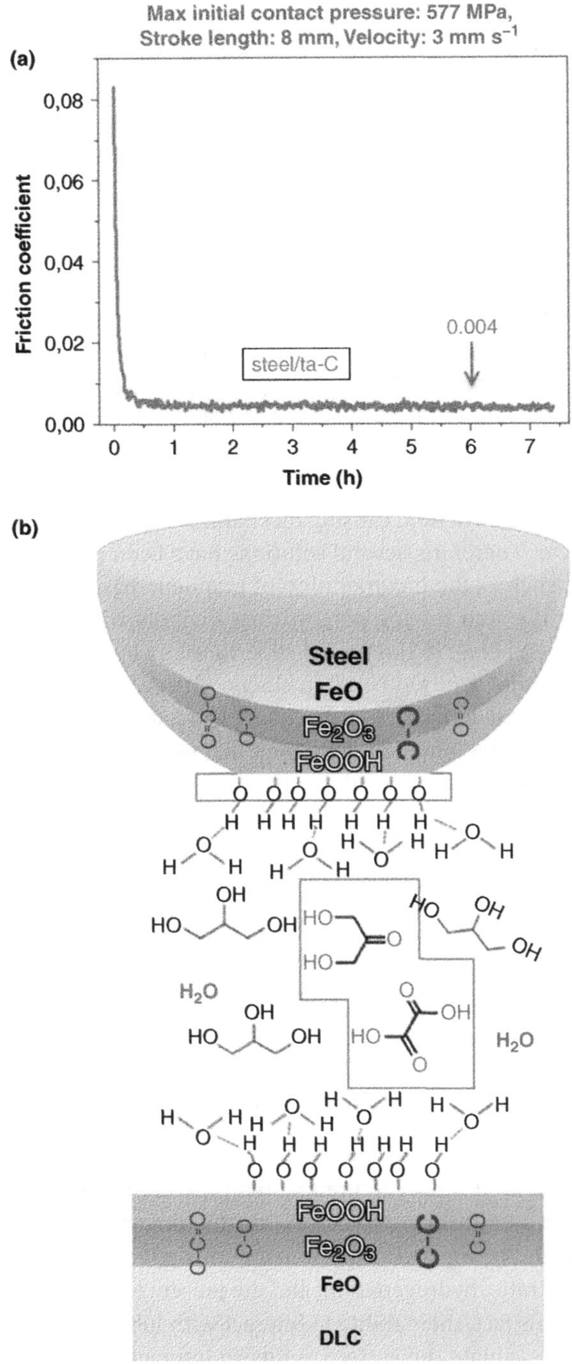

Figure 6.11 Tribological evaluation of glycerol between steel and tetrahedral amorphous DLC: (a) variation in COF as a function of time by glycerol at 50 °C, (b) schematic representation of lubrication mechanism indicating the oxide surface is covered with FeOOH providing hydroxyl-terminated groups, while nanometre thick self-sustained lubricant film ensures the elastohydrodynamic regime. *Source:* [26] / Springer Nature / CC BY 4.0.

surfaces, liquids, environments, and additives, surface energy is generally a very useful parameter. Since lubricants and lubricant additives are adapted to steel surfaces, the steel surface is always an adequate reference material for comparison with DLC coatings and their ability to interact with lubricants and additives.

As shown in Figure 6.12, DLC coatings (except for fluorine-doped DLC [F-DLC]) under ambient conditions do not have much lower total surface energy (γ) than steel surface – it varies only from 7 to 15%. Furthermore, the difference in dispersive surface energy (γ_d) is even smaller, namely less than 5%. On the other hand, the differences in polar surface energy (γ_p) are significant – between 30 and 65%. However, the surface energy of DLC coatings is further affected by the presence of hydrogen and different doping elements. Namely, ta-C has higher surface energy than steel surface. On the other hand, a-C:H provides lower surface energy than steel surface. This is due to the fact that hydrogen passivates the DLC dangling bonds and thus reduces the reactivity of the coating [75, 76]. The surface energy can be additionally tailored by adding different doping elements (Si, N, and F) into hydrogenated DLC coating. Si and N cause an increase in surface energy, especially the polar component (higher surface energy than a-C:H coating), while F causes a decrease in surface energy (lower surface energy than a-C:H coating). In particular, F-DLC possesses properties that are much different from other DLC coatings, with extremely low surface energy (below 20 mJ/m^2) and negligible polar components. The F-DLC surface energy is thus similar to Polytetrafluoroethylene (PTFE),

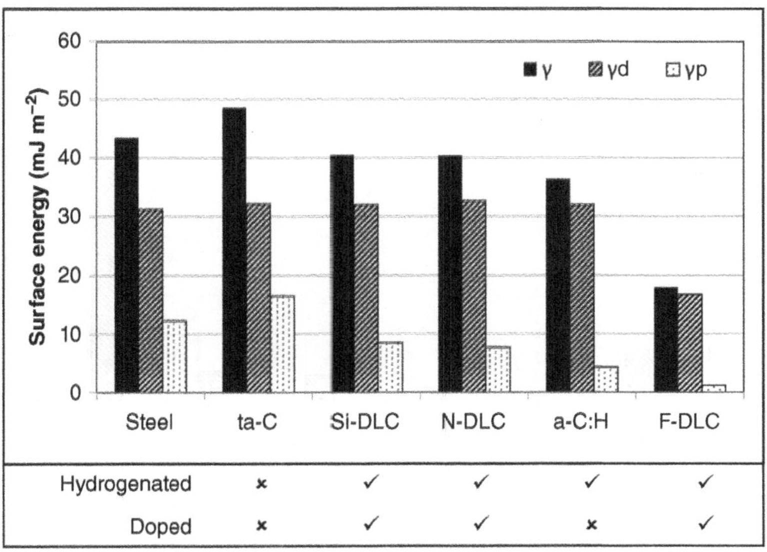

	ta-C	Si-DLC	N-DLC	a-C:H	F-DLC
Hydrogenated	×	✓	✓	✓	✓
Doped	×	✓	✓	×	✓

Figure 6.12 Surface energies of different DLC coatings (nonhydrogenated and hydrogenated, undoped and doped).

which is known as a nonsticky material. The main question is how these DLC coatings, with different surface energies, interact with lubricants and additives and how this affects their lubricating properties.

6.3.2 Influence of Dopants on the Wetting Properties of DLC Coatings

Lower surface energy of hydrogenated DLC coatings compared to steel surface (see Figure 6.12) would suggest that these DLC coatings should provide weaker interactions and poorer wetting properties with base oils than steel surface. However, using inappropriate parameters for wetting evaluation can lead to mismatching conclusions regarding the wetting properties of DLC coatings. Contact angle, the most common parameter for wetting evaluation, does not work well for vast DLC coatings [77]. As shown in Figure 6.13, contact angles for all DLC coatings (except F-DLC) show similar or lower contact angles with two different synthetic PAO oils compared to a steel surface, suggesting better interactions with oils that are contradictory to surface energy results in Figure 6.12. The reason behind this is the mismatching observation in the wetting nature of lubricating oils. Lubricating oils have low surface tension and, due to this fact, spread over steel as well as over DLC coatings (Figure 6.14a). The only exception is F-DLC, where oil drop sustain its shape when deposited on the surface (Figure 6.14b). Due to this reason, the spreading parameter should be used for the wetting evaluation of DLC

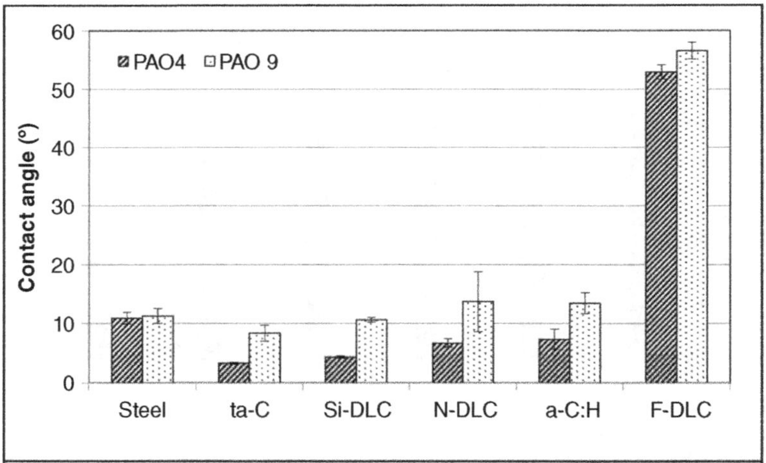

	ta-C	Si-DLC	N-DLC	a-C:H	F-DLC
Hydrogenated	×	✓	✓	✓	✓
Doped	×	✓	✓	×	✓

Figure 6.13 Steady-state contact angles of synthetic oils (PAO4 and PAO9) on steel and DLC coatings.

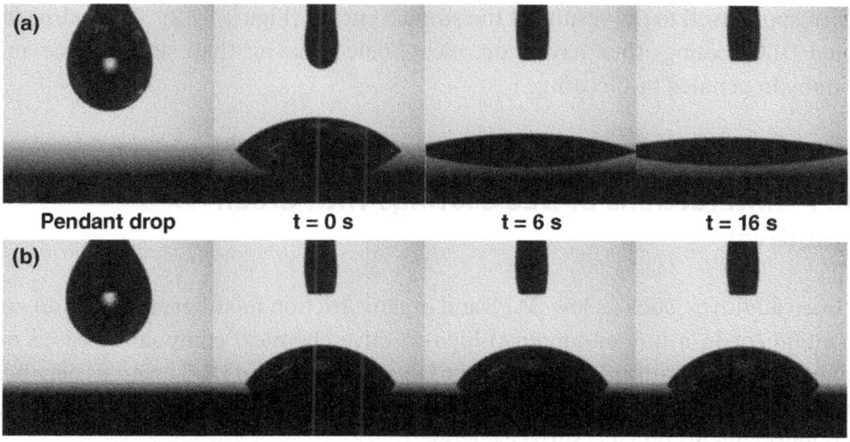

Figure 6.14 Difference between (a) spreading wetting of base oil observed on steel and vast DLC coatings and (b) adhesional wetting of base oils observed on low-surface energy F-DLC coating. *Source:* Adapted with permission from Ref. [78].

coatings instead of the contact angle. When spreading parameters are applied to evaluate the wetting properties of base oils on DLC coatings, the poorer wetting properties of hydrogenated DLC coatings compared to steel surface are revealed, as shown in Figure 6.15. What is more, the results of the spreading parameters

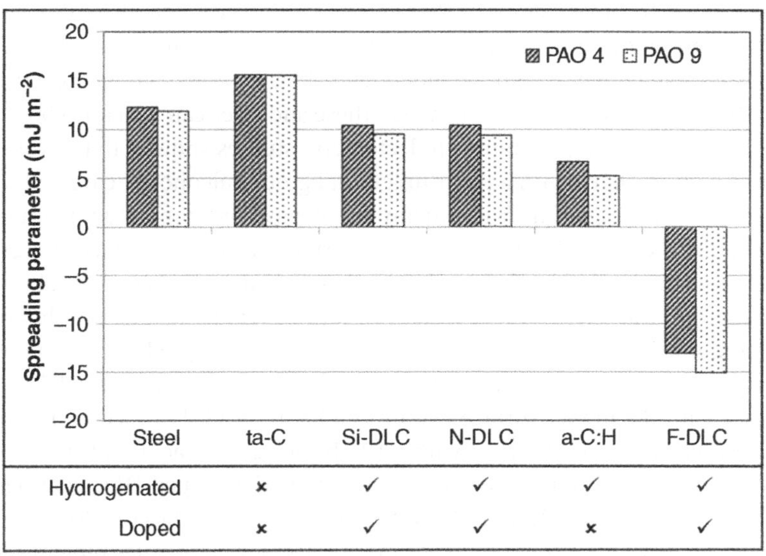

Figure 6.15 Spreading parameters of synthetic oils (PAO4 and PAO9) on steel and DLC coatings.

correspond well to the results of the surface energy (Figure 6.12). The hydrogenated DLC coatings thus have more oleophobic behavior than steel surface and nonhydrogenated ta-C coating.

6.4 Interactions of DLC Coatings with Green Additives

Green additives, such as low-SAPS and organic friction modifiers (OFMs), do not contain (or in a limited amount) high-reactive chemical compounds based on phosphorus and sulfur. The reactivity of these additives on DLC coatings depends on the type of DLC coating and the previously elaborated effect of hydrogen and doping elements on DLC surface energy.

As shown in Figure 6.16, the additive tribofilm thickness (Figure 6.16a) and wear track tribofilm coverage (Figure 6.16b) formed in boundary reciprocating sliding lubricating conditions depends on the reactivity of additives, being the lowest for neat low-SAPS additives, compared to harsher high-SAPS and ZDDP additive. Furthermore, tribofilm thickness and wear track coverage are further affected by surface energy, where a reduction in surface energy results in thinner tribofilm with lower wear track coverage. However, even the low reactive low-SAPS additive on a-C:H coating that has the second lowest surface energy among DLC coatings (see Figure 6.12) forms around 30 nm thick boundary film with 37% of wear track coverage, showing that low-SAPS additives can also interact with DLC coatings with significantly lower surface energy than the steel surface.

Another, even greener additive than low-SAPS are OFMs, where their reactivity with surfaces is even lower. However, these additives can also form boundary layers on DLC coatings. Neutron reflectometry studies showed that a very thin layer of hydrocarbons with a polar functional group (alcohol or carboxylic) can be formed on DLC coatings without any tribological or temperature stimulation [42, 43]. Furthermore, in this case, the importance of DLC's surface energy was revealed as an important factor affecting their additive adsorption ability on DLC coating. Namely, OFM hexadecanol formed around 1 nm thin boundary film on nonhydrogenated DLC and Si-doped DLC coatings (DLC coatings with the highest surface energy, see Figure 6.12), but not on a-C:H and F-DLC (DLC coatings with the lowest surface energy, see Figure 6.12) [43]. Same film thickness results were also observed with OFM hexadecanoic acid, but in this case, the film formed not only on nonhydrogenated and Si-doped DLC coatings but also on a-C:H coating [42]. Further XPS and ToF-SIMS studies of adsorbed additives on a-C:H coating revealed that polar molecules of hexadecanol and hexadecanoic acid can adsorb relatively strongly onto the a-C:H coatings by bonding

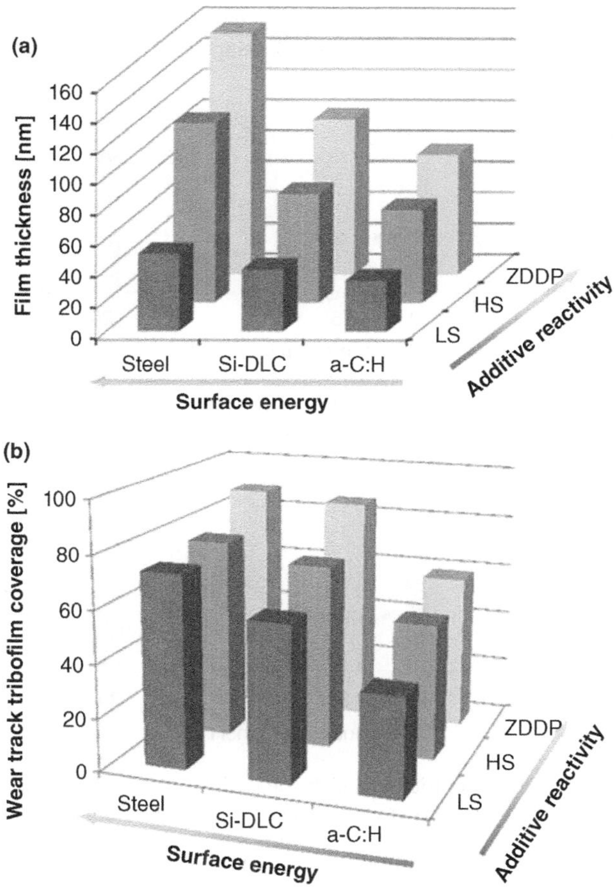

Figure 6.16 (a) Film thickness and (b) wear track tribofilm coverage for LS, HS, and ZDDP additives on steel and DLC coatings. *Source:* Reproduced from [17] / with permission of Springer Nature.

to the native surface oxides and hydroxides, however, in this case, the OFM hexadecanoic acid showed a greater adsorption ability than the OFM hexadecanol on a-C:H coating [79].

Furthermore, coupled tribological and AFM studies showed that under reciprocating sliding boundary lubricating conditions, the adsorption of hexadecanol and hexadecanoic acid onto the DLC surfaces forms a protective layer that prevents direct mechanical contacts and thus reduces the wear of the a-C:H coatings, but still, it is less effective in reducing the friction (see Figure 6.17) because of the already inherently low-friction properties of hydrogenated DLC coatings [40, 41].

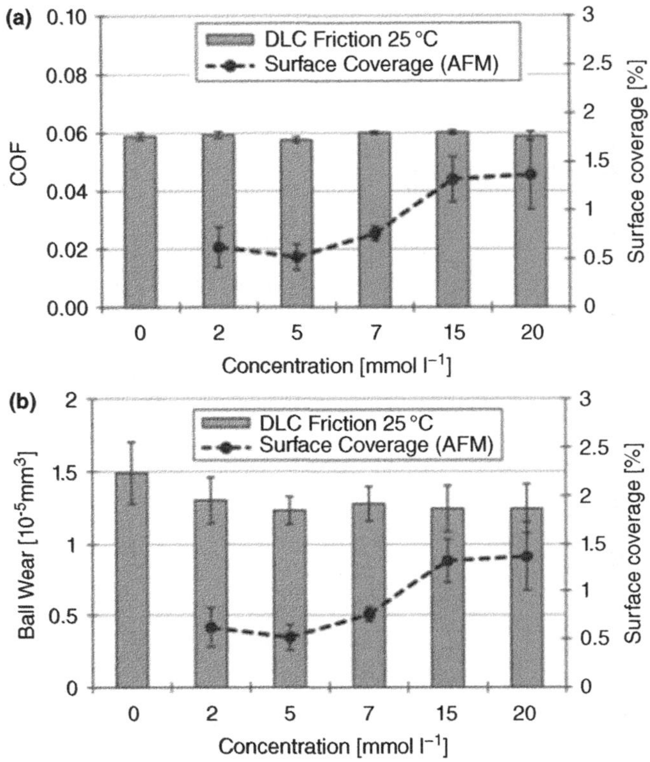

Figure 6.17 (a) Coefficient of friction and (b) wear of DLC/DLC contacts for different hexadecanoic acid concentrations at 25 °C. *Source:* Reproduced from [40] / with permission of ELSEVIER.

All aforementioned examples of successful lubrication of DLC coatings with green additives were based predominantly on chemical-based interactions; however, as mentioned in the review article about designing tribological interface for efficient and green DLC lubrication [80], this can also be achieved by physical-based lubrication technologies. Good practice of physical-based lubrication technologies are nanoparticles. Several studies confirmed that these particles can also perform well on a-C:H coating with low surface energy (see Figure 6.12). MoS_2 nanotubes have thus shown a reduction in COF up to 50% compared to base oil on a-C:H coating [50, 81]. Analyses of the formed boundary film further revealed that about 20% of the wear track of DLC surface was covered with the MoS_2-based tribofilm, and the thickness of the tribofilm was about 3–5 nm [82]. Similar observation to MoS_2 particles was also found for GNPs, where patchy tribofilm of graphene was formed on a-C:H surface, resulting in up to 50% friction reduction compared to base oil [55].

6.5 Lubrication Behavior of DLC Coatings with Base Lubricants

The role of doping elements and their effect on DLCs' surface energy is also revealed in DLC contacts lubricated with base oils under full-film conditions. As shown in Figure 6.18, the COF under full-film conditions is significantly lower for DLC coatings than steel surface [83]. What is more, there is also a significant difference between different DLC coatings, depending on the type of doping element. These significant differences can be explained by the surface energies of these coatings and their corresponding wetting behavior. Namely, the lower the (polar) surface energy and the wetting (evaluated by the spreading parameter), the lower the friction (Figure 6.18). The proposed mechanism behind observed friction reduction is lubricant slip over DLC coatings that reduce shear stresses within the lubricating film and thus full-film friction. More than 30% reduction in full-film friction is achieved compared to the steel contact by selecting DLC coatings of different doping elements that possess poorer wetting (lower spreading parameter) and the polar surface energy [83]. As seen from Figure 6.18, the highest friction reduction is achieved for F-DLC coating that has the lowest surface energy and negligible polar component (see Figure 6.12).

The effect of polar surface energy of DLC coating on lubricant interactions and lubrication on DLC coatings was also investigated under dynamic conditions by employing high-speed optical imaging of lubricant film flow over

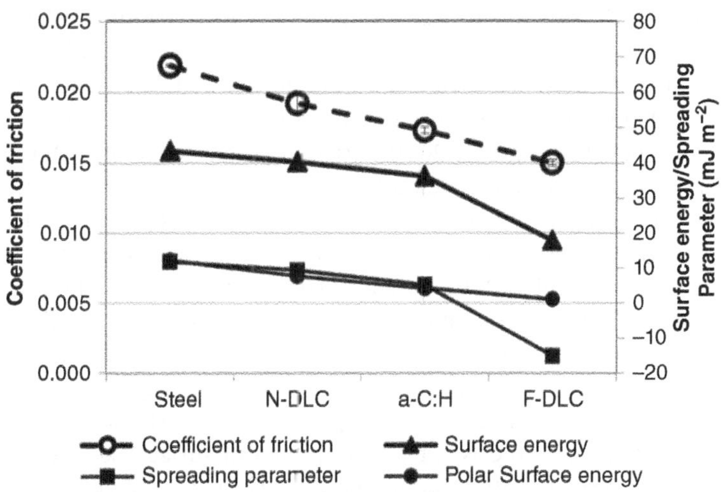

Figure 6.18 Coefficient of friction, surface energy, and wetting (spreading parameters) of steel and different DLC coatings under full-film lubricating conditions.

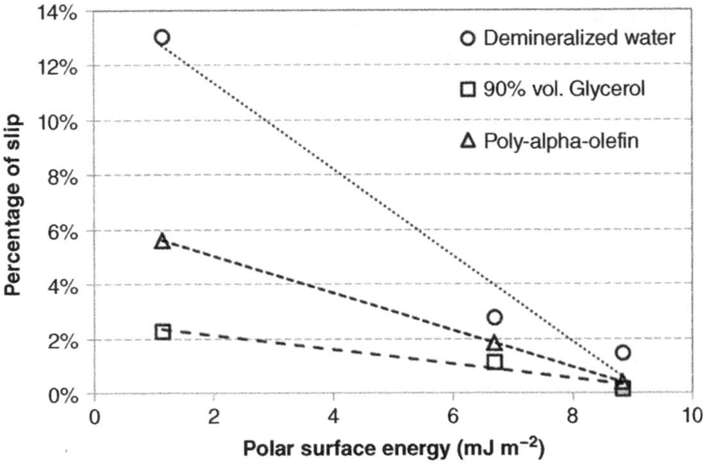

Figure 6.19 Percentage of lubricant slip as a function of the polar component of the surface energy. *Source:* Reproduced from [84] / with permission of John Wiley & Sons.

free-rotating DLC surfaces [84]. Results showed that lubricants can slip over DLC coatings, but the amount of slip depends on their viscosity and surface tension (see Figure 6.19). The amount of slip was very low or even negligible for steel surface, but it was up to 3% for DLC (a-C:H) and over 13% for F-DLC coating. As shown in Figure 6.19, in this case, a good correlation between DLC polar surface energy and the amount of slip was established.

In addition to decreasing the polar component of DLC coating, further friction reduction under full-film lubricating conditions can also be achieved with proper contact design. Namely, many tribological studies showed that friction reduction is always higher if both contact surfaces are coated with DLC instead of only one [83, 85]. As shown in Figure 6.20, the friction performance of F-DLC coating under full-film conditions is much different, depending on the contact surface coated with F-DLC and slide-to-roll ratio (SRR) [86]. At SRR of 50%, where rolling prevails in the contact, there is no significant difference in the frictional performance between steel/F-DLC2 (where F-DLC coating is applied on the faster surface) and F-DLC2/steel contacts (where F-DLC coating is applied on the slower surface; see Figure 6.20a). However, with a further increase of SRR up to 180%, where sliding prevails in the contact, the COF was significantly lower for steel/F-DLC2 contact (where F-DLC is applied on the faster surface) than for F-DLC2/steel (where F-DLC is applied on the slower surface; see Figure 6.20c). Therefore, with proper contact design, the COF was reduced by up to 48% compared to steel contact [86].

All studies mentioned earlier were done under high contact pressure conditions in elastohydrodynamic contacts; however, the beneficial tribological performance

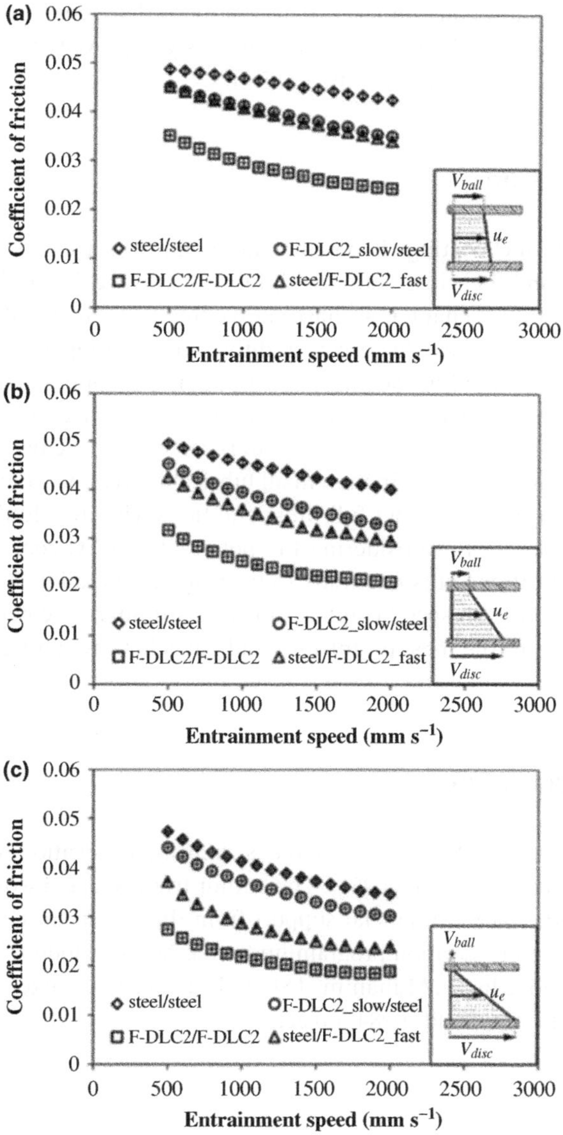

Figure 6.20 Coefficient-of-friction results for contacts of F-DLC coating at (a) SRR = 0.5, (b) SRR = 1.0, and (c) SRR = 1.8. *Source:* Reproduced from [86] / with permission of Springer Nature.

of low surface energy F-DLC coating was also shown for low-pressure hydrodynamic contacts. In a recent study [87], it has been shown that for base oils, the F-DLC coating reduced the COF up to 16% compared to the steel surface and, at the same time, due to the slip effect, increased the lubricant film thickness up to

13%. This result shows that under certain conditions, DLC coatings with very low surface energy (such as F-DLC) can simultaneously improve two crucial lubricating properties, i.e. film thickness and COF.

6.6 Conclusion

Lubrication of DLC coatings is possible, although they are considered to have low reactivity, and their interaction with lubricants is weak and difficult to predict or control. In this chapter, we have shown that it is also possible to lubricate them even with low reactive/green additives (e.g. nanoparticles, low SAPS, OFMs, ILs, water-based lubricants, etc.). In addition, doping elements play an important role in the reactivity of DLC coatings, as they affect the surface energy and thus the wetting behavior with the lubricants. In particular, DLC coatings exhibit significant friction reduction in all lubrication regimes: boundary regime (with additives) and also under full-film conditions without additives (by tailoring wetting properties). Considering the potential of DLC coatings in various industrial applications, it is necessary to investigate and tailor the properties of DLC coatings to match specific applications. In summary, we believe that DLC coatings are one of the most promising candidates for environmentally friendly lubrication.

Acknowledgements

This work is supported by the Slovenian Research and Innovation Agency (ARIS), Slovenia (core funding no. P2-0231). The author Irfan Nadeem would also like to acknowledge the partial financial support from the European Union's Horizon 2020 research and innovation program under the Marie Skłodowska-Curie grant agreement no. 860246. Dr. Muhammad Shahid Arshad would like to acknowledge partial funding from the PR-0084 project.

References

1 Holmberg, K., Siilasto, R., Laitinen, T. et al. (2013). Global energy consumption due to friction in paper machines. *Tribology International* 62: 58–77. doi: 10.1016/j.triboint.2013.02.003

2 (2013). International Energy Agency. *Key World Energy Statistics 2012*. Paris: International Energy Agency. https://doi.org/10.1787/key_energ_stat-2012-en (accessed 15 September 2023).

3 Holmberg, K., Andersson, P., and Erdemir, A. (2012). Global energy consumption due to friction in passenger cars. *Tribology International* 47: 221–234. doi: 10.1016/j.triboint.2011.11.022

4 Henze, V. and Thomas, C. (2017). Electric vehicles to accelerate to 54% of new car sales by 2040, BloombergNEF, NY, USA. https://about.bnef.com/blog/electric-vehicles-accelerate-54-new-car-sales-2040/ (accessed 3 June 2024).

5 Kalin, M., Polajnar, M., Kus, M. et al. (2019). Green tribology for sustainable engineering of the future. *Strojniški Vestnik – Journal of Mechanical Engineering* 65: 709–727. doi: 10.5545/sv-jme.2019.6406

6 Espejo, C., Wang, C., Thiébaut, B. et al. (2020). The role of MoDTC tribochemistry in engine tribology performance. A Raman microscopy investigation. *Tribology International* 150: 106366. doi: 10.1016/j.triboint.2020.106366

7 Cen, H., Morina, A., and Neville, A. (2018). Effect of slide to roll ratio on the micropitting behaviour in rolling-sliding contacts lubricated with ZDDP-containing lubricants. *Tribology International* 122: 210–217. doi: 10.1016/j.triboint.2018.02.038

8 Spikes, H. (2008). Low- and zero-sulphated ash, phosphorus and sulphur anti-wear additives for engine oils. *Lubrication Science* 20: 103–136. doi: 10.1002/ls.57

9 Brown, P., Cardenas, L., Del Vento, S. et al. (2023). UK Greenhouse Gas Inventory 1990 to 2021: annual report for submission under the Framework Convention on Climate Change. Natural Environment Research Council UK. https://nora.nerc.ac.uk/id/eprint/537272/.

10 Nosonovsky, M. and Bhushan, B. (2010). Green tribology: Principles, research areas and challenges. *Philosophical Transactions of the Royal Society A: Mathematical, Physical and Engineering Sciences* 368: 4677–4694. doi: 10.1098/rsta.2010.0200

11 Kalin, M., Vižintin, J., Vercammen, K. et al. (2006). The lubrication of DLC coatings with mineral and biodegradable oils having different polar and saturation characteristics. *Surface and Coatings Technology* 200: 4515–4522. doi: 10.1016/j.surfcoat.2005.03.016

12 Kalin, M. and Vižintin, J. (2005). The tribological performance of DLC-coated gears lubricated with biodegradable oil in various pinion/gear material combinations. *Wear* 259: 1270–1280. doi: 10.1016/j.wear.2005.02.028

13 Barriga, J., Kalin, M., Acker, K.V. et al. (2006). Tribological performance of titanium doped and pure DLC coatings combined with a synthetic bio-lubricant. *Wear* 261: 9–14. doi: 10.1016/j.wear.2005.09.004

14 Long, Y., Galipaud, J., Weihnacht, V. et al. (2022). Achieving superlubricity using selected tribo-pairs lubricated by castor oil and unsaturated fatty acids. *Tribology International* 169: 107462. doi: 10.1016/j.triboint.2022.107462

15 Yang, L., Neville, A., Brown, A. et al. (2015). Effect of lubricant additives on the WDLC coating structure when tested in boundary lubrication regime. *Tribology Letters* 57: 14. doi: 10.1007/s11249-015-0464-y

16 Durham, J. and Kidson, A. (2014). The effects of low sulfated ash, phosphorus and sulfur oils on camshaft/tappet tribocouples with various diamond-like-carbon coated tappets in motored and fired engines. *Lubrication Science* 26: 411–427. doi: 10.1002/ls.1251

17 Oblak, E. and Kalin, M. (2015). Relationship between the nanoscale topographical and mechanical properties of tribochemical films on DLC coatings and their macroscopic friction behavior. *Tribology Letters* 59: 49. doi: 10.1007/s11249-015-0575-5

18 Kalin, M., Oblak, E., and Akbari, S. (2016). Evolution of the nano-scale mechanical properties of tribofilms formed from low- and high-SAPS oils and ZDDP on DLC coatings and steel. *Tribology International* 96: 43–56. doi: 10.1016/j.triboint.2015.12.013

19 Bhushan, B. and Liu, H. (2005). Self-assembled monolayers for controlling adhesion, friction and wear. In: *Nanotribology and Nanomechanics: An Introduction*, (ed. B. Bhushan), 885–928. Berlin, Heidelberg: Springer. doi: 10.1007/3-540-28248-3_18

20 Watson, S., Nie, M., Wang, L. et al. (2015). Challenges and developments of self-assembled monolayers and polymer brushes as a green lubrication solution for tribological applications. *RSC Advances* 5: 89698–89730. doi: 10.1039/C5RA17468F

21 Sun, S., Li, J., Li, J. et al. (2022). Enhanced superlubricity on a-C films by lubrication with 3-hydroxypropionic acid. *Carbon* 199: 161–169. doi: 10.1016/j.carbon.2022.07.062

22 Kogovšek, J. and Kalin, M. (2023). Comparison of graphene as an oil additive with conventional automotive additives for the lubrication of steel and DLC-coated surfaces. *Tribology International* 180: 108220. doi: 10.1016/j.triboint.2023.108220

23 Tang, J., Chen, S., Jia, Y. et al. (2020). Carbon dots as an additive for improving performance in water-based lubricants for amorphous carbon (a-C) coatings. *Carbon* 156: 272–281. doi: 10.1016/j.carbon.2019.09.055

24 Kalin, M., Kogovšek, J., and Remškar, M. (2012). Mechanisms and improvements in the friction and wear behavior using MoS_2 nanotubes as potential oil additives. *Wear* 280–281: 36–45. doi: 10.1016/j.wear.2012.01.011

25 Tannous, J., Dassenoy, F., Lahouij, I. et al. (2011). Understanding the tribochemical mechanisms of IF-MoS_2 nanoparticles under boundary lubrication. *Tribology Letters* 41: 55–64. doi: 10.1007/s11249-010-9678-1

26 Long, Y., Bouchet, M.-I.D.B., Lubrecht, T. et al. (2019). Superlubricity of glycerol by self-sustained chemical polishing. *Scientific Reports* 9: 6286. doi: 10.1038/s41598-019-42730-9

27 Majdič, F., Velkavrh, I., and Kalin, M. (2013). Improving the performance of a proportional 4/3 water–hydraulic valve by using a diamond-like-carbon coating. *Wear* 297: 1016–1024. doi: 10.1016/j.wear.2012.11.060

28 Tokoro, M., Aiyama, Y., Masuko, M. et al. (2009). Improvement of tribological characteristics under water lubrication of DLC-coatings by surface polishing. *Wear* 267: 2167–2172. doi: 10.1016/j.wear.2009.04.009

29 Yamamoto, K. and Matsukado, K. (2006). Effect of hydrogenated DLC coating hardness on the tribological properties under water lubrication. *Tribology International* 39: 1609–1614. doi: 10.1016/j.triboint.2006.01.004

30 Uchidate, M., Liu, H., Iwabuchi, A. et al. (2007). Effects of water environment on tribological properties of DLC rubbed against stainless steel. *Wear* 263: 1335–1340. doi: 10.1016/j.wear.2006.10.024

31 Arshad, M.S., Čoga, L., Geue, T. et al. (2023). The W-cluster reactive sites interaction model for WDLC coatings with ionic liquids. *Tribology International* 185: 108550. doi: 10.1016/j.triboint.2023.108550

32 Arshad, M.S., Kovač, J., Cruz, S. et al. (2020). Physicochemical and tribological characterizations of WDLC coatings and ionic-liquid lubricant additives: potential candidates for low friction under boundary-lubrication conditions. *Tribology International* 151: 106482. doi: 10.1016/j.triboint.2020.106482

33 Feng, X. and Xia, Y. (2012). Tribological properties of Ti-doped DLC coatings under ionic liquids lubricated conditions. *Applied Surface Science* 258: 2433–2438. doi: 10.1016/j.apsusc.2011 10.066

34 Sierra-Cantor, J.F. and Guerrero-Fajardo, C.A. (2017). Methods for improving the cold flow properties of biodiesel with high saturated fatty acids content: a review. *Renewable and Sustainable Energy Reviews* 72: 774–790. doi: 10.1016/j.rser.2017.01.077

35 Sia, C.B., Kansedo, J., Tan, Y.H. et al. (2020). Evaluation on biodiesel cold flow properties, oxidative stability and enhancement strategies: a review. *Biocatalysis and Agricultural Biotechnology* 24: 101514. doi: 10.1016/j.bcab.2020.101514

36 Seifert, W.K. (1975). Carboxylic acids in petroleum and sediments. In: *Fortschritte Der Chemie Organischer Naturstoffe/Progress in the Chemistry of Organic Natural Products* (eds. L. Zechmeister, W. Herz, H. Grisebach et al.), 1–49. Vienna: Springer. doi: 10.1007/978-3-7091-7083-0_1

37 Kano, M., Martin, J.M., Yoshida, K. et al. (2014). Super-low friction of ta-C coating in presence of oleic acid. *Friction* 2: 156–163. doi: 10.1007/s40544-014-0047-1

38 De Barros Bouchet, M.I., Martin, J.M., Avila, J. et al. (2017). Diamond-like carbon coating under oleic acid lubrication: evidence for graphene oxide formation in superlow friction. *Scientific Reports* 7: 46394. doi: 10.1038/srep46394

39 Kuwahara, T., Romero, P.A., Makowski, S. et al. (2019). Mechano-chemical decomposition of organic friction modifiers with multiple reactive centres induces superlubricity of ta-C. *Nature Communications* 10: 151. doi: 10.1038/s41467-018-08042-8

40 Simič, R. and Kalin, M. (2013). Adsorption mechanisms for fatty acids on DLC and steel studied by AFM and tribological experiments. *Applied Surface Science* 283: 460–470. doi: 10.1016/j.apsusc.2013.06.131

41 Kalin, M. and Simič, R. (2013). Atomic force microscopy and tribology study of the adsorption of alcohols on diamond-like carbon coatings and steel. *Applied Surface Science* 271: 317–328. doi: 10.1016/j.apsusc.2013.01.192

42 Simič, R., Kalin, M., Hirayama, T. et al. (2014). Fatty acid adsorption on several DLC coatings studied by neutron reflectometry. *Tribology Letters* 53: 199–206. doi: 10.1007/s11249-013-0257-0

43 Kalin, M., Simič, R., Hirayama, T. et al. (2014). Neutron-reflectometry study of alcohol adsorption on various DLC coatings. *Applied Surface Science* 288: 405–410. doi: 10.1016/j.apsusc.2013.10.047

44 Loehlé, S., Matta, C., Minfray, C. et al. (2016). Mixed lubrication of steel by C18 fatty acids revisited. Part II: Influence of some key parameters. *Tribology International* 94: 207–216. doi: 10.1016/j.triboint.2015.08.036

45 Erdemir, A. (2005). Review of engineered tribological interfaces for improved boundary lubrication. *Tribology International* 38: 249–256. doi: 10.1016/j.triboint.2004.08.008

46 Dai, W., Kheireddin, B., Gao, H. et al. (2016). Roles of nanoparticles in oil lubrication. *Tribology International* 102: 88–98. doi: 10.1016/j.triboint.2016.05.020

47 Kotia, A., Chowdary, K., Srivastava, I. et al. (2020). Carbon nanomaterials as friction modifiers in automotive engines: recent progress and perspectives. *Journal of Molecular Liquids* 310: 113200. doi: 10.1016/j.molliq.2020.113200

48 Rosentsveig, R., Gorodnev, A., Feuerstein, N. et al. (2009). Fullerene-like MoS_2 nanoparticles and their tribological behavior. *Tribology Letters* 36: 175–182. doi: 10.1007/s11249-009-9472-0

49 Moshkovith, A., Perfiliev, V., Verdyan, A. et al. (2007). Sedimentation of IF-WS2 aggregates and a reproducibility of the tribological data. *Tribology International* 40: 117–124. doi: 10.1016/j.triboint.2006.02.067

50 Kalin, M., Kogovšek, J., and Remškar, M. (2013). Nanoparticles as novel lubricating additives in a green, physically based lubrication technology for DLC coatings. *Wear* 303: 480–485. doi: 10.1016/j.wear.2013.03.009

51 Liu, Y., Ge, X., and Li, J. (2020). Graphene lubrication. *Applied Materials Today* 20: 100662. doi: 10.1016/j.apmt.2020.100662

52 Chen, X. and Li, J. (2020). Superlubricity of carbon nanostructures. *Carbon* 158: 1–23. doi: 10.1016/j.carbon.2019.11.077

53 Joshi, P.N., Kundu, S., Sanghi, S.K. et al. (2016). Graphene quantum dots – from emergence to nanotheranostic applications. In: *Smart Drug Delivery System*, 7: 159–195. London: IntechOpen. doi: 10.5772/61932.

54 Singh, R.K., Dixit, A.R., Sharma, A.K. et al. (2018). Influence of graphene and multi-walled carbon nanotube additives on tribological behaviour of lubricants. *International Journal of Surface Science and Engineering* 12: 207–227. doi: 10.1504/IJSURFSE.2018.094773

55 Kogovšek, J. and Kalin, M. (2019). Lubrication performance of graphene-containing oil on steel and DLC-coated surfaces. *Tribology International* 138: 59–67. doi: 10.1016/j.triboint.2019.05.026

56 Fan, K., Chen, X., Wang, X. et al. (2019). Correction to "Toward excellent tribological performance as oil-based lubricant additive: particular tribological behavior of fluorinated graphene". *ACS Applied Materials & Interfaces* 11: 36296–36296. doi: 10.1021/acsami.9b15885

57 Georgakilas, V., Otyepka, M., Bourlinos, A.B. et al. (2012). Functionalization of graphene: covalent and non-covalent approaches, derivatives and applications. *Chemical Reviews* 112: 6156–6214. doi: 10.1021/cr3000412

58 Sun, Y.-P., Zhou, B., Lin, Y. et al. (2006). Quantum-sized carbon dots for bright and colorful photoluminescence. *Journal of the American Chemical Society* 128: 7756–7757. doi: 10.1021/ja062677d

59 Nadeem, I., Malok, M., Kovač, J. et al. (2023). Superior macro-scale tribological performance of steel contacts based on graphene quantum dots in aqueous glycerol. *Tribology International* 181: 108328. doi: 10.1016/j.triboint.2023.108328

60 Nadeem, I., Finšgar, M., Dražić, G. et al. (2024). Robust and durable superlubricity enabled by carboxylated graphene quantum dots in lubricated steel contacts. *Carbon* 226: 119226. doi: 10.1016/j.carbon.2024.119226

61 Xu, P., Hong, J., Xu, Z. et al. (2021). Novel aminated graphene quantum dots (GQDs-NH2)-engineered nanofiltration membrane with high Mg^{2+}/Li^{+} separation efficiency. *Separation and Purification Technology* 258: 118042. doi: 10.1016/j.seppur.2020.118042

62 Yi, S., Chen, X., Li, J. et al. (2021). Macroscale superlubricity of Si-doped diamond-like carbon film enabled by graphene oxide as additives. *Carbon* 176: 358–366. doi: 10.1016/j.carbon.2021.01.147

63 Zeng, Q., Yu, F., and Dong, G. (2013). Superlubricity behaviors of Si_3N_4/DLC Films under PAO oil with nano boron nitride additive lubrication. *Surface and Interface Analysis* 45: 1283–1290. doi: 10.1002/sia.5269

64 Tang, J., Ding, Q., Zhang, S. et al. (2016). Improved tribological performance of amorphous carbon (a-C) coating by ZrO_2 nanoparticles. *Materials* 9: 795. doi: 10.3390/ma9100795

65 Chen, S., Ding, Q., Gu, Y. et al. (2022). Study of tribological properties of fullerenol and nanodiamonds as additives in water-based lubricants for amorphous carbon (a-C) coatings. *Nanomaterials* 12: 139. doi: 10.3390/nano12010139

66 Erdemir, A. (2021). Chapter 12 – Diamond-like carbon films and their superlubricity. In: *Superlubricity*, 2e (eds. A. Erdemir, J.M. Martin, and J. Luo), 215–230. Elsevier. doi: 10.1016/B978-0-444-64313-1.00012-0

67 Sato, T., Besshi, T., Sato, D. et al. (2001). Effect of water based lubricants on wear of coated material. *Wear* 249: 50–55. doi: 10.1016/S0043-1648(01)00522-1

68 Shi, Y., Minami, I., Grahn, M. et al. (2014). Boundary and elastohydrodynamic lubrication studies of glycerol aqueous solutions as green lubricants. *Tribology International* 69: 39–45. doi: 10.1016/j.triboint.2013.08.013

69 Luo, J. and Zhou, X. (2020). Superlubricitive engineering—future industry nearly getting rid of wear and frictional energy consumption. *Friction* 8: 643–665. doi: 10.1007/s40544-020-0393-0

70 Matta, C., Joly-Pottuz, L., De Barros Bouchet, M.I. et al. (2008). Superlubricity and tribochemistry of polyhydric alcohols. *Physical Review B* 78: 085436. doi: 10.1103/PhysRevB.78.085436

71 Han, T., Yi, S., Zhang, C. et al. (2020). Superlubrication obtained with mixtures of hydrated ions and polyethylene glycol solutions in the mixed and hydrodynamic lubrication regimes. *Journal of Colloid and Interface Science* 579: 479–488. doi: 10.1016/j.jcis.2020.06.095

72 Ge, X., Li, J., and Luo, J. (2019). Macroscale superlubricity achieved with various liquid molecules: a review. *Frontiers of Mechanical Engineering* 5: 2. doi: 10.3389 /fmech.2019.00002

73 Donnet, C. (1998). Recent progress on the tribology of doped diamond-like and carbon alloy coatings: a review. *Surface and Coatings Technology* 100–101: 180–186. doi: 10.1016/S0257-8972(97)00611-7

74 Bewilogua, K. and Hofmann, D. (2014). History of diamond-like carbon films—from first experiments to worldwide applications. *Surface and Coatings Technology* 242: 214–225. doi: 10.1016/j.surfcoat.2014.01.031

75 Erdemir, A. (2004). Genesis of superlow friction and wear in diamond-like carbon films. *Tribology International* 37: 1005–1012. doi: 10.1016/j.triboint.2004. 07.018

76 Erdemir, A. (2001). The role of hydrogen in tribological properties of diamond-like carbon films. *Surface and Coatings Technology* 146–147: 292–297. doi: 10. 1016/S0257-8972(01)01417-7

77 Kalin, M. and Polajnar, M. (2013). The correlation between the surface energy, the contact angle and the spreading parameter, and their relevance for the wetting behaviour of DLC with lubricating oils. *Tribology International* 66: 225–233. doi: 10.1016/j.triboint.2013.05.007

78 Kalin, M. and Polajnar, M. (2014). The wetting of steel, DLC coatings, ceramics and polymers with oils and water: the importance and correlations of surface energy, surface tension, contact angle and spreading. *Applied Surface Science* 293: 97–108. doi: 10.1016/j.apsusc.2013.12.109

79 Simič, R., Kalin, M., Kovač, J. et al. (2016). Adsorption of alcohols and fatty acids onto hydrogenated (a-C:H) DLC coatings. *Applied Surface Science* 363: 466–476. doi: 10.1016/j.apsusc.2015.12.066

80 Kalin, M. (2012). Designing tribological interface for efficient and green DLC lubrication: the role of coatings and lubricants. *Tribology Online* 7: 112–118. doi: 10.2474/trol.7.112

81 Kogovšek, J., Remškar, M., and Kalin, M. (2013). Lubrication of DLC-coated surfaces with MoS_2 nanotubes in all lubrication regimes: surface roughness and running-in effects. *Wear* 303: 361–370. doi: 10.1016/j.wear.2013.03.033

82 Kalin, M., Kogovšek, J., Kovač, J. et al. (2014). The formation of tribofilms of MoS_2 nanotubes on steel and DLC-coated surfaces. *Tribology Letters* 55: 381–391. doi: 10.1007/s11249-014-0366-4

83 Kalin, M. and Polajnar, M. (2013). The effect of wetting and surface energy on the friction and slip in oil-lubricated contacts. *Tribology Letters* 52: 185–194. doi: 10.1007/s11249-013-0194-y

84 Polajnar, M., Bizjan, B., Širok, B. et al. (2017). High-speed optical imaging of liquid film flow and liquid macro-slip over free surfaces with different surface energies. *Lubrication Science* 29: 557–566. doi: 10.1002/ls.1388

85 Kalin, M., Velkavrh, I., and Vižintin, J. (2009). The Stribeck curve and lubrication design for non-fully wetted surfaces. *Wear* 267: 1232–1240. doi: 10.1016/j.wear.2008.12.072

86 Polajnar, M. and Kalin, M. (2015). Effect of the slide-to-roll ratio and the contact kinematics on the elastohydrodynamic friction in diamond-like-carbon contacts with different wetting behaviours. *Tribology Letters* 60: 8. doi: 10.1007/s11249-015-0593-3

87 Jing, Z., Guo, F., Jin, W. et al. (2022). Study on the influence of interfacial slip on the lubrication performance of a step slider bearing. *Tribology International* 176: 107822. doi: 10.1016/j.triboint.2022.107822

Section III

Diamond-like Carbon Coatings for Medical Applications

7

Diamond-like Carbon Coatings for Artificial Implants

Benedict Rothammer[1], Max Marian[2,3]

[1] *Engineering Design, Friedrich-Alexander-Universität Erlangen-Nuremberg (FAU), Erlangen, Germany*
[2] *Department of Mechanical and Metallurgical Engineering, School of Engineering, Pontificia Universidad Católica de Chile, Macul, Región Metropolitana, Chile*
[3] *Institute of Machine Design and Tribology (IMKT), Leibniz University Hannover, Garbsen, Germany*

CONTENTS

7.1 Introduction, 193
7.2 Requirements for DLC Coatings When Employed to Artificial Implants, 195
7.2.1 Coating-Substrate Adhesion, 196
7.2.2 Biological Compatibility, 198
7.3 Application for Load-bearing Implants, 198
7.3.1 Biotribological Studies on the Model Level, 201
7.3.2 Biotribological Studies on the Component Level, 213
7.4 Application for Dental Implants and Braces/Retainers, 216
7.4.1 Dental Implants, 216
7.4.2 Braces/Retainers, 219
7.5 Concluding Remarks, 222

7.1 Introduction

The use of artificial implants has revolutionized modern medicine. From hip replacements over dental implants to cardiovascular stents, these devices have greatly improved the quality of life for millions of people worldwide. However, the success of these implants depends on their ability to integrate with the human body and function properly without causing adverse reactions or complications [1]. One way to improve the performance of artificial implants is through the use of diamond-like carbon (DLC) coatings. DLC is a unique material that possesses excellent mechanical, chemical, and biocompatible properties, making it a prosperous candidate for various biomedical applications [2–7] (see Figure 7.1). Therefore, DLC coatings have been extensively studied for their use in orthopedic applications, including

Applications of Diamond-like Carbon Coatings, First Edition. Abdul Wasy Zia.

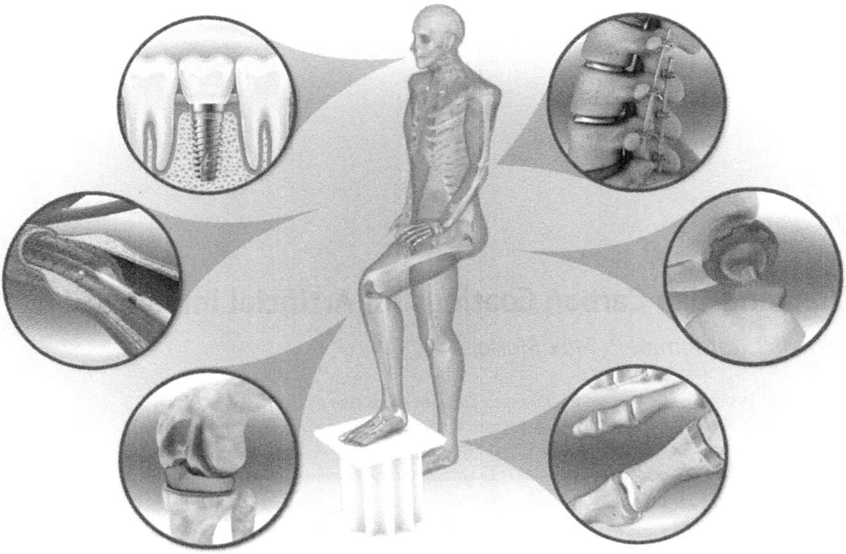

Figure 7.1 Application of DLC coatings to improve the biotribological behavior in the areas of orthopedic, dental, and cardiovascular implants. *Source:* Reproduced from [1] / with permission of ELSEVIER.

joint replacements, with the aim of improving their wear resistance, leading to a longer lifespan of the implant [8, 9]. DLC coatings have also been shown to promote osteointegration, or the integration of the implant with surrounding bone tissue, by enhancing cell adhesion and proliferation as well as drug delivery carriers. In dental implants, DLC coatings can be used to improve the biocompatibility of the implant and reduce the risk of infection. DLC coatings can also be used to improve the durability and wear resistance of orthodontic brackets, which are subject to high stresses and wear during treatment.

As briefly indicated, the use of DLC coatings has the potential to:

- improve patient outcomes and reduce the need for revision surgeries in orthopedic applications, and
- improve patient comfort and reduce the need for frequent replacements in dental medicine.

The present chapter provides a detailed overview of the requirements for DLC coatings when employed to artificial implants (Section 7.2) as well as the state of the art of DLC coatings for load bearing (Section 7.3) and dental implants (Section 7.4), while the main findings and remaining challenges for future research are summarized in the Section 7.5.

7.2 Requirements for DLC Coatings When Employed to Artificial Implants

Nowadays, DLC coatings are well established in many commercially available, industrial applications. However, they still are only occasionally employed for artificial implants that are subject to tribological stressing, which might be due to the fact that they need to be applied to surfaces with tendentially low adhesion [4]. The cell biological and tribological challenges in artificial implants require a comprehensive surface modification of the articulating components. This brings some challenges and requirements toward applied coatings, which are summarized in Table 7.1.

DLC coatings have been successfully employed in mechanical engineering and production technology to reduce friction and increase wear resistance. Additionally, DLC coatings feature good biocompatibility [4, 10], which is why coating artificial implants is considered to have potential for considerably increasing service life and prevent patients from premature surgical interventions [11]. Due to the cell biological peri-implant environment, fundamentally different reactions take place compared to mechanical engineering, which prevents a simple transfer of coating solutions from the mechanical engineering field [4, 12]. In particular, human tissue

Table 7.1 Requirements for coatings applied to artificial implants.

Requirement	Description
Good coating-substrate adhesion	Good adhesion of the coating system to soft/hard substrate material
	Avoidance of brittle mixed phases in the adhesion area of the layer
	Reproducible, optimum roughness to promote mechanical coating adhesion with low impact on friction
High biological compatibility – biopassivity	Low ion release, biochemically stable
	No cytotoxic, infectious, or allergenic interactions in tissue due to material solubility, corrosion, or wear particles
	Long-term stable and biocompatible surfaces
Good mechanical properties of the coating	High deformation resistance and ductility
	Preservation of the fatigue strength of the substrate material
Adjustable biotribological behavior	High wear resistance and low friction
	Full-coverage, dense, nanoscale coating systems
	Influencing the geometry and morphology of the wear particles. Avoidance of mean particle diameters between 0.1 and –1 μm, otherwise macrocytic reactions leading to osteolysis are stimulated

with its multitude of cells, bacteria, and proteins must be considered [12]. The goal of carbon-based coatings in medical technology is to modify the interface between the implant and the tissue biologically, chemically, and physically in such a way that the functional compatibility within the biological environment is continuously improved [13]. Coating the articulating surfaces of artificial implants primarily serves to minimize wear as well as the resulting abrasion [4, 10]. The excellent properties of DLC coatings, such as their favorable friction behavior as well as their high corrosion and wear resistance, have led to the investigation of DLC coatings for biomedical applications [3]. Before DLC coatings can be discussed with regard to their biotribological suitability, the two main requirements of good coating-substrate adhesion and cell biological compatibility have to be addressed.

7.2.1 Coating-Substrate Adhesion

The adhesion of DLC coatings can be improved by ion etching the substrate in order to remove oxide layers prior to the actual coating process [14]. However, one disadvantage of DLC coatings is the intrinsic residual stresses induced by the high-energy coating process [15]. Thus, a thicker DLC layer leads to higher residual stresses, which influence adhesion and can lead to local spalling or even complete delamination of the coating [16]. It is assumed that, apart from coating structure and morphology, the coating thickness primarily influences the cohesive coating failure, which can promote the formation of wear particles [17]. Residual stresses can also be reduced by increasing the hydrogen content in the DLC layer, which improves the adhesion but decreases the eventually desired high hardness [18]. Therefore, multilayer coating systems [18, 19] or individual intermediate layers [20] have been studied.

Different coating designs are typically used to ensure a sufficiently high adhesion of the coatings on metal substrates, such as Ti6Al4V, titanium (Ti), or cobalt-chromium (CoCr) alloys [21–23]. Usually, Ti [21, 24] or chromium (Cr) [25] are used in the form of a metal-carbide [4, 26] or metal-nitride [27–29] transition, frequently followed by gradient from the intermediate layer into the desired DLC layer to improve adhesion and prevent carbon accumulation at the transitions of the individual layers and increases the cohesion of the coating-substrate composite [28, 29]. Another way to achieve a strong adhesion to individual layers is to introduce doping elements, such as tungsten (W) [30], silver (Ag) [6], and Ti [26, 31], which reduces the diamond bonding fractions, improving the ductility of the coating system and thus the adhesion to the substrate. Alternatively, nanometer-scale crystals (nanocomposites), such as TiC particles [32], can be embedded in the amorphous matrix.

With regard to increasing the service life of artificial implants, not only the metallic components but also polymeric materials, for example ultrahigh molecular weight polyethylene (UHMWPE), can be coated with DLC [33, 34]. Before the deposition of the coating itself, an argon- (Ar) [34, 35], nitrogen- (N) [29], or oxygen- (O_2) [36, 37] ion etching process is usually performed to remove surface impurities, improve surface wettability, and selectively modify and activate the surface without changing the properties of the substrate [38]. Subsequently, hydrogenated amorphous carbon (a-C:H) layers – due to their chemically comparable structure with the substrate – are suitable to be directly deposited directly on UHMWPE surfaces by physical vapor deposition (PVD) or plasma-enhanced chemical vapor deposition (PECVD) [34]. Compared to metals, several polymer-specific challenges have to be considered: Although PVD and PECVD processes can be used to coat comparatively temperature-sensitive polymers [39–41] (for example, the melting temperature of UHMWPE is around 135°C [42]), it is necessary to avoid overheating caused by bombardment of high-energy particles and simultaneously enable good coating adhesion [43]. This conflict of objectives can be solved by permanently monitoring the coating process and by a specific adjustment of the coating process parameters, such as bias voltage, gas flow, cathode power, and voltage [44]. Due to the low conductivity of polymers, the bias voltage hardly contributes to substrate heating [43], but it does contribute to the level of ion energy, which influences the resulting coating and consequently its adhesion [45]. Thus, adhesion can be favored by the modulation of a sufficiently high bias voltage. In order to keep the heating of the polymer substrate as low as possible during the coating process, a sufficiently low cathode power or coil current should be selected [43]. Trakhtenberg et al. [46] recommended process temperatures between 80 and 100°C for optimal coating-substrate adhesion. Nevertheless, this type of coating connection involves an increased risk of spontaneous, stress-induced coating delamination due to large differences in hardness and hence in stiffness between the coating and the substrate. According to Teixeira [47], coating-intrinsic stresses can be attributed to the following reasons: On the one hand, thermal stresses are induced by the temperature change during the coating process when the layer grows (so-called layer growth stresses), coupled with the different coefficients of thermal expansion of substrate and coating material. On the other hand, external stresses occur as a result of tribological loading. The stronger elastic deformation of the polymeric substrate material may result in spalling or delamination of the coating up to complete coating failure [48]. This can lead to increased wear and, in turn, to tribologically induced, undesirable in vivo cell-biological interactions in vivo [4]. However, He et al. [35] showed that such a coating design can already exhibit acceptable adhesion. Analogous to the doping of DLC coatings on metallic substrates, doping elements, such as N or Si, are used to improve adhesion, adjust medically relevant properties, or influence tribological behavior [49].

7.2.2 Biological Compatibility

In principle, implants are foreign bodies for the immune system and promote infections as well as allergenic reactions [13]. In order to prevent such an undesirable inflammatory or long-term rejection reaction to an implanted material, the latter must be biocompatible and also fulfill specific functions, such as rapid integration of the bone into the implant [50]. During implantation, the material interacts with body fluids [6], most of which represent corrosive media [51]. The adsorption of proteins on the material surface starts already a few milliseconds after implantation [52] and significantly affects the subsequent cellular response to the introduced foreign body [6]. In this context, the physiological conditions and the affinity of the proteins for the surface determine the type of adherent proteins [6]. Accordingly, biomedical implant materials, such as Ti and CoCr alloys, possess temporarily a good application behavior due to passivation of their surfaces. However, the risk of foreign body reactions increases with increasing service life due to the release of metal ions [53] and allergenic sensitization due to friction-induced corrosion. Coatings can combine good mechanical properties of the substrate with a more advantageous tribological behavior and excellent biocompatibility [13]. The good biocompatibility of DLC coatings has been confirmed in vitro and in vivo studies [54, 55], whereas no negative effects on cell behavior [56, 57] or inflammatory reactions could be observed [50]. Allen et al. [55] investigated the behavior of osteoblastic cells on uncoated and DLC-coated plates. Using specific marker proteins, the influence of DLC on cell metabolism was examined and found to be biologically inconspicuous. Also, DLC-coated CoCr cylinders were implanted into muscular tissues of rats and sheep, and no negative effect was found [55]. Mostly, some bulk materials exhibit inert behavior per se and induce undesirable responses mainly in particulate form. Therefore, Bruinink et al. [58] simulated the delamination of a-C:H coatings and investigated the effects of the particles on cell behavior. The addition of the a-C:H particles did not have a negative effect on cell survival, cell proliferation, or cell differentiation. Again, it was assumed that the coating particles did not induce an inflammatory response in the human body. However, the biocompatibility confirmed by these short-term tests remains to be tested in long-term tribological studies.

7.3 Application for Load-bearing Implants

The failure of a natural joints, e.g. due to arthrosis, requires the implantation of an artificial joint. The aim of endoprosthetic replacement of the joint is to restore its functionality and enable patients to lead a pain-free, more mobile life [59]. For the selection of a total joint replacement, different endoprosthesis systems are available to the surgeons, which are selected on the basis of the indication, in consideration of the severity or the propagation of the osteoarthritis, and in consideration of the ligament stability [60]. Total knee replacements (TKR) mainly involve a femoral component rubbing against the bearing surface of the tibial plateau, whereas total hip replacements (THR) generally consist of a femoral stem and head moving in the acetabular cup embedded in the pelvis or sometimes a backing [61]. While hard head-on-soft cup (hard-on-soft), as well as hard head-on-hard cup (hard-on-hard) pairings, can be found for total hip arthroplasty (THAs), total knee arthroplasty (TKAs) always comprise a hard femur-on-soft tibia couple [61]. Hard total joint replacement components are frequently made of CoCr alloys, such as $Co_{28}Cr_6Mo$, titanium alloys, such as Ti6Al4V (Ti64) or Ti6Al7Nb (Ti67), or oxide ceramics, such as alumina (Al_2O_3) or zirconia (ZrO_2), while soft components can be made of UHMWPE or, increasingly, highly cross-linked polyethylene (HXLPE). Currently, CoCr dominates as a material for femoral and tibial components for TKR. Ti alloys exhibit a considerably better biocompatibility than CoCr, but the unfavorable tribological behavior still prohibits its use as articulating interfaces [60]. Friction and wear in the latter, generally have a great influence on the service life of load-bearing implants [62]. Even though the service life of artificial joints can be up to 25 years [63, 64], the lifetime of revision implants or implants of younger patients with a more active lifestyle can be substantially shorter [65–67]. The bone to which the prosthesis is attached to generally represents a weak area. Apart from postoperative infections [66], one of the most common reasons for the need for revision surgery is aseptic prosthesis loosening [65, 68], which generally has a multifactorial etiology [66]. However, wear particles removed from the implant materials can be considered to be largely responsible [69–71].

Therefore, wear-resistant material combinations or surface coatings can be used [72], whereby the application of tribologically effective coating systems on the articulating implant surfaces is a particularly promising approach [9, 73] (see Figure 7.2). At the same time, the outstanding properties of metallic and polymeric implant materials, including high ductility, can be combined with the advantages of ceramic materials, such as high hardness and chemical inertness. A high-cost, fracture-prone ceramic prosthesis can thus be replaced by a damage-tolerant and also more economical metallic/polymeric prosthesis [60] if, for example, DLC is applied to at least one of the surfaces [73].

DLC coatings for load-bearing implants have been studied with different degrees of simplifications of testing specimens as well as stress spectra and boundary conditions and thus transferability to the actual in vivo system (see Figure 7.3). Generally, the in situ measurement of friction as well as the ex situ derivation of gravimetric or volumetric wear rates is of interest. In contrast to classical tribotechnical systems, such as machine elements, there is no clear demarcation

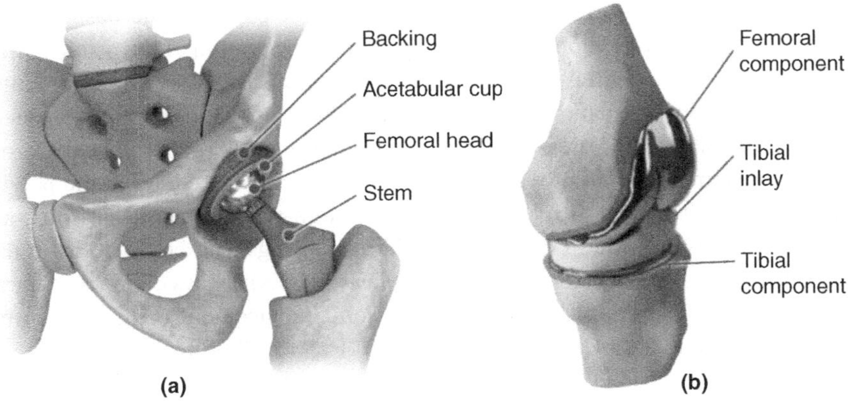

Figure 7.2 Structure of (a) total hip and (b) TKR as well as potential components for application of DLC coatings (highlighted in pink) to improve mechanical properties, durability, and wear resistance. Copyright Marian.

		Category	Measurements, test technique	Example and illustration
Operational tests and operation-like tests: Original system structure and simplified loading	Clinical *in vivo* testing	I	Operational test	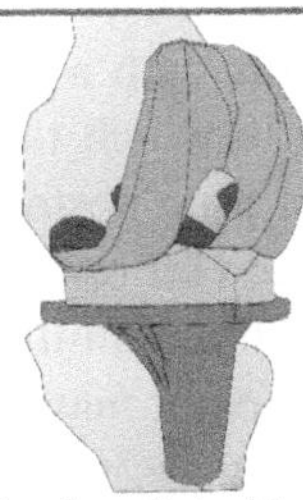 In vivo clinical test
		II	Bench test	
		III	Aggregate test-rig	
	Pre-clinical in vitro testing	IV	Component test-rig	Knee wear simulator 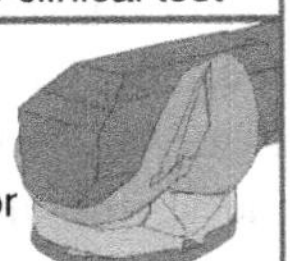
Model structure and simplified loading		V	Sample test-rig	Rolling-sliding test rig 
		VI	Model experiment	Pin-on-disk tribometer 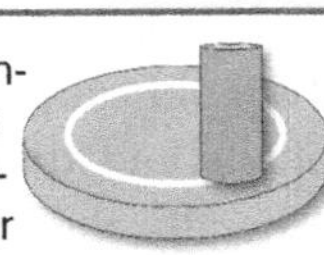

Figure 7.3 Categories of tribological testing technique according to Ref. [79] applied to TKA. Copyright Rothammer.

between operational test, bench test and aggregate test for biomedical tribological systems, such as a TKR. This is due to the fact that the original system – after implantation in the human body – is used under real environmental conditions. Yet, specimen tests play an important role since they involve testing specimens that are similar to components under similar stress and loading conditions. An important component test for determining and evaluating the wear behavior of TKR is the knee wear simulator (knee simulator) according to ISO 14243-1 [74]. Thereby, different, typical load cases, such as walking or climbing stairs, are simulated and the wear is subsequently examined [75], allowing for precise conclusions about the number and size distribution of wear particles and their cell biological effect [76]. Clinically relevant ligament situations, such as an insufficient posterior cruciate ligament, can also be simulated on the simulator [77]. Testing is usually performed in a closed chamber and diluted bovine calf serum (DBCS) tempered to 37°C is used as a substitute for synovium. Generally, 5×10^6 loading cycles are simulated, which corresponds to a duration of use in patients between 2 [78] and 5 [75] years. Finally, model tests with simplified test specimens (multi-ball, ball/disk, pin/disk, disk/disk or cylinder/cylinder test) and under modified loads are used for fundamental biotribological investigations [79]. These do not provide a direct reference to the biotribological conditions in the actual application, but are essential for the determination of comparable wear rates and for the investigation of individual wear mechanisms [80]. Thus, tribometer tests are suitable for a rapid and economical comprehensive screening of coating systems with respect to their tribological effectiveness [80]. With the aid of a suitable tribological test chain, the advantages of the respective test category can be combined [79]. The rolling-sliding test rig for knee components [81] can be seen as a connection between simplified tribometers and standardized knee simulators. Thereby, the wear behavior of various material combinations can be simulated and reproduced on relatively low-cost specimens with simple geometries under rolling-sliding conditions for up to 3×10^6 cycles and thus considerably more application-oriented motion for representative load cases of the knee joint [81]. In the following, the reported literature on DLC coatings when employed for load-bearing implants will be discussed clustered by the level of the conducted biotribological measurements.

7.3.1 Biotribological Studies on the Model Level

Xie et al. [82] successfully deposited a-C:H coatings on UHMWPE using electron cyclotron resonance microwave PECVD (ECR-PECVD). Thus, the hardness (0.139 GPa compared to 0.010 GPa of the UHMWPE substrate), the scratch resistance as well as the wear resistance in ball-on-disk tribometer tests were considerably improved. Using X-ray photoelectron (XPS) and Raman spectroscopy, the authors demonstrated an influence of the Ar/C_2H_2 gas flow ratio on the structure and properties of the a-C:H coatings. Thereby, a higher proportion of sp^3-hybridized carbon in the coating and thus improved mechanical properties as well as increased wear resistance was attributed to an increased Ar/C_2H_2 ratio.

Similarly, Dorner-Reisel et al. [34, 83] used an Ar/C_2H_2 gas mixture and radio frequency PECVD (RF-PECVD), analyzing the influence of individual process parameters (plasma etching, coating time as well as Ar/C_2H_2 ratio and flow) on coating thickness, structure and adhesion as well as wetting behavior. Regarding the latter, an improvement, i.e. a decrease in the contact angle by the application of a-C:H coatings, could be verified. An increasing Ar content in the gas mixture positively influenced the surface wetting with polar liquids. The increase in the coating time from 15 to 95 minutes resulted in an increase in the coating thickness from 1 to 3 μm. Scanning electron microscope (SEM) and atomic force microscopy (AFM) were employed to demonstrate that the coating power, Ar/C_2H_2 ratio, and Ar gas flow did not affect the coating (micro-) structure, which was cauliflower- and clod-like coarse. Raman spectroscopy indicated relatively low I_D/I_G ratios for all coatings, suggesting a low proportion of sp^2-hybridized carbon. Also, cross-sectional SEM images confirmed good coating-substrate adhesion for all coatings. Furthermore, Dorner-Reisel et al. [34] studied the effect of immersion of the coated and uncoated UHMWPE specimens in simulated body fluid for a period of twelve months on the wetting behavior, sp^2/sp^3 ratio, and the coating adhesion. Both uncoated and a-C:H coated UHMWPE specimens showed improved wetting behavior after twelve months in simulated body fluid. The reduced contact angles of coated compared to uncoated UHMWPE was also visible after twelve months of immersion in simulated body fluid. The adhesion of the coatings to the substrate was not affected by immersion in simulated body fluid and still turned out to be good for all specimens. The immersion also had no effect on the position of the D and G peaks measured by Raman spectroscopy, i.e. did not affect sp^2/sp^3 ratio. However, the authors found a reduction in the number of C–H bonds in the a-C:H coatings after immersion, which was attributed to a substitution of hydrogen by the ions contained in the simulated body fluid.

Puértolas et al. [84] fabricated a-C:H coatings by RF-PECVD on UHMWPE in their research and treated the specimens with an Ar plasma etching and the deposition of a thin hydrogen-rich layer (<20 nm) prior to the actual coating process. Subsequently, they used an $Ar/C_2H_2/H_2$ gas mixture for the deposition. In mechanical and tribological tests, they examined the effect of crosslinking the UHMWPE specimens using γ-radiation. The a-C:H coatings deposited on γ-irradiated UHMWPE showed higher hardness than the coatings deposited on non-cross-linked UHMWPE only at low indentation depths (<0.1 μm). With increasing indentation depth, the hardness values of the coatings deposited on both substrates converged. The coatings featured good adhesion on both materials and showed improved wear resistance against an Al_2O_3 ball counter-body under DBCS lubrication at 37 °C despite showing higher coefficients of friction (COF).

According to Puértolas et al. [84], the a-C:H coating with a thickness of 0.7 μm represented the optimum in terms of wear resistance.

After ultrasonic cleaning in ethanol followed by drying in a desiccator as well as Ar plasma pretreatment, Cuong et al. [41] deposited of a-C:H coatings by means of RF-PECVD in a CH_4/H_2 gas mixture atmosphere. In this case, polycarbonate (PC) served as the substrate material and the authors investigated the effect of CH_4 content (25–100%) on friction and wear in steel ball-on-disk tribometer tests under ambient conditions. The coefficient of friction was higher (≥0.7) for the uncoated, non-plasma-treated and the uncoated, plasma-treated control specimens than for the specimens coated with a-C:H coating. Thereby, friction was lower (COF < 0.3) for coatings with a medium or high CH_4 content (50 and 100%) than for coatings deposited using a low CH_4 content of 25%. The authors attributed the improved frictional behavior for the high CH_4 content during deposition to the lower H_2 content in the coating, which resulted in an increased coating hardness and a transition from soft, polymer-like to more graphite-like coatings [41].

Similarly, Shi et al. [85] applied a-C:H coatings by RF-PECVD with a CH_4/H_2 gas mixture, however, they used UHMWPE as the substrate material and employed a H_2 plasma pretreatment prior to coating deposition. SEM images showed significant folds in the deposited coating, which resulted in a high mean roughness value of 0.8 μm. According to the authors, the folding in the coating was due to the interaction between high residual stresses in the coating and low Young's modulus of the substrate material. In the tribological experiments (ZrO_2 ball-on-disk, DBCS lubrication, 35°C), the high roughness of the a-C:H coatings led to an increased COF. Nevertheless, the wear rate was lower for coated UHMWPE compared to the uncoated contact pairing. Ball indentation tests showed a 42% increase in hardness (0.025 GPa) compared to uncoated UHMWPE (0.017 GPa). Contact angle measurements with distilled water also indicated improved wettability for the a-C:H coating due to the polar functional groups, which is in agreement with Dorner-Reisel et al. [83].

Fischer et al. [86] likewise deposited a-C:H coatings by RF-PECVD using C_2H_2 as a precursor after O_2 pretreatment. They fabricated flexible (f-DLC) and robust a-C:H coatings (r-DLC) on high-density polyethylene (HDPE) in two different thicknesses, namely 10 and 100 nm. Since the f-DLC and r-DLC films differed in their deposition rates, they varied the process time to allow deposition of both types of coatings with the same coating thickness. During the deposition process, the parameters of ion energy, ion flux, and angle of incidence of the ions on the surface affected the different hybridization of the carbon, i.e. the sp^2/sp^3 ratio. The r-DLC coating was formed at high ion energy and flux. The ions impinged directly on the surface, resulting in an increased probability of ion supplantation. Varying the angle of incidence enabled the deposition of f-DLC coatings, which grew mainly by diffusion processes. Consequently, the sp^3 content in the r-DLC coating turned out to be higher than that in the f-DLC coating. As demonstrated by AFM, SEM, and near-edge X-ray absorption spectroscopy, all coatings exhibited a grain-like structure, presumably including that of the low-contrast specimens with 10 nm thick r-DLC coating. Scratches as well as irregularities of the substrate material were still visible after the coating deposition, but less pronounced. In addition, the authors demonstrated the formation of an intermediate layer and good layer adhesion to the substrate material for all coatings [86]. Catena et al. [36, 87, 88] deposited a-C:H coatings (r-DLC, f-DLC) on HDPE, polyethylene terephthalate (PET), and polyoxymethylene (POM). Thereby, the roughness was barely changed by the coatings and the wetting behavior on HDPE and PET while the authors assumed this was not the case for POM due to the absence of a solid interlayer between coating and substrate in contrast to HDPE and PET. According to Catena et al. [88], an improvement in the adhesion on POM can be achieved by adjusting the plasma pretreatment and the process and process parameters.

He et al. [35] investigated the mechanical properties and tribological behavior of amorphous carbon (a-C) coatings on UHMWPE prepared by closed field unbalanced magnetron sputtering (CF-UBMS). Prior to the deposition, the substrates were pretreated with Ar plasma, with and without the deposition of a process parameter induced intermediate layer. The hardness of UHMWPE (0.047 GPa) was considerably increased by the deposited coatings (0.720 GPa) and the UHMWPE wear rate (Si_3N_4 ball-on-disk under DBCS lubrication) was also substantially reduced. In contact angle measurements, the authors reported noticeable improvement in the surface wetting for a-C coated specimens compared to uncoated, non-plasma-treated and uncoated, plasma treated UHMWPE specimens. No significant effect on the adhesion strength due to the additional deposited interlayer was observed in scratch tests. Regardless of the existence of an intermediate layer, the a-C coatings showed no cracks, spalling or delamination at the applied scratch marks. However, in Rockwell penetration tests, the coatings without interlayer displayed distinct radial cracks, which He et al. [35] attributed to a poorer adaptation of the coating to substrate deformation. In a follow-up study, Xie et al. [48] applied a two-step linear increase to the bias voltage during the five hour lasting deposition of a-C coatings on UHMWPE by means of CF-UBM sputtering a graphite target. In addition, they deposited reference coatings of the same thickness at three different constant bias voltages. A positive effect of the linearly increasing bias voltage was observed with respect to coating adhesion, showing no fractures or delamination around the Rockwell indentations. The enhanced adhesion was also confirmed by scratch testing. The authors correlated this to a good adaptation of the coating to substrate deformations. Also, an increase in the hardness by depositing a-C coatings

(2.16 GPa) on UHMWPE (0.044 GPa) was demonstrated. When coating both contacting bodies in a Si_3N_4 ball-on-UHMWPE-disk tribometer setup under dry conditions as well as under lubrication by a Hank's solution resulted in a reduction of in COF and wear rate compared the pairing of uncoated Si_3N_4 and coated UHMWPE [48].

Yurjev et al. [89] studied the influences of process parameters on the mechanical and tribological properties of a-C coatings, which were deposited on stainless steel (SS; AISI 321) by means of medium frequency (MF) dual magnetron sputtering (DMS) in mirror and closed-field magnetic field configurations (MF-MFMS and MF-CFMS, respectively). A 100 nm thick titanium layer was applied to enhance adhesion, on which a 1 μm thick a-C functional layer was deposited. The different functional layers were realized by varying the process pressure (0.1–1.6 Pa) and using the different magnetic field configurations (MFMS and CFMS). The indentation hardness (5–20 GPa) was found to increase with decreasing process pressure (0.1–0.13 Pa). The magnetic field configuration had no significant effect on the coating hardness. The COFs determined in ball-on-disk tests were in the range of 0.08–0.12 for the mirrored magnetic field configuration, independent of the process pressure. Lower COFs were obtained for the closed magnetic field configuration, which, however, increased with higher process pressure. The a-C coatings showed no relevant coating damage in scratch testing and were recommended for use in biomedical applications.

Initially, Sheeja et al. [90] investigated hydrogen-free tetrahedral amorphous carbon (ta-C) coatings on different medical substrates with respect to the dependence of coating hardness and adhesion on the applied bias voltage as well as the resulting tribological performance. The ta-C coatings were deposited by filtered cathodic vacuum arc (FCVA) on technically pure Ti, Ti6Al4V, and 316L SS. The bias voltage varied between 0 and 5.5 kV. For the coating hardness, a decrease with increasing bias voltage was observed and explained by low formation of sp^3-hypridized carbon bonds at very high negative bias voltages. The ta-C coatings exhibited a stable adhesion to 316L regardless of the bias voltage. In contrast, stable film adhesion on Ti and Ti6Al4V was observed only at negative bias voltages of 1 and 2.5 kV, respectively. Premature coating delamination on pure Ti and Ti6Al4V was attributed to high intrinsic residual stresses of the coating at low bias voltages. In a ball-on-disk tribometer, short-term screening of the ta-C coatings (each deposited at a bias voltage of 2.5 kV) against sapphire balls at ambient conditions demonstrated a COF of roughly 0.8 for all coated substrate materials. Long-term tests revealed that, after 5000 cycles on Ti6Al4V and after about 20 000 cycles on pure Ti, premature coating failure occurred. In contrast, the ta-C coating on SS maintained a constant low friction level for the entire test duration of 80 000 cycles due to the superior coating adhesion to the carbon-rich 316L due to carbon–carbon bonding.

Sheeja et al. [91] also investigated ta-C coatings deposited on UHMWPE pins and CoCrMo disks by a FCVA technique while varying the pretreatment of the polymer pins in terms of C-implantation prior to the actual coating process. In pin-on-disk tribometer tests under DBCS lubrication, the C-implantation did not significantly improve the wear behavior. Furthermore, the authors showed that coating only one of the two components did not result in any considerable improvement in the tribological behavior compared to the conventional UHMWPE/CoCrMo pairing. Although the wear resistance of UHMWPE was improved by applying a ta-C coating, it led to wear of the uncoated CoCrMo counter-body due to the high hardness of the DLC coating, resulting in the release of undesirable metal particles. Even though coating both contact partners did not positively influence friction, the ta-C/ta-C pairing led to a substantial enhancement of the wear resistance.

Hinüber et al. [92] deposited ta-C coatings at low temperatures (<150°C) by PVD on medical grade CoCrMo and Ti6Al4V disks. After an Ar high voltage direct current (DC) discharge plasma etching process, an adhesive layer of about 200 nm thickness was applied by DC cathodic arc evaporators (DC-CVA) and slowly transferred into a ta-C layer (with thicknesses between 2 and 10 μm) by laser-controlled pulsed high current cathodic vacuum arc evaporation (LCP-HCCVA). A comprehensive characterization of ta-C was performed by measuring roughness, contact angle, coating adhesion, and wear behavior. In addition, biocompatibility was evaluated by osteoblast attachment and cell viability by live/dead test. No statistical difference in wettability was observed as contact angle measurements for the ta-C coated and the uncoated CoCrMo or Ti6Al4V specimens. Using Rockwell-C penetration testing and dynamic scratch testing of ta-C coatings of different thicknesses on CoCrMo, excellent adhesion was demonstrated with adhesion strength class HF1 and up to 48 N for critical load L_{C2} in scratch tests. The ta-C coating decreased wear from 3.5×10^{-5} mm^3 Nm^{-1} for an uncoated pairing (CoCrMo/316L) to 1.1×10^{-7} mm^3 Nm^{-1} (ta-C/Cr/CoCrMo/316L) in reciprocating pin-on-disk testing. The lowest wear factor of 3.9×10^{-10} mm^3 Nm^{-1} was measured with a ta-C coated steel ball sliding against a polished and ta-C coated CoCrMo disk. Using student's *t*-test, it was found that the ta-C coating had not any statistically significant effect ($p < 0.05$) effect on osteoblast attachment compared to the uncoated specimens. Also, there appeared to be no significant difference ($p < 0.05$) in terms of cell death between ta-C coated CoCrMo and Ti6Al4V as well as uncoated specimens. As a result, these ta-C coatings exhibited enhanced wear resistance and good in vitro cytocompatibility compared to the currently used uncoated CoCrMo and Ti6Al4V implant materials.

Wang et al. [93] analyzed the tribological behavior of three different thin coatings (a-C:H, a-C and TiN/Ti) deposited on Ti6Al4V by RF-PECVD and IBAD/MS, respectively. Prior to the actual deposition of the coatings, the substrates were pretreated using carbon ion implantation to enable a good coating-substrate adhesion. The adhesion of the TiN coating was achieved by a sputtered Ti transition. The coating properties were characterized by X-ray diffraction, Raman measurements, nanohardness and nanoscratch tests, and cross-sectional SEM images. The tribological behavior of these coatings was investigated using ZrO_2 and UHMWPE ball-on-disk tribo-tests under dry and DBCS-lubricated conditions at about 37°C. The values for the coating hardness of the a-C coatings were nearly doubled and of the TiN coating quadrupled compared to the untreated specimen while exhibiting high adhesion strengths. In the DBCS-lubricated wear test with ZrO_2 balls, the TiN coating exhibited the highest wear resistance, followed by the a-C and further by the a-C:H coating. All coatings resulted in a considerable wear reduction. The failure of the coatings was mainly attributed to plowing of the hard ZrO_2 ball, which was accompanied by abrasive and adhesive wear. In the wear test with UHMWPE balls, all coatings exhibited a significant improvement in wear resistance in both dry and lubricated conditions. Both, the a-C:H and a-C coatings, showed less signs of wear than the TiN coating, which was related to the self-lubrication phenomenon during dry sliding. The TiN coating had the highest COF for all tested conditions, which was attributed to the large TiN grains and resulting higher roughness. The self-lubricating transfer film formed on the a-C:H and a-C coatings may also reduce their friction. The results indicated that the three coatings had different wear mechanisms, with the authors recommending a thick a-C:H or a-C coating to be promising for use under moderate loading conditions.

The influences of zirconium as a doping element for a-C and a-C:H coatings regarding the mechanical and tribological properties were examined by Escudeiro et al. [94]. Thereby, a-C, a-C:H, a-C:Zr, and a-C:H:Zr coatings, with various Zr contents, were deposited by DC-DMS on pure Ti and Ti6Al4V. Prior to the actual coating deposition, an Ar plasma etching process was used to clean and activate the surfaces. In order to improve the coating-substrate adhesion, an about 450 nm thick interlayer system consisting of Ti, TiN, and TiNC(Zr) coatings was used. The various Zr contents of the a-C(:H):Zr coatings were realized by a different amount of Zr pellets within the erosion zone of the used graphite target. Based on the indentation hardness measurement performed, a higher hardness as well as an increasing Young's modulus of the a-C coatings could be determined compared to the uncoated substrates. The hydrogen-free a-C or a-C:Zr coatings showed a comparable hardness of about 10 GPa regardless of the Zr content of the coating, with the Young's modulus increasing (from 94 to 133 GPa) with increasing Zr content. In contrast, an increase in hardness from about 9 to 12 GPa and Young's modulus from 70 to 114 GPa was observed with increasing Zr content of the a-C:H:Zr coatings. The roughness of the coatings was enhanced by doping with Zr, although this positive effect decreased with increasing Zr content. The scratch test adhesion was mostly independent of the Zr doping and tended to be better on Ti6Al4V than on Ti. A tribological characterization was performed by ball-on-disk tests, using Ti6Al4V as a counter-body, whereby a substantial improvement in friction and wear was observed for all coatings. The influence of the Zr doping was mainly shown for milder wear of the counter-body with increasing Zr contents. In a follow-up study, Escudeiro et al. [20] studied the tribological behavior of the same a-C coatings when sliding against UHMWPE and polyetheretherketone (PEEK). Therefore, a multi-directional pin-on-disk wear tester under DBCS lubrication at 37 °C and up to two million cycles was employed. Against UHMWPE, no significant difference of the tribological performance was found between the coated surfaces compared to uncoated Ti6Al4V and CoCr. In contrast, the frictional and wear behavior when sliding against PEEK was substantially deteriorated by Zr-containing a-C coatings, which failed prematurely. Although the incorporation of Zr into the C-matrix increased the adhesion strength and decreased the residual stress of the coating, delamination was observed after a few hundred thousand cycles. According to the authors, this pointed toward surface defects under the influence of body fluids and high shear stresses, which can ultimately lead to interfacial failure and delamination. Nevertheless, the integrity of the coating was unexpectedly preserved.

Kobayashi et al. [37] employed a RF-PECVD process and CH_4 reactive gas to fabricate a-C:H coatings on polymethyl methacrylate (PMMA) and UHMPWE and studied the influence of O_2 plasma pretreatment on the coating adhesion and wear resistance. Scratch testing revealed that the coating adhesion on PMMA as well as UHMWPE was increased by the plasma pretreatment prior to the coating process. The wear resistance in SS ball-on-disk experiments (dry conditions, 25 °C) was also shown to be enhanced.

Martínez-Nogués et al. [95] compared different deposition techniques, namely RF-MS, FCVA, and ion beam assisted evaporation (IBAD), to fabricate a-C coatings as well as PECVD for a-C:H coatings on UHMWPE substrates. The thickness of the coatings varied between 0.1 and 0.7 μm. The a-C:H coating deposited by PECVD showed the highest hardness in nanoindentation tests, which was attributed to the higher coating thickness ($t = 0.7$ μm) compared to the other coatings and the highest amount of sp^3-hybridized carbon as represented by the lowest I_D/I_G ratio in Raman spectroscopy. All specimens demonstrated a decrease of friction with increasing sliding distance in Al_2O_3 ball-on-disk tests under DBCS lubrication at 37 °C. The a-C coating deposited by FCVA featured the lowest COF, whereas the a-C coating deposited by sputtering showed the highest COF. Despite the high friction of the sputtered a-C coating, it exhibited the highest wear resistance. Neither coating delamination nor measurable coating wear was observed. According to Martínez-Nogués et al. [95], this indicated that magnetron sputtered a-C coatings have superior adhesion to the substrate and are therefore suitable for the usage as implant coating.

Rather recently, Rothammer et al. [40, 96] studied the effects of tungsten-doped hydrogen-containing amorphous carbon coatings (a-C:H:W) deposited on CoCr and Ti64 pins as well as a pure a-C:H coating on UHMWPE by means of PVD (MF-UBMS) (see Figure 7.4). The deposited coatings exhibited typical characteristics of a-C coatings in terms of morphology and composition.

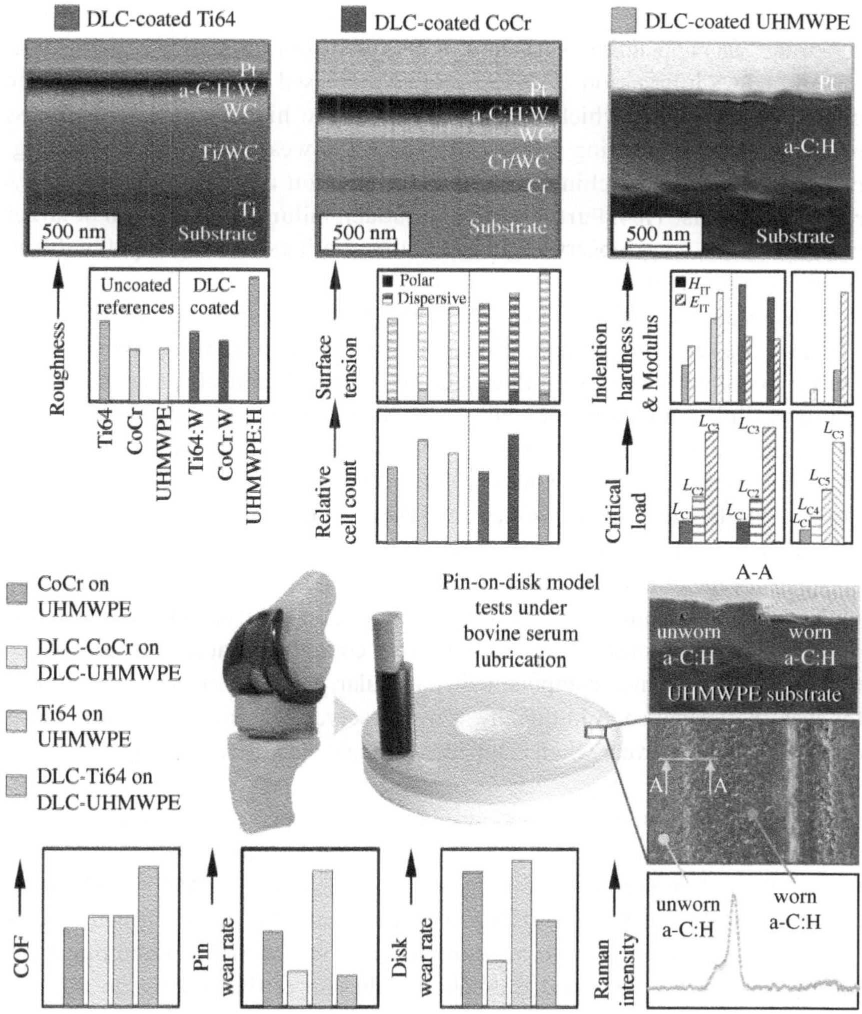

Figure 7.4 Comparison of the morphology, roughness, surface tension, relative cell count, indentation hardness, and modulus, as well as critical scratch loads for a-C:H:W coatings deposited on Ti64 and CoCr and a-C:H on UHMWPE as well as COF and wear behavior of coated compared to uncoated pairings. *Source:* Adapted from Refs. [40, 96] with permission by CC BY 4.0 (MDPI).

The coatings displayed higher roughness compared to the substrates, particularly in the case of UHMWPE. Initial evaluations with respect to contact angles and surface tensions as well as in vitro indirect and direct cell testing indicated that the DLC coatings were comparable to the substrates and did not cause any cytotoxic effects. Furthermore, the developed coatings demonstrated excellent mechanical properties, with a significant improvement of hardness-to-elasticity ratios. The adhesion of the coatings to the Ti64, CoCr, and UHMWPE substrates was considered sufficient. Subsequent pin-on-disk tribometer testing und DBCS lubrication at 37 °C revealed increased friction compared to the uncoated references, which was attributed to the higher surface roughness resulting from the coating process. However, the wear of UHMWPE was significantly reduced, reaching up to a 49% reduction against CoCr and a 77% reduction against Ti64. Furthermore, no sudden failures or spallation of larger wear particles were observed. Instead, the coatings exhibited gradual and steady signs of wear.

In a subsequent study conducted by Rothammer et al. [97], the authors examined how the selection of the coated component (metallic and/or polymeric) and the variations in load cases influenced the results. Specifically, they compared non-cross-linked and conventionally cross-linked UHMWPE under higher and lower load conditions. The coating systems studied demonstrated excellent mechanical properties, with a notable improvement in ratios of indentation hardness and elastic modulus. The adhesion of the coatings, as assessed through modified scratch tests, was deemed highly excellent for polymeric substrates and adequate for metallic substrates. Using a UHMWPE ball-on-three metal pin setup under DBCS lubrication at 37 °C, the authors again observed increased friction due to higher roughness of the coating, wear could be reduced. It was revealed that coating the polymer component, in particular, played a crucial role in improving wear performance. Additionally, it was observed that uncoated counterparts experienced higher wear when compared to single-sided coated components (see Figure 7.5).

Recently, Rothammer et al. [98] investigated the biocompatibility, mechanical properties, adhesion, and wear resistance of ta-C coatings produced through plasma-filtered pulsed laser-arc PVD on CoCr, Ti64, and UHMWPE substrates. The coatings demonstrated satisfactory cytocompatibility, as confirmed by contact angle and surface tension measurements, as well as indirect and direct cell testing. Scratch testing revealed excellent adhesion of the coatings to the substrates. Nanoindentation measurements indicated a remarkable increase in hardness, up to 13-fold for hard materials and 182-fold for soft materials. In sliding

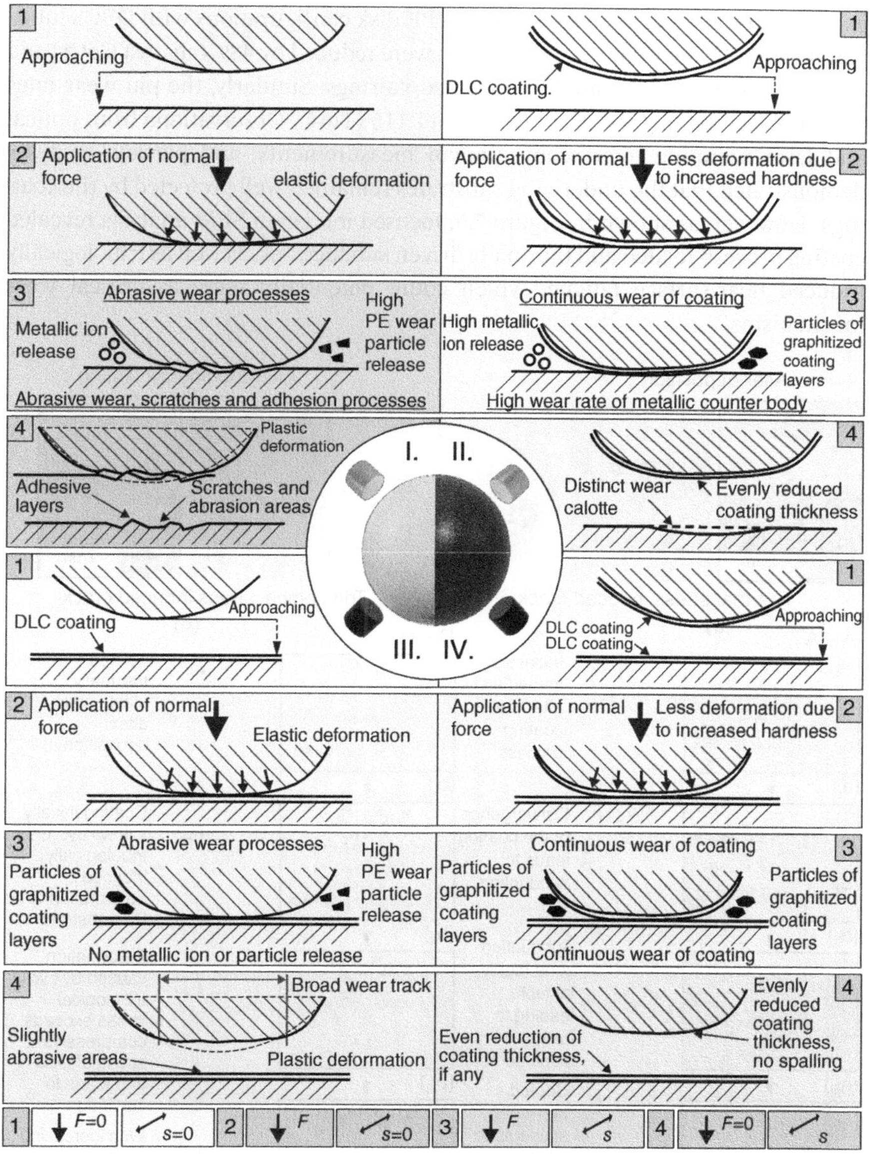

Figure 7.5 Schematics of the wear processes for the uncoated versus uncoated (I), coated versus uncoated (II), uncoated versus coated (III), and coated versus coated (IV) UHMWPE ball versus metal pin pairings. *Source:* Reproduced from [97] / with permission of ELSEVIER.

experiments using metal pin-on-UHMWPE disk configurations with DBCS lubrication at 37 °C, the wear rates of the disk were reduced by 48% (against CoCr) and 73% (against Ti64) compared to uncoated pairings. Similarly, the pin wear rates were reduced by factors of 20 (CoCr) and 116 (Ti64). Observations from optical and laser scanning microscopy, Raman measurements, and particle analyses demonstrated that the underlying substrates remained well protected by the coatings. However, as shown in Figure 7.6, focused ion beam SEM analysis revealed coating process-related and thermally driven subductions, as well as tribologically induced near-surface fatigue, which could potentially serve as critical wear mechanisms.

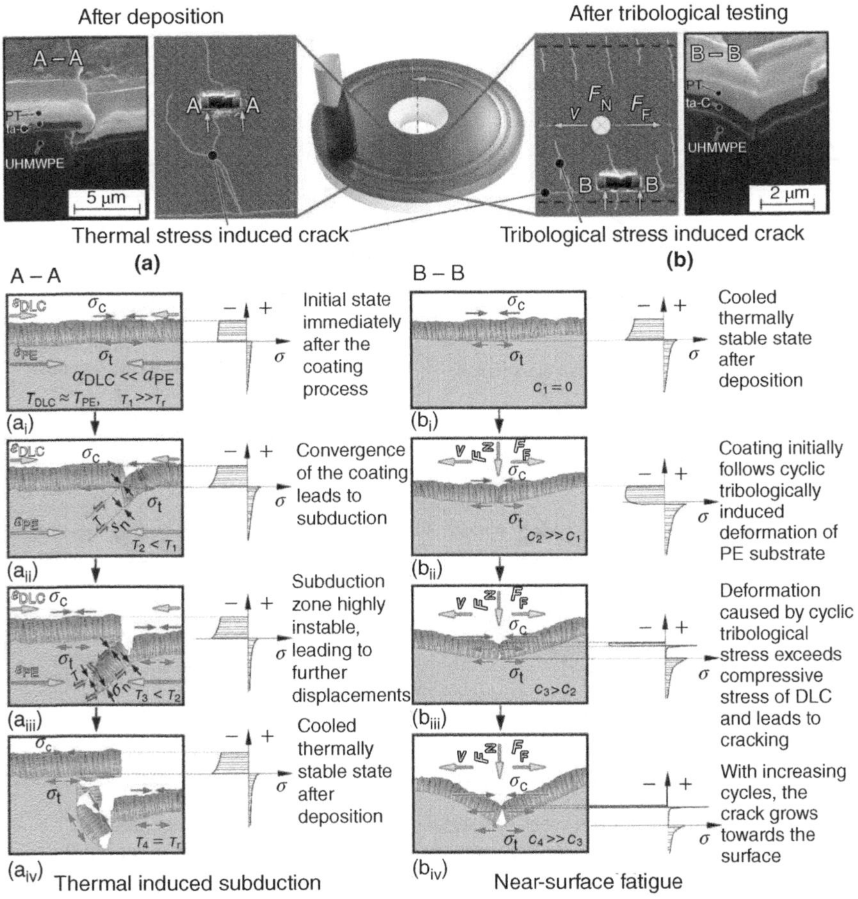

Figure 7.6 Mechanisms for (a) coating process-related thermal and intrinsic compressive stress-induced subduction and (b) tribologically-related near-surface fatigue of superhard ta-C-coated UHMWPE substrates. *Source:* [98] / John Wiley & Sons / CC BY 4.0.

7.3.2 Biotribological Studies on the Component Level

Compared to the investigations on the model level discussed before, there's a considerably smaller number of studies focusing on the actual component level. Hauert et al. [99, 100] conducted an explantation study on DLC-coated metatarsophalangeal (MTP) prosthesis after four years in vivo. Both articulating components of the two-piece toe joint consisted of DLC-coated CoCrMo alloy. They found some coating delamination, which was due to the slow interlayer crevice corrosion. Weak mechanical adhesion between substrate material and applied Si interlayer led to penetration of body fluid beneath the interlayer. This clearly demonstrates the importance interlayers for the biotribological effectiveness of a-C coatings in load-bearing implants. Taeger et al. [101] conducted a THR explant study and reported significantly higher failure rate for DLC-coated Ti6Al4V femoral heads compared to uncoated Al_2O_3 components against UHMWPE cups. The sp^3-rich DLC coatings (Adamante®, Biomecanique, France) deposited by IBAD showed a survival rate of only 54%, whereas the Al_2O_3/UHMWPE pairing was at 88% ($p < 0.001$) while there was no correlation of the results with age, sex, or weight of patients. The DLC-coated femoral heads removed in revision surgeries exhibited microscopic coating delamination in varying quantities, which resulted in high UHMWPE wear and, in most cases, aseptic implant failure. In few cases, even the metallic substrate was subjected to wear.

Saikko et al. [102] investigated the wear of THR consisting of UHMWPE cups and femoral heads made of Al_2O_3, CoCr, and a-C coated CoCr in a biaxial rocking motion simulator in DBCS at temperatures between approximately 33 and 36 °C. The 3 μm thick coating was deposited by a hybrid PVD/PECVD process (Inasmet, Spain). As an articulating surface in combination with polyethylene, the a-C coating showed quite similar biotribological performance to Al_2O_3 and CoCr. Although the coated variant (about 58 mg per 10^6 cycles) had the highest average polyethylene wear rate compared to CoCr (about 57 mg per 10^6 cycles) and Al_2O_3 (about 49 mg per 10^6 cycles), the differences between the three head materials were small. The authors concluded that the wear behavior in terms of size and shape of wear particles as well as wear rates were in a good accordance with clinical studies.

Scholes et al. [103] investigated the frictional behavior and lubrication condition of five different contact pairings for the use in THR. The components (head/cup) consisted of SS/UHMWPE, ZrO_2/UHMWPE, DLC on SS/UHMWPE, Al_2O_3/Al_2O_3, and CoCrMo/CoCrMo and were tested in a Durham hip function simulator with various artificial nonbiological and biological substitute synovial fluids as well as with real synovial fluid. The friction results showed that Al_2O_3/Al_2O_3 had the lowest COF (0.002) for the nonbiological substitute synovial fluids. Once biological lubricants were used, the friction (0.065) increased to the same level as the materials tested against UHWMPE. The CoCrMo/CoCrMo pairing exhibited the highest friction (0.21); however, a general reduction in friction (0.16) was observed with the use of biological fluids. The DLC/UHMWPE contact pairing showed inconsistent results with COFs between 0.055 and 0.150, which could be attributed to different surface topographies and defects of the coating. Lower friction was obtained for smoother and more homogeneous coatings. The authors therefore concluded that such coatings and their deposition processes need to be further developed.

Lappalainen et al. [104] deposited 40–100 μm thick amorphous diamond (AD) coatings on 316L SS using filtered pulsed plasma arc (FPPA) vapor deposition for potential application on acetabular cups of THR. By using high ionic energy, an intermediate layer was first deposited with the aim of ensuring high coating-substrate adhesion. The subsequently deposited functional layer was generated at sufficiently low ion energy to achieve a predominance of sp^3 hybridizations (80–85%) of carbon. After testing in a hip joint wear simulator according to ISO 9225 under DBCS lubrication at 23 °C for 15 million cycles, the coating remained nearly unworn with microscopic nanometer-sized cracks that could hardly be evaluated. The authors estimated the wear particle amount of the coated metal-metal contact compared to conventional metal-polymer prostheses or purely metallic prostheses to be at least a factor of 10^5–10^6 smaller.

Ortega-Saenz et al. [105] investigated the wear of THR, more specifically CoCr/CoCr, multilayer titanium nitride/chromium nitride coatings, namely $(TiN/CrN)_3$, on CoCr/CoCr, CrN on CoCr/CoCr and a-C:H on CoCr/CoCr in a hip joint wear simulator under DBCS lubrication at ca. 25°C. The nitride coatings were deposited by hybrid PVD and the a-C coating was deposited by PECVD. The thicknesses were 3.25 μm for the multilayer coating, 3.75 μm for the monolayer CrN coating and 2 μm for the a-C:H coating. The coatings almost doubled the hardness from about 8 GPa in case of a-C:H (14 GPa) and even tripled it for the nitride coatings (up to about 27 GPa). The surface of the a-C:H coating was smooth and almost without defects in contrast to the nitride coatings. However, the adhesion of the a-C:H coating to CoCr was about 5 N until the beginning of material removal, which was considerably lower than that of CrN with about 23 N or that of $(TiN/CrN)_3$ with about 13 N. The coated femoral heads did not exhibit pronounced running-in wear, and the volumetric wear for the coated specimens of <3 mm^3 was substantially reduced by roughly a factor of three. In contrast, the uncoated counterparts (cups) showed the highest volumetric wear for contact with the nitride coatings and an 85 times lower wear for pairing with the uncoated or the a-C:H coated surface. Overall, the wear of a-C:H on CoCr/CoCr was the lowest.

Choudhury et al. [106, 107] investigated the tribological performance of smooth and micro-textured (dimples) prosthetic heads coated with a-C coatings against ceramic cups in a pendulum hip joint simulator. In this context, silver-doped a-C:H and ta-C coatings with a chromium adhesion layer were deposited on SS (316L) using a hybrid RF-MS technique. The results showed a considerable reduction in friction for the ta-C/ceramic as well as the a-C:H/ceramic combination with dimples compared to a metal/ceramic pairing due to the improved material and surface properties and hydrodynamic effect. Ex situ analysis showed that the a-C:H-coated dimples caused a slight change in roughness and produced a larger amount of wear particles (equivalent diameter = 40–200 nm). According to the authors, the size of the abrasive particles could be stored in the dimples, reducing potential third body wear. Therefore, the authors considered the use of dimpled a-C:H as a "metal-on-ceramic hip interface" as feasible for further research.

Finally, Oñate et al. [29] examined TiN and a-C:H coatings, as well as nitrogen ion implantation surface treatments, as potential ways to reduce wear of UHMWPE when applied to the CoCr alloy articulating surface. The effects of nitrogen ion implantation of UHMWPE were also analyzed. To evaluate the wear performance, a modified knee wear simulator with a combined roll-slide motion was used to mimic the most severe situation in the knee. Simplified coated and uncoated CoCr and Al_2O_3 femoral components were contacted with simplified UHMWPE tibial inlays. Testing was performed at a pressure of 50 MPa in distilled water at ca. 37 °C. Wear measurements were performed using UHMWPE mass loss and laser profilometry according to ASTM F732. The results after up to five million cycles revealed positive effects of a-C:H coatings on CoCr and N^+ ion implantation on UHMWPE on reducing the UHMWPE wear. Analogous results were also obtained for Al_2O_3 ceramic material. The former combination reduced the wear by a factor of five compared to the uncoated control specimens. However, the TiN coating considerably increased the wear of the UHMWPE. The adhesion of UHMWPE to the uncoated as well as to the TiN coated counter-bodies was clearly seen in SEM. UHMWPE transfer to the a-C:H coated or the ceramic counter-bodies was observed only to a much lower extent. Micro-delamination at the worn UHMWPE surface could only be detected for uncoated CoCr and was not observed for the coated specimens. A potential combination of N^+ ion implantation and a-C:H coatings represented a favorable approach to reduce the wear of UHMWPE in CoCr total joint replacements and to replace a more expensive ceramic material. According to the authors, a-C:H passed all biocompatibility tests in accordance with FDA regulations and ISO 10993 standards for implantable devices.

7.4 Application for Dental Implants and Braces/Retainers

7.4.1 Dental Implants

Dental implants are widely used for the replacement of damaged or missing teeth (see Figure 7.7a). The employed materials have to be biocompatible, stable, and resistant against fretting and wear [108]. Since aesthetics play a major role, nowadays ceramics or (ceramic or resin) composites are mainly used for restoration purposes while polymers have low wear-resistance and metals tend to have an undesired color [108]. Similarly, DLC coatings are rather not employed due to their darker color compared to human teeth. Apart from the "visible" parts of dental implants, i.e. the crown, metallic alloys based on Ti, zirconium, or tantalum are frequently used for fixtures, screws, and abutments. These materials have good biocompatibility, but wear due to micro-movements and associated tribo-corrosion can be an major issue, potentially leading to the release toxic elements like aluminum or vanadium, inflammatory reactions of the body as well as decreased stability or even implant loosening [108]. In this case, DLC coatings can be employed and tailored via doping elements to enhance bone ingrowth/osseointegration [109–111], prevent bacteria [112] as well as especially to improve the mechanical stability as well as wear and corrosion resistance [113, 114].

Generally, the fast and strong adhesion of the dental implant to the jawbone is of utmost importance. When coatings are applied, they have to withstand the high stresses during screwing without cohesive or adhesive fracturing and spalling in order to avoid particles in the bone and biological immune responses. Waldhauser et al. [115] deposited various thin (thickness ca. 150 nm) coatings on silicon (Si) wafers to analyze the failure stresses. Thereby, despite featuring lower cohesion and adhesion strength, ta-C:H displayed a low-adhesive, low-friction behavior in dry as well as Ringer solution lubricated alumina ball-on-disk experiments compared to tantalum oxide, zirconium oxide, aluminum oxide, silicon oxide and titanium oxide coatings. Furthermore, in contrast to the oxides, no salt deposits were formed in the wear track of the DLC coating, which seemed to be crucial for low friction and wear. However, this study was limited by the fact that Si substrates were used instead of a material relevant for dental implants.

Kim et al. [116] studied the extent of abutment/fixture loosening under cyclic loading for uncoated as well as a-C:H:Si-coated commercially pure (CP) Ti screws. Thereby, before assembly, the coating was applied to the surface of a self-tapping external hexagon type fixture that contacts with a nonsegmented abutment. The obtained data clearly indicated a (on average) more than five times higher number

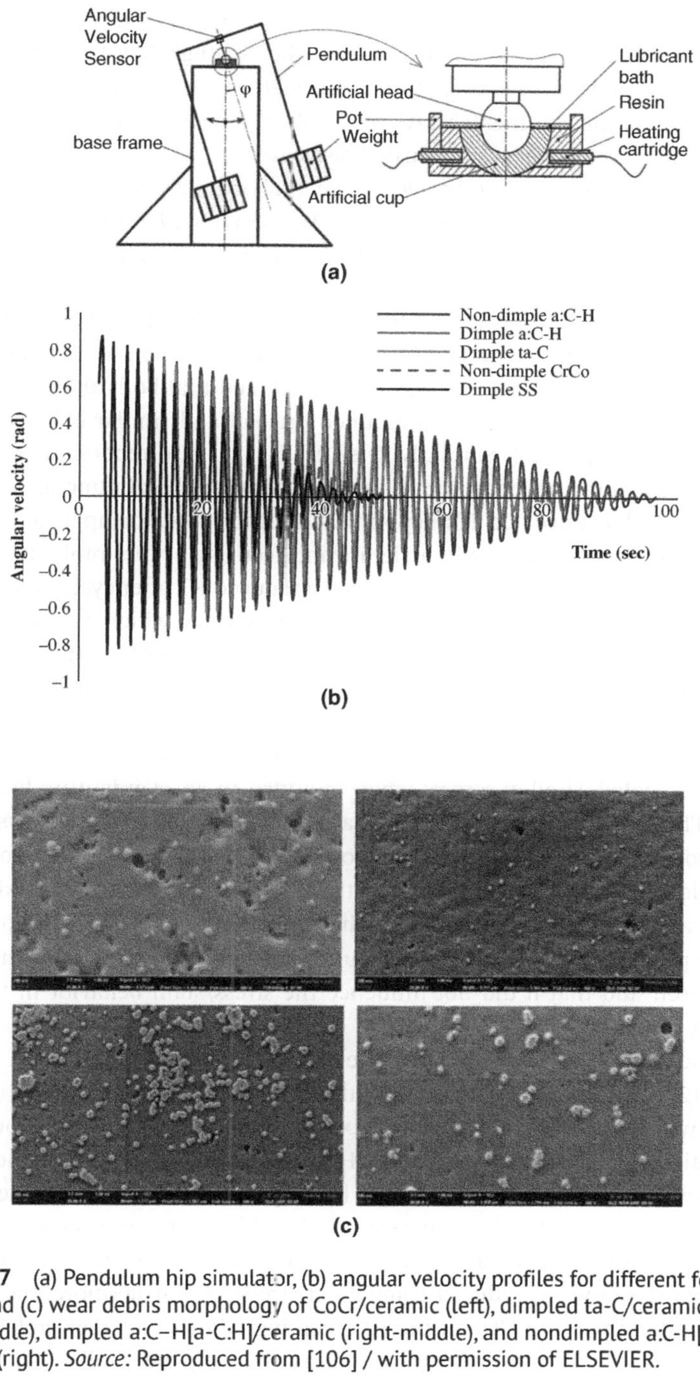

Figure 7.7 (a) Pendulum hip simulator, (b) angular velocity profiles for different femoral heads and (c) wear debris morphology of CoCr/ceramic (left), dimpled ta-C/ceramic (left-middle), dimpled a:C–H[a-C:H]/ceramic (right-middle), and nondimpled a:C-H[a-C:H]/ceramic (right). *Source:* Reproduced from [106] / with permission of ELSEVIER.

of required cycles for loosening at constant loading for the DLC coating compared to the references. The authors argued that the coating's low friction and high hardness allowed for releasing stresses through micro-slippage and prevented deformation of the abutment. Moreover, SEM imaging also showed that the coating acted as a protective layer against abrasive wear.

In contrast, some studies reported detrimental or at least no positive effects for DLC coatings. De Almeida Basílio et al. [117] compared Ti6Al4V alloy screws with a tungsten-doped DLC and uncoated screws in terms of the provided stability for zirconia (ZrO_2) ceramic abutments. They reported that the required removal torque (RT) before cyclic stressing was substantially higher for uncoated screws than for coated screws while after cyclic loading, the was no longer statistically significant difference. It was thus concluded that both abutment screws showed a reduction in the mean retained preload and similar effectiveness in maintaining the preload. Similarly, Diez et al. [118] also reported a loss of RT after cyclic loading for both, the connection between the abutments and DLC-coated as well as uncoated Ti alloy screws. Furthermore, the application of the DLC coating did not results in a significant decrease of the implant-abutment interface area. Bacchi et al. [119] observed a reduced RT after cyclic stressing for DLC-coated compared to uncoated Ti abutment screws, which was attributed to the lower friction of DLC. Da Silva-Neto et al. [120] reported that DLC-coated abutment screws did not contribute to the prevention to implant-abutment microleakage.

Dos Santos et al. [121] evaluated the axial tightening for applied to uncoated and DLC-coated Ti alloy screws when mounted to an overdenture bar framework. They observed greater preload values for the DLC-coated screws, which was also attributed to lower frictional losses during tightening. In a subsequent finite element methods analysis using the experimentally measured axial force data as an input, the authors showed that the higher preload also led to higher surface pressures on the framework, but to lower von Mises stresses inside the screw itself and that it did not influence the stress/stain behavior in the peri-implant bone tissue.

Furthermore, DLC coatings have been deposited on the abutments. Corazza et al. [122] deposited DLC coatings with a thin silicon adhesion layer on abutments, which were then mounted on hexagonal dental implants. They found that the coating did not affect the initial RT. However, the RT of pairings without coating significantly decreased after cyclic loading, which was not the case for the DLC-coated abutments. The authors stated that the coatings protected the surface and reduced the maladaptation between abutment and implant, which in turn improved stability and reduced component stressing. Furthermore, the authors assumed that the lower COF of the DLC coatings allowed a larger preload using the same torque during tightening, thus also enhancing the stability.

Hirata et al. [123] conducted a comparative study on the effects of DLC coatings when applied to abutments and/or abutment screws on the reliability and strength of implant supported single crowns In the performed mechanical bending tests under dynamic conditions and water lubrication, all screws failed irrespective of whether a DLC coating was applied or not. Thereby, no statistically significant difference in the probability of survival could be observed. The authors attributed this to the fact that failure was mainly driven by the bulk properties and not by the applied coating. However, the DLC coating on the abutment, the screw, or both, led to an increase in the Weibull modulus, i.e. a decrease in the scatter of load at which failure occurred. This is an indication for less variability and a higher predictability of the systems with the coatings. Furthermore, Hirata et al. [123] hypothesized that the benefits from the DLC coating in terms of its tribological properties might rather come into play when the abutment screw gets torqued several times, e.g. when the screw is subjected to insertion and removal processes to adapt the crown during surgery.

Apart from the usage as the outer and functional layer, DLC has also been explored as an interlayer. Using a multiple-target pulsed laser deposition process, Bell et al. [124] fabricated a gradient DLC/DLC:Ag/TiC/TiCN/TiN coating to – as verified in scratch testing – enhance the adhesion between a hydroxyapatite (HA) coating and the Ti6Al4V substrate.

7.4.2 Braces/Retainers

Finally, DLC coatings were investigated when employed to of orthodontic brackets or wires of braces/retainers (see Figure 7.8b). Thereby, nickel-titanium (Ni-Ti) alloys are frequently employed due to their superelasticity and shape

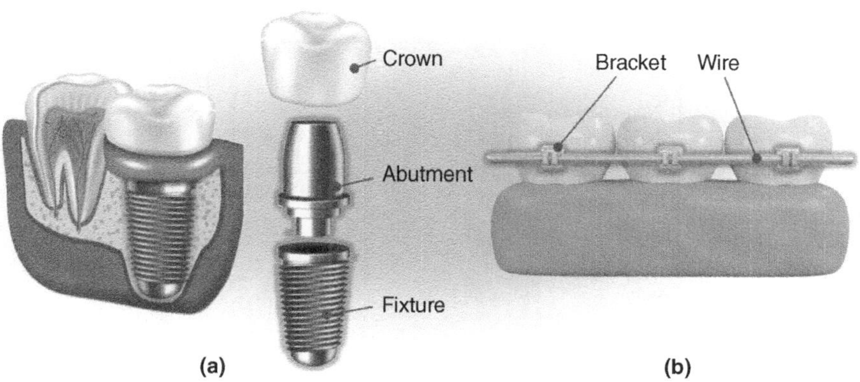

Figure 7.8 Structure of (a) dental implants and (b) braces/retainers as well as potential components for application of DLC coatings (highlighted in pink) to improve mechanical stability, durability, wear resistance, and toxic particle release. ©Marian.

memory effect. However, there is a growing prevalence allergy against nickel, which can be released due to Ni-Ti corrosion when immersed in oral saliva. Kobayashi et al. [125, 126] deposited a-C:H coatings with a SiC adhesion layer using arc discharge ion plating and verified excellent mechanical and adhesion properties in a mechanical brushing tests. Moreover, microwave induced plasma mass spectrometry after immersion into physiological saline revealed a reduction of Ni ion release by a factor of six compared to uncoated wires (see Figure 7.9). Moreover, cytotoxicity tests, the direct cell growth rate on the DLC-coated wires was shown to be higher than the reference wires. Since friction between the bracket and wire can be a relevant factor for orthodontic tooth movement, Muguruma et al. [127] compared the static friction force required for displacing various bracket/wire combinations including SS and Ni-Ti wires coated with a-C:H by plasma-based ion implantation/deposition (PBIID). It was reported that the roughness wasn't changed due to the coating, but the

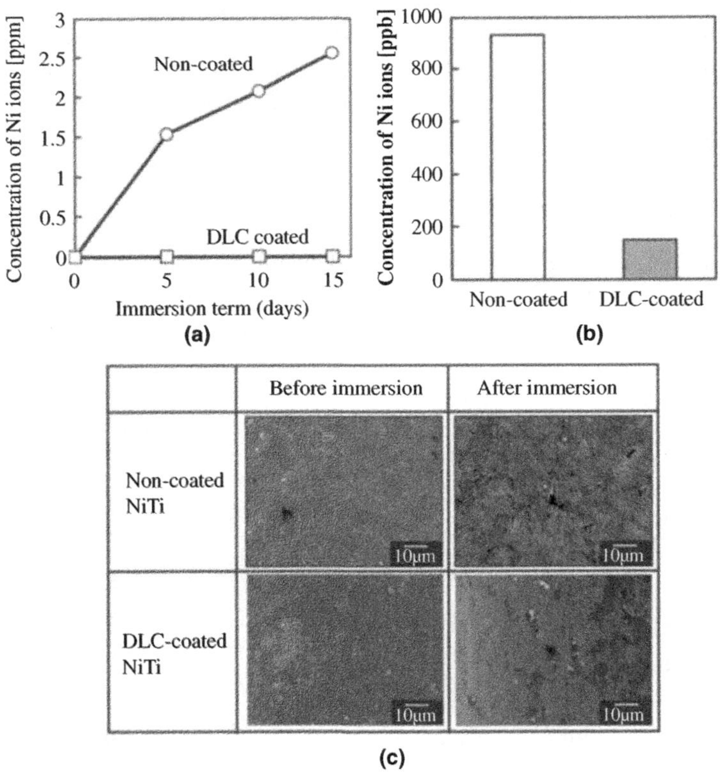

Figure 7.9 Concentrations of the released Ni ions (a) over time up to 15 days and (b) after six months as well as SEM images of before and after immersion in physiological saline for six months at 37 °C. *Source:* Reproduced from [125] / with permission of ELSEVIER.

hardness was increased and generally the friction was decreased compared to uncoated pairings. The authors thus concluded that this might enhance the efficiency of tooth movement.

Regarding SS archwires, Zhang et al. [128] showed that a nanostructured DLC coating deposited by RF-PECVD revealed no significant toxicity while allowing for an increase in hardness by a factor of 1.5, a decrease in elastic modulus, a reduction of friction by 41% and a decrease in corrosion current density by a factor of 2 compared to uncoated controls. The DLC coating thereby also outperformed wires that underwent nitrocarburization, which also led to higher hardness and lower friction, but weakened the corrosion resistance.

Kang et al. [129] deposited 200 nm thin a-C coatings with sp^2 carbon dominated bonding structures on SS archwires and brackets using a mirror-confinement-type electron cyclotron resonance (MCECR) plasma sputtering and studied the friction as well as fretting behavior under dry as well as saliva lubricated conditions. It was observed that the lowest/more stable friction and wear was achieved when the wire was coated (either against coated or uncoated brackets). However, the tribological behavior wasn't improved when only the brackets were coated. In contrast, irrespective of SS or uncoated titanium-molybdenum wires (uncoated), wire sizes, angulation, and dry or wet conditions, Akaike et al. [130] reported a reduction of static friction for a-C:H/a-C:H:Si:O-coated SS brackets. Thereby, the authors emphasized the importance of homogenous coatings in the entire slot surface of the bracket, whereby the employed direct current-pulsed (DC-pulsed) PECVD process might have been superior for components with complex shapes. In a follow-up study, Akaike et al. [131] demonstrated that doping the coating by fluorine or silicon led to an increase in hardness and further lowered the static friction (see Figure 7.10). Thereby, the fluorine-doped coating

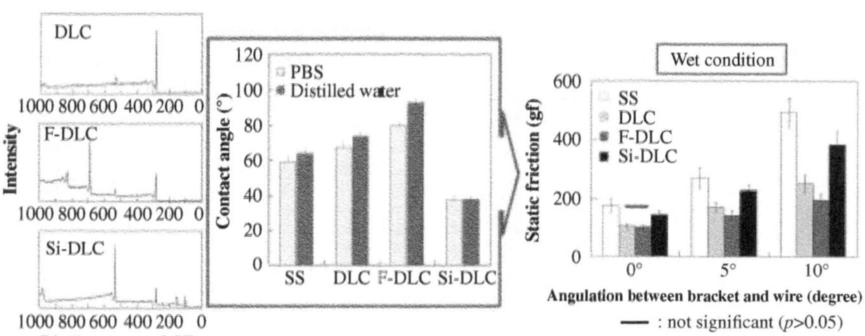

Figure 7.10 XPS spectra, contact angles as well as static friction of undoped, F- and Si-doped DLC compared to SS. *Source:* Reproduced from [131] / with permission of Elsevier.

(a-C:H:F, F-DLC) outperformed the silicon-doped (a-C:H:Si:O, Si-DLC) and undoped DLC (a-C:H) coatings, thus offering great potential for effective tooth movement and shortening orthodontic treatment time.

7.5 Concluding Remarks

Despite significant advancements in the design and materials of artificial implants, challenges persist regarding the interaction between the articulating surfaces as well as the interaction of released and worn products with the human body, necessitating further development. As discussed in Section 7.2, the utilization of DLC coatings holds promise in meeting the biotribological demands of implant surfaces. This is primarily attributed to the outstanding physical, chemical, and structural properties of DLC coatings allowing for biocompatibility and wear protection when sufficient adhesion to the substrate is ensured. In Sections 7.3 and 7.4, we demonstrated that DLC coatings have been successfully employed for load bearing as well as dental implants. The applications, studied substrates as well as conducted analyses and main findings are summarized in Table 7.2. It becomes evident that for load-bearing implants such as THR and TKR, DLC coating processes and systems have been successfully developed and studied especially on the model level, i.e. under simplified conditions, indicating the promise for wear protection while featuring no cytotoxic effects as well as good adhesion. However, there are very few studies – and the ones that are there are dating from some years back – actually reporting positive effects on the component level. Future work therefore should now aim to transfer the progress made at the model level in recent years to actual components and more application-relevant testing conditions. In contrast, investigations aiming at dental implants or braces/retainers mainly employ original components thus providing higher transferability to actual application. Yet, the published studies often contain comparatively little information on the coatings as well as the deposition parameters, so that reproducibility is strongly hampered. Partly, there is a lack of a mechanism-oriented understanding, which would be necessary to optimize the coatings and deposition processes.

Table 7.2 Summary of literature findings on DLC coatings employed for load-bearing implants, dental implants, and cardiovascular devices.

Coating with thickness t	Deposition method	Substrate material(s)	Evaluation on the model (ML) or component level (CL)	Main effect due to the DLC coating	Ref.
Load-bearing implants					
a-C:H ($t \approx 1$–2.4 μm)	ECR-PECVD	UHMWPE	Ball-on-disk (ML), scratch testing, hardness, coating structure	Increased wear resistance and hardness, increasing Ar/C_2H_2 ratio leads to higher sp^3 proportion, better mechanical properties, and wear resistance	[82]
a-C:H ($t \approx 1$–3 μm)	RF-PECVD	UHMWPE	Coating thickness, structure and adhesion, wetting behavior (ML and CL)	Improved wetting behavior and decreased contact angles by application of a-C:H; increasing Ar content positively influences the wetting behavior, process parameters did not affect coating structure, coatings exhibited cauliflower-like microstructure and clod-like coarse structure, low I_D/I_G ratio and low proportion of sp^2-hybridized carbon for all coatings, good coating-substrate adhesion	[34,83]
a-C:H ($t \approx 0.25$–0.70 μm)	RF-PECVD	UHMWPE; HXLPE	Ball-on-disk, coating thickness, hardness, structure (ML)	Improved wear performances of a-C:H coated UHMWPE and HXLPE, higher COF due to a-C:H, good adhesion and high hardness, coating thickness of 0.70 μm showed best wear resistance	[84]

(Continued)

Table 7.2 (Continued)

Coating with thickness t	Deposition method	Substrate material(s)	Evaluation on the model (ML) or component level (CL)	Main effect due to the DLC coating	Ref.
a-C:H ($t \approx 0.25$ μm)	RF-PECVD	PC	Ball-on-disk, coating structure (ML)	Decrease of friction for a-C:H ($CH_4 > 50\%$), decrease of roughness a-C:H ($CH_4 = 50\%$), no effect of H_2	[41]
a-C:H ($t \approx 6$–8 μm)	RF-PECVD	UHMWPE	Ball-on-disk, coating thickness, ball indentation, wetting behavior, structure (ML)	Folding, higher roughness, higher hardness due to coating, increase of friction and decrease of wear due to coating improve of wettability for the a-C:H coating	[85]
a-C:H ($t \approx 10$ and 100 nm)	RF-PECVD	HDPE	Coating structure, adhesion and thickness (ML)	Mainly grain-like structure, good adhesion	[86]
a-C:H ($t \approx 2$–1000 nm)	RF-PECVD	HDPE, PET, POM	Coating structure, adhesion and thickness, wetting behavior (ML)	No considerable effect on roughness due to coatings, improve of wetting behavior on HDPE and PET due to a-C:H coatings, good coating adhesion, except for POM	[36, 87, 88]
a-C ($t \approx 0.9$ μm)	CF-UBMS	UHMWPE	Ball-on-disk, coating adhesion, wetting behavior (ML)	Increase of hardness, decrease of wear rate and improve of wetting behavior by the a-C coatings, good adhesion	[35]
a-C ($t \approx 1$ μm)	CF-UBMS	UHMWPE, Si_3N_4	Ball-on-disk, coating adhesion (ML)	Increase of hardness, decrease of wear rate due to coating (gradient bias voltage performed better), decrease of wear and friction due to a-C/a-C contact	[48]

a-C/Ti ($t \approx 1/0.1$ μm)	MF-MFMS, MF-CFMS	AISI 321	Ball-on-disk, coating hardness and adhesion (ML)	No significant effect on coating hardness and adhesion due to different magnetic field configurations, increase in hardness due to decreasing process pressures, good coating adhesion and low friction	[89]
ta-C ($t \approx 0.36$ μm)	FCVA	Ti, Ti6Al4V, 316L	Ball-on-disk, coating adhesion and hardness (ML)	No considerably effect of variation in bias voltage on microstructure, mechanical and tribological coating properties, no premature coating failure when applied to 316L	[90]
ta-C ($t \approx 2$ and 6 μm)	FCVA	UHMWPE, CoCrMo	Pin-on-disk (ML)	No effect on wear due to C-implantation of UHMWPE, no improvement of wear performance due to one-sided coating compared to reference UHMWPE/CoCrMo bearing couple, increase in wear resistance of UHMWPE due ta-C coating, but increase of uncoated CoCrMo counter-body wear and particle release, increase of wear resistance due to ta-C/ta-C contact partners	[91]

(Continued)

Table 7.2 (Continued)

Coating with thickness t	Deposition method	Substrate material(s)	Evaluation on the model (ML) or component level (CL)	Main effect due to the DLC coating	Ref.
ta-C/Cr ($t \approx 2$–10/0.2 μm)	LCP-HCCVA/ DC-CVA	CoCrMo, Ti6Al4V	Pin-on-disk, coating structure, friction and wear, thickness, hardness, adhesion, wetting behavior, chemical properties and cytological Interactions (ML)	No cytotoxically effects due to coatings, increase in hardness due to coatings, excellent adhesions of coatings, decrease of friction due to the use of coatings, decrease of wear due to the use of coatings	[92]
a-C:H ($t \approx 0.46$ μm) a-C ($t \approx 0.33$ μm)	RF-PECVD	Ti6Al4V	Ball-on-disk, coating structure and topography, hardness, adhesion (ML)	Increase of hardness due to coatings, while exhibiting high adhesion strengths, considerably increase of wear resistance due to coatings and low friction due to self-lubrication of a-C:H and a-C, promising for use under moderate loading conditions, such as an artificial cervical disk	[93]
a-C:H/TiNC/TiN/Ti ($t \approx 1/0.45$ μm) a-C:H:Zr/TiNCZr/TiN/Ti ($t \approx 1/0.45$ μm) a-C/TiNC/TiN/Ti ($t \approx 1/0.45$ μm) a-C:Zr/TiNCZr/TiN/Ti ($t \approx 1/0.45$ μm)	DC-DMS	Ti, Ti6Al4V	Ball-on-disk, hardness, adhesion, roughness, coating structure (ML)	Increase of hardness due to coatings, Enhancement of the roughness of the coatings by doping with Zr, although this positive effect decreased with increasing Zr concentration, better coating adhesion to Ti6Al4V than to Ti, considerably decrease of friction and wear due to coatings, Zr doping was mainly shown in a milder wear of the counter-body	[94]

a-C:H/TiNC/TiN/Ti ($t \approx 1/0.45$ μm) a-C:H:Zr/TiNCZr/TiN/Ti ($t \approx 1/0.45$ μm) a-C/TiNC/TiN/Ti ($t \approx 1/0.45$ μm) a-C:Zr/TiNCZr/TiN/Ti ($t \approx 1/0.45$ μm)	DC-DMS	Ti6Al4V	multi-directional pin-on-disk wear testing, hardness, adhesion, roughness, coating structure (ML)	No significant difference of the tribological performance was found between the pairings investigated when sliding against UHMWPE, tribological behavior with PEEK as sliding partner was significantly negatively influenced by Zr-containing a-C coatings	[20, 94]
a-C:H ($t \approx 0.5$ μm)	RF-PECVD	UHMWPE, PMMA	Ball-on-disk, scratch testing (ML)	Increase of coating adhesion by the plasma pretreatment, increase of wear resistance by coated UHMWPE, lowest wear depth due to plasma treated and coated UHMWPE	[37]
a-C ($t \approx 0.1$–0.5 μm), a-C:H ($t \approx 0.7$ μm)	RF-MS, FCVA, IBAD, PECVD	UHMWPE	Ball-on-disk, hardness and structure (ML)	Highest hardness was exhibited by a-C:H_{PECVD} supported by lowest I_D/I_G ratio and consequently the highest amount of sp^3-hybridized carbon, decrease of the coefficient of friction with increasing sliding distance, lowest friction for the a-C_{FCVA}, whereas a-C_{MS} showed highest friction, high decrease of wear due to a-C_{MS}	[95]

(Continued)

Table 7.2 (Continued)

Coating with thickness t	Deposition method	Substrate material(s)	Evaluation on the model (ML) or component level (CL)	Main effect due to the DLC coating	Ref.
a-C:H, a-C:H:N, a-C:H:Si, a-C:H:N/a-C:H, a-C:H:N/a-C:H:Si	RF-PECVD	LDPE	Cytotoxicity tests, contact angle, hardness (ML)	Improvement in wetting behavior due to coating, decrease in contact angle and best wetting behavior due to a-C:H:N, increase in hardness due to coatings, especially due to a-C:H:N/a-C:H coating system, no negative effect on in vitro cytotoxicity due to coating	[49]
a-C:H:W/WC/ CrWC/Cr ($t \approx 0.3/0.2/0.4/0.2$ μm), a-C:H:W/WC/ TiWC/Ti ($t \approx 0.2/0.2/0.6/0.2$ μm), a-C:H ($t \approx 1.4$ μm)	Hybrid UBMS	UHMWPE, CoCrMo, Ti6Al4V	Pin-on-disk, friciton, wear and wear particles, coating structure, thickness, hardness, adhesion, wetting behavior, chemical properties and cytological Interactions (ML)	No cytotoxically effects due to coatings, increase in hardness due to coatings, sufficient to excellent adhesions of the coatings, increase of friction due to the use of coatings, decrease of wear due to the use of coatings, no delamination or spalling of the coatings during tribological testing	[40, 96]
a-C:H:SiO/a-C:H:Ti:N:SiO/ TiN/Ti ($t \approx 1.4$–1.5 μm), a-C:H/a-C:H:SiO/ a-C:H:Ti:N:SiO/ TiN/Ti ($t \approx 1.3$ μm), a-C:H/a-C:H:W/ WC/Cr ($t \approx 1.7$ μm), a-C:H ($t \approx 0.9$–1.8 μm)	Hybrid UBMS and MF-PECVD	UHMWPE, CoCrMo, Ti6Al4V	Ball-on-three-pin, friciton, wear, coating structure, thickness, hardness, adhesion (ML)	Increase in hardness due to coatings, satisfactory to excellent adhesions of the coatings, increase of friction due to the use of coatings, decrease of wear due to the use of coatings, no delamination or spalling of the coatings during tribological testing, single sided coating resulted in higher wear of the uncoated counter-part	[97]

ta-C/Cr ($t \approx 0.5$–3.6/0.05–0.5 μm)	Plasma-filtered pulsed laser-arc	UHMWPE, CoCrMo, Ti6Al4V	Pin-on-disk, friciton, wear and wear particles, coating structure, thickness, hardness, adhesion, wetting behavior, chemical properties and cytological Interactions (ML)	No cytotoxically effects due to coatings, increase in hardness due to coatings, excellent adhesions of the coatings, increase of friction due to the use of coatings, decrease of wear due to the use of coatings, but potential of near-surface fatigue	[98]
DLC ($t \approx 0.35$ μm)	No information	CoCrMo (MTP prostheses (toe joint)	Wear study of explants (CL)	Delamination of DLC due to corrosion of the interlayer	[99, 100]
DLC (Adamante®, Biomecanique, France, $t \approx 2$–3 μm)	IBAD	Ti6Al4V (femoral heads of THA)	Wear study of explants (CL)	In the THA explant study by significantly higher failure rate due to DLC-coated Ti6Al4V femoral heads compared to uncoated Al_2O_3 components, DLC-coated femoral heads exhibited microscopic coating delamination, which resulted in high UHMWPE wear and also aseptic implant failure, sometimes DLC wear led to wear of metallic substrate	[101]
DLC (Inasmet, Spain, $t \approx 3$ μm)	Hybrid PVD/PECVD	CoCr (femoral heads of THR)	Biaxial rocking motion simulator testing (CL)	Not considerably increased wear rate due to DLC contact pairing compared to CoCr and Al_2O_3. Good agreement of wear behavior with clinical studies	[102]

(Continued)

Table 7.2 (Continued)

Coating with thickness t	Deposition method	Substrate material(s)	Evaluation on the model (ML) or component level (CL)	Main effect due to the DLC coating	Ref.
DLC (no information)	No information	SS (femoral heads of THR)	Hip joint simulator testing (CL)	Inconsistent results of friction behavior for DLC/UHMWPE contact pairings due to different surface topographies and defects of the coating	[103]
AD coating ($t \approx 40$–$100\ \mu m$	FPPA	316L (acetabular cups of THR)	Hip joint simulator testing (CL)	Considerably decrease of wear due to sp^3-rich AD coatings in comparison to metal-polymer or metal-metal prostheses	[104]
a-C:H (BALINIT® DLC, Oerlikon Balzers, Liechtenstein, $t \approx 2\ \mu m$)	PECVD	CoCr (femoral heads of THR)	Hip joint simulator testing, hardness, scratch testing, coating structure (CL)	Increase in hardness, satisfactory coating adhesion, smooth and almost defect-free surfaces, Considerably decreased wear due to the use of a-C:H coatings	[105]
aC:H:Ag/Cr/316L ($t \approx 1/0.15\ \mu m$), ta-C:Ag/Cr/316L ($t \approx 0.9/0.07\ \mu m$)	Hybrid RF-MS	316L (femoral heads of THR)	Pendulum hip joint simulator testing, hardness and structure (CL)	Increase of hardness due to a-C:H and ta-C coatings, considerably decreased friction and wear due to a-C:H and ta-C coatings	[106, 107]
a-C:H ($t \approx 3$–$4\ \mu m$) and N^+ ion implantation	MS and RF-PECVD/ N^+ ion implantation	CoCr (femoral component of TKR) UHMWPE (tibial inlays of TKR)	Modified knee wear simulator (CL)	Considerably decreased wear due to the use of a-C:H coatings and N^+ ion implantation, Decrease in UHMWPE Adhesion on counter-bodies	[29]

Dental implants					
ta-C:H ($t \approx 0.15$ μm)	PECVD	Si	Ball-on-disk, scratch testing (ML)	Low friction in dry and lubricated conditions compared to other coatings, no salt deposit formation in the wear track	[115]
a-C:H:Si/Si ($t \approx 1/0.5$μm)	RF-PECVD	CP Ti (screw)	Cyclic loading (CL)	Increased resistance against loosening	[116]
DLC:W	No information	Ti6Al4V alloy (screw)	Bending strength and RT test before and after cyclic loading (CL)	Decrease of RT before dynamics stressing, no prevention of RT decrease due to cycling loading	[117]
No information	No information	Ti alloy (screw)	RT test before and after cyclic loading (CL)	Decrease of RT before dynamics stressing, no prevention of RT decrease due to cycling loading	[118]
No information	No information	Ti alloy (screw)	RT test before and after cyclic loading (CL)	Reduced RT after cyclic loading	[119]
No information	No information	Ti alloy (screw)	Axial tightening test (CL)	Higher preload due to lower friction during tightening	[121]
Undoped DLC/Si ($t \approx 3$ μm)	PECVD	Ti alloy (abutment)	Cyclic loading (CL)	No effect on initial RT, but prevention of RT decrease RT due to dynamical stressing; increased stability; protection against wear	[122]

(Continued)

Table 7.2 (Continued)

Coating with thickness t	Deposition method	Substrate material(s)	Evaluation on the model (ML) or component level (CL)	Main effect due to the DLC coating	Ref.
No information	No information	Ti alloy (screw and/or abutment)	Cyclic bending	No effect on probability of survival, but less variability	[123]
HA/DLC/DLC:Ag/TiC/TiCN/TiN ($t \approx 0.5$ μm)	DLP	Ti alloy	Scratch testing (ML)	Enhanced adhesion between HA and substrate	[124]
Braces/retainers					
a-C:H/SiC ($t \approx 1$ μm)	Ion beam plating	Ni-Ti alloy (wire)	Mechanical brushing, immersion, and cytotoxicity test (CL)	Increased wear resistance and reduced Ni ion release	[125, 126]
a-C:H ($t \approx 0.5$ μm)	PBIID	Ni-Ti alloy (wire)	Nanoindentation and wire drawing friction test (CL)	Friction reduction through	[127]
Undoped DLC ($t \approx 1$ μm)	RF-PECVD	SS (wire)	Cytotoxicity, hardness, friction, and corrosion test (CL)	No toxicity, increased hardness, reduced friction, enhanced corrosion resistance	[128]
a-C ($t \approx 0.2$ μm)	MCECR plasma sputtering	SS (wire and brackets)	Friction and fretting wear test (CL)	Reduced friction and fretting wear for coated wires; no improvement for coated brackets	[129]
a-C:H/a-C:H:Si:O, a-C:H, a-C:H:F, a-C:H:Si:O ($t \approx 1$ μm)	(DC-pulsed) PECVD	SS (brackets)	Static friction test (CL)	Reduced static friction, especially for F-doping	[130, 131]

References

1 Marian, M., Berman, D., Nečas, D. et al. (2022). Roadmap for 2D materials in biotribological/biomedical applications – a review. *Advances in Colloid and Interface Science* 307: 102747. https://doi.org/10.1016/j.cis.2022.102747.

2 Malisz, K., Świeczko-Żurek, B., and Sionkowska, A. (2023). Preparation and characterization of diamond-like carbon coatings for biomedical applications—a review. *Materials (Basel)* 16: 3420. https://doi.org/10.3390/ma16093420.

3 Grill, A. (2003). Diamond-like carbon coatings as biocompatible materials—an overview. *Diamond and Related Materials* 12: 166–170. https://doi.org/10.1016/S0925-9635(03)00018-9.

4 Hauert, R. and Müller, U. (2003). An overview on tailored tribological and biological behavior of diamond -like carbon. *Diamond and Related Materials* 12: 171–177. https://doi.org/10.1016/S0925-9635(03)00019-0.

5 Tiainen, V. (2001). Amorphous carbon as a bio-mechanical coating—mechanical properties and biological applications. *Diamond and Related Materials* 10: 153–160. https://doi.org/10.1016/S0925-9635(00)00462-3.

6 Love, C.A., Cook, R.B., Harvey, T.J. et al. (2013). Diamond like carbon coatings for potential application in biological implants—a review. *Tribology International* 63: 141–150. https://doi.org/10.1016/j.triboint.2012.09.006.

7 Stoica, A., Manakhov, A., Polčák, J. et al. (2015). Cell proliferation on modified DLC thin films prepared by plasma enhanced chemical vapor deposition. *Biointerphases* 10: 29520. https://doi.org/10.1116/1.4920978.

8 Birkett, M., Zia, A.W., Devarajan, D.K. et al. (2023). Multi-functional bioactive silver - and copper -doped diamond -like carbon coatings for medical implants. *Acta Biomaterialia* 167: 54–68. https://doi.org/10.1016/j.actbio.2023.06.037.

9 Shah, R., Gashi, B., Hoque, S. et al. (2021). Enhancing mechanical and biomedical properties of protheses – surface and material design. *Surfaces and Interfaces* 27: 101498. https://doi.org/10.1016/j.surfin.2021.101498.

10 Griesser, H.J. (ed.) (2016). *Thin Film Coatings for Biomaterials and Biomedical Applications*. Duxford, UK: Woodhead Publishing.

11 Dorner-Reisel, A., Schürer, C., Müller, E. (2004). The wear resistance of diamond -like carbon coated and uncoated $Co_{28}Cr_6Mo$ knee prostheses. *Diamond and Related Materials* 13: 823–827. https://doi.org/10.1016/j.diamond.2003.11.080.

12 Boyan, B.D., Hummert, T.W., Dean, D.D. et al. (1996). Role of material surfaces in regulating bone and cartilage cell response. *Biomaterials* 17: 137–146. https://doi.org/10.1016/0142-9612(96)85758-9.

13 Repenning, D. (2006). Beschichtungen auf implantaten. In: *Ossäre Integration*, 53–61. Berlin, Heidelberg: Springer.

14 Fontaine, J., Donnet, C., and Erdemir, A. (2008). Fundamentals of the tribology of DLC coatings In: *Tribology of Diamond-like Carbon Films* (eds. C. Donnet and A. Erdemir), 139–154. Boston, MA: Springer.

15 Weikert, T., Wartzack, S., Baloglu, M.V. et al. (2021). Evaluation of the surface fatigue behavior of amorphous carbon coatings through cyclic nanoindentation. *Surface and Coatings Technology* 407: 126769. https://doi.org/10.1016/j.surfcoat.2020.126769.

16 Bai, W.Q., Cai, J.B., Wang, X.L. et al. (2014). Mechanical and tribological properties of a-C/a-C:Ti multilayer films with various bilayer periods. International Conference on Metallurgical Coatings, San Diego, 1989. *Thin Solid Films* 558: 176–183. https://doi.org/10.1016/j.tsf.2014.03.039.

17 Dorner, A., Schürer, C., Reisel, G. et al. (2001). Diamond-like carbon-coated Ti6Al4V: Influence of the coating thickness on the structure and the abrasive wear resistance. *Wear* 249: 489–497. https://doi.org/10.1016/S0043-1648(01)00587-7.

18 Carbon Films – Fundamentals, Film Types, and Properties (2023). VDI Standard 2840, Engl. Düsseldorf, Germany: VDI-Gesellschaft Produktion und Logistik.

19 Bai, W.Q., Li, L.L., Wang, X.L. et al. (2015). Effects of Ti content on microstructure, mechanical and tribological properties of Ti-doped amorphous carbon multilayer films. *Surface and Coatings Technology* 266: 70–78. https://doi.org/10.1016/j.surfcoat.2015.02.024.

20 Escudeiro, A., Wimmer, M.A., Polcar, T. et al. (2015). Tribological behavior of uncoated and DLC-coated CoCr and Ti-alloys in contact with UHMWPE and PEEK counterbodies. *Tribology International* 89: 97–104. https://doi.org/10.1016/j.triboint.2015.02.002.

21 Bai, W.Q., Li, L.L., Li, R.L. et al. (2017). Deposition and characterization of a ZrN/Zr/a-C multilayer : implication on bio-tribological and corrosion behaviors. *Surface and Coatings Technology* 324: 509–517. https://doi.org/10.1016/j.surfcoat.2017.05.058.

22 Cai, J.B., Wang, X.L., Bai, W.Q. et al. (2013). Bias-graded deposition and tribological properties of Ti-contained a-C gradient composite film on Ti6Al4V alloy. *Applied Surface Science* 279: 450–457. https://doi.org/10.1016/j.apsusc.2013.04.136.

23 Dorner-Reisel, A., Schürer, C., Klemm, V. et al. (2003). Nano- and microstructure of diamond -like carbon films modified by Ca–O incorporation. *Diamond and Related Materials* 12: 1030–1033. https://doi.org/10.1016/S0925-9635(02)00293-5.

24 Cai, J.B., Wang, X.L., Bai, W.Q. et al. (2013). Microstructure, mechanical and tribological properties of a-C/a-C:Ti nanomultilayer film. *Surface and Coatings Technology* 232: 403–411. https://doi.org/10.1016/j.surfcoat.2013.05.042.

25 Bujak, J. and Michalczewski, R. (2011). Characterization and properties of low-friction , multilayered Cr-doped diamond -like carbon coatings prepared by pulse biased filtered cathodic arc deposition. *Proceedings of the IMechE* 225: 875–882. https://doi.org/10.1177/1350650111411380.

26 Rothammer, B., Weikert, T., Tremmel, S. et al. (2019). Tribological behavior of amorphous carbon coatings on metals for total knee arthroplasty. *Tribologie und Schmierungstechnik* 66: 15–24. https://doi.org/10.30419/TUS-2019-0002.

27 Döring, J., Crackau, M., Nestler, C. et al. (2019). Characteristics of different cathodic arc deposition coatings on CoCrMo for biomedical applications. *Journal of the Mechanical Behavior of Biomedical Materials* 97: 212–221. https://doi.org/10.1016/j.jmbbm.2019.04.026.

28 Avelar-Batista, J.C., Spain, E., Fuentes, G.G. et al. (2006). Triode plasma nitriding and PVD coating: a successful pre-treatment combination to improve the wear resistance of DLC coatings on Ti6Al4V alloy. *Surface and Coatings Technology* 201: 4335–4340. https://doi.org/10.1016/j.surfcoat.2006.08.070.

29 Oñate, J.I., Comin, M., Braceras, I. et al. (2001). Wear reduction effect on ultra-high-molecular-weight polyethylene by application of hard coatings and ion implantation on cobalt chromium alloy, as measured in a knee wear simulation machine. *Surface and Coatings Technology* 142–144: 1056–1062. https://doi.org/10.1016/S0257-8972(01)01074-X.

30 Strondl, C., Carvalho, N.M., de Hosson, J. et al. (2005). Influence of energetic ion bombardment on W-C:H coatings deposited with W and WC targets. *Surface and Coatings Technology* 200: 1142–1146. https://doi.org/10.1016/j.surfcoat.2005.02.182.

31 Bai, W.Q., Wang, X.L., Gu, C.D. et al. (2015). Influence of duty cycle on microstructure, tribological and corrosion behaviors of a-C/a-C:Ti multilayer films. International Conference on Metallurgical Coatings, San Diego, 1989. *Thin Solid Films* 584: 214–221. https://doi.org/10.1016/j.tsf.2015.01.015.

32 Voevodin, A.A. and Zabinski, J.S. (1998). Superhard, functionally gradient, nanolayered and nanocomposite diamond -like carbon coatings for wear protection. *Diamond and Related Materials* 7: 463–467. https://doi.org/10.1016/S0925-9635(97)00214-8.

33 Poliakov, V., Siqueira, C.M., Veiga, W. et al. (2004). Physical and tribological properties of hard amorphous DLC films deposited on different substrates. *Diamond and Related Materials* 13: 1511–1515. https://doi.org/10.1016/j.diamond.2003.11.031.

34 Dorner-Reisel, A., Gärtner, G., Reisel, G. et al. (2008). Diamond-like carbon films for polyethylene femoral parts: Raman and FT-IR spectroscopy before and after incubation in simulated body liquid. *Analytical and Bioanalytical Chemistry* 390: 1487–1493. https://doi.org/10.1007/s00216-007-1744-7.

35 He, F.F., Bai, W.Q., Li, L.L. et al. (2016). Enhancement of adhesion by a transition layer: deposition of a-C film on ultrahigh molecular weight polyethylene (UHMWPE) by magnetron sputtering. *Applied Surface Science* 364: 280–287. https://doi.org/10.1016/j.apsusc.2015.12.150.

36 Catena, A., Agnello, S., Rösken, L.M. et al. (2016). Characteristics of industrially manufactured amorphous hydrogenated carbon (a-C:H) depositions on high-density polyethylene. *Carbon* 96: 661–671. https://doi.org/10.1016/j.carbon.2015.09.101.

37 Kobayashi, S., Ozeki, K., Hirakuri, K.K. et al. (2006). Improvements of wear properties DLC coated UHMWPE and PMMA for biomaterial. *Key Engineering Materials* 309–311: 1289–1292. https://doi.org/10.4028/www.scientific.net/KEM.309-311.1289.

38 Thanu, D.P., Srinadhu, E.S., Zhao, M. et al. (2019). Fundamentals and applications of plasma cleaning. In: *Developments in Surface Contamination and Cleaning: Applications of Cleaning Techniques*, 289–353. Elsevier.

39 Lake, M. (ed.) (2016). Beschichtungstechnik. In: *Oberflächentechnik in der Kunststoffverarbeitung*, 61–181. München: Carl Hanser Verlag GmbH & Co. KG.

40 Rothammer, B., Neusser, K., Marian, M. et al. (2021). Amorphous carbon coatings for total knee replacements-part I: deposition, cytocompatibility, chemical and mechanical properties. *Polymers* 13: 1952. https://doi.org/10.3390/polym13121952.

41 Cuong, N.K., Tahara, M., Yamauchi, N. et al. (2003). Diamond-like carbon films deposited on polymers by plasma-enhanced chemical vapor deposition. *Surface and Coatings Technology* 174–175: 1024–1028. https://doi.org/10.1016/S0257-8972(03)00360-8.

42 Geier, S., Schmitz, H., Göschel, U. et al. (2012). Synthetische kunststoffe. In: Kunststoffe (eds. H. Domininghaus, P. Elsner, P. Eyerer et al.), 115–1201. Berlin, Heidelberg: Springer Berlin Heidelberg.

43 Carley, S.M. (2017). Diamond-like carbon coatings for improved mechanical performance of polymer materials. PhD thesis, University of Leeds.

44 Sauer, C., Rothammer, B., Pottin, N. et al. (2022). Design of amorphous carbon coatings using Gaussian processes and advanced data visualization. *Lubricants* 10: 22. https://doi.org/10.3390/lubricants10020022.

45 Wang, Y., Ye, Y., Li, H. et al. (2011). A magnetron sputtering technique to prepare a-C:H films: effect of substrate bias. *Applied Surface Science* 257: 1990–1995. https://doi.org/10.1016/j.apsusc.2010.09.040.

46 Trakhtenberg, I., Bakunin, O., Korneyev, I. et al. (2000). Substrate surface temperature as a decisive parameter for diamond -like carbon film adhesion to polyethylene substrates. *Diamond and Related Materials* 9: 711–714. https://doi.org/10.1016/S0925-9635(99)00254-X.

47 Teixeira, V. (2001). Mechanical integrity in PVD coatings due to the presence of residual stresses. International Conference on Metallurgical Coatings, San Diego, 1989. *Thin Solid Films* 392: 276–281. https://doi.org/10.1016/S0040-6090(01)01043-4.

48 Xie, Y.J., Bai, W.Q., Li, L.L. et al. (2017). Adhesion and tribological properties of gradient designed a-C film on ultrahigh molecular weight polyethylene. International Conference on Metallurgical Coatings, San Diego, 1989. *Thin Solid Films* 624: 7–15. https://doi.org/10.1016/j.tsf.2016.12.054.

49 Kyzioł, K., Oczkowska, J., Kottfer, D. et al. (2018). Physicochemical and biological activity analysis of low-density polyethylene substrate modified by multi-layer coatings based on DLC structures, obtained using RF CVD method. *Coatings* 8: 135. https://doi.org/10.3390/coatings8040135.

50 Hauert, R. (2008). DLC films in biomedical applications. In: *Tribology of Diamond-Like Carbon Films* (eds. C. Donnet and A. Erdemir), 494–509. Boston, MA: Springer.

51 Thorwarth, K., Thorwarth, G., Figi, R. et al. (2014). On interlayer stability and high-cycle simulator performance of diamond -like carbon layers for articulating joint replacements. *International Journal of Molecular Sciences* 15: 10527–10540. https://doi.org/10.3390/ijms150610527.

52 Schmidt, D.R., Waldeck, H., and Kao, W.J. (2009). Protein adsorption to biomaterials. In: *Biological Interactions on Materials Surfaces* (eds. D.A. Puleo and R. Bizios), 1–18. New York, NY: Springer US.

53 Kretzer, J.P., Reinders, J., Sonntag, R. et al. (2014). Wear in total knee arthroplasty—just a question of polyethylene? Metal ion release in total knee arthroplasty. *International Orthopaedics* 38: 335–340. https://doi.org/10.1007/s00264-013-2162-4.

54 Hauert, R., Thorwarth, K., and Thorwarth, G. (2013). An overview on diamond -like carbon coatings in medical applications. *Surface and Coatings Technology* 233: 119–130. https://doi.org/10.1016/j.surfcoat.2013.04.015.

55 Allen, M., Myer, B., and Rushton, N. (2001). In vitro and in vivo investigations into the biocompatibility of diamond -like carbon (DLC) coatings for orthopedic applications. *Journal of Biomedical Materials Research* 58: 319–328. https://doi.org/10.1002/1097-4636(2001)58: 3<319::AID-JBM1024>3.0.CO;2-F.

56 Thomson, L.A., Law, F.C., Rushton, N. et al. (1991). Biocompatibility of diamond -like carbon coating. *Biomaterials* 12: 37–40. https://doi.org/10.1016/0142-9612(91)90129-X.

57 Allen, M., Law, F., and Rushton, N. (1994). The effects of diamond -like carbon coatings on macrophages, fibroblasts and osteoblast-like cells in vitro. *Clinical Materials* 17: 1–10. https://doi.org/10.1016/0267-6605(94)90041-8.

58 Bruinink, A., Schroeder, A., Francz, G. et al. (2005). In vitro studies on the effect of delaminated a-C:H film fragments on bone marrow cell cultures.

Biomaterials 26: 3487–3494. https://doi.org/10.1016/j.biomaterials.2004.09.030.

59 Seidlitz, C. and Kip, M. (2017). Einführung in das indikationsgebiet und verfahren. In: *Weißbuch Gelenkersatz: Versorgungssituation bei endoprothetischen Hüft- und Knieoperationen in Deutschland* (eds. H.-H. Bleß and M. Kip), 1–15. Berlin, Heidelberg: Springer.

60 Flören, M. and Reichel, H. (2011). Implantate. In: *AE-Manual der Endoprothetik* (ed. D.C. Wirtz), 57–71. Heidelberg, Dordrecht, London, New York, NY: Springer.

61 Nečas, D., Marian, M., and Sawae, Y. (2021). Lubrication of hip and knee joint replacements: the contribution of experiments and numerical modeling. In: *Biotribology: Emerging Technologies and Applications* (eds. T. Rao, S.B. Kasolang, X. Guoxin et al.), 33–61. Boca Raton, FL: CRC Press.

62 Hauert, R., Falub, C.V., Thorwarth, G. et al. (2012). Retrospective lifetime estimation of failed and explanted diamond -like carbon coated hip joint balls. *Acta Biomaterialia* 8: 3170–3176. https://doi.org/10.1016/j.actbio.2012.04.016.

63 Cook, R., Davidson, P., and Martin, R. (2019). More than 80% of total knee replacements can last for 25 years. *BMJ: British Medical Journal* 367: l5680. https://doi.org/10.1136/bmj.l5680.

64 Evans, J.T., Walker, R.W., Evans, J.P. et al. (2019). How long does a knee replacement last? A systematic review and meta-analysis of case series and national registry reports with more than 15 years of follow-up. *The Lancet* 393: 655–663. https://doi.org/10.1016/S0140-6736(18)32531-5.

65 Lombardi, A.V., Berend, K.R., and Adams, J.B. (2008). Patient-specific approach in total knee arthroplasty. *Orthopedics* 31: 927–930. https://doi.org/10.3928/01477447-20080901-21.

66 Sundfeldt, M., Carlsson, L.V., Johansson, C.B. et al. (2006). Aseptic loosening, not only a question of wear: a review of different theories. *Acta Orthopaedica* 77: 177–197. https://doi.org/10.1080/17453670610045902.

67 Liu, A., Jennings, L.M., Ingham, E. et al. (2015). Tribology studies of the natural knee using an animal model in a new whole joint natural knee simulator. *Journal of Biomechanics* 48: 3004–3011. https://doi.org/10.1016/j.jbiomech.2015.07.043.

68 Robertsson, O., Dahle, A., Lidgren, L. Sundberg, M., (2020). *Swedish Knee Arthroplasty Register*, Annual Report2020. Lund University, Sweden.

69 Kretzer, J.P., Krachler, M., Reinders, J. et al. (2010). Determination of low wear rates in metal-on-metal hip joint replacements based on ultra trace element analysis in simulator studies. *Tribology Letters* 37: 23–29. https://doi.org/10.1007/s11249-009-9486-7.

70 Gallo, J., Goodman, S.B., Konttinen, Y.T. et al. (2013). Osteolysis around total knee arthroplasty: a review of pathogenetic mechanisms. *Acta Biomaterialia* 9: 8046–8058. https://doi.org/10.1016/j.actbio.2013.05.005.

71 Miron, R.J. and Bosshardt, D.D. (2016). OsteoMacs: Key players around bone biomaterials. *Biomaterials* 82: 1–19. https://doi.org/10.1016/j.biomaterials.2015.12.017.

72 Wang, L., Zhou, S., and Pu, J. (2015). Functionally graded and multilayered carbon-based coatings: the microstructure and tribological performances. In: *Multifunctional Materials for Tribological Applications* (ed. R.J.K. Wood), 133–172. Jenny Stanford Publishing.

73 McGeough, J.A. (2013). *The Engineering of Human Joint Replacements*. Chichester, UK: John Wiley & Sons Ltd.

74 International Standard ISO ISO 14243-1: 2009-11 (2009). Implants for Surgery – Wear of Total Knee-joint Prostheses – Part 1: Loading and Displacement Parameters for Wear-testing Machines with Load Control and Corresponding Environmental Conditions for Test. Geneva, Switzerland: International Organization for Standardization.

75 Reinders, J., Sonntag, R., Vot, L. et al. (2015). Wear testing of moderate activities of daily living using in vivo measured knee joint loading. *PLoS ONE* 10: e0123155. https://doi.org/10.1371/journal.pone.0123155.

76 Schröder, C., Reinders, J., Zietz, C. et al. (2013). Characterization of polyethylene wear particle: the impact of methodology. *Acta Biomaterialia* 9: 9485–9491. https://doi.org/10.1016/j.actbio.2013.07.039.

77 Kretzer, J.P., Jakubowitz, E., Sonntag, R. et al. (2010). Effect of joint laxity on polyethylene wear in total knee replacement. *Journal of Biomechanics* 43: 1092–1096. https://doi.org/10.1016/j.jbiomech.2009.12.016.

78 Reinders, J., Sonntag, R., and Kretzer, J.P. (2015). Synovial fluid replication in knee wear testing: an investigation of the fluid volume. *Journal of Orthopaedic Research* 33: 92–97. https://doi.org/10.1002/jor.22736.

79 Czichos, H. and Habig, K.-H. (2015). *Tribologie-Handbuch*. Wiesbaden: Springer Fachmedien Wiesbaden.

80 Kretzer, J.P., Zietz, C., Schröder, C. et al. (2012). Grundlagen zur tribologischen analyse von endoprothesen. *Orthopade* 41: 844–852. https://doi.org/10.1007/s00132-012-1948-1.

81 Richter, B.I., Ostermeier, S., Turger, A. et al. (2010). A rolling-gliding wear simulator for the investigation of tribological material pairings for application in total knee arthroplasty. *BioMedical Engineering OnLine* 9: 24. https://doi.org/10.1186/1475-925X-9-24.

82 Xie, D., Liu, H., Deng, X. et al. (2009). Deposition of a-C:H films on UHMWPE substrate and its wear-resistance. *Applied Surface Science* 256: 284–288. https://doi.org/10.1016/j.apsusc.2009.08.017.

83 Reisel, G. and Dorner-Reisel, A. (2007). Hydrogen containing DLC coatings on UHMW-PE deposited by r.f.-PECVD. *Diamond and Related Materials* 16: 1370–1373. https://doi.org/10.1016/j.diamond.2006.11.095.

84 Puértolas, J.A., Martínez-Nogués, V., Martínez-Morlanes, M.J. et al. (2010). Improved wear performance of ultra high molecular weight polyethylene coated with hydrogenated diamond like carbon. *Wear* 269: 458–465. https://doi.org/10.1016/j.wear.2010.04.033.

85 Shi, X., Wang, Q., Xu, L. et al. (2009). Hydrogenated diamond -like carbon film deposited on UHMWPE by RF-PECVD. *Applied Surface Science* 255: 8246–8251. https://doi.org/10.1016/j.apsusc.2009.05.053.

86 Fischer, C.B., Rohrbeck, M., Wehner, S. et al. (2013). Interlayer formation of diamond -like carbon coatings on industrial polyethylene: thickness dependent surface characterization by SEM, AFM and NEXAFS. *Applied Surface Science* 271: 381–389. https://doi.org/10.1016/j.apsusc.2013.01.210.

87 Catena, A., Guo, Q., Kunze, M.R. et al. (2016). Morphological and chemical evolution of gradually deposited diamond-like carbon films on polyethylene terephthalate: from subplantation processes to structural reorganization by intrinsic stress release phenomena. *ACS Applied Materials & Interfaces* 8: 10636–10646. https://doi.org/10.1021/acsami.6b02113.

88 Catena, A., Kunze, M.R., Agnello, S. et al. (2016). Amorphous hydrogenated carbon (a-C:H) depositions on polyoxymethylene: substrate influence on the characteristics of the developing coatings. *Surface and Coatings Technology* 307: 658–665. https://doi.org/10.1016/j.surfcoat.2016.09.064.

89 Yurjev, Y.N., Kiseleva, D.V., Zaitcev, D.A. et al. (2016). Antifriction coatings based on a-C for biomedicine applications. *Journal of Physics: Conference Series* 669: 12066. https://doi.org/10.1088/1742-6596/669/1/012066.

90 Sheeja, D., Tay, B., Shi, X. et al. (2001). Mechanical and tribological characterization of diamond -like carbon coatings on orthopedic materials. *Diamond and Related Materials* 10: 1043–1048. https://doi.org/10.1016/S0925-9635(00)00413-1.

91 Sheeja, D., Tay, B.K., and Nung, L.N. (2005). Tribological characterization of surface modified UHMWPE against DLC-coated Co–Cr–Mo. *Surface and Coatings Technology* 190: 231–237. https://doi.org/10.1016/j.surfcoat.2004.02.051.

92 Hinüber, C., Kleemann, C., Friederichs, R.J. et al. (2010). Biocompatibility and mechanical properties of diamond -like coatings on cobalt-chromium-molybdenum steel and titanium-aluminum-vanadium biomedical alloys. *Journal of Biomedical Materials Research Part A* 95: 388–400. https://doi.org/10.1002/jbm.a.32851.

93 Wang, S., Liao, Z., Liu, Y. et al. (2015). The tribological behaviors of three films coated on biomedical titanium alloy by chemical vapor deposition. *Journal of Materials Engineering and Performance* 24: 4462–4474. https://doi.org/10.1007/s11665-015-1732-6.

94 Escudeiro, A., Polcar, T., and Cavaleiro, A. (2013). a-C(:H) and a-C(:H)_Zr coatings deposited on biomedical Ti-based substrates: tribological properties.

International Conference on Metallurgical Coatings, San Diego, 1989. *Thin Solid Films* 538: 89–96. https://doi.org/10.1016/j.tsf.2012.12.086.

95 Martínez-Nogués, V., Medel, F.J., Mariscal, M.D. et al. (2010). Tribological performance of DLC coatings on UHMWPE. *Journal of Physics: Conference Series* 252: 12006. https://doi.org/10.1088/1742-6596/252/1/012006.

96 Rothammer, B., Marian, M., Neusser, K. et al. (2021). Amorphous carbon coatings for total knee replacements – part II: tribological behavior. *Polymers* 13: 1–25. https://doi.org/10.3390/polym13111880.

97 Rothammer, B., Neusser, K., Bartz, M. et al. (2023). Evaluation of the wear-resistance of DLC-coated hard-on-soft pairings for biomedical applications. *Wear* 523: 204728. https://doi.org/10.1016/j.wear.2023.204728.

98 Rothammer, B., Schwendner, M., Bartz, M. et al. (2023). Wear mechanism of superhard tetrahedral amorphous carbon (ta-C) coatings for biomedical applications. *Advanced Materials Interfaces* 10: 2202370. https://doi.org/10.1002/admi.202202370.

99 Hauert, R., Thorwarth, G., Müller, U. et al. (2012). Analysis of the in-vivo failure of the adhesive interlayer for a DLC coated articulating metatarsophalangeal joint. *Diamond and Related Materials* 25: 34–39. https://doi.org/10.1016/j.diamond.2012.02.001.

100 Joyce, T.J. (2007). Examination of failed ex vivo metal-on-metal metatarsophalangeal prosthesis and comparison with theoretically determined lubrication regimes. *Wear* 263: 1050–1054. https://doi.org/10.1016/j.wear.2006.11.045.

101 Taeger, G., Podleska, L.E., Schmidt, B. et al. (2003). Comparison of diamond-like-carbon and alumina-oxide articulating with polyethylene in total hip arthroplasty. *Materialwissenschaft und Werkstofftechnik* 34: 1094–1100. https://doi.org/10.1002/mawe.200300717.

102 Saikko, V., Ahlroos, T., Calonius, O. et al. (2001). Wear simulation of total hip prostheses with polyethylene against CoCr, alumina and diamond -like carbon. *Biomaterials* 22: 1507–1514. https://doi.org/10.1016/s0142-9612(00)00306-9.

103 Scholes, S.C., Unsworth, A., and Goldsmith, A.A. (2000). A frictional study of total hip joint replacements. *Physics in Medicine & Biology* 45: 3721–3735. https://doi.org/10.1088/0031-9155/45/12/315.

104 Lappalainen, R., Selenius, M., Anttila, A. et al. (2003). Reduction of wear in total hip replacement prostheses by amorphous diamond coatings. *Journal of Biomedical Materials Research Part B: Applied Biomaterials* 66: 410–413. https://doi.org/10.1002/jbm.b.10026.

105 Ortega-Saenz, J.A., Alvarez-Vera, M., and Hernandez-Rodriguez, M. (2013). Biotribological study of multilayer coated metal-on-metal hip prostheses in a hip joint simulator. *Wear* 301: 234–242. https://doi.org/10.1016/j.wear.2013.01.024.

106 Choudhury, D., Urban, F., Vrbka, M. et al. (2015). A novel tribological study on DLC-coated micro-dimpled orthopedics implant interface. *Journal of the Mechanical Behavior of Biomedical Materials* 45: 121–131. https://doi.org/10.1016/j.jmbbm.2014.11.028.

107 Choudhury, D., Ay Ching, H., Mamat, A.B. et al. (2015). Fabrication and characterization of DLC coated microdimples on hip prosthesis heads. *Journal of Biomedical Materials Research Part B: Applied Biomaterials* 103: 1002–1012. https://doi.org/10.1002/jbm.b.33274.

108 Zhang, X., Zhang, Y., Jin, Z. (2022). A review of the bio-tribology of medical devices. *Friction* 10: 4–30. https://doi.org/10.1007/s40544-021-0512-6.

109 Mine, Y., Nakatani, T., Okamoto, K. et al. (2014). Impact of biomimetic diamond-like carbon coated titanium on osteoblast and osteoclast differentiation in vitro. *Journal of Photopolymer Science and Technology* 27: 373–378. https://doi.org/10.2494/photopolymer.27.373.

110 Wachesk, C.C., Trava-Airoldi, V.J., Da-Silva, N.S. et al. (2016). The influence of titanium dioxide on diamond-like carbon biocompatibility for dental applications. *Journal of Nanomaterials* 2016: 1–7. https://doi.org/10.1155/2016/8194516.

111 Filova, E., Vandrovcova, M., Jelinek, M. et al. (2017). Adhesion and differentiation of Saos-2 osteoblast-like cells on chromium-doped diamond -like carbon coatings. *Journal of Materials Science: Materials in Medicine* 28: 17. https://doi.org/10.1007/s10856-016-5830-2.

112 Velioglu, N., Akova, T., and Ozkomur, A. (2016). Effects of hard thin-film coatings on adhesion of early colonizer bacteria over titanium surfaces. *Implant Dentistry* 25: 114–121. https://doi.org/10.1097/ID.0000000000000355.

113 Roy, R.K. and Lee, K.-R. (2007). Biomedical applications of diamond -like carbon coatings: a review. *Journal of Biomedical Materials Research Part B: Applied Biomaterials* 83: 72–84. https://doi.org/10.1002/jbm.b.30768.

114 Nakatani, T., Shiraishi, K., and Kunitsugu, S. (2015). Development of new hydrogen-containing tetrahydral amorphous carbon thin films using cathodic arc plasma for dental implant. *Journal of Photopolymer Science and Technology* 28: 471–474. https://doi.org/10.2494/photopolymer.28.471.

115 Waldhauser, W., Lackner, J.M., Kot, M. et al. (2015). Dry and ringer solution lubricated tribology of thin osseoconductive metal oxides and diamond-like carbon films. *Archives of Metallurgy and Materials* 60: 2139–2144. https://doi.org/10.1515/amm-2015-0359.

116 Kim, S.K., Lee, J.B., Koak, J.Y. et al. (2005). An abutment screw loosening study of a diamond like carbon-coated CP titanium implant. *Journal of oral rehabilitation* 32: 346–350. https://doi.org/10.1111/j.1365-2842.2004.01475.x.

117 Basilio, M.D.A., Butignon, L.E., and Filho, J. (2012). Effectiveness of screw surface coating on the stability of zirconia abutments after cyclic loading. *International Journal of Oral and Maxillofacial Implants* 27: 1061–1067.

118 Diez, J., Brigagão, V.C., Cunha, L.G. et al. (2012). Influence of diamond-like carbon-coated screws on the implant-abutment interface. *International Journal of Oral and Maxillofacial Implants* 27: 1055–1060.

119 Bacchi, A., Regalin, A., Bhering, C.L.B. et al. (2015). Loosening torque of universal abutment screws after cyclic loading: influence of tightening technique and screw coating. *Journal of Advanced Prosthodontics* 7: 375–379. https://doi.org/10.4047/jap.2015.7.5.375.

120 Da Silva-Neto, J.P., Prudente, M.S., Dantas, T.S. et al. (2017). Microleakage at different implant-abutment connections under unloaded and loaded conditions. *Implant Dentistry* 26: 388–392. https://doi.org/10.1097/ID.0000000000000568.

121 Fernandes dos Santos, M.B., Bacchi, A., Consani, R.L.X. et al. (2015). Influence of diamond-like carbon coating of screws on axial tightening force and stress distribution on overdenture bar frameworks with different fit levels and materials. *International Journal of Oral & Maxillofacial Implants* 30: 1019–1027. https://doi.org/10.11607/jomi.3730.

122 Corazza, P.H., de Moura Silva, A., Cavalcanti Queiroz, J.R. et al. (2014). Carbon film coating of abutment surfaces: effect on the abutment screw removal torque. *Implant Dentistry* 23: 434–438. https://doi.org/10.1097/ID.0000000000000111.

123 Hirata, R., Machado, L.S. Bonfante, E.A. et al. (2015). Effect of diamond-like carbon coating on reliability of implant-supported crowns. *International Journal of Oral & Maxillofacial Implants* 30: 806–813. https://doi.org/10.11607/jomi.3750.

124 Bell, B., Scholvin, D., Jin, C. et al. (2006). Pulsed laser deposition of hydroxyapatite-diamond-like carbon multilayer films and their adhesion aspects. *Journal of Adhesion Science and Technology* 20: 221–231. https://doi.org/10.1163/156856106775897748.

125 Kobayashi, S., Ohgoe, Y., Ozeki, K. et al. (2005). Diamond-like carbon coatings on orthodontic archwires. *Diamond and Related Materials* 14: 1094–1097. https://doi.org/10.1016/j.diamond .2004.11.036.

126 Kobayashi, S., Ohgoe, Y., Ozeki, K. et al. (2007). Dissolution effect and cytotoxicity of diamond -like carbon coatings on orthodontic archwires. *Journal of Materials Science: Materials in Medicine* 18: 2263–2268. https://doi.org/10.1007/s10856-007-3118-2.

127 Muguruma, T., Iijima, M., Brantley, W.A. et al. (2011). Effects of a diamond -like carbon coating on the frictional properties of orthodontic wires. *Angle Orthodontist* 81: 141–148. https://doi.org/10.2319/052110-276.1.

128 Zhang, H., Guo, S., Wang, D. et al. (2016). Effects of nanostructured, diamond-like, carbon coating and nitrocarburizing on the frictional properties and biocompatibility of orthodontic stainless steel wires. *Angle Orthodontist* 86: 782–788. https://doi.org/10.2319/090715-602.1.

129 Kang, T., Huang, S.-Y., Huang, J.-J. et al. (2015). The effects of diamond-like carbon films on fretting wear behavior of orthodontic archwire-bracket contacts. *Journal of Nanoscience and Nanotechnology* 15: 4641–4647. https://doi.org/10.1166/jnn.2015.9788.

130 Akaike, S., Hayakawa, T., Kobayashi, D. et al. (2015). Reduction in static friction by deposition of a homogeneous diamond -like carbon (DLC) coating on orthodontic brackets. *Dental Materials Journal* 34: 888–895. https://doi.org/10.4012/dmj.2015-130.

131 Akaike, S., Kobayashi, D., Aono, Y. et al. (2016). Relationship between static friction and surface wettability of orthodontic brackets coated with diamond -like carbon (DLC), fluorine - or silicone-doped DLC coatings. *Diamond and Related Materials* 61: 109–114. https://doi.org/10.1016/j.diamond.2015.11.016.

8

Diamond-like Carbon Coatings for Orthopedic Implants and Surgical Tools

Witold Kaczorowski, Krzysztof Jastrzębski, Piotr Niedzielski

Division of Biomedical Engineering and Functional Materials, Institute of Material Science and Engineering, Lodz University of Technology, Lodz, Poland

CONTENTS

8.1 Introduction, 245
8.2 Orthopedic Substrates and DLC Coating Techniques, 246
8.3 Carbon Coatings as Part of Multilayer Systems or as a Base for Doping, 248
8.4 Properties of DLC Coatings and Their Use in Orthopedics and Surgical Instruments, 249
8.5 Examples of Applications of Carbon Coatings in Orthopedics and on Surgical Instruments, 261
8.6 Conclusion, 263

8.1 Introduction

Diamond-like carbon (DLC) coatings are one of the well-studied and vastly used types of surface modifications used for biomedical applications. Many years of exploration of that topic by various research centers worldwide have shown their potential in such demanding applications as orthopedics or surgical instruments, where substrates in the form of metallic materials are leading the way. Those coatings can be produced using a wide spectrum of techniques, including pulsed laser deposition (PLD) and plasma immersion ion implantation and deposition (PIIID). Nowadays, among the most prominent ways of their manufacturing are plasma techniques based on assisted chemical vapor deposition (CVD) or physical vapor deposition (PVD), which are most commonly described for industrial applications. It is important to note that the choice of manufacturing technique can determine the chemical structure and morphology of the coating produced, influencing the properties of the final product. Today, for biomedical

Applications of Diamond-like Carbon Coatings, First Edition. Abdul Wasy Zia.

applications, carbon coatings are used as composite coatings based on one or multiple interlayers used to improve their adherence to the substrate material [1–3]. In addition, in order to achieve bioactivity suitable for the intended application, DLC coatings are very often doped with additional elements to improve a range of their properties from mechanical to biological, e.g. thrombogenicity [4, 5] antibacterial [6, 7], or osteointegration [8, 9]. DLC coatings have a number of advantages that positively alter the properties of implants or surgical instrumentation. Many years of research into improving the biological properties of implants with carbon coatings, extremely important and critical for this type of application, confirm their high biocompatibility, bioinertness, or resistance to microbial colonization. Considering the metallic substrates used in orthopedics, especially stainless steel, it seems extremely important to protect these surfaces from releasing metal ions to surrounding tissues and prevent their accumulation in the patient's body. Details on the effectiveness of such barrier properties, as well as other features of DLC coatings, will be described later in this chapter. The mechanical behavior of carbon coatings produced on implants or orthopedic instruments is also important. Without adequate functional features like high adhesion to substrates, hardness, and tribological properties, the use of carbon coatings, for example, in the case of the modification of intramedullary nails or hip endoprosthesis components, would be impossible. The application of the carbon coating completely changes the composition and chemical structure of the implant surface, its chemical resistance (affecting corrosion resistance), wettability, refractive index, or electrical properties. All of those mentioned earlier can be controlled and further tuned through the adjustment of synthesis parameters of the manufacturing processes or interprocess treatments (e.g. premodification of substrates using plasma-chemical, electrochemical or laser treatment, generation of interlayers, doping of coatings, etc.).

8.2 Orthopedic Substrates and DLC Coating Techniques

The manufacturing of carbon coatings for orthopedics and surgical instruments affects the metallic substrates commonly used in this area: medical steels, titanium alloys, or CoCr alloys. It is worth noticing that this includes implants fabricated using a variety of techniques, also gaining popularity in recent years with a wide range of 3D printing techniques [10–13]. Today, the most commonly used alloys for implants in orthopedics include AISI 316LVM steel (characterized by a lower carbon content than typical 316 steels and increased corrosion resistance), cobalt-based, and titanium-based alloys, mainly with the addition of vanadium (Ti6Al4V) and niobium (Ti6Al7Nb) [14, 15]. From the point of view of the body's reaction after implantation, these materials belong to different groups: steel and cobalt-based alloys are materials that are biotolerated by the body, and titanium alloys are considered biologically inert. Regardless of the abovementioned group, all those materials have to exert good corrosion resistance in the environment of physiological fluids while maintaining high mechanical properties. These properties, however, must be viewed comprehensively, i.e. concerning the manufacturing technology used. Casting, forming (for example, forging, rolling, etc.), machining, 3D printing, or finishing (grinding, electrochemical or mechanical polishing, plasma modification) can significantly alter both the corrosive and mechanical or biological properties, depending on the parameters used and the manner in which they are performed. It should be remembered that also the way of using the element, manufactured from those materials, or even the region of the body where it is implanted can all affect the expectations placed on them. Neurosurgery, orthopedics, maxillofacial surgery, cardiac surgery, and other branches of medicine may require slightly different properties from implants due to specific characteristics of their work environment. A high degree of smoothness or roughness, resistance to contact with blood, saliva, or bone, and tendency, or its lack, to integrate with surrounding tissues are just a few examples of the contrasting properties that the same biomaterial can exert. In addition to the modification of metallic substrates for orthopedic purposes, researchers are also examining other groups of materials like polymers. DLC coatings are also useful in such applications, and for example, they are commonly used to modify substrates made of polyether ether ketone (PEEK) [16–19]. Among the techniques used for the modification of substrates for orthopedic implants or surgical instruments, the most common are PVD and CVD technologies [20, 21]. On one hand, by means of PVD processes, the production of nanocomposite materials based on multilayer or gradient systems can be easily controlled. Such techniques also give the additional possibility of simply introducing doping elements into the carbon matrix, for example, via magnetron sputtering, which is most commonly used. On the other hand, the use of CVD technology may allow similar functionalities of DLC coatings to be obtained without the problems of directionality of the processes carried out. It appears, however, that via a synergistic approach – achieved by combining PVD and CVD techniques in a single device – boosting process and limiting the disadvantages of each method can be achieved. Obviously, it is not the only solution. It is also possible to produce composite coatings on separate equipment, using two different coating manufacturing techniques [22] or by combining thermochemical (e.g. nitriding, oxidation, or carburising), electrochemical, or laser processing techniques with the manufacture of coatings [23–25]. Other techniques for producing DLC coatings for orthopedics include but are not limited to PLD [26, 27] and PIIID [28].

8.3 Carbon Coatings as Part of Multilayer Systems or as a Base for Doping

DLC coatings for orthopedic applications are nowadays often proposed as a material that allows excellent tribological properties, especially when combined with appropriately proposed interlayers between the substrate and the coating. Those are most often encountered in various variants of structural solutions for endoprostheses, particularly of the hip joint [1, 2, 29, 30]. The interlayer between the carbon coating and the substrate guarantees improvement of adhesion. In this case, internal stresses affect the durability of the substrate-coating connection [31–34]. Two approaches are used to reduce them in the coating: interlayers (single-layer [35], multilayer [36], and gradient [37–39] systems) or doping. Among the investigated solutions, we can find titanium interlayers [40], chromium carbide (CrC) [41], or multilayer systems based on Cr-Cr2N systems [3], Ti/TiN/TiCN [42], or Zr:ZrN [2]. The use of interlayers also has a beneficial effect on other properties of the modified substrates, improving not only their wear resistance but also their corrosion resistance or enabling the creation of a barrier blocking the release of substrate ions [38]. According to Grabarczyk [23], good results can also be obtained by modifying metallic substrates previously subjected to thermochemical treatment with carbon coatings. In this study, the best results were demonstrated (nanohardness 21.5 ± 1.0 GPa; the open circuit potential [OCP] values +0.04 V; low coefficient of friction (COF) about 0.08, for titanium alloy Ti6Al4V premodified using oxidation processes with a fabricated DLC coating by radio frequency plasma-assisted chemical vapor deposition (RFPACVD).

Doping of DLC coatings is realized so as to change the properties of such films (for example, improvement of tribological properties) or to tailor their properties toward new applications (shift in surface free energy, change of biological response in contact with examined material, modification of optical or electrical properties). In the literature, one can find reports on the use of at least 28 different admixtures of DLC coatings [43–52]. The studies focus not only on the introduction of a single element to the carbon matrix but also on the benefits of double or multi-doping phenomena [53, 54]. The choice of the type of doping element is commonly in firm connection with the deposition method selected for coating deposition. The limitations are usually the physical properties of the element in question: state of matter at given conditions, temperatures of phase transition, reactivity, flammability or toxicity, etc. Doping carbon coatings with CVD techniques requires the use of precursors whose vapors are relatively easy to obtain (mostly gaseous or liquids). In their structure, they should contain only the elements that are components of the coating, i.e. the admixture element, carbon, hydrogen, and oxygen – this means the use of organic precursors, e.g.

tetramethylsilane (TMS) [55, 56] and hexamethyldisiloxane (HDMSO) [57, 58] in the case of silicon precursors, isopropoxide (TIPOT) [59, 60], allowing the introduction of titanium or ferrocene [61], which is a precursor of iron. In the case of PVD methods, doping seems simpler since the element of interest should be in a solid state. Such a target made as a bulk powder of pelleted material turns into vapor and/or ionized gas in a manner analogous to that used to create a carbon matrix out of pyrolytic carbon or graphite [62, 63]. The application of dopants in favor of carbon coatings allows to tailor further the properties of such films. It is extremely important, especially from the perspective of the medical industry and implant manufacturers that are trying to fit the properties to the requirements of very divergent surroundings of various systems of the human body.

8.4 Properties of DLC Coatings and Their Use in Orthopedics and Surgical Instruments

Carbon coatings have several excellent properties that make them find applications in orthopedics and as coatings on surgical instruments, and importantly, they can be controlled by changing the parameters of the manufacturing processes. As the first of these, the thickness of DLC coatings is worth discussing. This property is closely related to the technology used, particularly the process parameters and the substrate used for modification. Some orthopedic applications use thin carbon coatings, usually not exceeding 100–150 nm, but in applications such as hip endoprosthesis modifications or orthopedic instrument modifications, they can be much larger, exceeding even 1 μm, so that they can guarantee the longest possible use. By varying the thickness of the coating, different results will be obtained in tribological or corrosion tests. It is also important to note that as the thickness of the coatings changes, their stress state changes. The thicker they are, the higher the values of the residual stresses increase until the critical values are exceeded, resulting in self-modeling of the coating, as described by Mosaner [64]. Thick carbon coatings, on the other hand, are possible to produce, especially with a doping process or by using interlayers between the modified substrate and the carbon coating. Both methods mentioned earlier are well known for reducing the stress state of DLC coatings [65]. Of course, the values of the stress state of DLC coatings are strongly influenced by their chemical structure and, more specifically, by the ratio of C–C sp^3 and C=C sp^2 bonds. The more sp^3 phase there is in the coatings, the higher their stress state will be [66]. When comparing the results of CVD and PVD carbon coatings, much higher (sometimes by an order of magnitude) values can be observed for CVD coatings. As a result, it is easy to understand why the thickness values of coatings produced by PVD processes often exceed 1 μm, a level that is unattainable for CVD processes. As the achievable thickness of DLC coatings is closely related to the substrate used for the modification, it seems most reliable to test it directly on the modified substrate (implant) using cross-grinding and cross-sectional measurements [67, 68] or calotes techniques [69]. Observations of carbon coating grinds or fractures also provide the opportunity to determine their typical morphology, which may depend on the technology used to produce them. In the case of CVD processes, they are amorphous in nature, whereas when PVD technology is used, their morphology can change to adopt, for example, an amorphous to columnar structure. For thin DLC coatings (even below 100 nm), other methods and testing techniques are used. The simplest of these is to place an obscuration (e.g. a fragment of a silicon wafer) on the test surface prior to the modification process so that once the DLC coating is formed, it can be easily removed and the thickness obtained can be determined by profilometry. With suitably defined substrates, ellipsometry techniques can also be used for this purpose [70, 71]. Unfortunately, measurements of the thickness of fabricated carbon coatings are often carried out on reference substrates (polished silicon wafers), but the results obtained in this way do not allow knowledge of the changes in the tested property on the target modified material. The reason for this is that the nucleation and growth of the carbon coating are linked to the material of the substrate and the way it is prepared. Another often overlooked, thickness-dependent property of carbon coatings is their color [72]. The color of the coating we see is related to the phenomenon of interference between reflected light from the surface of the DLC coatings and the surface of the substrate and the refractive index of the coating [72, 73]. In general, for amorphous carbon coatings, we can assume that the color of DLC films has a strong periodic oscillation with the film thickness increasing [74]. Above a certain thickness (depending on the technology used and the properties obtained), carbon films acquire a black color. Among all the features of the coatings in question, especially for the modification of the surface of metallic implants, the possibility for them to prevent metal ions from entering the body is of great importance. In the case of stainless steels used in medicine, we would necessarily like to reduce the amounts of chromium and nickel ions released, the elements most often responsible for allergic reactions [75], and in the case of Ti6Al4V titanium alloys, for example, the amount of vanadium ions, the presence of which can induce adverse hematological, biochemical, and toxic reactions in the body [76]. According to a few reports in the literature, the DLC coating can provide a barrier whose effectiveness is very high [77–80]. In many studies, we see a significant reduction in the amount of ions released from the metallic surfaces under test, although sometimes it is impossible to eliminate their presence completely [25, 26]. This effect is most often attributed to a lack of coverage of all surfaces of the test item or a lack of homogeneity of coverage (e.g. presence of cracks in the coating). The presence of metal ions found in tests using solutions taken after corrosion processes is also unsurprising. It is worth mentioning, however, that the presence of limited amounts of elemental ions in studies of substrates that have been protected with DLC coatings may be related to the way in which plasma modification processes are carried out, particularly using plasma and CVD systems. To produce a carbon coating with satisfactory properties (especially mechanical properties, e.g. adhesion), a surface pretreatment process (ion or plasma etching) is necessary, if only to clean or raise the temperature of the modified implant. During this process, ions of the modified material (or holder) enter the plasma and, under favorable conditions, can build up into the carbon coating being fabricated. On the other hand, too high a temperature of the DLC coating fabrication process can lead to the formation of specific metal whiskers, which can form channels up to the substrate [81]. Therefore, the observed low levels of metal ions entering the test solutions in tests with metallic samples coated with DLC coatings may be the result of spontaneous doping of the coatings by plasma processes or the resulting leaks in the DLC coatings. The sheer amount of ions released from the surface can vary over time, which is closely related to the composition of the solution and the phenomena occurring on the surface of the material under test [27]. As shown by Gutensohn [26], the amount of metal ion release from the surface of 316L steel without and with DLC coatings in contact with a plasma-based solution decreases with time, which is due to the adhesion of proteins to it and its coating with other blood morphotic elements. For the study of released ions, prefabricated implants or samples made from them immersed in a chosen solution (e.g. NaCl salt solution or Hank's) are used, from which samples are taken at a predetermined time for studies carried out using various techniques, e.g. inductively coupled plasma-mass spectrometry (ICP-MS), atomic adsorption spectrophotometry AAS, or graphite furnace atomic adsorption spectrophotometry GFAAS [78, 79, 82].

Barrier tests of DLC coatings are also confirmed in biological tests checking for a possible allergic reaction to the implant (or the sample prepared from it) without and with the carbon coating, especially in people suffering from metallosis. Patch tests conducted on the skin of the backs can provide evidence of adequate protection of the implant surface with the carbon coating. For example, Mitura presents the results of patch tests on a patient with severe metallosis using elements typical medical steel contains, i.e. Cr and Ni, and the results for carbon powders obtained from RFPACVD processes [83]. Unfortunately, due to the test model used and the brief contact of the implants or powders with the patient's body, these tests do not give us complete knowledge of the processes of ion release from the tested surfaces over time. Another important aspect for which carbon coatings find application in orthopedics is their mechanical properties. Here, the most important properties will be those that will give the materials, after modification with DLC coatings, the appropriate functional characteristics, such as adequate adhesion to the substrate, hardness, wear resistance, etc. One of the most important mechanical properties without which carbon coatings could not be used in orthopedics is their adhesion to implant surfaces. This property of DLC coatings can be controlled in various ways. It depends, among other things, on the plasma's energy parameters, coating thickness, initial surface modification procedures used, and the use or not of interlayer/interlayers or doping [67]. This property is also closely related to the DLC coating manufacturing technology. For example, the processes for producing carbon coatings on metallic substrates of AISI 316L steel implants using RFPACVD technology consist of two very important processes for adhesion: surface etching and coating production, carried out one after the other in a single reactor chamber. In the first of these, we decide on the development of the modified surface but also, and more importantly, on the formation of an interlayer right at the surface, made up of carbide-forming elements present in the modified substrate [81]. This procedure results from the fact that by varying the electrode potential value and the duration of this stage, we control the temperature of the surface and core of the implants, and therefore also the carbon diffusion processes and the formation of carbides, which translates into a change in the thickness and hardness distribution over the cross-section of the resulting interlayer. Ultimately, the resulting properties of the formed interlayer influence the adhesion of the carbon coatings produced in the second stage. However, the adhesion of the carbon coatings will also be related to the technological parameters of the DLC coating manufacturing processes. By changing the electrode potential, the composition of the gas mixture supplied to the process, or its duration, we decide on the final adhesion value of DLC coatings to metallic substrates, as well as their other final properties, by changing their chemical structure [84]. Also, the use of an interlayer reduces the mismatch between the coating and the substrate, thereby reducing stress and improving adhesion. Silicon and titanium are the elements that are commonly used in such adhesive interconnections, but the overall effect is the result of not only the interlayer itself but also deposition parameters. Several research programs on the manufacturing of silicon interlayers conducted by Cemin [85] show that the adhesion of carbon coatings to AISI 4140 steel depends on the temperature of deposition. Si-C bonds occurred above 573 K, which strengthened the interlayer-film connection. When considering the deposition time and thus the thickness of the interlayer, it is possible to determine the optimal parameters that balance purely mechanical aspects as well as chemical interactions – the lower strength of Si-Si bonds than C–C bonds. However, the results presented by Capote et al. [86] do not indicate a significant difference in the stress level of DLC coatings deposited using silicon interlayers with thicknesses from 100 to 300 nm. Moreover, regardless of the applied polarization of the substrate and the thickness of the interlayer, the stresses decreased by nearly 1 GPa compared to the coatings without silicon. The smooth transition of element concentration in the interlayer (gradient) is another key aspect of improving adhesion, especially for Ti-DLC films. As shown in the data presented by Nißen et al. [87], the critical delamination force of coatings deposited with an interlayer but without the use of a gradient was 97 mN, while a gradual change of admixture concentration resulted in improvement of those parameters by almost three times. Also, when using the Ti/TiN/TiCN transition between the substrate made of M2 tool steel and the carbon coating, the critical delamination force in the scratch test may increase by up to 350% [42]. It is worth remembering that not only substrate and coating have a direct influence on the adhesion but also the surrounding environment. It is especially important in medical applications that are hard to simulate by ordinary tests. A Swiss research team working on a cobalt-chromium-molybdenum alloy tested the mentioned alloying elements as an interlayer. The obtained results were influenced primarily by carbides formed in the transition layer, the corrosion of which affects the strength of the entire coating [88, 89]. The presence of cobalt carbides increased the delamination rate of the coating in phosphate-buffered saline (PBS) up to four times. Also, the analysis of the damaged heads of explanted hip endoprostheses covered with carbon coatings showed that the adhesive coatings made of silicon were the place of corrosion damage, leading to delamination of the DLC and Si-DLC coatings in the period from 2 to 10 years of implant use [90]. On the other hand, the doping element's direct effect on the coating's adhesion can be related to various aspects. Cornella et al. [52] associated the reduction of internal stresses with the columnar structure of the vanadium-doped carbon coating. For example, in the case of silicon, this effect is probably related to the presence of Si-C bonds that are longer than C–C bonds and, therefore, allow relaxation of some of the stresses [53] appearing in the thin films. In the case of some elements like silver, the influence of admixture on the adhesion of carbon coatings doped with it is ambiguous. Manien et al. [54] indicated the possibility of obtaining higher values of the critical delamination force than for DLC in the scratch test for silver doped diamond-like carbon (Ag-DLC) coatings; nevertheless, this relationship is not linear. The results presented by the research team from the Lodz University of Technology [55] indicate, however, that with the increase in the silver content, the adhesion of the doped coatings decreases (the value of the critical force of delamination may decrease even by more than 20% in relation to the undoped DLC coating). Among the most commonly used testing techniques to analyze the adhesion of carbon coatings on metallic substrates for orthopedics is the scratch test method, both in micro [67] and nanoscale. Of course, other methods are also used for thick PVD coatings, e.g. Rockwell C penetrant indentation trace analysis (Daimler-Benz method) [91].

Other important mechanical properties are hardness and their longitudinal modulus of elasticity. These two properties are usually reported in publications simultaneously due to the testing technique used to determine them – namely nanoindentation [1, 92, 93]. Admittedly, publications reporting the hardness of DLC coatings on the Vickers scale can also be found in the literature [94], but especially for thin carbon coatings, this testing technique becomes useless due to the depth of penetration of the indenter. In the case of carbon coatings, their hardness is in close correlation with their C–C sp^3 and C=C sp^2 bond content or hydrogen compactness. Thus, if the chemical structure of the DLC coatings is more diamond-like (with a higher content of C–C sp^3 bonds), the coatings become harder. Sheeja, using the filtered cathodic vacuum arc FCVA technique, showed that as the bias on modified metallic substrates increases, the hardness of DLC coatings decreases while the ID/IG ratio increases from Raman spectroscopy studies [93]. The hardness of DLC coatings is also strongly correlated with their hydrogen content. The less of it there is in the coatings, the more the coatings gain in hardness. Of course, depending on the techniques used to produce the carbon coatings, they can be obtained with some hydrogen content or no hydrogen at all (most easily with PVD techniques and the use of graphite targets and a hydrogen-free plasma atmosphere). The effect of hydrogen on hardness usually results in harder coatings being obtained with PVD techniques than with CVD.

DLC coatings of all varieties exhibit very good tribological properties in both dry and wet tests in physiological fluid environments. The good tribological properties of DLC coatings are particularly related to their C=C sp^2 phase and hydrogen content [84, 95, 96]. Carbon coatings not only cause a decrease in friction coefficients but also determine an increase in the wear resistance of the modified surface. The latter characteristic is best seen in comparisons of the wear index, which determines the loss of surface volume after tribological tests in relation to the applied load on the counterexample from its path along the tested surface. As reported by Oliveira [97], the use of DLC coatings can also reduce the amount of wear products formed during friction processes while deciding on the formation of larger particles, thus reducing the risk of phagocytosis. In addition, the tribological properties of DLC coatings and, in particular, their wear resistance can be improved by the use of an interlayer [2, 98] or by carrying out other technological treatments, such as surface structuring [99].

The reduction of friction coefficient of carbon coatings is widely investigated in relation to doping with silicon and silicon oxide [56–60], but interesting results are also obtained for coatings containing fluorine [61, 62]. Results at the level of 0.01–0.02 of coefficient of friction were achieved by the Wang team [42], but in the case of friction occurring under reduced pressure and coatings containing more than 11 at.% fluorine and also about 8 at.% sulfur. The tribological properties of a-C:S:F coatings are highly related to the phenomenon of electrostatic repulsion.

Ohana et al. [59] obtained a significantly higher coefficient of friction of the Si-DLC coatings in a frictional connection with AISI 440C steel (at the level of 0.07) than for F-DLC. Nevertheless, the presented research showed that such coatings ensure adhesion in tribological tests conducted in an aqueous environment. Jedrzejczak et al. [100], working with carbon coatings doped with silicon oxide obtained using the RF PACVD method, determined that the tribological properties of the produced material were influenced by both the oxide content and the magnitude of the polarization potential at which the deposition occurred. The change in the coefficient of friction was also related to the material of the counter-sample used in tribological tests. The presence of SiO_x reduced the coefficient of friction when using a steel ball but increased it when the measurements were conducted with the use of a ball made of zirconium oxide. The reason for such an effect may be related to the higher affinity of this material to the doped carbon coating. As an example, Figure 8.1 shows the course of tribological tests on a modified or unmodified Ti6Al4V titanium alloy substrate using carbon coatings doped with silicon (Si-DLC) or silicon and nitrogen (SiN-DLC), during which an Al_2O_3

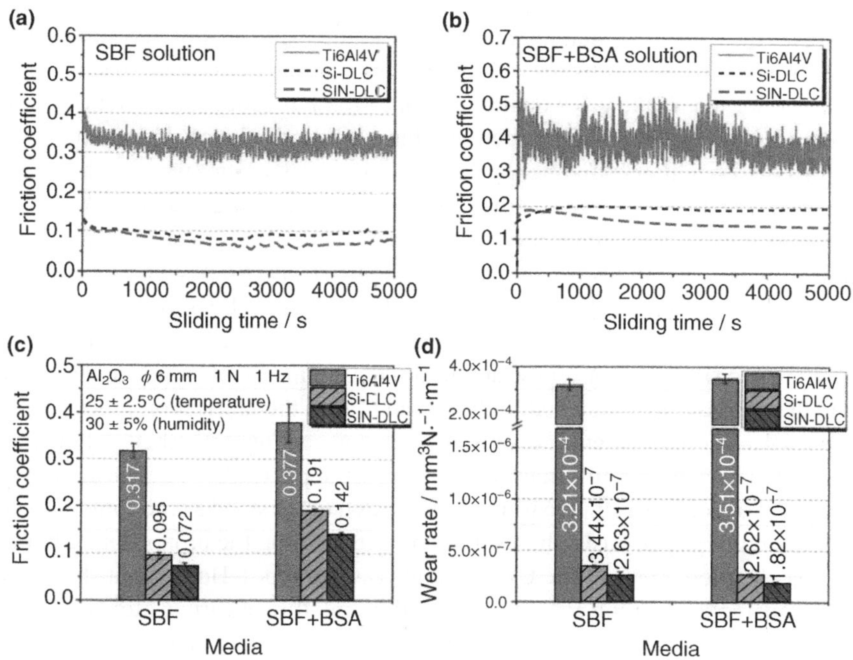

Figure 8.1 Comparison of the friction coefficients, their determined mean value and wear indices for Ti6Al4V substrates, and their modifications using Si-DLC and SiN-DLC coatings in tests with Al_2O_3 balls in SBF and SBF + BSA solutions. *Source:* Reproduced from [67] / with permission of American Chemical Society.

ceramic counterexample and a simulated body fluid (SBF) or SBF + bovine serum protein (BSA) environment were used. After the tests, the average values of friction coefficients and wear indices were determined for each of the associations and test conditions, and the results presented clearly show the potential for reducing COF values and wear indices using the proposed modifications [67].

Similar results were obtained by Jedrzejczak [92], who additionally proved that the tribological properties of Si-DLC coatings are influenced by the type of silicon precursor used in the tests (Si-DLC coatings produced using hexamethyldisiloxane–HMDSO in comparison to tetramethylsilane TMS exhibited better properties). Most often, DLC coatings are tested using tribotesters based on ball-on-disc association [3, 66]. However, due to their attempts to be used in the ball-on-disc association of hip endoprostheses, tribotesters with a special design reflecting natural joint movement are used for this purpose [1, 2, 66]. Increasingly, tribocorrosion tests are also being performed to provide additional information on the effect of tribological phenomena on the corrosion properties of the surfaces under investigation [23, 101–103].

Carbon coatings produced on orthopedic substrates alter their chemical composition and structure. The identification of DLC on metallic substrates usually involves the control of their chemical composition using mostly XPS and chemical structure studies with the additional contribution of Raman spectroscopy. The former technique gives quantitative information about the content of C–C sp^3 and C=C sp^2 bonds in carbon coatings, while the latter determines their change in a qualitative manner by analyzing the ID/IG ratio (when it increases, we know that in the studied coatings the compactness of C–C sp^3 bonds decreases simultaneously). Sometimes, in order to precisely analyze the hydrogen content in the produced coatings, other testing methods are also used, such as secondary ion mass spectrometry [104]. The study of the chemical composition enables determining the basic elements that make up the manufactured coating. In the case of pure DLC coatings, these are most often composed of carbon and oxygen, which are most often attached to the coatings after the fabrication process from the air. Additional elements are present in carbon coatings after their doping processes, subject to careful analysis for concentration on the surface [68]. The chemical structure of DLC coatings is most often understood as the content (usually the ratio) of two basic phases in them, determining the proportion of bonds between carbon atoms of the C–C sp^3 and C=C sp^2 types. However, it should be remembered that carbon coatings are usually a mixture of carbon atoms with the modes of hybridization mentioned earlier and a certain compactness of hydrogen, according to the triangular phase system proposed by Ali Erdemir and Christophe Donnet [73]. The chemical structure of DLC coatings is closely related to their fabrication techniques. For example, in RFPECVD methods, the obtained chemical structure of carbon coatings is particularly influenced by the

RF generator power used in the study, which determines the values of the negative self-polarization potential at the high-frequency electrode. Chunjaemsri showed that as the RF generator power increases from 60 to 160 W, the ratio of sp^3/sp^2 phase content in the coatings decreases [105]. Plasma energy conditions are just as important in magnetron PVD technologies. Using magnetron sputtering with an increase in sputtering power (possibly DC voltage), the ratio of C–C sp^3 bonds to C=C sp^2 bonds also changes. Kim determined that an increase in magnetron sputtering power from 15 to 30 W increases the ID/IG ratio determined from Raman spectroscopy spectra from 0.74 to 1.56 [106].In addition, all literature data confirm that the application of a DLC coating improves corrosion properties. For the most commonly used instruments or orthopedic implants made of metallic materials, it is important to maintain the relevant properties in saline environments, cleaning or sterilization agents, but especially in the human body environment with a high amount of aggressive ions (especially Cl-), a possible variable pH value, and a high oxygen content. Therefore, corrosion resistance tests are carried out using various solutions, most often potentiodynamic tests. The results obtained this way are presented using potentiodynamic polarization curves, from which the most important parameters such as corrosion potential (Ecorr), the density of the corrosion current (Jcorr), and polarization resistance (Rp). These tests are often combined with an initial OCP test. As shown by the research of Sharma et al., the DLC coating on AISI 304 SS steel provides effective protection even in a 3.5% NaCl solution (almost four times higher salt concentration than commonly found in the human body) [107]. As with the assessment of the tribological properties of DLC coatings, the interlayers used may be important for the corrosion safety of this coating. Experimental studies have shown, among other things, the validity of using a SiN_x interlayer [108] or titanium [109] on AISI 316 substrates. In the former case, a breakthrough potential of 1.4 mV was achieved compared to nitride layers, for which this value was about three times lower. Although the application of the DLC coating to the nitrogen- and/or silicon-modified material made it possible to equalize the value of the breakthrough potential, the magnitude of the corrosion current remained lowest for the carbon coating with the SiN_x interlayer [108]. Carbon coatings deposited on steel with a titanium interlayer were characterized by a threefold increase in the value of the breakthrough potential [109]. This parameter was further improved when the titanium coatings were doped, reaching a level of about 1.7 mV. In this case, the Rp parameter increase was as high as 1 $M\Omega cm^2$ compared to undoped DLC without interlayer.

In the literature, one can also find information about the beneficial effects of DLC coatings on other properties, among them surface wettability or refractive index. Carbon coatings change the wettability of modified surfaces. Also, doping has a great influence on the interactions of the coating with water or other liquids.

The hydrophilic or hydrophobic character of doped DLC film may be strictly related to the nature of the given element. Ostrovskaya et al. [63], who studied the doping of fullerene coatings, determined that the elements forming carbides increase the polar component of the surface energy. This effect is due to the presence of polar bonds between the metal and the carbon. On the other hand, elements that do not form carbides increase the hydrophobicity of the surface and reduce the surface energy. The reason for this is that such admixtures can only saturate the free bonds present in the overall structure. This approach confirms the increasing contact angle of coatings containing copper [64] and silver [65, 66] and the more hydrophilic nature of Si-DLC coatings [67–70]. In some cases, the effect is nonobvious, like in the case of titanium. Although this element easily forms carbides, Ti in the carbon matrix causes an increase in the contact angle for water and a decrease in surface energy [71, 72]. In this case, the overall wettability also has an effect on the Ti-O bonds present in these coatings.

In terms of orthopedic and surgical instrument applications, the biological properties of DLC coatings are extremely important. Medical application of carbon coatings requires, apart from mechanical properties that allow maintaining the structural integrity of the film and its basic properties, as well as the biological aspects. Today, most attention is given to making the coatings bioactive, especially by introducing appropriate admixtures into their structure. The use of various admixtures allows one to tailor the response of leaving matter in contact with the coating, i.e. affect and growth of surrounding cells, exert bacteriostatic and bactericidal properties, or hemocompatibility. Dopant elements can affect the overall biological properties of the DLC coating in numerous ways; however, some types of responses are dominant. Considering the above, it is possible to determine the groups of elements of higher interest due to a given activity. The most common example is the usage of elements of well-known bactericidal, bacteriostatic, antifungal effect, etc., to reduce microbial colonization of coatings dopped with them. Among the admixtures limiting microbial colonization well described are silver and copper [110], but scientific studies in such applications are also devoted to silicon [111], fluorine [112], and germanium [113]. Ag-DLC coatings limit the development of both model organisms, such as Escherichia coli (*E. coli*) bacteria [114], and members of the microbiota of the human body being opportunistic pathogens, e.g. affecting the respiratory tract and even bones, common in peri-implantation infections [115, 116]. The presence of silver admixture reduces the amount of bacteria in just a few hours. Baba et al. [117] indicated an antibacterial effect of more than 60% after just three hours of incubation, and Marciano et al. [114] for this time, even 68% reduction of microbial adhesion. The samples are usually subjected to a silver composition in the coating, reaching about 10% [118, 119]. In the case of Cu-DLC coatings, there are publications in which the copper content in the coating exceeds even 50 at.% [120], but already 9 at.% Cu limits the growth of microbial biofilm. Ivanov-Omskii et al. [121] indicate the fungicidal effect of the use of copper, which was confirmed in studies using *Aspergillus Niger, Chaetomium globosum*, and *Cladosporium cladosporioides*. Current research also focuses on the simultaneous introduction of both silver and copper to the carbon matrix [122]. In both silver and copper applications, the bacteriostatic or bactericidal effect is associated primarily with the impact of these metals or their ions on the cells of microorganisms. The mechanism of that action is complex. For example, research conducted on nanosilver shows that its bactericidal activity is related to the presence of an oxide layer on the surface of nanoparticles, which favors the formation of Ag+ ions [123]. Undoubtedly, silver ions can denature proteins by attaching to thiol groups [124] and COOH groups present in glutamic acid [125]. Moreover, such ions break the hydrogen bonds and disulfide bridges and, as a result, change the secondary structure of proteins [126]. In consequence, even the most basic functions of microorganisms are destabilized. Research conducted with the use of Vibrio cholerae indicates that Ag+ ions may interfere with ion exchange across the cell membrane (probably as a result of modification of membrane proteins or phospholipids) and, as a result, the metabolic processes of the cell [127]. Silver ions inhibit the action of, among others, cytochrome B, affecting the functioning of the respiratory chain. As a result, cells are destroyed by the accumulation of reactive oxygen species [128]. Dakal et al. [129] indicate that the action of silver is not limited only to interactions with proteins but also with nucleic acids. By forming bonds with pyrimidine bases, silver leads to DNA compaction and, in effect, inhibits replication. Ag+ ions attached to DNA can also perform catalytic functions [130]. The free radicals produced in these places lead to local damage to the genetic material. In the case of silicon and fluorine, the bacteriostatic effect is achieved indirectly, so it is associated with a change in roughness or surface energy. Ren et al. [111] showed that as the concentration of silicon or fluorine in the carbon coating increases, the surface energy decreases, and consequently, the bacterial adhesion to the surface drops. These results are also confirmed by studies by Ishihara et al. [112], although, in the case of F-DLC coatings, the decreasing number of bacteria with increasing fluorine content can also be explained as a result of reducing the roughness of the produced coatings. Research conducted on Ge-DLC coatings by Robertson et al. [113] does not indicate a statistical difference in the number of Pseudomonas aeruginosa cells inhabiting the doped and undoped surfaces, but the presence of germanium affects the destruction of their cell walls. On the other hand, Arslan et al. [131] achieved inhibition of that bacterial strain by 40% and, in the case of *Staphylococcus aureus*, by 25%. Despite this, the mechanism of action of this element on microorganisms has not yet been fully understood. Carbon coatings used for implants of the cardiovascular system, in general, should not be thrombogenic. Such a state effect is possible by controlling the surface energy and hydrophobicity. An increase in hydrophilicity and surface energy of Si-DLC films, and as a result, a decrease in thrombogenicity, was described by Okpalugo et al. [132] and Roy et al. [133]. However, Ong et al. [134] observed a reduction in surface energy, which also resulted in higher hemocompatibility of silicon-doped coatings. Explaining these seemingly contradictory relationships requires a deeper analysis of the factors responsible for blood coagulation. Cloth formation is controlled by two proteins: fibrinogen and albumin. The first one is responsible for the aggregation of activated thrombocytes, i.e. it participates in the coagulation process. The second, on the other hand, inhibits platelet adhesion. Although these proteins have a similar value of surface energy, fibrinogen has a higher polar component than albumin. In addition, these proteins are negatively charged, just like platelets. The Si-O bonds present in the case of Si-DLC are polarized in such a way that the presence of a negative charge on the surface of silicon-doped coatings can be assumed. The increase of the surface energy, as demonstrated in the work published by Okpalugo et al. [132] and Roy et al. [133], electrostatically blocks the adhesion of fibrinogen in contrast to albumin, which has a lower polar component. The polar component of the Si-DLC coatings produced by Ong et al. [134] is small and undergoes only slight changes. Therefore, it does not differentiate the adhesion of these proteins. However, due to its higher dispersive nature, albumin will interact more easily with the hydrophobic surfaces, limiting clot formation. The albumin-to-fibrinogen absorption ratio also increases when silver is used in the carbon matrix [135]. High hemocompatibility exerts also Ca-DLC and P-DLC coatings [136, 137]. However, calcium and phosphorus are not among the elements whose biological properties, when used as an admixture of carbon coatings, are frequently the subject of publications. As in the case of the previously described admixtures, this effect is closely related to the low surface energy, which is the result not only of the amount of phosphorus in the carbon matrix but also of the manufacturing process and its parameters. By stimulating osteoblasts, which are bone-forming cells, it is possible to accelerate the healing process and strengthen the integration of the biomaterial with the surrounding tissues. The osteoinduction effect is, in that way, another important feature of doped DLC coatings. The use of Ti-DLC is associated with the possibility of releasing titanium ions into the surrounding tissues, which, at concentrations above 20 ppm, may affect the viability and functionality of cells responsible for both resorption and bone reconstruction [138]. Bharathy et al. [139] claim that to maintain the biocompatibility of Ti-DLC coatings while improving osteoblast proliferation, the Ti content should not exceed 1.1 at.%. Randeniya et al. [140], while conducting research on DLC-TiOx, determined the optimal titanium content at 3 at.%. Several studies by Schroeder et al. [141] pay attention to no more than 7.2 at.% of titanium in the coatings since higher amounts may inhibit osteoclast cell activity, which can impair the healing process. The impact of Ti-DLC coatings on cell proliferation is not limited only to the direct impact of titanium and its ions but also to changes in surface morphology or the presence of an unbalanced charge. Higher surface roughness, and thus an increase in the available contact surface and the electrostatic attraction of calcium ions, may affect the adsorption of proteins to Ti-DLC coatings and, as a result, modify interactions with cells [142]. Silicon is an important building element of the skeleton, and its influence on the calcification process or densification of skeletal structures by cross-linking connective tissue [143]. That is why such admixture is of great interest, for example, in Thorwarth et al. [144] or the research team from Lodz University of Technology [145], indicating that Si-DLC coatings can promote osteoblast proliferation. Also, the mineralization assays show that Si-DLC increases the production of calcium and phosphate by Saos-2 cells, and the effect was even greater for N-DLC [146]. Studies conducted on MC3T3 cells, so osteoblast precursors from mice, have shown that silver-containing coatings can reduce the number of osteogenic cells within the first three days of incubation [115]. However, this effect disappears after seven days of contact of Ag-DLC coatings with those cells. Additionally, in the case of the results presented by Almaguer-Flores et al. [147] with the MG63 human osteoblast line, no negative effect of the presence of silver in the carbon coating was found. Doping carbon coatings with silicon can also promote attachment to the surface of human endothelial cells [148]. Okpalugo et al. [149] for DLC coatings containing from approx. 10–20 at.% Si achieved at least a fivefold increase in the amount of attached endothelial cells. Studies conducted using mouse fibroblasts indicate that the addition of nitrogen [150] or calcium oxide [151] can significantly affect the biological properties of DLC coatings. After both one and three days of incubation by Liao et al. [150], fibroblasts on N-DLC coatings were characterized by undisturbed morphology (spindle-shaped, not spherical cells), as well as more of them than on the undoped DLC. Similar results were obtained by Dorner-Reisel et al. [151] for cultures of fibroblasts growing for four days on the surface of carbon coatings containing CaO. Both cell density and morphological features (the shape and the presence of microvilli on the surface of the cell membrane) indicate their high viability.

8.5 Examples of Applications of Carbon Coatings in Orthopedics and on Surgical Instruments

So far, all reports show that DLC coatings are materials with enormous potential for applications in orthopedics and surgical instruments. According to literature reports, there are several advanced projects in this field [6]. Among the described applications for bone-anchored implant systems for amputation limb prostheses, Thesleff points to the ITAP system; the solution for this system in veterinary medicine is well established. This type of implant is based on a stem coated with a DLC coating, used to achieve low-friction properties on the one hand and to allow improved antimicrobial properties on the other, which was initially used in veterinary medicine [152, 153]. This system is encountered in clinical applications of individual implants in prosthetic reconstruction of extra-oral craniofacial defects [154].

DLC coatings work well on printed orthopedic implants. According to Kawaguchi's research, combining porous honeycomb-shaped structures of 3D titanium alloy Ti-6Al-7Nb implants with applied DLC coatings can accelerate the process of new bone formation [11]. The fabrication of DLC coatings on metallic 3D printouts for biomedical applications is an issue being addressed by many centers. Examples of their use on Ti6Al4V titanium alloy printouts can be found in the literature [10] AISI 316 L steel [12, 13].

The properties of carbon coatings are also used by manufacturers of surgical instruments. These are most often used to protect their working parts with DLC coatings in black. On the one hand, such solutions are supposed to determine the appropriate hardness, wear, and corrosion resistance of the modified surfaces, but also, through the antireflective character of the coating, to enable better visibility in the operating field. In addition, due to the low coefficient of friction, the coating can be used to produce nonsticking surgical tools [155]. Such applications for DLC coatings can be found on the websites of companies such as Oerlikon Balzers (under the name BALIMED DLC) or IHI Ionbond AG (under the name Medthin™), Microqore Medical GmbH. A great deal of research also addresses the possibility of creating surfaces on tools that are as resistant to bacteria as possible. In this case, silver-doped DLC coatings become the ideal solution [156, 157]. It is also important to be able to subject DLC-coated instruments to repeated sterilization without losing their properties [21].

Among the few commercially available applications of DLC coatings for orthopedic implants, one must mention the Carbioceram coating™ used in total ankle arthroplasty implants under the name AKILE™, on an industrial scale with good results [158]. There are also orthopedic implants on the market covered with antiallergic, silicon-doped diamond-like coatings usually produced using RF PACVD technology by different manufacturers such as Medgal Sp. z o.o. of Poland [159]. These coatings are produced on intramedullary nails, bone plates, or hip endoprosthesis stems and are a solution that is widely used not only in the domestic market.

It is worth noting that thick carbon coatings, as evidenced by their black color, are most often dominant in the applications described. This dominance is certainly due to the fact that most of them aim to achieve the longest possible wear resistance. Looking at the examples presented, one can deduce that carbon coatings are a good solution for orthopedic instruments and implants. However, the number of their commercial applications still seems to be quite low.

8.6 Conclusion

The significant potential of carbon coatings to provide high-quality performance features such as biocompatibility and anti-allergic properties, low coefficients of friction and wear, chemical inertness and the associated corrosion resistance, etc., has already granted these coatings a place on the market of both surgical tools and broadly understood implants. Of course, the development of DLC coatings in these applications is related to continuous work on overcoming their limitations and the technologies involved in their manufacturing. In this respect, important aspects include, for example, optimizing the adhesion of coatings to various materials and increasing their economic potential. By doping carbon coatings, it is possible to further tailor their properties toward given, highly specific needs like interaction with various tissues, altering the hydrophobicity or hydrophilicity of the surface, and so on. Ongoing research and development efforts in the commercialization of DLC coatings in biomedical applications offer promising prospects for improved patient outcomes in the future.

References

1 Rothammer, B., Neusser, K., Bartz, M. et al. (2023). Evaluation of the wear-resistance of DLC-coated hard-on-soft pairings for biomedical applications. *Wear* 523: 204728.

2 Choudhury, D., Lackner, J., Fleming, R. et al. (2017). Diamond-like carbon coatings with zirconium-containing interlayers for orthopedic implants. *Journal of the Mechanical Behavior of Biomedical Materials* 68: 51–61.

3 Choudhury, D., Morita, T., Sawae, Y. et al. (2016). A novel functional layered diamond like carbon coating for orthopedics applications. *Diamond and Related Materials* 61: 56–69.

4 Hasebe, T., Shimada, A., Suzuki, T. et al. (2006). Fluorinated diamond-like carbon as antithrombogenic coating for blood-contacting devices. *Journal of Biomedical Materials Research Part A* 76A: 86–94.

5 Alakoski, E., Tiainen, V.M., Soininen, A. et al. (2008). Load-bearing biomedical applications of diamond-like carbon coatings - current status. *The Open Orthopedics Journal* 2: 43–50.

6 Birkett, M., Zia, A.W., Devarajan, D. et al. (2023). Multi-functional bioactive silver- and copper-doped diamond-like carbon coatings for medical implants. *Acta Biomaterialia* 167: 54–68.

7 Jastrzebski, K., Bialecki, J., Jastrzebska, A. et al. (2021). Induced Biological Response in Contact with Ag-and Cu-Doped Carbon Coatings for Potential Orthopedic Applications. *Materials* 14: 1861.

8 Sasamoto, T., Kawaguchi, M., Yonezawa, K. et al. (2021). Antibacterial Fluorinated Diamond-like Carbon Coating Promotes Osteogenesis-Comparison with Titanium Alloy. *Applied Sciences-Basel* 11: 9451.

9 Katouno, J., Fujioka, K., Kidera, S. et al. (2017). Evaluation of the enhancement of osteogenesis by Zn-releasing diamond-like carbon film. *Diamond and Related Materials* 77: 131–136.

10 Yick, S., Reneman, J., Martin, P. et al. (2023). Enhancing the Biocompatibility of Additively Manufactured Ti-6al-4 V Eli with Diamond-Like Carbon Coating. *Advanced Materials Interfaces* 10: 2300225.

11 Kawaguchi, M., Segawa, A., Shintani, K. et al. (2021). Bone formation at Ti-6Al-7Nb scaffolds consisting of 3D honeycomb frame and diamond-like carbon coating implanted into the femur of beagles. *Journal of Biomedical Materials Research Part B-Applied Biomaterials* 109: 1283–1291.

12 Tillmann, W., Dias, N., Stangier, D. et al. (2021). Coatability of diamond-like carbon on 316L stainless steel printed by binder jetting. *Additive Manufacturing* 44: 102064.

13 Laskowska, D., Balasz, B., Kaczorowski, W. et al. (2022). The DLC Coating on 316L Stainless Steel Stochastic Voronoi Tessellation Structures Obtained by Binder Jetting Additive Manufacturing for Potential Biomedical Applications. *Coatings* 12: 1373.

14 Jambagi, S. and Malik, V. (2021). A Review on Surface Engineering Perspective of Metallic Implants for Orthopaedic Applications. *Jom* 73: 4349–4364.

15 Morsiya, C. (2022). A review on parameters affecting properties of biomaterial SS 316L. *Australian Journal of Mechanical Engineering* 20: 803–813.

16 Chen, J., Xie, L., Ruan, Q. et al. (2022). Diamond-like carbon coating and surface grafting of osteoprotegerin and alendronate on polyetheretherketone to ameliorate the mechanical performance and osseointegration simultaneously. *Composites Part B-Engineering* 236: 109815.

17 Mo, S., Zhao, F., Gao, A. et al. (2021). Simultaneous application of diamond-like carbon coating and surface amination on polyether ether ketone: Towards superior mechanical performance and osseointegration. *Smart Materials in Medicine* 2: 219–228.

18 Kaczorowski, W., Szymanski, W., Batory, D. et al. (2014). Tribological Properties and Characterization of Diamond Like Carbon Coatings Deposited by MW/RF and RF Plasma-Enhanced CVD Method on Poly(ether-ether-ketone). *Plasma Processes and Polymers* 11: 878–887.

19 Kaczorowski, W., Batory, D., Szymanski, W. et al. (2015). Evaluation of the surface properties of PEEK substrate after two-step plasma modification: Etching and deposition of DLC coatings. *Surface & Coatings Technology* 265: 92–98.

20 Zia, A.W., Hussain, S., and Baig, M. (2022). Optimizing diamond-like carbon coatings-From experimental era to artificial intelligence. *Ceramics International* 48: 36000–36011.

21 Geyao, L., Yang, D., Wanglin, C. et al. (2020). Development and application of physical vapor deposited coatings for medical devices: A review. *Procedia CIRP* 89: 250–262.

22 Yang, W., Gao, Y., Guo, P. et al. (2020). Adhesion, biological corrosion resistance and biotribological properties of carbon films deposited on MAO coated Ti substrates. *Journal of the Mechanical Behavior of Biomedical Materials* 101: 103448.

23 Grabarczyk, J., Gaj, J., Pazik, B. et al. (2021). Tribocorrosion behavior of Ti6Al4V alloy after thermo-chemical treatment and DLC deposition for biomedical applications. *Tribology International* 153: 106560.

24 Gabor, R., Cvrcek, L., Doubková, M. et al. (2022). Hybrid coatings for orthopaedic implants formed by physical vapour deposition and microarc oxidation. *Materials & Design* 219: 110811.

25 Dufils, J., Faverjon, F., Héau, C. et al. (2017). Combination of laser surface texturing and DLC coating on PEEK for enhanced tribological properties. *Surface & Coatings Technology* 329: 29–41.

26 Lu, Y., Huang, G., Wang, S. et al. (2021). A review on diamond-like carbon films grown by pulsed laser deposition. *Applied Surface Science* 541: 148518.

27 Loir, A., Garrelie, F., Donnet, C. et al. (2005). Mechanical and tribological characterization of tetrahedral diamond-like carbon deposited by femtosecond pulsed laser deposition on pre-treated orthopaedic biomaterials. *Applied Surface Science* 247: 225–231.

28 Lu, T., Qiao, Y., and Liu, X. (2012). Surface modification of biomaterials using plasma immersion ion implantation and deposition. *Interface Focus* 2: 325–336.

29 Skjöldebrand, C., Tipper, J., Hatto, P. et al. (2022). Current status and future potential of wear-resistant coatings and articulating surfaces for hip and knee implants. *Materials Today Bio* 15: 100270.

30 Rafiq, N.M., Wang, W., Liew, S.L. et al. (2023). A review on multifunctional bioceramic coatings in hip implants for osteointegration enhancement. *Applied Surface Science Advances* 13: 100353.

31 Peng, X. and Clyne, T. (1998). Residual stress and debonding of DLC films on metallic substrates. *Diamond and Related Materials* 7: 944–950.

32 Ban, M., Hasegawa, T., Fujii, S. et al. (2003). Stress and structural properties of diamond-like carbon films deposited by electron beam excited plasma CVD. *Diamond and Related Materials* 12: 47–56.

33 Lacerda, R., Hammer, P., Lepienski, C. et al. (2001). Hard graphitic-like amorphous carbon films with high stress and local microscopic density. *Journal of Vacuum Science & Technology a-Vacuum Surfaces and Films* 19: 971–975.

34 Morshed, M., McNamara, B., Cameron, D. et al. (2003). Stress and adhesion in DLC coatings on 316L stainless steel deposited by a neutral beam source. *Journal of Materials Processing Technology* 141: 127–131.

35 Kang, S., Lim, H., and Lee, K. (2015). Effects of TiCN interlayer on bonding characteristics and mechanical properties of DLC-coated Ti-6Al-4V ELI alloy. *International Journal of Refractory Metals & Hard Materials* 53: 13–16.

36 Schwarz, C., Heeg, J., Rosenberg, M. et al. (2008). Investigation on wear and adhesion of graded Si/SiC/DLC coatings deposited by plasma-enhanced-CVD. *Diamond and Related Materials* 17: 1685–1688.

37 Batory, D., Szymanski, W., and Clapa, M. (2013). Mechanical and tribological properties of gradient a-C:H/Ti coatings. *Materials Science-Poland* 31: 415–423.

38 Popa, A., Stan, G., Husanu, M. et al. (2013). Multi-layer haemocompatible diamond-like carbon coatings obtained by combined radio frequency plasma enhanced chemical vapor deposition and magnetron sputtering. *Journal of Materials Science-Materials in Medicine* 24: 2695–2707.

39 Joska, L., Fojt, J., Cvrcek, L. et al. (2014). Properties of titanium-alloyed DLC layers for medical applications. *Biomatter* 4: e29505.

40 Wongpanya, P., Wongpinij, T., Photongkam, P. et al. (2022). Improvement in corrosion resistance of 316L stainless steel in simulated body fluid mixed with antiplatelet drugs by coating with Ti-doped DLC films for application in biomaterials. *Corrosion Science* 208: 110611.

41 Zhu, H., Ma, L., Liu, N. et al. (2017). Improvement in anti-corrosion property of hydrogenated diamond-like carbon film by modifying CrC interlayer. *Diamond and Related Materials* 72: 99–107.

42 Chang, C. and Wang, D. (2001). Microstructure and adhesion characteristics of diamond-like carbon films deposited on steel substrates. *Diamond and Related Materials* 10: 1528–1534.

43 Safaie, P., Eshaghi, A., and Bakhshi, S. (2016). Optical properties of oxygen doped diamond-like carbon thin films. *Journal of Alloys and Compounds* 672: 426–432.

44 Khun, N. and Liu, E. (2014). Effects of platinum content on tribological properties of platinum/nitrogen doped diamond-like carbon thin films deposited via magnetron sputtering. *Friction* 2: 64–72.

45 Wong, H., Foong, Y., and Chua, D. (2011). Improving the conductivity of diamond-like carbon films with zinc doping and its material properties. *Applied Surface Science* 257: 9616–9620.

46 Corbella, C., Vives, A., Pinyol, A. et al. (2004). Preparation of metal (W, Mo, Nb, Ti) containing a-C:H films by reactive magnetron sputtering. *Surface & Coatings Technology* 177: 409–414.

47 Wan, S., Wang, L., and Xue, Q. (2010). Electrochemical deposition of sulfur doped DLC nanocomposite film at atmospheric pressure. *Electrochemistry Communications* 12: 61–65.

48 Omer, A., Adhikari, S., Adhikary, S. et al. (2006). Iodine doping in amorphous carbon thin-films for optoelectronic devices. *Physica B-Condensed Matter* 376: 316–319.

49 Benchikh, N., Garrelie, F., Donnet, C. et al. (2006). Nanostructured coatings of metal containing diamond-like carbon films deposited by femtosecond pulsed laser ablation. *Surface & Coatings Technology* 200: 6272–6278.

50 Li, H., Hu, J., Lin, G. et al. (2010). V and N co-doped diamond-like carbon films deposited by pulsed bias arc ion plating. *International Journal of Refractory Metals & Hard Materials* 28: 544–549.

51 Wu, Y., Zhou, S., Zhao, W. et al. (2018). Comparative corrosion resistance properties between (Cu, Ce)-DLC and Ti co-doped (Cu, Ce)/Ti-DLC films prepared via magnetron sputtering method. *Chemical Physics Letters* 705: 50–58.

52 Jelinek, M., Kocourek, T., Zemek, J. et al. (2015). Chromium-doped DLC for implants prepared by laser-magnetron deposition. *Materials Science & Engineering C-Materials For Biological Applications* 46: 381–386.

53 Swiatek, L., Olejnik, A., Grabarczyk, J. et al. (2016). Multi-doped diamond like-carbon coatings (DLC-Si/Ag) for biomedical applications fabricated using the modified chemical vapour deposition method. *Diamond and Related Materials* 67: 54–62.

54 Wang, F., Wang, L., and Xue, Q. (2016). Fluorine and sulfur co-doped amorphous carbon films to achieve ultra-low friction under high vacuum. *Carbon* 96: 411–420.

55 Nelson, N., Rakowski, R., Franks, J. et al. (2014). The effect of substrate geometry and surface orientation on the film structure of DLC deposited using PECVD. *Surface & Coatings Technology* 254: 73–78.

56 Jongwannasiri, C., Krasaesin, A., Pinijsuwan, S. et al. (2023). Diamond-like carbon (DLC)-coated titanium surface inhibits bacterial growth and modulates human alveolar bone cell responses in vitro. *Diamond and Related Materials* 136: 110022.

57 Beake, B., Liskiewicz, T., Vishnyakov, V. et al. (2015). Development of DLC coating architectures for demanding functional surface applications through nano- and micro-mechanical testing. *Surface & Coatings Technology* 284: 334–343.

58 Batory, D., Jedrzejczak, A., Kaczorowski, W. et al. (2015). Influence of the process parameters on the characteristics of silicon-incorporated a-C:H:SiOx coatings. *Surface and Coatings Technology* 271: 112–118.

59 Caschera, D., Federici, F., Kaciulis, S. et al. (2007). Deposition of Ti-containing diamond-like carbon (DLC) films by PECVD technique. *Materials Science & Engineering C-Biomimetic and Supramolecular Systems* 27: 1328–1330.

60 Jedrzejczak, A., Batory, D., Prowizor, M. et al. (2020). Titanium(IV) isopropoxide as a source of titanium and oxygen atoms in carbon based coatings deposited by Radio Frequency Plasma Enhanced Chemical Vapour Deposition method. *Thin Solid Films* 693: 137697.

61 Zhang, H., Wu, W., Wang, W. et al. (2009). Influence of Fe-doped on structural, electronic structural and optical properties of hydrogenated amorphous carbon films prepared by plasma enhanced chemical vapor deposition.*Journal of Alloys and Compounds* 487: 522–526.

62 Yetim, A., Kovaci, H., Kasapoglu, A. et al. (2021). Influences of Ti, Al and V metal doping on the structural, mechanical and tribological properties of DLC films. *Diamond and Related Materials* 120: 108639.

63 Depla, D. (2021). Sputter deposition with powder targets: An overview. *Vacuum* 184: 109892.

64 Mosaner, P., Bonelli, M., and Miotello, A. (2003). Pulsed laser deposition of diamond-like carbon films: reducing internal stress by thermal annealing. *Applied Surface Science* 208: 561–565.

65 Sun, H., Yang, L., Wu, H. et al. (2023). Effects of Element Doping on the Structure and Properties of Diamond-like Carbon Films: A Review. *Lubricants* 11: 186.

66 Love, C., Cook, R., Harvey, T. et al. (2013). Diamond like carbon coatings for potential application in biological implants-a review. *Tribology International* 63: 141–150.

67 Wei, X., Zhang, Y., Feng, H. et al. (2022). Bio-Tribology and Corrosion Behaviors of a Si- and N-Incorporated Diamond-like Carbon Film: A New Class of Protective Film for Ti6Al4V Artificial Implants. *ACS Biomaterials Science & Engineering* 8: 1166–1180.

68 Wei, X., Chen, L., Zhang, M. et al. (2020). Effect of dopants (F, Si) material on the structure and properties of hydrogenated DLC film by plane cathode PECVD. *Diamond and Related Materials* 110: 108102.

69 Cano, D., Lousa, A., Esteve, J. et al. (2018). Low Wear and Low Friction DLC Coating With Good Adhesion to CoCrMo Metal Substrates. *Physica Status Solidi B-Basic Solid State Physics* 255.

70 Kato, N., Mori, H., and Takahashi, N. (2008). Spectroscopic ellipsometry of silicon-containing diamond-like carbon (DLC-Si) films. *Physica Status Solidi C* 5: 1117–1120.

71 Kanasugi, K., Eguchi, H., Ohgoe, Y. et al. (2022). Correlation of Cell Proliferation with Surface Properties of Polymer-like Carbon Films of Different Thicknesses Prepared by a Radio-Frequency Plasma CVD Process. *Materials* 15: 4466.

72 Ikeyama, M. and Sonoda, T. (2013). Transparent Si-DLC coatings on metals with high repetition bi-polar pulses of a PBII system. *Nuclear Instruments & Methods in Physics Research Section B-Beam Interactions With Materials and Atoms* 307: 340–343.

73 Erdemir, A. and Donnet, C. (2006). Tribology of diamond-like carbon films: recent progress and future prospects. *Journal of Physics D-Applied Physics* 39: R311–R327.

74 Li, Q., Wang, F., and Zhang, L. (2013). Study of colors of diamond-like carbon films. *Science China-Physics Mechanics & Astronomy* 56: 545–550.

75 Mosas, K., Chandrasekar, A., Dasan, A. et al. (2022). Recent Advancements in Materials and Coatings for Biomedical Implants. *Gels* 8: 323.

76 Tardelli, J., Valente, M., de Oliveira, T. et al. (2021). Influence of chemical composition on cell viability on titanium surfaces: A systematic review. *Journal of Prosthetic Dentistry* 125: 421–425.

77 Kobayashi, S., Ohgoe, Y., Ozeki, K. et al. (2007). Dissolution effect and cytotoxicity of diamond-like carbon coatings on orthodontic archwires. *Journal of Materials Science-Materials in Medicine* 18: 2263–2268.

78 Gutensohn, K., Beythien, C., Bau, J. et al. (2000). In vitro analyses of diamond-like carbon coated stents: Reduction of metal ion release, platelet activation, and thrombogenicity. *Thrombosis Research* 99: 577–585.

79 Mu, Y., Kobayashi, T., Sumita, M. et al. (2000). Metal ion release from titanium with active oxygen species generated by rat macrophages in vitro. *Journal of Biomedical Materials Research* 49: 238–243.

80 Ohgoe, Y., Kobayashi, S., Ozeki, K. et al. (2006). Reduction effect of nickel ion release on a diamond-like carbon film coated onto an orthodontic archwire. *Thin Solid Films* 497: 218–222.

81 Jastrzebski, K. and Grabarczyk, J. (2018). Radio-frequency plasma-enhanced chemical vapour deposition of carbon films on AISI 316LVM steel: Formation of a transition layer and metal whiskers, and their influence on coating properties. *Thin Solid Films* 659: 16–23.

82 Konkhunthot, N., Photongkam, P., and Wongpanya, P. (2019). Improvement of thermal stability, adhesion strength and corrosion performance of diamond-like carbon films with titanium doping. *Applied Surface Science* 469: 471–486.

83 Mitura, S., Mitura, K., Niedzielski, P. et al. (2006). Nanocrystalline diamond, its synthesis, properties and applications. *Journal of Achievements in Materials and Manufacturing Engineering* 16: 9–15.

84 Li, C., Huang, L., and Yuan, J. (2020). Effect of sp3 Content on Adhesion and Tribological Properties of Non-Hydrogenated DLC Films. *Materials* 13: 1911.

85 Cemin, F., Bim, L., Menezes, C. et al. (2015). The influence of different silicon adhesion interlayers on the tribological behavior of DLC thin films deposited on steel by EC-PECVD. *Surface & Coatings Technology* 283: 115–121.

86 Capote, G., Bonetti, L., Santos, L. et al. (2008). Adherent amorphous hydrogenated carbon films on metals deposited by plasma enhanced chemical vapor deposition. *Thin Solid Films* 516: 4011–4017.

87 Nissen, S., Heeg, J., Warkentin, M. et al. (2017). The effect of deposition parameters on structure, mechanical and adhesion properties of a-C:H on Ti6Al4V with gradient Ti-a-C:H:Ti interlayer. *Surface & Coatings Technology* 316: 180–189.

88 Falub, C., Müller, U., Thorwarth, G. et al. (2011). In vitro studies of the adhesion of diamond-like carbon thin films on CoCrMo biomedical implant alloy. *Acta Materialia* 59: 4678–4689.

89 Müller, U., Falub, C., Thorwarth, G. et al. (2011). Diamond-like carbon coatings on a CoCrMo implant alloy: A detailed XPS analysis of the chemical states at the interface. *Acta Materialia* 59: 1150–1161.

90 Hauert, R., Falub, C., Thorwarth, G. et al. (2012). Retrospective lifetime estimation of failed and explanted diamond-like carbon coated hip joint balls. *Acta Biomaterialia* 8: 3170–3176.

91 Ghio, E., Bolelli, G., Bertè, A. et al. (2023). Diamond-Like Carbon (DLC) and AlCrN films onto Ti-6Al-4V substrates by Laser-Powder Bed Fusion (L-PBF): Effect of substrate heat treatment and surface finish. *Surface & Coatings Technology* 475: 130128.

92 Jedrzejczak, A., Szymanski, W., Kolodziejczyk, L. et al. (2022). Tribological Characteristics of a-C:H:Si and a-C:H:SiOx Coatings Tested in Simulated Body Fluid and Protein Environment. *Materials* 15: 2082.

93 Sheeja, D., Tay, B., Shi, X. et al. (2001). Mechanical and tribological characterization of diamond-like carbon coatings on orthopedic materials. *Diamond and Related Materials* 10: 1043–1048.

94 Catledge, S.A., Thomas, V., Vohra, Y.K. et al. (2013). 5-Nanostructured diamond coatings for orthopaedic applications. In: *Diamond-Based Materials for Biomedical Applications*, 105–150. Sawston: Woodhead Publishing.

95 Ould, C., Tite, T., Héau, C. et al. (2024). Tribological behavior of DLC film models in base oils: Analysis of influence of sp3/sp2 ratio and hydrogen content. *Tribology International* 194: 109446.

96 Rao, X., Yang, J., Chen, Z. et al. (2020). Tribological behavior of DLC film models in base oils: Analysis of influence of sp3/sp2 ratio and hydrogen content. *Bioactive Materials* 5: 192–200.

97 Oliveira, L., Kuromoto, N., and Siqueira, C. (2014). Treating orthopedic prosthesis with diamond-like carbon: minimizing debris in Ti6Al4V. *Journal of Materials Science-Materials in Medicine* 25: 2347–2355.

98 Yang, W., Xu, D., Gao, Y. et al. (2019). Microstructure, adhesion, in vitro corrosion resistance and tribological behavior of (Si:N)-DLC coated pure Ti. *Diamond and Related Materials* 92: 109–116.

99 Liu, Y., Zhang, K., Han, J. et al. (2022). One-step method to enhance biotribological properties and biocompatibility of DLC coating by ion beam irradiation. *Friction* 10: 1114–1126.

100 Jedrzejczak, A., Kolodziejczyk, L., Szymanski, W. et al. (2017). Friction and wear of a-C:H:SiOx coatings in combination with AISI 316L and ZrO2 counterbodies. *Tribology International* 112: 155–162.

101 Bayón, R., Igartua, A., González, J. et al. (2015). Influence of the carbon content on the corrosion and tribocorrosion performance of Ti-DLC coatings for biomedical alloys. *Tribology International* 88: 115–125.

102 Oliveira, S.M.M., Barzotto, I.L.M., Vieira, L. et al. (2019). Tribocorrosion studies on diamond-like carbon film deposited by PECVD on 304 stainless steel in simulated body fluid. *Biomedical Physics & Engineering Express* 5: 045012.

103 He, D., He, C., Li, W. et al. (2020). Tribological behaviors of in-situ textured DLC films under dry and lubricated conditions. *Applied Surface Science* 525: 146581.

104 de Oliveira, A., Placias, F., Sobrinho, A. et al. (2021). Secondary ion mass spectrometry and atomic force microscopy analysis of silver-doped diamond-like carbon films on titanium alloy (Ti6Al4V) for possible biomedical application. *Thin Solid Films* 719: 138487.

105 Chunjaemsri, T., Chongsereecharoen, E., Chanlek, N. et al. (2020). Influence of RF power and CH4 flow rate on properties of diamond-like carbon films deposited by PECVD technique. *Radiation Physics and Chemistry* 176: 109073.

106 Kim, J., Kang, S., Park, J. et al. (2021). Tailored Hydrogen-Free Carbon Films by Tuning the sp2/sp3 Configuration. *Acs Applied Electronic Materials* 3: 1771–1779.

107 Shanna, R., Barhai, R., and Kumari, N. (2008). Corrosion resistant behaviour of DLC films. *Thin Solid Films* 516: 5397–5403.

108 Azzi, M., Amirault, P., Paquette, M. et al. (2010). Corrosion performance and mechanical stability of 316L/DLC coating system: Role of interlayers. *Surface & Coatings Technology* 204: 3986–3994.

109 Wongpanya, P., Pintitraratibodee, N., Thumanu, K. et al. (2021). Improvement of corrosion resistance and biocompatibility of 316L stainless steel for joint replacement application by Ti-doped and Ti-interlayered DLC films. *Surface & Coatings Technology* 425: 127734.

110 Ohta, T. and Kamiya, Y. (2023). Antimicrobial coating using copper-doped diamond-like carbon film deposited by dual magnetron sputtering. *Japanese Journal of Applied Physics* 62: 078002.

111 Ren, D., Zhao, Q., and Bendavid, A. (2013). *Surface & Coatings Technology* 226: 1–6.

112 Ishihara, M., Kosaka, T., Nakamura, T. et al. (2006). Anti-bacterial property of Si and F doped diamond-like carbon coatings. *Diamond and Related Materials* 15: 1011–1014.

113 Robertson, S., Gibson, D., MacKay, W. et al. (2017). Investigation of the antimicrobial properties of modified multilayer diamond-like carbon coatings on 316 stainless steel. *Surface & Coatings Technology* 314: 72–78.

114 Marciano, F., Bonetti, L., Santos, L. et al. (2009). Antibacterial activity of DLC and Ag-DLC films produced by PECVD technique. *Diamond and Related Materials* 18: 1010–1014.

115 Endrino, J., Galindo, R., Zhang, H. et al. (2008). Structure and properties of silver-containing a-C(H) films deposited by plasma immersion ion implantation. *Surface & Coatings Technology* 202: 3675–3682.

116 Zimmerli, W. and Moser, C. (2012). Pathogenesis and treatment concepts of orthopaedic biofilm infections. *Fems Immunology and Medical Microbiology* 65: 158–168.

117 Baba, K., Hatada, R., Flege, S. et al. (2013). Preparation and antibacterial properties of Ag-containing diamond-like carbon films prepared by a combination of magnetron sputtering and plasma source ion implantation. *Vacuum* 89: 179–184.

118 Wu, Y., Chen, J., Li, H. et al. (2013). Preparation and properties of Ag/DLC nanocomposite films fabricated by unbalanced magnetron sputtering. *Applied Surface Science* 284: 165–170.

119 Mazare, A., Anghel, A., Surdu-Bob, C. et al. (2018). Silver doped diamond-like carbon antibacterial and corrosion resistance coatings on titanium. *Thin Solid Films* 657: 16–23.

120 Chan, Y., Huang, C., Ou, K. et al. (2011). Mechanical properties and antibacterial activity of copper doped diamond-like carbon films. *Surface & Coatings Technology* 206: 1037–1040.

121 Ivanov-Omskii, V., Panina, L., and Yastrebov, S. (2000). Amorphous hydrogenated carbon doped with copper as antifungal protective coating. *Carbon* 38: 495–499.

122 Sanzone, G., Field, S., Lee, D. et al. (2022). Antimicrobial and Aging Properties of Ag-, Ag/Cu-, and Ag Cluster-Doped Amorphous Carbon Coatings Produced by Magnetron Sputtering for Space Applications. *Acs Applied Materials & Interfaces* 14: 10154–10166.

123 Sotiriou, G.A., Meyer, A., Knijnenburg, J.T.N. et al. (2012). Quantifying the Origin of Released Ag+ Ions from Nanosilver. *Langmuir* 28: 15929–15936.

124 Matsumura, Y., Yoshikata, K., Kunisaki, S. et al. (2003). Mode of bactericidal action of silver zeolite and its comparison with that of silver nitrate. *Applied and Environmental Microbiology* 69: 4278–4281.

125 Jung, W., Koo, H., Kim, K. et al. (2008). Antibacterial activity and mechanism of action of the silver ion in Staphylococcus aureus and Escherichia coli. *Applied and Environmental Microbiology* 74: 2171–2178.

126 Jing, P., Su, Y., Li, Y. et al. (2021). Mechanism of protein biofilm formation on Ag-DLC films prepared for application in joint implants. *Surface & Coatings Technology* 422.

127 Dibrov, P., Dzioba, J., Gosink, K. et al. (2002). Chemiosmotic mechanism of antimicrobial activity of Ag+ in Vibrio cholerae. *Antimicrobial Agents and Chemotherapy* 46: 2668–2670.

128 Yamanaka, M., Hara, K., and Kudo, J. (2005). Bactericidal actions of a silver ion solution on *Escherichia coli*, studied by energy-filtering transmission electron microscopy and proteomic analysis. *Applied and Environmental Microbiology* 71: 7589–7593.

129 Dakal, T., Kumar, A., Majumdar, R. et al. (2016). Mechanistic Basis of Antimicrobial Actions of Silver Nanoparticles. *Frontiers in Microbiology* 7: 1831.

130 Hossain, Z. and Huq, F. (2002). Studies on the interaction between Ag+ and DNA. *Journal of Inorganic Biochemistry* 91: 398–404.

131 Arslan, M., Kurt, M., Aslan, N. et al. (2022). Structural, biocompatibility, and antibacterial properties of Ge-DLC nanocomposite for biomedical applications. *Journal of Biomedical Materials Research Part B-Applied Biomaterials* 110: 1667–1674.

132 Okpalugo, T., Ogwu, A., Maguire, P. et al. (2004). Platelet adhesion on silicon modified hydrogenated amorphous carbon films. *Biomaterials* 25: 239–245.

133 Roy, R., Choi, H., Yi, J. et al. (2009). Hemocompatibility of surface-modified, silicon-incorporated, diamond-like carbon films. *Acta Biomaterialia* 5: 249–256.

134 Ong, S., Zhang, S., Du, H. et al. (2007). Influence of silicon concentration on the haemocompatibility of amorphous carbon. *Biomaterials* 28: 4033–4038.

135 Choi, H., Dauskardt, R., Lee, S. et al. (2008). Characteristic of silver doped DLC films on surface properties and protein adsorption.*Diamond and Related Materials* 17: 252–257.

136 Chu, P. (2007). Plasma surface treatment of artificial orthopedic and cardiovascular biomaterials. *Surface & Coatings Technology* 201: 5601–5606.

137 Kwok, S.C.H., Jin, W., and Chu, P.K. (2005). Surface energy, wettability, and blood compatibility phosphorus doped diamond-like carbon films. *Diamond and Related Materials* 14: 78–85.

138 Mine, Y., Makihira, S., Nikawa, H. et al. (2010). Impact of titanium ions on osteoblast-, osteoclast- and gingival epithelial-like cells. *Journal of Prosthodontic Research* 54: 1–6.

139 Bharathy, P., Nataraj, D., Chu, P. et al. (2010). Effect of titanium incorporation on the structural, mechanical and biocompatible properties of DLC thin films prepared by reactive-biased target ion beam deposition method. *Applied Surface Science* 257: 143–150.

140 Randeniya, L., Bendavid, A., Martin, P. et al. (2010). Thin-film nanocomposites of diamond-like carbon and titanium oxide; Osteoblast adhesion and surface properties. *Diamond and Related Materials* 19: 329–335.

141 Schroeder, A., Francz, G., Bruinink, A. et al. (2000). Titanium containing amorphous hydrogenated carbon films (a-C: H/Ti): surface analysis and evaluation of cellular reactions using bone marrow cell cultures in vitro. *Biomaterials* 21: 449–456.
142 Cheng, H., Chiou, S., Liu, C. et al. (2009). Effect of plasma energy on enhancing biocompatibility and hemocompatibility of diamond-like carbon film with various titanium concentrations. *Journal of Alloys and Compounds* 477: 931–935.
143 Calomme, M., Sindambiwe, J., Cos, P. et al. (2004). Effect of choline stabilized orthosilicic acid on bone density in ovariectomized rats. *Journal of Bone and Mineral Research* 19: S449–S449.
144 Thorwarth, G., Saldamli, B., Schwarz, F. et al. (2007). Biocompatibility of Doped Diamond-Like Carbon Coatings for Medical Implants. *Plasma Processes and Polymers* 4: S364–S368.
145 Batory, D., Czerniak-Reczulska, M., Kolodziejczyk, L. et al. (2013). Gradient titanium and silver based carbon coatings deposited on AISI316L. *Applied Surface Science* 275: 303–310.
146 Subramanian, B., Rajan, S., Martin, P. et al. (2018). Biomineralization of osteoblasts on DLC coated surfaces for bone implants. *Biointerphases* 13.
147 Argelia, A.-F., Sandra, E.R., and René, O.-N. (2011). Oral bacterial adhesion and biocompatibility of silver-amorphous carbon films: A surface modification for dental implants (Chapter 12). *Implant Dentistry*. Rijeka: IntechOpen.
148 Ogwu, A., Okpalugo, T., Ali, N. et al. (2008). Endothelial cell growth on silicon modified hydrogenated amorphous carbon thin films. *Journal of Biomedical Materials Research Part B-Applied Biomaterials* 85B: 105–113.
149 Okpalugo, T., Murphy, H., Ogwu, A. et al. (2006). Human microvascular endothelial cellular interaction with atomic N-doped DLC compared with Si-doped DLC thin films. *Journal of Biomedical Materials Research Part B-Applied Biomaterials* 78B: 222–229.
150 Liao, W., Lin, C., Wei, D. et al. (2012). Concurrent improvement in biocompatibility and bioinertness of diamond-like carbon films with nitrogen doping. *Journal of Biomedical Materials Research Part a* 100A: 3151–3156.
151 Dorner-Reisel, A., Schürer, C., Irmer, G. et al. (2002). Diamond-like carbon coatings with Ca-O-incorporation for improved biological acceptance. *Analytical and Bioanalytical Chem*istry 374: 753–755.
152 Thesleff, A., Brånemark, R., Håkansson, B. et al. (2018). Biomechanical Characterisation of Bone-anchored Implant Systems for Amputation Limb Prostheses: A Systematic Review. *Annals of Biomedical Engineering* 46: 377–391.

153 Fitzpatrick, N., Smith, T., Pendegrass, C. et al. (2011). Intraosseous Transcutaneous Amputation Prosthesis (ITAP) for Limb Salvage in 4 Dogs. *Veterinary Surgery* 40: 909–925.

154 Kang, N., Morritt, D., Pendegrass, C. et al. (2013). Use of ITAP implants for prosthetic reconstruction of extra-oral craniofacial defects. *Journal of Plastic Reconstructive and Aesthetic Surgery* 66: 497–505.

155 Shum, P., Zhou, Z., and Li, K. (2014). To increase the hydrophobicity, non-stickiness and wear resistance of DLC surface by surface texturing using a laser ablation process. *Tribology International* 78: 1–6.

156 Peng, Y., Peng, J., Wang, Z. et al. (2022). Diamond-like Carbon Coatings in the Biomedical Field: Properties, Applications and Future Development. *Coatings* 12: 1088.

157 Vishwakarma, V., Kaliaraj, G., and Mosas, K. (2023). Multifunctional Coatings on Implant Materials-A Systematic Review of the Current Scenario. *Coatings* 13: 69.

158 Hernandez, J., Laffenêtre, O., Toullec, E., Darcel, V., Chauveaux, D. (2014). AKILE™ total ankle arthroplasty: Clinical and CT scan analysis of periprosthetic cysts. Orthopedics & Traumatology-Surgery & Research 100: 907–915.

159 Medgal Sp. z o. o., Poland https://medgal.com.pl/. Accessed 17/12/2024.

Section IV

Diamond-like Carbon Coatings for Electrical and Optical Devices

Section IV

[illegible]

9

Diamond-like Carbon Coatings for Electric Storage Batteries

Muhammad Ihsan-Ul-Haq[1,2,3]

[1] *Energy Storage and Distributed Resources Division, Lawrence Berkeley National Laboratory, Berkeley, CA, USA*
[2] *Interdisciplinary Research Center for Sustainable Energy Systems (IRC-SES), King Fahd University of Petroleum and Minerals, Dhahran, Saudi Arabia*
[3] *Materials Science and Engineering Department, King Fahd University of Petroleum and Minerals, Dhahran, Saudi Arabia*

CONTENTS

9.1 Introduction, 279
9.2 Applications of DLC Coatings in Secondary Batteries, 280
9.2.1 DLC Coatings on Alkali Metal Anodes, 282
9.2.2 DLC Coatings on Carbon or Alloy-based Anodes, 283
9.2.3 DLC Coatings on Cathodes, 284
9.2.4 DLC Coatings on the Separator, 286
9.3 Conclusions and Future Research, 287

9.1 Introduction

Better battery electrodes are pivotal for advancing energy storage technology. These electrodes must possess a range of key properties to optimize battery performance. High electrical conductivity of electrode materials is paramount, ensuring efficient charge and discharge processes, reducing energy losses, and enabling fast charging. Additionally, a high capacity is crucial, as it allows electrodes to store and release substantial energy, extending device runtime or electric vehicle driving range.

Stability and durability are equally essential, ensuring that electrodes maintain their structural integrity and electrochemical performance over numerous charge–discharge cycles. High-rate capability is another critical property, facilitating rapid charging and discharging by supporting fast ion and electron transport [1]. Furthermore, chemical stability is vital to resist undesirable reactions

Applications of Diamond-like Carbon Coatings, First Edition. Abdul Wasy Zia.

with the electrolyte, preventing capacity fade and safety concerns [2, 3]. Cost-effectiveness and sustainability also play a pivotal role in making advanced electrode materials and production methods accessible and environmentally friendly [4]. Finally, electrode materials must be adaptable and optimized to work with new battery technologies as they emerge, such as solid-state batteries or evolving anode and cathode materials [5, 6]. Achieving the right balance of these properties is fundamental to developing better battery electrodes that meet the diverse needs of various applications and contribute to a cleaner, more efficient energy future.

In particular, the electrode–electrolyte interface and interphases play a critical role in achieving the cyclic stability and high-rate capability of advanced battery electrodes [7]. Therefore, rationale design strategies such as modification of chemistry and chemical composition of the electrolyte and electrode as well as adding protective functional coatings on the electrodes have been used. Functional coatings like diamond-like carbon (DLC) on battery electrodes have a synergistic effect on enhancing their electrochemical performance [8, 9]. These coatings serve several critical purposes. Unlike crystalline carbon materials, such as graphite, carbon nanotubes, graphene, and fullerene, DLC can be formed at room temperature without a catalyst or a pretreatment of the substrate surface [10]. Firstly, they act as barriers to prevent direct contact between the electrode material and the electrolyte, reducing undesirable side reactions and corrosion. This preservation of the electrode's integrity and the reduction of the formation of the solid–electrolyte interface (SEI) layer leads to improved cyclic stability and extended battery life.

Additionally, functional coatings can facilitate more efficient ion and electron transport, enhancing the rate capability of the battery [7, 11]. By modifying the electrode–electrolyte interface, these coatings can reduce interfacial impedance, allowing for faster charge and discharge rates [12, 13]. Overall, these coatings contribute to improved capacity retention, reduced capacity fade over time, and increased power output, all of which are crucial for optimizing battery performance in various applications, from consumer electronics to electric vehicles and renewable energy storage.

9.2 Applications of DLC Coatings in Secondary Batteries

DLC coatings have garnered significant attention in the literature for their potential to enhance the electrochemical performance of battery electrodes due to their chemical inertness and mechanical hardness [14]. Various synthesis strategies have been used to deposit DLC coatings on electrodes/substrates [15, 16]. For example, chemical vapor deposition (CVD) is a commonly used method for DLC coating synthesis. It involves the chemical reaction of gaseous hydrocarbons to deposit DLC films with precise control over film thickness, composition, and uniformity by tuning the sp^3/sp^2 carbon bonding ratio and hydrogen content [17]. However, high operating temperatures and the complexity of equipment are the main barriers to the scalability of this approach. It is suitable for high-performance applications but can be energy-intensive and expensive.

Physical vapor deposition (PVD) methods, including sputtering and evaporation, vaporize a solid carbon source to condense onto a substrate [18]. These methods offer better film density and adhesion. The advantages include good adhesion and compatibility with various substrates as well as a high deposition rate, but they may require high-energy processes. Plasma-enhanced chemical vapor deposition (PECVD) introduces plasma to CVD, enabling lower deposition temperatures [19]. This method provides good adhesion and film quality, making it suitable for various applications. However, control over the film structure can be complex and challenging. Other approaches include magnetron sputtering [20] and pulsed laser deposition [21] of DLC coatings.

In summary, DLC coatings for battery electrodes present a diverse array of synthesis methods, each with its unique advantages and disadvantages. The choice of approach depends on the specific application and requirements, such as film quality, adhesion, and cost-effectiveness. Comprehensive structural and electrochemical characterization is crucial to understanding the coating's impact on battery performance. These findings help guide the selection of the most suitable DLC synthesis method for a given application, ultimately contributing to the development of high-performance batteries. In Figure 9.1, an overview is presented, summarizing diverse applications of DLC coatings aimed at

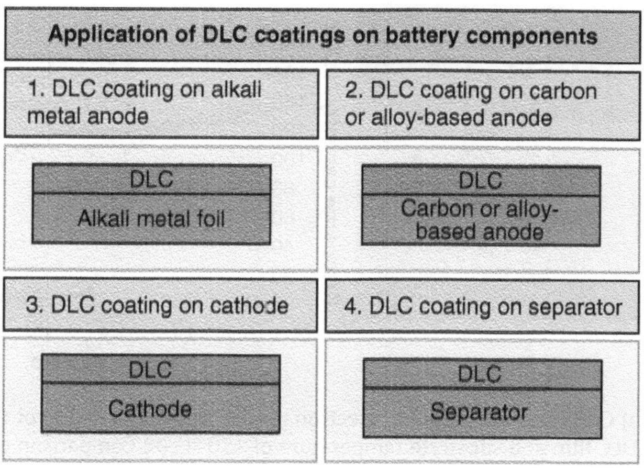

Figure 9.1 Various strategies to apply DLC coatings on battery components.

stabilizing the electrochemical performance of batteries. These applications encompass DLC coatings on alkali metal anodes like lithium, carbon, and alloy-based anodes such as Sn, Sb, and SiO_x-graphite composites, as well as cathodes and separators.

9.2.1 DLC Coatings on Alkali Metal Anodes

The rapid capacity fade and potential safety concerns associated with the interfacial instability and dendritic growth of lithium and sodium metals in batteries underscore the need for a protective coating film [22, 23]. Therefore, a protective coating film is necessary to tackle these issues and thus improve their cyclic performance. For example, Arie and Lee [24] and Shim et al. [25] used radio frequency (RF) PECVD to coat DLC thin film on lithium metal to protect lithium metal and inhibit the growth of lithium dendrites, thus improving the specific capacity, and reduce the interfacial impedance. DLC film formed at a temperature of 150 °C exhibited better performance than the films formed at 30 and 100 °C as well as bare lithium anodes, as shown in Figure 9.2a,b. The enhanced performance is attributed to a relatively high content of sp^3-bonded carbon, which could provide a more conductive layer for insertion/extraction of lithium upon cycles. These results signify the importance of tuning the ratio of sp^2 to sp^3 carbon bonds in DLC coatings to achieve the optimal electrochemical performance of battery electrodes.

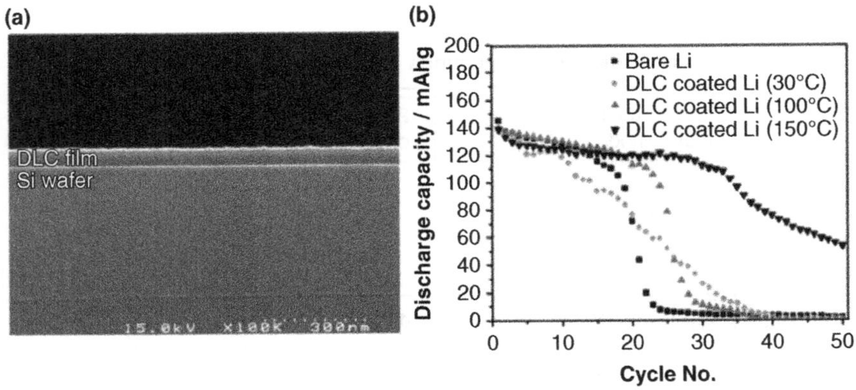

Figure 9.2 (a) Cross-section scanning electron microscope (SEM) image of the as-deposited DLC film at a substrate temperature of 150 °C. (b) Comparison of cyclic performance of bare lithium and DLC-coated lithium electrodes deposited at various substrate temperatures. *Source:* Reproduced from [24] / with permission of ELSEVIER.

9.2.2 DLC Coatings on Carbon or Alloy-based Anodes

The rapid capacity fade in the electrochemical performance of carbon and high-capacity alloy-based anodes, attributed to the formation of an unstable electrode–electrolyte interface and significant volume changes, underscores the need for protective coatings [26]. Therefore, protective coatings that have chemical inertness and high mechanical strength are necessary to ameliorate the cyclic life and high-rate capability of carbon and other alloy-based electrodes. In this pursuit, Nam et al. deposited direct current (DC) sputtered carbon thin films and studied the influence of post-annealing temperature on the lithium storage performance and revealed that films formed at room temperature without any post-annealing show the best performance, as shown in Figure 9.3a,b. The enhanced electrochemical performance is attributed to the more defective nature of films due to lower sp^2 to sp^3 bond ratios [27]. Lee et al. used PECVD to form a DLC coating on a SiO_x-graphite composite electrode with a subsequent chromium coating via

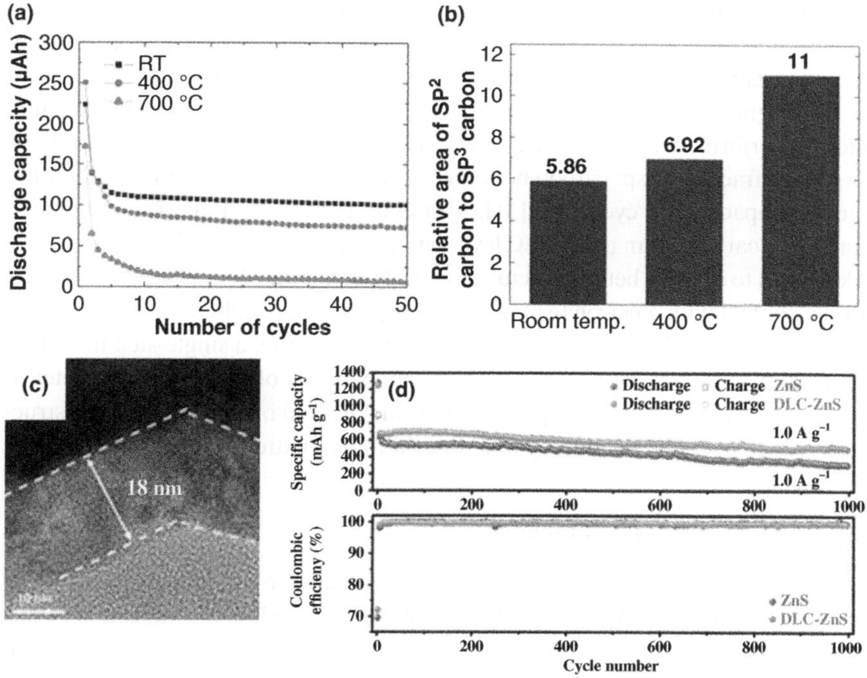

Figure 9.3 (a) Cyclic performance and (b) relative peak area of sp^2 to sp^3 of the carbon films annealed at various temperatures. *Source:* Reproduced from [27] / with permission of ELSEVIER. (c) High-resolution transmission electron microscopy image of DLC-ZnS film, and (d) comparison of cyclic performance and Coulombic efficiency plots of DLC-ZnS and bare ZnS electrodes. *Source:* Reproduced from [30] / with permission of American Chemical Society.

PVD to achieve a higher initial capacity of 591 mAh g^{-1} and improved capacity retention of 83% after 50 cycles compared to bare electrodes which delivered an initial capacity of 517 mAh g^{-1} and 59% capacity retention after 50 cycles [28]. The DLC-Cr-coated electrodes also presented an over ninefold reduction in the impedance caused by the SEI layer after the first cycle signifying the importance of this coating to form an improved interface.

Tang et al. used ambient pressure PECVD to grow DLC nanoparticles on SnSb/C nanofibers to effectively alleviate the large volume changes in the alloy anodes and thus achieve a capacity retention of 94.4% after 100 cycles [29]. Jiang et al. used RF magnetron sputtering to achieve DLC-ZnS double-layer-structured thin-film electrodes with ~18 nm thick DLC film and thus achieved superior high-rate capability and enhanced cyclic stability for up to 1000 cycles with a higher capacity retention of 66% compared to 40.2% of bare ZnS thin-film electrodes, as shown in Figure 9.3c,d [30]. The ameliorated electrochemical performance is attributed to (i) reinforced interfacial mechanical strength with a higher Young's modulus of ~95 GPa of DLC thin films than ~70 GPa of ZnS counterparts and (ii) enhanced kinetics of electrochemical reactions in DLC-ZnS thin films (diffusion coefficient = 1.37×10^{-14} cm^2 s^{-1}) than ZnS thin films (diffusion coefficient = 2.18×10^{-14} cm^2 s^{-1}).

DC magnetron sputtering was used to deposit DLC thin films, and their lithium storage performance was investigated for all-soli-state Li-ion microbatteries showing that a fraction of sp^2 carbon bonding is necessary to achieve enhanced lithium storage capacity and cyclic life [31]. Kim et al. used to RF-PECVD technique to partially coat a ~60 nm thick DLC layer on ~250 nm thick amorphous Si thin-film electrodes to achieve better structural integrity of Si thin films, protect them from direct contact with electrolyte and achieve superior electrochemical performance than bare Si thin films upon cycles [32]. Zhang et al. used a single-step induction thermal plasma process to synthesize Si nanoparticles of an average diameter of 67–78 nm with carbon coating of 2–8 nm thickness to obtain a core–shell structure, which could be potentially used in lithium-ion batteries [33].

9.2.3 DLC Coatings on Cathodes

It is imperative to apply protective coatings on cathodes to enhance their interfacial stability, maintain structural integrity, and inhibit the dissolution of transition metals into the electrolyte [34, 35]. For example, Visbal et al. used CVD to form a 2.6–4.8 nm thin DLC film on $LiNi_{0.8}Co_{0.15}Al_{0.05}O_2$ (NCA), which acted as a buffer layer to hinder side reactions between the NCA cathode and Li_2S-P_2S_5 solid-state electrolyte [36]. Endo et al. used a DC plasma CVD of ethylene to coat a conformal DLC layer on the $LiCoO_2$ electrode to suppress the electrolyte decomposition and thus achieved a high discharge capacity of 29 mAh g^{-1} [37]. Oka et al. coated $LiNi_{1/3}Co_{1/3}Mn_{1/3}O_2$ (NCM) cathode with DLC film through a hybrid process of plasma-based ion implantation and deposition. The electrodes with around 5 nm thick DLC layer effectively reduced the side reactions at the cathode–electrolyte interface and thus improved the cyclic stability of electrodes for 100 cycles compared to bare NCM electrode and 2 nm thick DLC coated NCM cathode, as shown in Figure 9.4a,b [38]. Lee et al. coated DLC on LiV_3O_8 cathode and achieved an initial discharge capacity of 219 mAh g^{-1} and a capacity retention of 92% after 50 cycles, as shown in Figure 9.4c [39].

Cho et al. used PECVD to form a 30 nm thick Si-DLC coating on carbon-based electrodes, which extended their cyclability from 7 to 50 cycles in Li-O_2 cells and

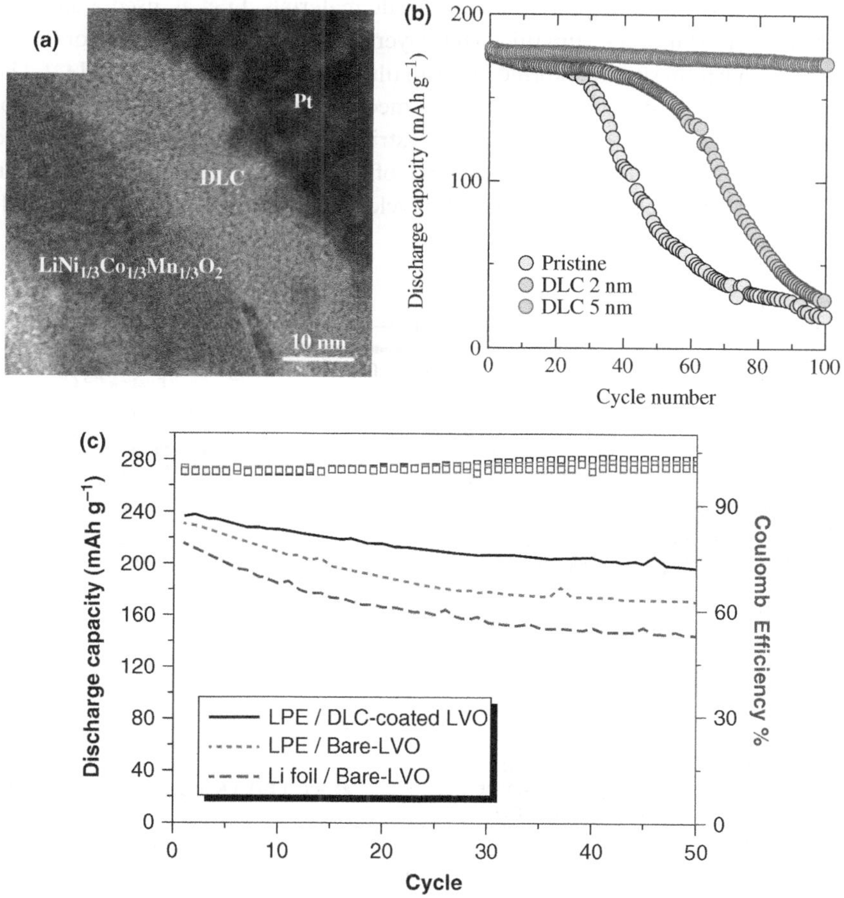

Figure 9.4 (a) SEM image and (b) cyclic performance of the pristine and DLC-coated NCM cathodes. *Source:* Reproduced from [38] / with permission of IOP Publishing. (c) Comparison of cyclic performances of Li foil/bare-LVO, LPE/bare-LVO, and LPE/DLC-coated LVO cells. *Source:* [39] / ELSEVIER / CC BY 4.0.

was attributed to chemical inertness, high hardness, and good adhesion of Si-DLC coating to the carbon electrodes [40]. Besides, the Si-DLC coating remained stable in the electrolyte and prevented direct contact between Li_2O_2 and carbon, thus suppressing the formation of side reaction products, such as Li_2CO_3 and HCO_2Li. Yang et al. investigated a 385 nm thick DLC thin film, prepared by RF sputtering, for lithium-air batteries and proposed that Sp^3-bonded carbon atoms could play a key role in improving the electrochemical performance of lithium-air batteries [41].

9.2.4 DLC Coatings on the Separator

Instead of coating DLC film on the electrode materials, Li et al. used a different approach of forming a ~3 μm thick DLC layer on a polypropylene separator (DLC/PP), which was in situ transformed into an ultra-strong Li-ion conductor [42]. The high Young's modulus of DLC ~100 GPa mechanically suppressed the growth of lithium dendrites and enabled uniform distribution of Li-ions and thus achieved more than 1000 stable cycles at a high rate of 5 C with ~71% capacity retention in Li || $LiFePO_4$ full cell and more than 200 cycles in Li || $LiNi_{0.5}Co_{0.3}Mn_{0.2}O_2$ (NCM) full cells as shown in Figure 9.5.

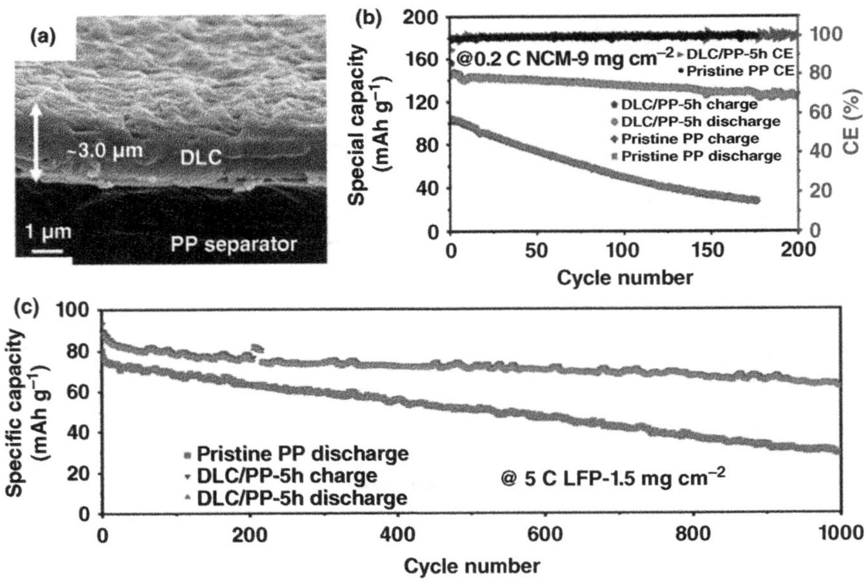

Figure 9.5 (a) Side view SEM images of DLC coating on PP separator, (b) long-cycling performance of cells based on the DLC/PP separator and the pristine PP separator, and their Coulombic efficiency at 5 C in EC/DEC electrolytes, (c) long-term cycling properties of Li || NCM pouch cell with high cathode mass loading by using pristine PP or DLC/PP-5h separators. *Source:* Reproduced from [42] / with permission of John Wiley & Sons.

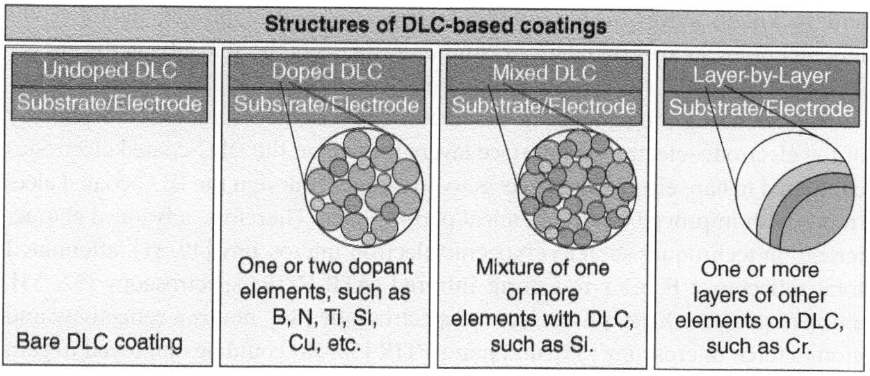

Figure 9.6 Structural configurations of various DLC-based coatings for improving the performance of secondary batteries.

Figure 9.6 provides a summary of the structural configurations of different types of DLC-based coatings, highlighting their potential as promising candidates for improving the electrochemical performance and cyclic stability of secondary batteries.

9.3 Conclusions and Future Research

DLC coatings on battery electrodes are formed using several techniques including PECVD, ambient pressure PECVD, PVD, CVD, and RF magnetron sputtering. The electrodes with DLC coatings exhibited improved electrochemical performance compared to the bare electrodes. However, efforts are required to optimize the thickness, nanostructure, and chemistry of DLC films to improve the electrochemical performance of lithium-based and beyond lithium secondary batteries, such as sodium-, potassium-, calcium-, magnesium-, aluminum-, and zinc-based batteries. For example, Hevia et al. studied the influence of DLC film thickness on P-doped silicon substrates ranging from 10 to 100 nm in thickness of films on their electrochemical response and revealed a quasi-reversible behavior and a lower charge transfer resistance in the films having thickness of ~30 nm [43]. Therefore, the influence of the thickness of DLC coatings on cell impedance and interfacial stability for the improved electrochemical performance of specific secondary battery electrodes must be systematically investigated. In addition, the impact of dopant elements, such as boron [44], nitrogen [45], titanium [46], silicon [47], and copper [48], in DLCs to tune their nanostructure to optimize the electrochemical performance of battery electrodes must be investigated.

Current challenges in DLC coating technologies for batteries include scalability issues from laboratory to large-scale production and ensuring durability through numerous charge–discharge cycles. Achieving uniform coating quality and thickness, addressing environmental concerns, and adapting to next-gen battery technologies are crucial for maximizing DLC's potential in advancing energy storage.

Understanding the changes in the chemistry, nanostructure, and morphology of the electrode–electrolyte interface layers formed on the DLC-coated electrodes compared to bare electrodes is necessary to rationally design the DLC-coated electrodes with improved electrochemical performance. Therefore, advanced characterization techniques such as cryogenic electron microscopy [49–51], attenuated total reflectance Fourier transform infrared (ATR-FTIR) spectroscopy [52, 53], ambient pressure X-ray photoelectron spectroscopy [54], neutron reflectivity and atomic force microscopy [55], and nano-FTIR [56–58] could be employed to gain insights into the electrode–electrolyte interfaces.

DLC coatings and their interactions with electrolytes, along with addressing barriers hindering ion diffusion, especially lithium ions, necessitate the utilization of advanced computational tools [59, 60]. Additionally, there is an opportunity to integrate artificial intelligence into the synthesis and characterization processes for materials, acting as a bridge between computational screening and the experimental realization of novel materials [61]. This approach could pave the way for the synthesis of optimal DLC coatings tailored for battery applications.

References

1 Liu, Y., Zhu, Y., and Cui, Y. (2019). Challenges and opportunities towards fast-charging battery materials. *Nature Energy* 4 (7): 540–550.

2 Xiao, Y., Wang, Y., Bo, S.H. et al. (2020). Understanding interface stability in solid-state batteries. *Nature Reviews Materials* 5 (2): 105–126.

3 He, X., Bresser, D., Passerini, S. et al. (2021). The passivity of lithium electrodes in liquid electrolytes for secondary batteries. *Nature Reviews Materials* 6 (11): 1036–1052.

4 Vaalma, C., Buchholz, D., Weil, M. et al. (2018). A cost and resource analysis of sodium-ion batteries. *Nature Reviews Materials* 3 (4): 18013.

5 Bachman, J.C., Muy, S., Grimaud, A. et al. (2016). Inorganic solid-state electrolytes for lithium batteries: mechanisms and properties governing ion conduction. *Chemical Reviews* 116 (1): 140–162.

6 Trahey, L., Brushett, F.R., Balsara, N.P. et al. (2020). Energy storage emerging: a perspective from the joint center for energy storage research. *Proceedings of the National Academy of Sciences of United States of America* 117 (23): 12550–12557.

7 Dang, H., Peng, Y., Wang, L. et al. (2023). Designing interface coatings on anode materials for lithium-ion batteries. *Journal of Energy Storage* 74 (6010): 109526.

8 Ma, Y., Li, L., Qian, J. et al. (2021). Materials and structure engineering by magnetron sputtering for advanced lithium batteries. *Energy Storage Materials* 39: 203–224.

9 Zia, A.W., Hussain, S.A., Rasul, S. et al. (2023). Progress in diamond-like carbon coatings for lithium-based batteries. *Journal of Energy Storage* 72 (1): 108803.

10 Zeng, A., Neto, V.F., Gracio, J.J. et al. (2014). Diamond-like carbon (DLC) films as electrochemical electrodes. *Diamond and Related Materials* 43: 12–22.

11 Zhou, H., Yu, S., Liu, H. et al. (2020). Protective coatings for lithium metal anodes: recent progress and future perspectives. *Journal of Power Sources* 450: 227632.

12 Xu, Z., Wang, X., Wang, Z. et al. (2023). Interface problems, modification strategies and prospects of Ni–rich layered oxide cathode materials in all–solid–state lithium batteries with sulfide electrolytes. *Journal of Power Sources* 571: 233079.

13 Li, N., Luo, J., Zhu, J. et al. (2023). Cathodic interface in sulfide-based all-solid-state lithium batteries. *Energy Storage Materials* 63: 103034.

14 Zeng, A., Liu, E., Tan, S.N. et al. (2002). Cyclic voltammetry studies of sputtered nitrogen doped diamond-like carbon film electrodes. *Electroanalysis: An International Journal Devoted to Fundamental and Practical Aspects of Electroanalysis* 14 (15–16): 1110–1115.

15 Zia, A.W. and Birkett, M. (2021). Deposition of diamond-like carbon coatings: conventional to non-conventional approaches for emerging markets. *Ceramics International* 47 (20): 28075–28085.

16 Rajak, D.K., Kumar, A., Behera, A. et al. (2021). Diamond-like carbon (DLC) coatings: classification, properties, and applications. *Applied Sciences* 11 (10): 4445.

17 Jatti, V.S., Sefene, E.M., Jatti, A.V. et al. (2023). Synthesis and characterization of diamond-like carbon coatings for drill bits using plasma-enhanced chemical vapor deposition. *International Journal of Advanced Manufacturing Technology* 127 (9–10): 4081–4096.

18 Vysotina, E.A., Kazakov, V.A., Polyansky, M.N. et al. (2017). Investigation of the structure and functional properties of diamond-like coatings obtained by physical vapor deposition. *Journal of Surface Investigation* 11 (6): 1177–1184.

19 Santra, T.S., Liu, C.H., Bhattacharyya, T.K. et al. (2010). Characterization of diamond-like nanocomposite thin films grown by plasma enhanced chemical vapor deposition. *Journal of Applied Physics* 107 (12): 124320.

20 Yeldose, B.C. and Ramamoorthy, B. (2008). Characterization of DC magnetron sputtered diamond-like carbon (DLC) nano coating. *International Journal of Advanced Manufacturing Technology* 38 (7–8): 705–717.

21 Lackner, J.M., Stotter, C., Waldhauser, W. et al. (2003). Pulsed laser deposition of diamond-like carbon coatings for industrial tribological applications. *Surface and Coatings Technology* 174–175: 402–407.

22 Lee, B., Paek, E., Mitlin, D. et al. (2019). Sodium metal anodes: emerging solutions to dendrite growth. *Chemical Reviews* 119 (8): 5416–5460.

23 Ma, L., Cui, J., Yao, S. et al. (2020). Dendrite-free lithium metal and sodium metal batteries. *Energy Storage Materials* 27: 522–554.

24 Arie, A.A. and Lee, J.K. (2011). Electrochemical characteristics of lithium metal anodes with diamond like carbon film coating layer. *Diamond and Related Materials* 20 (3): 403–408.

25 Shim, H.T., Lee, J.K., and Cho, B.W. (2007). DLC film coating on a lithium metal as an anode of lithium secondary batteries. *Solid State Phenomena* 124–126: 919–922.

26 Wu, J., Ihsan-Ul-Haq, M., Ciucci, F. et al. (2021). Rationally designed nanostructured metal chalcogenides for advanced sodium-ion batteries. *Energy Storage Material* 34: 582–628.

27 Nam, S.C., Lee, J.M., Pukha, V.E. et al. (2014). Carbon anode thin films for lithium batteries. *Current Applied Physics* 14 (8): 1010–1015.

28 Lee, J.K., Lee, J.H., Kim, B.K. et al. (2014). Electrochemical characteristics of diamond-like carbon/Cr double-layer coating on silicon monoxide-graphite composite anode for Li-ion batteries. *Electrochimica Acta* 127: 1–6.

29 Yin, J., Wada, M., Tanase, S. et al. (2020). Multidimensional jagged SnSb/C/DLC nanofibers fabricated by AP-PECVD method for Li-ion battery anode. *Nanotechnology* 31 (20): 205401.

30 Jiang, H., Peng, H., Guo, H. et al. (2020). Interfacial mechanical strength enhancement for high-performance ZnS thin-film anodes. *American Chemical Society Applied Materials & Interfaces* 12 (46): 51344–51356.

31 Lu, Z.G. and Chung, C.Y. (2008). Electrochemical characterization of diamond like carbon thin films. *Diamond and Related Materials* 17 (11): 1871–1876.

32 Kim, S.O., Shim, H.T., and Lee, J.K. (2010). Electrochemical performance of silicon thin film anodes covered by diamond-like carbon with various surface coating morphologies. *Journal of Solid State Electrochemistry* 14 (7): 1247–1253.

33 Zhang, X., Hayashida, R., Tanaka, M. et al. (2020). Synthesis of carbon-coated silicon nanoparticles by induction thermal plasma for lithium ion battery. *Powder Technology* 371: 26–36.

34 Wu, Y., Liu, X., Wang, L. et al. (2021). Development of cathode-electrolyte-interphase for safer lithium batteries. *Energy Storage Materials* 37: 77–86.

35 Maleki Kheimeh Sari, H. and Li, X. (2019). Controllable cathode–electrolyte interface of $Li[Ni_{0.8}Co_{0.1}Mn_{0.1}]O_2$ for lithium ion batteries: a review. *Advanced Energy Materials* 9 (39): 1901597.

36 Visbal, H., Aihara, Y., Ito, S. et al. (2016). The effect of diamond-like carbon coating on $LiNi_{0.8}Co_{0.15}Al_{0.05}O_2$ particles for all solid-state lithium-ion batteries based on Li_2S–P_2S_5 glass-ceramics. *Journal of Power Sources* 314: 85–92.

37 Endo, E., Yasuda, T., Kita, A. et al. (2000). A $LiCoO_2$ cathode modified by plasma chemical vapor deposition for higher voltage performance. *Journal of the Electrochemical Society* 147 (4): 1291.

38 Oka, Y., Obata, T., Nishimura, Y. et al. (2015). High-temperature cycling performance of $LiNi_{1/3}Co_{1/3}Mn_{1/3}O_2$ cathode with DLC protective film. *Journal of the Electrochemical Society* 162 (2): A3032–A3037.

39 Lee, J.H., Lee, J.K., and Yoon, W.Y. (2013). Electrochemical characteristics of a diamond-like-carbon-coated LiV_3O_8 cathode when used in a Li-metal battery with a Li-powder anode. *Japanese Journal of Applied Physics* 52 (10 PART2): 10MB05.

40 Cho, S.M., Lee, J.K., and Yoon, W.Y. (2015). Electrochemical effects of silicon/diamond-like carbon layered composite on oxygen cathodes in lithium–oxygen batteries. *Electrochimica Acta* 158: 246–252.

41 Yang, Y., Sun, Q., Li, Y.-S. et al. (2011). Nanostructured diamond like carbon thin film electrodes for lithium air batteries. *Journal of the Electrochemical Society* 158 (10): B1211.

42 Li, Z., Peng, M., Zhou, X. et al. (2021). In situ chemical lithiation transforms diamond-like carbon into an ultrastrong ion conductor for dendrite-free lithium-metal anodes. *Advanced Materials* 33 (37): 2100793.

43 Hevia, S.A., Bejide, M., Duran, B. et al. (2018). Nanometric thin films of non-doped diamond-like carbon grown on n-type (P-doped) silicon substrates as electrochemical electrodes. *Journal of Solid State Electrochemistry* 22 (9): 2845–2853.

44 Khadro, B., Sikora, A., Loir, A.S. et al. (2011). Electrochemical performances of B doped and undoped diamond-like carbon (DLC) films deposited by femtosecond pulsed laser ablation for heavy metal detection using square wave anodic stripping voltammetric (SWASV) technique. *Sensors and Actuators B: Chemical* 155 (1): 120–125.

45 Menegazzo, N., Kahn, M., Berghauser, R. et al. (2011). Nitrogen-doped diamond-like carbon as optically transparent electrode for infrared attenuated total reflection spectroelectrochemistry. *Analyst* 136 (9): 1831–1839.

46 Jo, Y.J., Zhang, T.F., Son, M.J. et al. (2018). Synthesis and electrochemical properties of Ti-doped DLC films by a hybrid PVD/PECVD process. *Applied Surface Science* 433: 1184–1191.

47 Zou, Y., Wang, X., and Wang, L. (2024). Microstructure and high-temperature tribological properties of Ti/Si co-doped diamond-like carbon films fabricated by twin-targets reactive HiPIMS. *Diamond and Related Materials* 141: 110573.

48 Wang, H., Qiu, Z., Lin, S. et al. (2022). A comparative study of copper-doped and copper, nitrogen co-doped DLC film electrode and its electrochemical properties. *Journal of the Electrochemical Society* 169 (7): 077503.

49 Ihsan-Ul-Haq, M., Huang, H., Wu, J. et al. (2020). Thin solid electrolyte interface on chemically bonded Sb_2Te_3/CNT composite anodes for high performance sodium ion full cells. *Nano Energy* 71: 104613.

50 Ihsan-Ul-Haq, M., Huang, H., Wu, J. et al. (2021). Unveiling solid electrolyte interface morphology and electrochemical kinetics of amorphous Sb_2Se_3/CNT composite anodes for ultrafast Na storage. *Carbon N Y* 171: 119–129.

51 Wang, J., Huang, W., Pei, A. et al. (2019). Improving cyclability of Li metal batteries at elevated temperatures and its origin revealed by cryo-electron microscopy. *Nature Energy* 4 (8): 664–670.

52 Song, S.W. and Baek, S.W. (2009). Surface layer formation on Sn anode: ATR FTIR spectroscopic characterization. *Electrochimica Acta* 54 (4): 1312–1318.

53 Shi, F., Ross, P.N., Zhao, H. et al. (2015). A catalytic path for electrolyte reduction in lithium-ion cells revealed by in situ attenuated total reflection-Fourier transform infrared spectroscopy. *Journal of the American Chemical Society* 137 (9): 3181–3184.

54 Salmeron, M. (2018). From surfaces to interfaces: ambient pressure XPS and beyond. *Topics in Catalysis* 61 (20): 2044–2051.

55 Steinhauer, M., Stich, M., Kurniawan, M. et al. (2017). In situ studies of solid electrolyte interphase (SEI) formation on crystalline carbon surfaces by neutron reflectometry and atomic force microscopy. *ACS Applied Materials & Interfaces* 9 (41): 35794–35801.

56 Huth, F., Govyadinov, A., Amarie, S. et al. (2012). Nano-FTIR absorption spectroscopy of molecular fingerprints at 20 nm spatial resolution. *Nano Letters* 12 (8): 3973–3978.

57 Yoon, I., Larson, J.M., and Kostecki, R. (2023). The effect of the SEI layer mechanical deformation on the passivity of a Si anode in organic carbonate electrolytes. *ACS Nano* 17 (7): 6943–6954.

58 He, X., Larson, J.M., Bechtel, H.A. et al. (2022). In situ infrared nanospectroscopy of the local processes at the Li/polymer electrolyte interface. *Nature Communications* 13 (1): 1398.

59 Louie, S.G., Chan, Y.H., da Jornada, F.H. et al. (2021). Discovering and understanding materials through computation. *Nature Materials* 20 (6): 728–735.

60 Jain, A., Shin, Y., and Persson, K.A. (2016). Computational predictions of energy materials using density functional theory. *Nature Reviews Materials* 1 (1): 15004.

61 Szymanski, N.J., Rendy, B., Fei, Y. et al. (2023). An autonomous laboratory for the accelerated synthesis of novel materials. *Nature* 624 (7990): 86–91.

10

DLC Coatings for Optical and Optoelectronic Applications

Mümin Mehmet Koç[1,2], Naim Aslan[3,4]

[1] *Department of Medical Services and Techniques, School of Medical Service, Kırklareli University, Kırklareli, Turkiye*
[2] *Department of Physics, Faculty of Science and Art, Kırklareli University, Kırklareli, Turkiye*
[3] *Faculty of Engineering, Department of Metallurgical and Materials Engineering, Munzur University, Tunceli, Turkiye*
[4] *Rare Earth Elements Application and Research Center, Munzur University, Tunceli, Turkiye*

CONTENTS

10.1 Introduction, 293
10.2 Production and Characterization Methods, 295
10.3 Optical and Optoelectronic Properties, 299
10.4 Effect of Production Methods/Parameters on DLC Film Properties, 301
10.5 DLC Films in Optical and Optoelectronic Applications, 303
10.6 Conclusion, 307

10.1 Introduction

Carbon is a marvelous material, which can be found in both living organisms and different materials. Various forms of carbon were addressed in the literature, which were amorphous carbon, diamond-like carbon (DLC), graphite, highly oriented pyrolytic graphite, carbon nanotubes, bulky balls, fullerene, graphene, etc. Among those, DLC materials are one of the most prominent materials, which were widely used in different fields and various applications. DLC is one of the most important amorphous carbon materials, which is composed of a mixture of sp^2- and sp^3-hybridized carbon atoms [1, 2]. DLC is widely studied for its chemical and physical properties. It is known that such structures are inert, highly stable, and shear-resistant. DLC also exhibits high hardness, outstanding surface characteristics, good tribological behaviors, and high thermal stability [3–8]. DLC is also

Applications of Diamond-like Carbon Coatings, First Edition. Abdul Wasy Zia.

a special material with good biocompatibility [9]. The mentioned properties mesmerize researchers from different fields. However, the mechanical properties of DLC-based materials are often a step forward since the literature mostly focuses on the mechanical characteristics of the material [10]. On the other hand, DLC films have great potential due to the special structure of the material. The aforementioned properties have been commonly defined by the ratio of sp^2- and sp^3-hybridized carbon atoms. Alteration in such a ratio may alter the physical and chemical properties of the material. Moreover, the addition of different molecules or metals to the carbon structures also alters the overall structure and characteristics of the material [11, 12]. Therefore, different reports are addressed in the literature where different materials and metals can be added to the structure of DLC materials. Moreover, the addition of different molecules or metals to the carbon structures also alters the overall structure and properties of the DLC-based materials [2, 13, 14]. In addition, production characteristics and methods may have an essential role in thin film properties.

DLC-based materials are quite functional and can be used in different industries and industrial applications. Therefore, the characterization of such materials is quite important to illustrate the potential opportunities and to understand the intrinsic properties of the materials. Different characterization techniques were proposed for the characterization of such films. Microscopic techniques, such as scanning electron microscopy (SEM), atomic force microscopy (AFM), etc., were the foremost microscopic techniques used for the physical and surface characterization of the DLC-based materials. Besides, spectroscopical techniques, such as Raman spectroscopy, Fourier transform infrared spectroscopy (FTIR), UV-vis (UV visible range) spectroscopy, X-ray photoelectron spectroscopy (XPS), etc., are important spectroscopic techniques used in the chemical and structural characterization of the films [15–18].

While characterization techniques are important to illustrate the thin film quality, production methods and production parameters are also important. Various characterization techniques are addressed in the literature where DLC-based thin films could be obtained. Magnetron sputtering (MS) and ion beam deposition (IBD)-based techniques are advanced techniques, which were commonly used in DLC thin film production. Besides MS-based techniques, vapor deposition techniques such as plasma-enhanced vapor deposition and chemical vapor deposition (CVD) were also addressed in the literature. Such techniques are advanced and precise where high control was achieved on thin film structure and thickness. Some reports pertaining to pulsed laser deposition (PLD) and cathodic vacuum arc techniques were also found in the literature [3, 4, 16, 19, 20].

It is a known fact that DLC materials are favored for their mechanical properties. DLC films are durable, inert, and resistant to scratches [3, 16]. However, DLC films also exhibit interesting optical and optoelectronic properties [21]. Optical characteristics of the thin films were discussed in different reports and illustrated that DLC films have good absorbance and transmittance in infrared (IR) zones and have the potential to be used in IR-related applications [18]. It was reported that optical and optoelectronic properties alter depending on fabrication methods and parameters where the intrinsic structures change. Moreover, doping DLC thin films with metals and different molecules is often used to tune the optical and optoelectronic properties. In this regard, DLC films find implications for different applications.

In this chapter, production and investigation methods used for the DLC thin film production and evaluation were addressed in the chapter. It is known that understanding such characterization and production techniques is important to comprehend the optical properties of the DLC films and therefore, the importance of such materials for the optical applications. Thus, the optical and optoelectronic applications of DLC films are discussed in detail.

10.2 Production and Characterization Methods

DLC-based materials are outstanding materials and sophisticated methods were needed to produce such a practical material. DLC-based materials are often coated on a surface as a thin film for protective, optical, and optoelectronic purposes; thus, various methods were proposed to obtain good-quality films. It is important to achieve high control over the films during the production process where DLC films in the desired structure and formation could be obtained. The methods used in the DLC film production process are high-tech methods and therefore, they mostly require complicated and expensive infrastructure and setup [7]. MS-based methods are some of them. The MS system is a quite complex system to achieve a pure and high-quality DLC film, and a clean prediction environment was required. Hence, the production of DLC films is often conducted under high vacuum (HV) or ultra-high vacuum (UHV) conditions [7]. However, the deposition rate of the MS deposition is found to be low. To enhance the deposition rate, plasma-enhanced deposition techniques are proposed. Plasma-enhanced chemical vapor deposition (PECVD) techniques were found to be quite similar to MS technique where a reactive gas increasing the deposition rate is used to achieve high deposition rates [22]. Similarly, CVD and physical vapor deposition techniques are also addressed in the literature. However, it was reported that DLC films produced using these methods were found to be less transparent. Thus, DLC films produced using such methods are not found to be favorable for optical applications. Besides these methods, pulsed laser-based techniques were mentioned for DLC film production. Lasers are strong light sources and can be used to vaporize materials under UHV and HV conditions. Laser-related methods are often referred to as laser vapor deposition (LVD) and/or pulsed laser vapor deposition (PLVD). Since the method is applied in UHV and HV conditions, high-quality films with minimum contamination can be obtained. Therefore, nonhydrogenated inert DLC films with outstanding hardness characteristics could be produced. A similar scenario is also valid for IBD techniques. This technique allows researchers to fabricate pure and noncontaminated films. The arc deposition technique is another common technique and frequently reported in the literature where DLC films with good mechanical and tribological properties can be produced. Alternative methods were also addressed in the literature such as electron deposition and electrophoretic deposition techniques [23, 24]. Table 10.1 is prepared to compare and illustrate the details of the production techniques.

DLC-based materials were often applied in thin film form. Hence, some essential thin film characterization techniques could be used in the characterization of the DLC films. Advanced microscopical characterization techniques such as SEM and AFM are common microscopic techniques addressed in the literature for DLC film characterizations. SEM is a nondestructive method and is able to provide high-resolution data about the surface characteristics of the DLC thin films. As can be known, SEM scans the surface with free electrons and detects the secondary electrons bouncing back from the surface of the investigated area. Thus, such a method requires conducting surface area to achieve results in high resolution. In this regard, better results could be obtained for DLC films, which were doped with metals and conductive elements. Producing metal-doped films

Table 10.1 Deposition techniques used for the production of DLC films.

Technique	Abbreviation	Result
Magnetron sputtering	MS	High-quality films, low contamination
Laser vapor deposition	LVD	High-quality films, low contamination, good hardness
Plasma-enhanced chemical vapor deposition	PECVD	High-quality films, low contamination, less transparent, and not suitable for optical applications
Ion beam deposition	IBD	High-quality films, low contamination
Arc deposition	–	Good mechanical and tribological properties
Electrophoretic deposition	EPD	Slightly lower quality DLC films, contamination in the film, good for the production of composite films

could also alter the optic and optoelectronic properties of the materials as well. SEMs are often equipped with energy dispersive X-ray spectroscopy (EDX) apparatus. The method uses X-rays released by excited atoms, which were triggered by electrons released by the filament of SEM hitting the investigation area. EDX apparatus could give elemental analysis where detailed results could be achieved for especially dopants in DLC structure. Such dopants were added to DLC structure to alter the overall characteristics of DLC thin films. Further characterization techniques can be used to assess the alteration in the DLC structure. Another important tool used for the investigation of the surface structure of the DLC films is AFM. AFM is a precise and important microscopic method used for the surface characterization where important results about surface topography and surface roughness could be achieved. AFM is a nanotool, which can be used to obtain 3D surface topography. Different modes of AFM could be used for the surface investigations. The noncontact mode does not have direct contact with the surface and is often acknowledged as a nondestructive method. However, hybrid modes (tapping mode) and contact mode have intermittent contact or total contact with the surface, which may cause distortion or destruction on the surface. Such a distortion could also be used to understand the tribological properties of the surface. The mentioned drawback can be an advantage for using and investigating the mechanical characteristics of DLC thin films. Touching the surface with AFM tip or scratching the DLC surface with AFM tip by raster movement can be used to understand the mechanical and tribological properties of the DLC films. Using a similar approach, the nanoindentation tool can be used to investigate the tribological properties, surface roughness, hardness, shear behaviors, and stiffness. Understanding and characterization of mechanical properties are important parameters for DLC films since DLC films are often preferred for their mechanical superiority. DLC films illustrate superior durability, inertness, and scratch resistance [16, 20, 25]. Thus, the characterization of mechanical properties is essential to use proper material in a specific application. The nanoindentation technique is a successful method to determine such characteristics; however, the technique is a destructive method and may damage the surface and investigated area [26]. Some thin film characteristics can also be assessed using ellipsometry. The method is a nondestructive method where thin film quality and thickness can be obtained [27].

Spectroscopic methods are able to give far more details about the thin film structure and characteristics where information pertaining to the intrinsic properties of the thin films could be achieved. Spectroscopic methods are often referred to as nondestructive methods. FTIR spectroscopy and Raman spectroscopy are essential methods that were used to understand and characterize the DLC films [28–30]. FTIR and Raman spectroscopy are often used to characterize and verify the chemical composition of DLC films. Since the ratio of sp^2- and sp^3-hybridized carbon atoms is associated with the identification of DLC films, such methods can give an insight into the identification of the quality and chemical composition of DLC-based films. In Raman spectra, DLC-related peaks could be obtained to confirm the thin film structure. Especially doped DLC-based films could be characterized using FTIR spectroscopy where vibration and stretching-related bands about dopants could easily be identified. Similarly, XPS also gives an insight into sp^2- and sp^3-hybridized carbon atoms. XPS is a nondestructive method where X-ray is highly penetrable and does not pose a distortion on the applied surface [21, 31, 32]. Thus, XPS is often used to assess sp^2- and sp^3-hybridized carbon atoms where the quality of the DLC films could be obtained. Further spectroscopic methods could also be used to analyze the different characteristics of the DLC films. Among those, UV-vis spectroscopy could be used to assess the optical and optoelectronic properties of the DLC films. In UV-vis spectroscopy investigations, transmittance, reflectance, and absorbance characteristics of the thin films could be obtained [27]. Such investigations could be used to calculate band energies where optical and optoelectronic characteristics were obtained. Table 10.2 illustrates the investigation methods used for the characterization of the DLC films.

Table 10.2 Common methods used for the characterization of DLC films.

Method	Purpose	Destructive/nondestructive
SEM	Surface topography	ND
EDX	Chemical composition	ND
AFM	Surface topography, tribological properties, mechanical properties, roughness	D/ ND
Ellipsometry	Film thickness, optical constants	ND
UV-vis	Optical and optoelectronic properties	ND
Nanoindentation	Mechanical properties, toughness, mechanical properties	D
Raman spectroscopy	sp^2- and sp^3-hybridized carbon atoms	ND
FTIR	Chemical composition	ND
XPS	sp^2- and sp^3-hybridized carbon atoms	ND

D, destructive; ND, nondestructive.

10.3 Optical and Optoelectronic Properties

DLC films are often preferred for their tribological and mechanical properties. However, it is possible to produce transparent DLC films, which find a wide area of implication in optical and optoelectronic applications. For this purpose, understanding the optical and optoelectronic characteristics of the DLC films such as optical constant, refractive index, optical bandgap, transmittance, reflectance, absorption, etc., is essential. Suitable, reliable, tunable, and high-quality materials may help us to obtain high-quality and efficient optical and optoelectronic devices.

The optical constant is one of the most fundamental parameters used to define the optical characteristics of a material. Optical constant consists of two different parameters: k (extinction coefficient) and n (refractive index). The extinction coefficient is often referred to as the imaginary refractive index. In this regard, the complex refractive index can be defined as follows:

$$n_c(\lambda) = n(\lambda) - ik(\lambda), \tag{10.1}$$

where n_c is the complex refractive index, n is the refractive index, ik is the imaginary part of the refractive index, and λ is the wavelength. It can be assumed that the refractive index is highly dependent on wavelengths where light in different wavelength may have different refractive characteristics in the same material. The refractive index basically gives an intuition about the light bending capacity of a material.

The refractive index is highly dependent on the thickness of the material. It is important to know the thickness of the material. In some cases, such as fine structures and/or thin films, the thin film thickness can only be determined by special tools such as ellipsometry, transmittance spectra, etc. Different theories were proposed to determine the optical constant and refractive index such as Hartmann, Sellmeier, Drude, Schott Glass, Snell rules, Cauchy, etc. [33]. Besides, different optical parameters, such as transmittance, reflectance, absorbance, etc., could be calculated using optical constants. Such optical parameters are affected by the DLC structures. In this regard, hydrogenated and nonhydrogenated DLC structures may have different optical characteristics. Moreover, the addition of different molecules to the DLC film structure also alters the intrinsic properties of the DLC films, so the optical parameters. The main reason for such alteration is alteration in atomic structure. Atoms forming hydrogenated and nonhydrogenated DLC films have different molecular sizes; such a case alters the deposition conditions and therefore the optical properties. A similar case is also valid for the doped DLC films. The doping process alters the structure of the DLC film, which results in the alteration of the optical properties of the films. Flaws and surface defects may also alter the optical constants of the DLC films. In most cases, the refractive index of the DLC films was measured around 1.4–1.5 for 20–100 nm film thickness. Such a refractive index is found to be quite low compared to the refractive index of common semiconductors [33]. For example, the refractive index of semiconductors such as silicon and germanium was found to be 3.5 and 4, respectively. It was seen that DLC films have quite low refractive index; such a low refractive index makes DLC-based materials suitable candidates for optical and optoelectronic applications and devices.

Optical bandgap (E_g) is another optoelectronic parameter, which is essential for optical and optoelectronic applications. Optical bandgap is a quite complex issue, and it is important to understand microelectronic and optoelectronic devices. In Tauc definition, the bandgap is defined with absorption coefficient, which indicates that optoelectronic properties of the materials are highly dependent on optical parameters.

The relation is defined as follows:

$$\alpha h\nu = A\ (h\nu - E_g)\ n, \tag{10.2}$$

where $\alpha = F(R)/d$, A is a constant that refers to the absorption coefficient, d is the thickness of thin films, E_g is bandgap energy, and n is a constant that determines the optic transfer. Two different kinds of transfer may be observed: (i) direct transfers and (ii) indirect transfers. Indirect transfers were determined as $n = 1/2, 3/2, 5/2...$, while direct transfers were determined as $n = 1, 2, 3....$

It was seen that different production methods result in DLC films with different qualities where hybridization ratios alter and therefore, thin film structure. Such alterations change the optical parameters. It was seen that different bandgap energies were reported for different production techniques. In the literature, the optical bandgap of DLC films was found between 0.5 and 3.0 eV. In the report of Lu et al., it was claimed that increased sp^3/sp^2 hybridization ratio enhances the calculated optical bandgap [33]. They claim that the sp^3 ratio is important, but it is not a single criterion. The production and structure of the films are also important for the optical characteristics. For example, C–H bond content, film density, grain size of the films, deposition rates, and surface roughness may also have an impact on the bandgap energy. Different production methods have different production environments and, therefore, result in DLC films in different structures. In this regard, it is reasonable to say that adding different dopants and different materials in thin films inevitably alters the thin film structure. It was seen in the literature that DLC films doped with different molecules and/or metals exhibit different bandgap energies (Table 10.3).

Table 10.3 Refractive index and/or bandgap energy of DLC-based thin films.

Film type	Production method	Refractive index	Binding energy (eV)	Reference
Boron-doped DLC	Low methane/ argon sputtering	1.7–3.2	3.65–2.38	[34]
DLC	Electrodeposition	–	3.87–3.89	[35]
Hydrogenated DLC	PECVD	2.06–2.15	1.60–1.70	[22]
Silicon-incorporated DLC	PECVD	–	0.62–2.22	[36]
Nickel-incorporated DLC	Low-voltage electrode position	–	1.58–2.31	[37]
Nonhydrogenated DLC	DC MS	1.6–2.4	0.80–2.06	[11]
Phosphorous-doped DLC	PLD	–	0.75–0.95	[38]
Iodine-doped DLC	PECVD	–	1.9–3.5	[30]
Silver-doped DLC	Radio frequency reactive sputtering	–	3.25–3.97	[39]
Cu-doped DLC	PLD	–	1.71–1.93	[40]
Zinc-doped DLC	PECVD	–	1.71–1.79	[41]

10.4 Effect of Production Methods/Parameters on DLC Film Properties

As mentioned earlier, different reports illustrate that the production parameters, production method, and addition of foreign atoms to the DLC structure alter the chemical and physical properties of the DLC films through which properties of the DLC films could be adjusted. It was mentioned in the production method that DLC films have outstanding properties and therefore, they are produced in a very clean environment and under HV conditions to reduce contamination and obtain high-quality films. Therefore, different production methods are used for the fabrication of the films. However, almost all of the methods follow high standards. Ar and He are favorable gasses used in the production processes to achieve high-quality films. To enhance the deposition rate and reduce the contamination, different types of gasses are used in the fabrication process. Ar is a large molecule; in ion-based deposition techniques, ionized Ar^+ enhances the bombardment and increases the deposition rate [7]. Inert gases are widely used for the thin film production and are commonly addressed in the literature. Since inert gasses do not react with materials, using inert gasses in the production of DLC films reduces contamination and the addition of different molecules. Such a case enables researchers to obtain nonhydrogenated DLC films. Application of inert gasses in the production process also reduces the oxidation and therefore C=O bonds in the DLC film structure. However, ionized Ar^+ atom bombardment increases the formation rate of sp^2-hybridized atoms. Moreover, thin films produced under inert gas tend to be more granular form, which could be detected by probe microscopy-related methods [33].

Reactive gasses such as hydrogen, oxygen, and nitrogen were also addressed in the literature for DLC production. The addition of such gasses results in DLC thin films in different formations. For example, the addition of hydrogen gas enhances hydrogenated molecules and is used to produce hydrogenated DLC films. The process is commonly used in physical vapor deposition (PVD) and CVD-based production methods. The authors use such methods to tune the optical and optoelectronic properties of the DLC films. It was reported that such methods enhance the number of molecules with C–H bonds and sp^3 hybridization. It was reported that such a case enhances the optical bandgap of standard DLC films [12, 34, 42, 43]. Increased hydrogenated molecules enhance the transparency of the films in the visible region [21, 44, 45]. As expected, hydrogenation process slightly reduces the hardness of the DLC films.

Similarly, nitrogen gas is used in CVD and PVD methods. The addition of nitrogen alters the chemical structure and bonding between C–C atoms and enhances the number of molecules with sp^1 hybridization. It was seen that the presence of nitrogen gas in the production process enhances the transmittance in the visible region and increases the optical bandgap. It was also reported that the addition of nitrogen gas increases the antireflective characteristics of the DLC films [15, 46, 47].

The addition of oxygen gas is a favorable method to obtain nonhydrogenated DLC films since oxygen is a highly reactive gas and likes to make bonds with C-based atoms. DLC films produced in the existence of oxygen gas tend to contain C–O and C=O bonds. It was reported that the addition of oxygen gas in the production increases the optical bandgap and diminishes the refractive index [15, 46, 47]. As expected, the mechanical properties of the DLC film are also affected by the existence of oxygen molecules in the DLC structure where decreased mechanical properties (such as decreased hardness) were observed. Since oxygen destructs C=C and C–C bonding, it should be expected to obtain decreased mechanical properties.

The addition of different materials in DLC structures is becoming increasingly popular. It can be proposed that adding different materials in the DLC structure can be used to tune or adjust electrical, electronic, optical, and optoelectronic properties. Among these materials, metals are quite popular. Metals exhibit good electrical properties and are used in different microelectronic and optoelectronic applications. Therefore, the addition of metal molecules in DLC structures can enhance the electrical, electronic, and optoelectronic characteristics of the DLC films. Silver, titanium, copper, and zinc are favorable materials used in different optoelectronic applications [32, 48–50]. Hence, such materials are preferred to produce metal/composite DLC films or metal-doped DLC films. MS, PVD, and CVD techniques are widely used techniques, which have high precision where dopant content and rate can strictly be controlled. Doping DLC films with metals enhances the tribological, mechanical, and electronic properties. It was also seen in the literature that metal doping can be used to tune optical bandgap, electrical, and electronic properties; reduced electrical conductivity was addressed in the literature [51, 52]. Different materials may have different effects on DLC films. For example, Ag doping increases the absorption coefficient of DLC films, Cu and Ni doping decreases the bandgap energy [1, 40, 53].

Besides metals, different molecules are also used to dope DLC structures. Reports pertaining to nonmetallic elements like boron and phosphorous were evidenced in the literature [34, 54]. In addition, metalloids like Si and Ge are commonly addressed to dope DLC films. The atomic structure of the boron is quite similar to carbon, and it can replace the carbon atoms with great success and get into the structure of DLC films. Thus, high-quality boron-doped DLC films were previously reported in the literature. However, reports claim that boron doping increases the graphite-related formations in the film structure [55]. Silicon and germanium are metalloids vastly used in semiconductor applications. Reports pertaining to the investigation of the optical and optoelectronic properties of the Si- and Ge-doped DLC films can be found in the literature [9, 23, 36]. It was shown in the literature that the addition of Si atoms to the DLC structure enhances the sp^3/sp^2 rate and also increases the optical properties where optical bandgap increases and the transmittance in IR zone enhances [56]. The addition of silicon atoms also enhances the elasticity of the DLC films and reduces the film density [44]. However, a slight decrease in the refractive index was also addressed in the reports. Using Ge atoms as a dopant increases the electrical and optoelectrical characteristics of the DLC films where reduced bandgap was reported. A similar case was also valid for phosphorous doping; it was seen that the addition of phosphorous atoms to the DLC structure reduces the optical bandgap but increases the refractive index [54, 57, 58].

10.5 DLC Films in Optical and Optoelectronic Applications

Today, it is a well-known fact that DLC films exhibit outstanding mechanical performance including shear resistance and good tribological properties. They are quite hard and inert; thus, they can resist corrosive and abrasive effects. Such facts make DLC films a suitable material for protective purposes; hence, DLC films were used in mechanical parts, biomedical applications, prosthetic applications, micromechanical applications, the food industry (inner wall of food containers), the automotive industry, aerospace industry, microelectronic industry, construction equipment, etc. Besides their mechanical excellence, DLC films have huge potential for optical implications. Optical properties like transparency, refractive index, absorbance, and reflectance can be adjusted by altering production parameters. DLC films have relatively high transmittance, and such characteristics enable them to be used in various optical and optoelectronic applications. For example, solar cells were coated with DLC thin films where a high transmission rate was observed (approx. 80–95%) [27, 59]. Compared to standard silicon-based solar cells, the refractive index of the DLC films was found to be low. It should also be noted that DLC is inert and has high resistance to chemical and physical artefacts. The reflectance rate of the DLC materials was also found to be quite low where enhanced absorbance and transmittance rate was observed. Such results indicate that DLC-based materials are suitable candidates for optical and optoelectronic applications. In this regard, it is possible to say that enhanced transmittance and absorbance rates could enhance the efficiency of the optoelectronic performance of solar harvesting devices. The DLC films could also protect the device from external factors.

Light-emitting microelectronic devices, like light-emitting diodes (LEDs), are essential for the modern world and advanced technologies. They can also find applications in display technologies [60]. Different protective materials can be used in display technologies for protective purposes. Plastics and polycarbonate-based materials are quite popular in display technologies for protective purposes. They are transparent and flexible and illustrate good optical characteristics. However, they are soft and can easily be scratched. Also, the optical properties of polycarbonate-based materials and plastics could not be tuned. DLC thin films were applied with polycarbonate and plastic materials to enhance tribological properties. For example, the inner walls of plastic bottles were coated for this purpose. In this regard, DLC films were applied to the polycarbonate films, which were used in display technologies. It was proposed that such applications increase scratch resistance. High transparency characteristics could enhance visual perfection.

Antireflectance characteristics of the DLC films are reported in the literature and in our report. Coating a surface with DLC films can dramatically increase the antireflective characteristics and enhance the transmittance of the surface. It was previously reported in the literature that coating a germanium surface with DLC could increase the transmittance by around 60% in the mid-IR zone [18, 61]. Similar reports pertaining to the far IR zone were also reported in the literature [61]. Cases indicate that coating a semiconductor-based surface with DLC could enhance the transmittance and decrease reflectance in the IR zone, which shows that DLC films have the potential to enhance the performance of IR-based detectors and cameras. For this purpose, multilayered DLC films have been proposed. It was reported that adding Zn-based materials to the DLC structure enhances the transparency of the films [33]. Such a multilayered structure can have high mechanical tolerance and performance; they can also provide mechanical protection for the detectors and cameras.

The mechanical properties of DLC films were vastly addressed in the literature while there is a limited effort in the literature to understand the optical properties. Besides, there is slight interest in the literature to assess and investigate the electrical properties of DLC films. DLC films are carbon-based structures and may not exhibit outstanding electrical characteristics. However, doping DLC with metal-based atoms could enhance the electrical characteristics dramatically. Since DLC films have considerable optical properties, enhancing the electrical properties may help them to be used in optoelectronic and microelectronic applications. Organic electronic applications basically count on organic materials with conductive properties. Flexibility is the fundamental advantage of such materials. Organic materials are flexible, and the molecules used in organic electronics can be applied to surfaces in different shapes and geometries. Similarly, DLC films consist of carbon- and hydrogen-based molecules and exhibit organic characteristics. DLC films are durable, scratch resistant, inert, and resistant to humidity. They are transparent and have quite interesting absorbance characteristics. In this perspective, metal-doped DLC films have the potential to be used in organic electronic applications as microelectronic or optoelectronic components. In this regard, enhancing the electrical properties of the DLC films can make a considerable contribution to optoelectronic applications. Different transparent substrates and base materials were used in organic electronic applications. Indium tin oxide (ITO) is one of the most favorable materials, which are used in organic electronic applications. It has high transparency and low resistivity. Optical and optoelectronic characteristics of the ITO films are found to be quite similar to DLC films. Different reports in the literature illustrate that DLC films can be used as an essential material for solar harvesting applications such as solar cells, photodiodes, and photodetectors. It should be noted that the electrical properties of DLC films are relatively poor and the resistivity of the DLC films is quite high. Therefore, it is essential to doping DLC films with metals and conductive materials. In the literature, metals like Zn, Ti, Ag, Ni, and Cu are important materials that enhance the electrical characteristics [32, 48–50, 62]. It was evidenced in the literature that enhancing the electrical properties of DLC films with metals enables them to be used in solar harvesting applications. It was illustrated that Zn-doped, Ti-doped, Ag-doped, Ni-doped, and Cu-doped films can be used as active layers for photodetector and photodiode applications [32, 48–50, 62]. Besides metal doping, metalloids have the potential and can be used to dope DLC thin films for optoelectronic applications. Ge and Si are common metalloids, which are expected to be used in microelectronic and optoelectronic applications. Reports showed that Ge- and Si-doped DLC films exhibit good photodiode characteristics [23, 27, 63].

Since the mechanical properties of the DLC films are well known and well defined, DLC films were applied to different surfaces for decorative and protective purposes where indirect benefits of the optical characteristics of the DLC films were seen. In this regard, using DLC films in the fashion industry has great value since the fashion market is quite big and DLC films can be used for decorative purposes. Optic accessories like glasses and sunglasses are leading products, which are quite expensive and profitable for the fashion industry. Outstanding optical properties (low refractivity, absorbance, reflectance, transmittance characteristics) of DLC films enables them to be applied to such products for decorative and protective purposes. As DLC films have good mechanical endurance, DLC films provide scratch and shear resistance on lenses applied in sunglasses [3]. DLC films are inert to chemicals; thus, coating the surface of the lenses of the glasses significantly enhances the resistance against corrosion, dirt, and chemicals. Similarly, protective goggles are also treated with DLC films. Protective goggles cover the eye of the user, which were vastly used in tough conditions. Protective goggles are made of polycarbonate and/or plastics. Polymer-based materials are often found to be soft and they are not resistant to scratches and mechanical factors. Protective goggles aim to protect the eye of the user and provide protection against external factors like chemicals, dust, grains, scratching objects, etc. DLC films are applied to the surface of the protective goggles to reduce reflection and enhance visibility. DLC films are resistant to chemicals and mechanical effects. They are hard and scratch resistant. Hence, they improve the mechanical and chemical resistance of the surface of the safety goggles against external factors [3].

Zinc sulphite is a special material, which was used in different optical applications such as optical windows. It is highly transparent and has high clarity. Therefore, zinc sulphite-based materials were used in different optical and optoelectronic applications especially on IR detectors and IR instrumentation windows [64]. DLC films have good optical properties where good transparency and absorbance characteristics in IR regions were reported in different works. Due to the optical properties and mechanical resistance, DLC films were applied on zinc sulphite-based IR windows. Similar cases were also valid for aluminum mirrors. Such mirrors were used in IR detectors and IR-based cameras. Such material is found to be fragile and is not resistant to chemical and physical effects. Due to the optical properties of DLC films in IR regions, DLC films have huge potential as a protective layer, which is applied to aluminum mirrors [64]. Similarly, laser bar code scanner windows are produced using different materials. Bar code scanners often use a special light, which is around 630 nm wavelength. Thus, scanners need to be highly transparent for the light at around 630 nm, which is close to near-IR zone. Laser barcode scanners are used in different sectors and fields. Therefore, they need to be durable and reliable. DLC thin films are special films, which are durable, inert, and resistant to physical factors. DLC films have high transparency around near-IR zone. Such films are suitable films, which can be applied to the bar code scanner windows.

10.6 Conclusion

The mechanical properties of the DLC films are commonly discussed in the literature. However, it should be noted that DLC films exhibit interesting optical and optoelectronic properties. Intrinsic and optical properties of DLC films can be tuned with different techniques where DLC films in desired characteristics are obtained. Optical properties of DLC films such as absorbance, transmittance, reflectance, etc., enabled DLC films to be used in different optic and optoelectronic applications. In this chapter, the optical and optoelectronic applications of DLC films are discussed in detail. It is illustrated that DLC films have great potential and can be used in different optical and optoelectronic applications.

References

1 Meškinis, S., Čiegis, A., Vasiliauskas, A. et al. (2015). Optical properties of diamond like carbon films containing copper, grown by high power pulsed magnetron sputtering and direct current magnetron sputtering: structure and composition effects. *Thin Solid Films* 581: 48–53.

2 Moghadam, R.Z., Ehsani, M.H., Dizaji, H.R. et al. (2021). Oxygen doping effect on wettability of diamond-like carbon films. *Materials Research Express* 8 (3): 035601.

3 Kimock, F.M. and Knapp, B.J. (1993). Commercial applications of ion beam deposited diamond-like carbon (DLC) coatings. *Surface and Coatings Technology* 56 (3): 273–279.

4 Joshi, P., Riley, P.R., Denning, W. et al. (2022). Laser-patterned carbon coatings on flexible and optically transparent plastic substrates for advanced biomedical sensing and implant applications. *Journal of Materials Chemistry C* 10 (8): 2965–2975.

5 Moolsradoo, N., Abe, S., and Watanabe, S. (2011). Thermal stability and tribological performance of DLC-Si-O films. *Advances in Materials Science and Engineering* 2011 (1): 483437.

6 Schlatter, M. (2002). DLC-based wear protection on magnetic storage media. *Diamond and Related Materials* 11 (10): 1781–1787.

7 Grill, A. (1999). Diamond-like carbon: state of the art. *Diamond and Related Materials* 8 (2–5): 428–434.

8 Khan, M.I. Adil, F., Majeed, S. et al. (2019). Structural, morphological, electrical and optical properties of Cu doped DLC thin films. *Materials Research Express* 6 (12): 126420.

9 Arslan, M.E., Kurt, M.Ş., Aslan, N. et al. (2022). Structural, biocompatibility, and antibacterial properties of Ge–DLC nanocomposite for biomedical applications. *Journal of Biomedical Materials Research Part B: Applied Biomaterials* 110 (7): 1667–1674.

10 Suasnavas, C.V., Makeev, M.O., Osipkov, A.S. et al. (2019). Relationship between hardness and optical properties of diamond-like carbon coatings. *IOP Conference Series: Materials Science and Engineering* 675 (1): 012051.

11 Coşkun, Ö.D. and Zerrin, T. (2015). Optical, structural and bonding properties of diamond-like amorphous carbon films deposited by DC magnetron sputtering. *Diamond and Related Materials* 56: 29–35.

12 Sheng, H., Xiong, W., Zheng, S. et al. (2021). Evaluation of the sp^3/sp^2 ratio of DLC films by RF-PECVD and its quantitative relationship with optical band gap. *Carbon Letters* 31 (5): 929–939.

13 Jo, Y.J., Zhang, T.F., Son, M.J. et al. (2018). Synthesis and electrochemical properties of Ti-doped DLC films by a hybrid PVD/PECVD process. *Applied Surface Science* 433: 1184–1191.

14 Basman, N., Uzun, R., Gocer, E. et al. (2018). Electrodeposition of Si–DLC nanocomposite film and its electronic application. *Microsystem Technologies* 24 (5): 2287–2294.

15 Solomon, I., Shukla, K., Bhatnagar, M. et al. (2020). Effect of nitrogen concentrations on optical, structural and mechanical properties of self organized a-C:N films. *Ceramics International* 46 (9): 13743–13751.

16 Lin, Z., Wang, F., Gao, D. et al. (2013). Frictional and optical properties of diamond-like-carbon coatings on polycarbonate. *Plasma Science and Technology* 15 (7): 690.

17 Dwivedi, N., Kumar, S., Carey, J.D. et al. (2013). Influence of silver incorporation on the structural and electrical properties of diamond-like carbon thin films. *ACS Applied Materials & Interfaces* 5 (7): 2725–2732.

18 Ankit, K., Varade, A., Reddy, N. et al. (2017). Synthesis of high hardness IR optical coating using diamond-like carbon by PECVD at room temperature. *Diamond and Related Materials* 78: 39–43.

19 Wang, F.M., Chen, M.W., and Lai, Q.B. (2010). Metallic contacts to nitrogen and boron doped diamond-like carbon films. *Thin Solid Films* 518 (12): 3332–3336.

20 Patnaik, L., Maity, S.R., and Kumar, S. (2020). Comprehensive structural, nanomechanical and tribological evaluation of silver doped DLC thin film coating with chromium interlayer (Ag-DLC/Cr) for biomedical application. *Ceramics International* 46 (14): 22805–22818.

21 Das, D. and Banerjee, A. (2015). Anti-reflection coatings for silicon solar cells from hydrogenated diamond like carbon. *Applied Surface Science* 345: 204–215.

22 Sun, Y., Huang, X.Y., and Wang, H. (2016). Influence of hydrogen content on optical and mechanical performances of diamond-like carbon films on glass substrate. *Journal of Materials Engineering and Performance* 25 (4): 1570–1577.

23 Aslan, N., Kurt, M.Ş., and Mehmet Koç, M. (2022). Morpho-structural and optoelectronic properties of diamond like carbon–germanium (DLC-Ge) composite thin films produced by magnetron sputtering. *Optical Materials* 126: 112229.

24 Aslan, N., Ceylan, B., Koç, M.M. et al. (2020). Metallic nanoparticles as X-ray computed tomography (CT) contrast agents: a review. *Journal of Molecular Structure* 1219: 128599.

25 Jing, P.P., Ma, D.L., Gong, Y.L. et al. (2021). Influence of Ag doping on the microstructure, mechanical properties, and adhesion stability of diamond-like carbon films. *Surface and Coatings Technology* 405: 126542.

26 Modabberasl, A., Kameli, P., Ranjbar, M. et al. (2015). Fabrication of DLC thin films with improved diamond-like carbon character by the application of external magnetic field. *Carbon N.Y.* 94: 485–493.

27 Banerjee, A. and Das, D. (2014). Fabrication of highly transparent diamond-like carbon anti-reflecting coating for Si solar cell application. *AIP Conference Proceedings* 1591 (1): 856–857.

28 Zeng, Q., Yu, F., and Dong, G. (2013). Superlubricity behaviors of Si3N4/DLC Films under PAO oil with nano boron nitride additive lubrication. *Surface and Interface Analysis* 45 (8): 1283–1290.

29 Zemek, J., Jiricek, P., Houdkova, J. et al. (2021). On the origin of reduced cytotoxicity of germanium-doped diamond-like carbon: role of top surface composition and bonding. *Nanomaterials* 11 (3): 567.

30 Omer, A.M.M., Adhikari, S., Adhikary, S. et al. (2006). Electrical conductivity improvement by iodine doping for diamond-like carbon thin-films deposited by microwave surface wave plasma CVD. *Diamond and Related Materials* 15 (4–8): 645–648.

31 Ray, S.C., Pong, W.F., and Papakonstantinou, P. (2016). Iron, nitrogen and silicon doped diamond like carbon (DLC) thin films: a comparative study. *Thin Solid Films* 610: 42–47.

32 Koç, M., Aslan, N., Erkovan, M. et al. (2019). Electrical characterization of solar sensitive zinc oxide doped-amorphous carbon photodiode. *Optik* 178: 316–326.

33 Lu, Y., Wang, S., Huang, G. et al. (2022). Fabrication and applications of the optical diamond-like carbon films: a review. *Journal of Materials Science* 57 (6): 3971–3992.

34 Ahmad, A.A. (2017). Optical and electrical properties of synthesized reactive RF sputter deposited boron-rich and boron-doped diamond-like carbon thin films. *Journal of Materials Science: Materials in Electronics* 28 (2): 1695–1705.

35 Aslan, N., Başman, N., Uzun, O. et al. (2019). The effects of deposition potential on the optical, morphological and mechanical properties of DLC films produced by electrochemical deposition technique at low. *Materials Science* 37 (2): 166–172.

36 Ahmed, S.F., Banerjee, D., Mitra, M.K. et al. (2011). Visible photoluminescence from silicon-incorporated diamond like carbon films synthesized via direct current PECVD technique. *Journal of Luminescence* 131 (11): 2352–2358.

37 Pandey, B., Mukherjee, J., Das, B. et al. (2014). Nickel concentration dependent structural and optical properties of electrodeposited diamond like carbon thin films. *European Physical Journal Applied Physics* 66 (1): 10302.

38 Rusop, M., Soga, T., and Jimbo, T. (2006). Photovoltaic characteristics of phosphorus-doped amorphous carbon films grown by r.f. plasma-enhanced CVD. *Solar Energy Materials and Solar Cells* 90 (18–19): 3214–3222.

39 Huang, Q.F., Yoon, S.F., Zhang, Q. et al. (2002). Dielectric properties of molybdenum-containing diamond-like carbon films deposited using electron cyclotron resonance chemical vapor deposition. *Thin Solid Films* 409 (2): 211–219.

40 Khan, M.I., Adil, F., Majeed, S. et al. (2019). Structural, morphological, electrical and optical properties of Cu doped DLC thin films. *Materials Research Express* 6 (12): 126420.

41 Chattopadhyay, K.K. and Banerjee, D. (2016). Non-linear conduction behaviour of chemical vapour deposited zinc doped diamond like carbon. *Advanced Science, Engineering and Medicine* 8 (11): 881–887.

42 Risplendi, F., Bernardi, M., Cicero, G. et al. (2014). Structure-property relations in amorphous carbon for photovoltaics. *Applied Physics Letters* 105 (4): 043903.

43 Zhou, X. Zheng, Y., Shimizu, T. et al. (2020). Colorful diamond-like carbon films from different micro/nanostructures. *Advanced Optical Materials* 8 (11): 1902064.

44 Baek, S.M., Shirafuji, T., Saito, N. et al. (2011). Fabrication of transparent protective diamond-like carbon films on polymer. *Japanese Journal of Applied Physics* 50 (8 PART 2): 08JD08.

45 Mancini, D., Percot, A., Bellot-Gurlet, L. et al. (2021). On-site contactless surface analysis of modern paintings from Galleria Nazionale (Rome) by reflectance FTIR and Raman spectroscopies. *Talanta* 227: 122159.

46 Mieszawska, A.J., Gianella, A., Cormode, D.P. et al. (2012). Engineering of lipid-coated PLGA nanoparticles with a tunable payload of diagnostically active nanocrystals for medical imaging. *Chemical Communications* 48 (47): 5835–5837.

47 Yimin, L., Guojun, H., Yanlong, G. et al. (2018). Experiment research on the double-layer diamond-like carbon film prepared by double laser beams. *Infrared Laser Engineering* 47 (11): 1121003.

48 Aslan, N., Kurt, M.Ş., and Koç, M.M. (2023). Silver-doped diamond-like carbon (DLC:Ag) nanocomposite films for solar tracking applications. *Journal of Electronic Materials* 52 (4): 2807–2818.

49 Aslan, N., Koç, M.M., Dere, A. et al. (2018). Ti doped amorphous carbon (Al/Ti-a:C/p-Si/Al) photodiodes for optoelectronic applications. *Journal of Molecular Structure* 1155: 813–818.

50 Koc, M.M. (2020). Photoelectrical properties of solar sensitive CuO doped carbon photodiodes. *Journal of Molecular Structure* 1208: 127872.

51 Jelinek, M., Zemek, J., Kocourek, T. et al. (2016). Dual laser deposition of Ti:DLC composite for implants. *Laser Physics* 26 (10): 105605.

52 Grigoriev, S.N., Fominski, V.Y., Volosova, M.A. et al. (2016). Surface modification of diamond-like carbon coatings to control over run-in processes in friction pair. *Journal of Friction and Wear* 37 (1): 7–14.

53 Meškinis, Š., Peckus, D., Vasiliauskas, A. et al. (2017). Photovoltaic properties and ultrafast plasmon relaxation dynamics of diamond-like carbon nanocomposite films with embedded Ag nanoparticles. *Nanoscale Research Letters* 12 (1): 1–10.

54 Dey, R., Dolai, S., Hussain, S. et al. (2018). Phosphorus doping of diamond-like carbon films by radio frequency CVD-cum-evaporation technique. *Diamond and Related Materials* 82: 70–78.

55 Xia, X.H., Lu, L., Walton, A.S. et al. (2012). Origin of significant visible-light absorption properties of Mn-doped TiO_2 thin films. *Acta Materialia* 60 (5): 1974–1985.

56 Barve, S.A., Chopade, S.S., Kar, R. et al. (2017). SiOx containing diamond like carbon coatings: effect of substrate bias during deposition. *Diamond and Related Materials* 71: 63–72.

57 Hevia, S.A., Bejide, M., Duran, B. et al. (2018). Nanometric thin films of non-doped diamond-like carbon grown on n-type (P-doped) silicon substrates as electrochemical electrodes. *Journal of Solid State Electrochemistry* 22 (9): 2845–2853.

58 Da Silva, D.S., Côrtes, A.D.S., Oliveira, M.H. et al. (2011). Application of amorphous carbon based materials as antireflective coatings on crystalline silicon solar cells. *Journal of Applied Physics* 110 (4): 043510.

59 Leonard, R.L., Hasan, S.A., Terekhov, A.Y. et al. (2012). Protective coatings for enhanced performance in biomedical applications. *Surface Engineering* 28 (7): 473–479.

60 Chabinyc, M.L. and Salleo, A. (2004). Materials requirements and fabrication of active matrix arrays of organic thin-film transistors for displays. *Chemistry of Materials* 16 (23): 4509–4521.

61 Zambrano, D.F., Villarroel, R., Espinoza-González, R. et al. (2021). Mechanical and microstructural properties of broadband anti-reflective TiO_2/SiO_2 coatings for photovoltaic applications fabricated by magnetron sputtering. *Solar Energy Materials and Solar Cells* 220: 110841.

62 Kurt, M.Ş., Yıldırım, F., Orhan, Z. et al. (2021). Influence of thickness of the sputtered diamond-like carbon (DLC) on electronic and dielectric parameters of the Au/DLC/n-Si heterojunction. *Journal of Materials Science: Materials in Electronics* 32 (20): 25214–25224.

63 Basman, N., Uzun, R., Gocer, E. et al. (2018). Electrodeposition of Si–DLC nanocomposite film and its electronic application. *Microsystem Technologies* 24 (5): 2287–2294.

64 Lettington, A.H. and Smith, C. (1992). Optical properties and applications of diamond-like carbon coatings. *Diamond and Related Materials* 1 (7): 805–809.

Section V

Emerging Applications of Diamond-like Carbon Coatings

11

Diamond-like Carbon Coatings for Polymers and Textiles

Witold Kaczorowski

Division of Biomedical Engineering and Functional Materials, Institute of Material Science and Engineering, Lodz University of Technology, Lodz, Poland

CONTENTS

11.1 Introduction, 315
11.2 Polymer Substrates and Techniques for Developing DLC Coatings on Them, 317
11.3 Properties of DLC Coatings for Which They Find Application in Polymer Modification, 318
11.4 Application Areas for Carbon Coatings for Polymers and Textiles, 331
11.5 Conclusion, 335

11.1 Introduction

The issues related to the production of diamond-like carbon (DLC) coatings on polymer substrates or textiles are widely reported in the literature today. Developments in this field are closely linked to the achievable properties of carbon coatings, such as, among others, excellent tribological and biological properties or a high wetting angle of the modified surface. Compared to other engineering materials, polymeric materials in the form of solid materials – fibers, powders, or textiles – can be modified by plasma processes, provided that the temperature conditions are suitable for them. Therefore, in this case, in order to obtain a functional diamond-like coating, it is necessary to know not only about the structure or properties of the substrates but also about the interaction of the plasma with their surface. However, like other engineering materials, we can etch polymers with plasma, functionalize their surface, or apply coatings to them. However, one phenomenon encountered only in polymer-plasma considerations is the possibility of cross-linking of polymer chains, and this process can be used to achieve the best performance of modified substrates.

Applications of Diamond-like Carbon Coatings, First Edition. Abdul Wasy Zia.

Among the plasma techniques used to produce DLC coatings on the described group of materials, one can mention not only those based on low (e.g. radio frequency plasma-enhanced chemical vapor deposition [RF-PECVD]) or high vacuum (e.g. pulsed laser deposition [PLD] or magnetron sputtering physical vapor deposition [MS-PVD]), but also techniques based on atmospheric plasma (e.g. dielectric barrier discharge [DBD]). The usability of carbon coatings on flexible polymeric materials is particularly influenced by their elastic modulus or stiffness, which must be selected depending on the substrate used for modification. The effects of limitations due to the properties of DLC coatings can be reduced by applying interlayers (e.g. through the polymer cross-linking processes mentioned above) that allow a smooth change in properties across the cross-section between the substrate and the coating. That is why the production of DLC coatings is so often preceded by plasma pretreatment processes, during which the surface layer of the polymeric materials is modified to form an interlayer with individual properties that depend on the substrate material used [1]. On the other hand, having the possibility to control the properties of carbon coatings by controlling the proportion of C–C sp^3, C=C sp^2 bonds, and hydrogen content in the coatings, it is possible to match the properties of DLC coatings to such demanding materials as polymer substrates. It should also not be forgotten that, as with other coatings, the stress state of DLC coatings on polymeric materials increases with thickness [2]. Therefore, among the promising coatings for polymer substrates, those with thicknesses in the order of a few to tens of nanometers are often indicated, which is closely related to their low-stress state and highest flexibility. Of course, the problem of the high state of residual stress can also be solved in other ways, most often by interlayer or doped DLC coatings.

By controlling the process parameters, DLC and doped DLC coatings are most commonly produced on polymer substrates, either without or with an interlayer. Carbon coatings on polymer substrates are usually subjected to identification using two basic techniques – Raman spectroscopy and X-ray photoelectron spectroscopy (XPS) [3–5]. Each of these allows the assessment of the presence of C–C sp^3 and C=C sp^2 bonds in the coatings, with the proviso that the XPS technique enables this process to be carried out quantitatively and Raman spectroscopy qualitatively. The analysis of the occurrence of the phases above-mentioned in the coatings provides the basis for considering virtually all the properties of DLC coatings on various substrates.

From the point of view of the various applications of carbon coatings on polymers, the most common property conditioning their use is the achievement of good tribological properties, especially wear resistance [6, 7]. Some modifications are also subject to evaluation of the way the coating loses continuity during tensile or compression tests on flexible polymer substrates [8–10]. The mentioned, but also other, mechanical properties of DLC coatings on polymeric materials, such as their hardness and Young's modulus, are closely related to the substrate material and the content of C–C sp^3 and

C=C sp^2 bonds in the coatings [11, 12]. The deposited carbon coatings strongly influence the morphology and topography of the surface, which can be practically exploited in the formation of wrinkles on their surface [13–17]. In many applications, the wettability of the surface is also very important, which, thanks to the use of carbon coatings, can be successfully altered to obtain parameters that are not achievable with unmodified polymers. On the one hand, it is possible to achieve hydrophobicity and wetting angles of modified surfaces even above 130° [3, 18], and on the other hand, hydrophilicity [2, 11] is attributed to the extensive controllability of changes in the chemical structure of the coatings and/or their doping processes. Studies of changes in wettability, over time, of polymer surfaces modified with DLC coatings demonstrate their high stability and prove the barrier nature of such coatings [3, 12, 16]. Other properties of the modifications discussed are closely related to their intended application. Much attention has been paid to biological research showing the usefulness of DLC coatings on polymers in biomedicine [11, 19–21]. Their various properties are analysed to determine the effect on altering the permeability of polymers to water or gases [5, 22, 23], chemical resistance [24–26], optical [4, 27, 28], or electrical [11, 29] parameters. The properties of DLC coatings for which they find application in polymer modification will be discussed in detail in the following subsection.

11.2 Polymer Substrates and Techniques for Developing DLC Coatings on Them

There is a wide range of polymeric materials whose surfaces are attempted to be modified with DLC coatings. It is also important to note that these range from bulk materials to fabrics or powders. Looking at their application, these are most often used materials for biomedicine (e.g. polyurethanes, polydimethylsiloxane [PDMS], polyether-ether-ketone [PEEK], polyethylenes [PEs], polyamide 12 [PA], poly(methyl methacrylate) [PMMA], polycarbonates [PC]), packaging (PE, polyethylene terephthalate [PET]), sealing (nitrile butadiene rubber [NBR], fluoroelastomer) or for AMOLED-based flexible display (polyimide [PI]).

A wide variety of plasma techniques are used to produce carbon coatings on polymer substrates, and among these, devices based on technologies are most commonly described: RF-PECVD [1, 10, 24, 25, 30–33], microwave plasma-enhanced chemical vapor deposition (MW-PECVD) [12], direct current plasma-enhanced chemical vapor deposition (DC-PECVD) [34], MS-PVD [3, 4, 35–37], plasma immersion ion implantation and deposition [19, 38, 39], PLD [40–42], but also DBD [43] or high-power impulse magnetron sputtering (HiPIMS) [44]. Techniques based on chemical reaction-based phenomena use various hydrocarbons (e.g. methane CH_4, acetylene CH_2) or their mixtures with other gases as the carbon precursor, while techniques using standard physical phenomena use graphite targets.

11.3 Properties of DLC Coatings for Which They Find Application in Polymer Modification

Assessment of the properties of carbon coatings usually begins with tests to identify the manufactured material in terms of chemical and structural composition. For this purpose, Raman spectroscopy and XPS studies are most commonly carried out. Particularly, the latter method allows the precise characterization of the chemical composition of the modification on the nanometer scale, additionally enabling (through deconvolution and C1s peak analysis) the quantification of the C–C sp^3, C=C sp^2 bond content in the coating [33, 38, 41, 45]. Raman spectroscopy, on the other hand, allows qualitative determination of changes in the chemical structure by deconvolution of the obtained spectra (usually into D and G peaks) and analysis of the I_D/I_G ratio [9, 27, 46–48], an increase or decrease of which indicates a changing content of the C–C sp^3 phase in the produced DLC coatings. As an example, the chemical composition of DLC coatings deposited on PI substrates using variable plasma parameters is presented in Figure 11.1 [4]. It clearly shows that the analysis of the C1s peak obtained from XPS tests allows a direct (percentage) analysis of the content of C–C sp^3 bonds in the coatings, the amount of which increases with increasing bias applied to the sample holder. On the other hand, it can be seen in Figure 11.1 that the same change in bias causes a decrease in the I_D/I_G ratio (the areas of the two characteristic peaks D and G obtained by deconvolution of the Raman spectra). As reported in the literature, the smaller the I_D/I_G ratio, the higher the C–C sp^3 bond content in the studied coatings [49, 50]. Thus, it can be noted that the results obtained from XPS and Raman spectroscopy are in agreement, but in the case of the latter technique, the chemical structure assessment was carried out only through qualitative changes.

One of the main properties for which DLC coatings find application in the polymer world is the possibility of using them as coatings to increase the wear resistance of modified surfaces. In order to verify their resistance, various tribological studies are most often used, from which it is clear that diamond-like coatings on polymers have an effect on reducing the coefficient of friction and specific wear resistance [8, 9, 35, 36, 51–55]. In addition to tribological tests, scratch resistance tests [4], tensile or compression tests, and other technological tests [3, 10, 20, 34, 44] are also used, which can verify their suitability for the intended application. They make it possible to obtain information about the scratch resistance, the mode of cracking of the coating, the stress relaxation in the coating, or the adhesion of the coating to polymer substrates [2, 4, 10, 56]. For example, carbon coatings on polymeric materials for the packaging industry are being tested to analyze fracture behavior under increasing loads. Among others, such studies were carried out by Yamamoto, who evaluated the defragmentation process (cracking) of

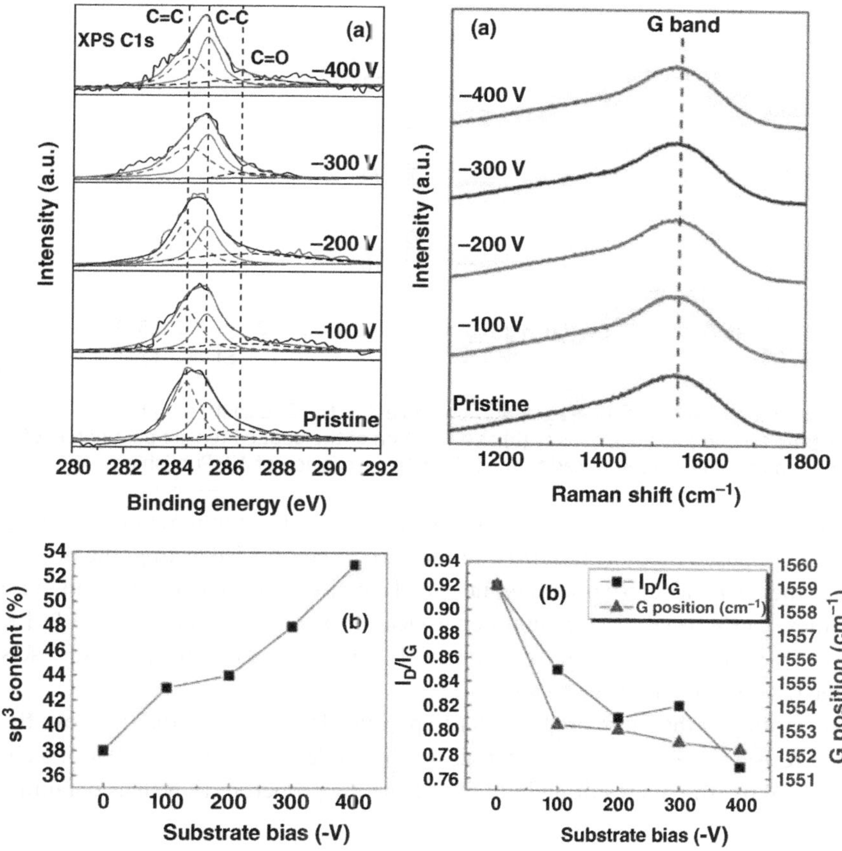

Figure 11.1 Chemical structure analysis of DLC coatings on PI substrates deposited using a bias from −100 to −400 V. *Source:* Reproduced from [4] / with Permission of ELSEVIER.

DLC coatings on PET substrates and the crack density, as shown in Figure 11.2 [37]. When modifying substrates for food packaging, it is important not only the modified material's behavior under varying loads but also its resistance to freezing and defrosting or external thermal loads. The latter aspect is most commonly addressed when using DLC coatings on hot drinks [22].

A different approach is used to analyze the wear resistance of components of the artificial heart modified with DLC coatings: blood chambers or heart valve discs. For this purpose, special test benches are built, e.g. to enable the chamber implants to be operated at appropriate pressures, where the diffusion of ions through them is analyzed. Alanazi's study presents results from the diffusion of Na ions through a polyurethane substrate before and after the formation of a DLC

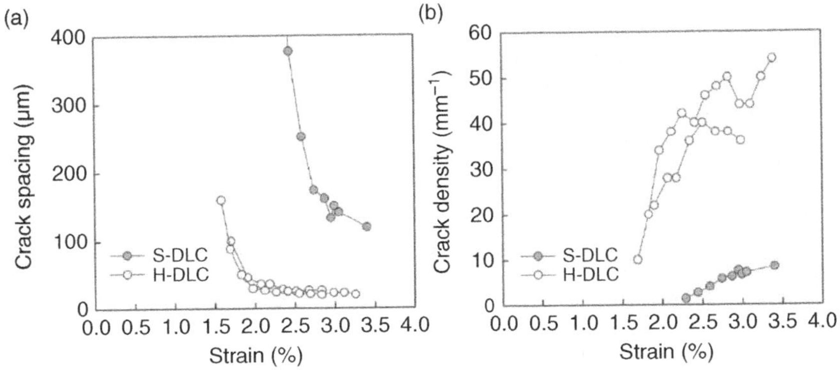

Figure 11.2 Analysis of the fragmentation process of the DLC coating on the PET substrate (a) and the crack density (b) as a function of the applied tensile parameters of the samples. *Source:* Reproduced from [37] / with permission of AIP Publishing.

coating on it, demonstrating the possibility of limiting the phenomenon described by the proposed modification [57]. In the case of heart valve testing, the devices force the heart valve to work under near-real conditions. As Szuber shows in these studies, DLC-coated discs exhibit better wear resistance than unmodified PEEK discs [58]. Figure 11.3 presents a view of the unmodified and modified disc sets after 120 days of fatigue test testing.

According to the literature, in order to produce DLC coatings that improve the wear resistance of polymeric materials, it is possible to rely on a number of technological treatments resulting in different coating properties as well as obtaining

Figure 11.3 View of the Moll mechanical valves after the 120-day test fatigue: unmodified PEEK disc (on the left), DLC-coated PEEK disc (on the right). *Source:* Reprinted from Ref. [58] with permission from Polish Society for Biomaterials (Copyright 2013).

changes in the properties of the material being modified. On the coating side, improvements in wear resistance can be achieved by controlling its composition and chemical structure. As it is a mixture of C=C sp^2 and C–C sp^3 bonds and, most often, hydrogen, its tribological properties are optimized for the content of the mentioned components. For example, Bai et al. [59] conducted tests on an NBR substrate and showed a correlation between the amount of hydrogen in the DLC coating and the obtained averaged friction coefficient. The best of the proposed modifications with approximately 29% hydrogen content achieved a 74% reduction in friction coefficient values compared to an unmodified NBR substrate [55]. Another way of altering the tribological properties of carbon coatings on polymer substrates is to introduce additional elements into their chemical structure, the most common being Ti [60], Si [61, 62], F [63], or W [64]. In general, the occurrence of a C=C sp^2 phase in coatings favors the reduction of friction coefficients and, thus, surface wear. In the literature, we can find examples confirming this theory in the modification of polymeric materials such as PEEK or PDMS [7, 8, 15, 54]. However, it is worth bearing in mind that the chemical structure of DLC coatings produced on polymer substrates depends on the manufacturing process conditions used, which can also play an important role in the cross-linking of the surface layer of the substrates by changing their tribological properties. In addition, the achievement of low friction coefficient values may also be related to the graphitization of the DLC coating (i.e. the transformation of C–C type sp^3 bonds into C=C type sp^2) during tribological tests [65].

In the case of tribological wear resistance testing of DLC coatings produced on polymer substrates, surface development is an additional important factor. After DLC coating manufacturing processes on polymeric substrates, an increase in surface roughness parameters is observed, which depends on the modification conditions applied regardless of the technology used [6, 30, 37, 66, 67]. Figure 11.4 presents the influence of the time of the production of DLC coatings on PE substrates by the RF-PECVD method and two ways of setting the substrates in the reactor chamber (allowing the obtaining of coatings designated as r-DLC and f-DLC) on the changes of the roughness profile of the modified surfaces [30]. As can be seen, the obtained surface roughness of DLC coatings on polymer substrates depends on how the modification processes are carried out. It will also be influenced by plasma pretreatment processes [7].

Going further into the material, we should consider aspects related to the stress state of DLC coatings and their adhesion to the polymer substrate [6, 8, 12, 68]. These parameters strongly depend on the coating manufacturing conditions used in the study. Yamamoto et al., for example, made a comparison of carbon coatings produced on PET substrates using the magnetron method at two different argon injection pressures, thus obtaining coatings with different adhesion [37]. The adhesion strength of DLC coatings to polymer substrates also depends on whether or not a plasma pretreatment process is used. For example, pretreatment of polymers in

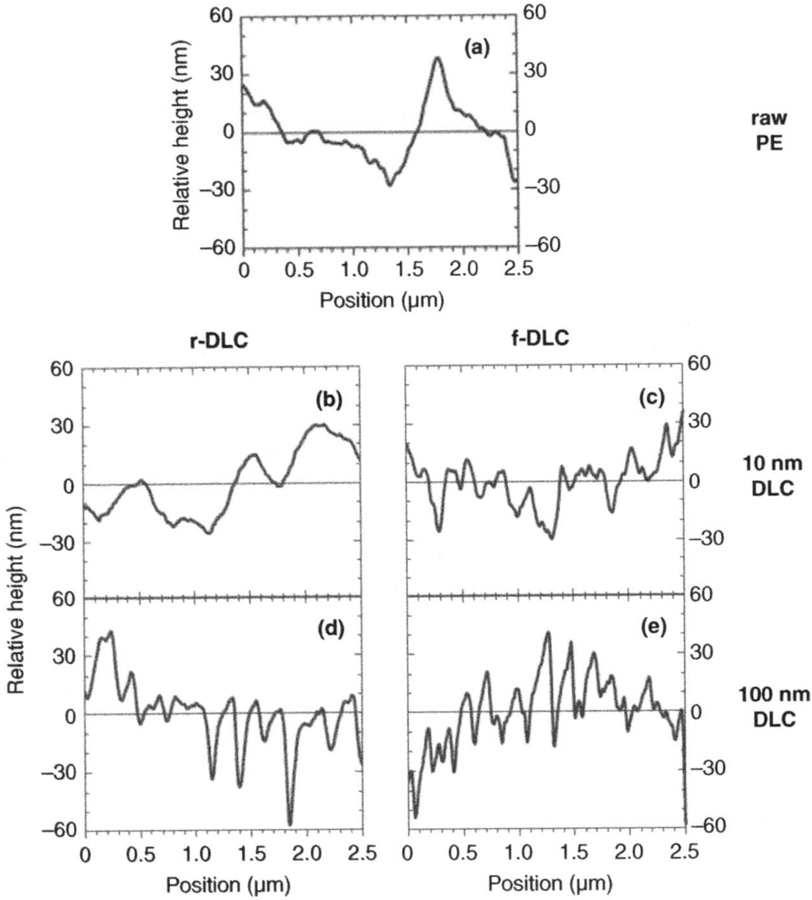

Figure 11.4 Surface roughness profiles from atomic force microscopy studies of r-DLC and f-DLC coatings of different thicknesses. *Source:* Reproduced from [30] / with permission of ELSEVIER.

oxygen plasma increases the adhesion of DLC coatings to polymer substrates. Matsumoto et al. have demonstrated its positive effect using PMMA, PC, PU, and polytetrafluoroethylene (PTFE) in studies [68]. The adhesion of DLC coatings to polymer substrates will be influenced by the duration of the plasma pretreatments and the type of gases or gas mixtures used in the process [27, 46, 47]. It is during the plasma pretreatment process that an interlayer is formed between the substrate and the coating, which increases adhesion. Similarly, as in the case of the adhesion obtained, using different pretreatment durations, and changing the type of atmosphere during these processes, we will determine the change in friction coefficients (COF) of the modified surface [46, 47]. For example, we can use argon or its

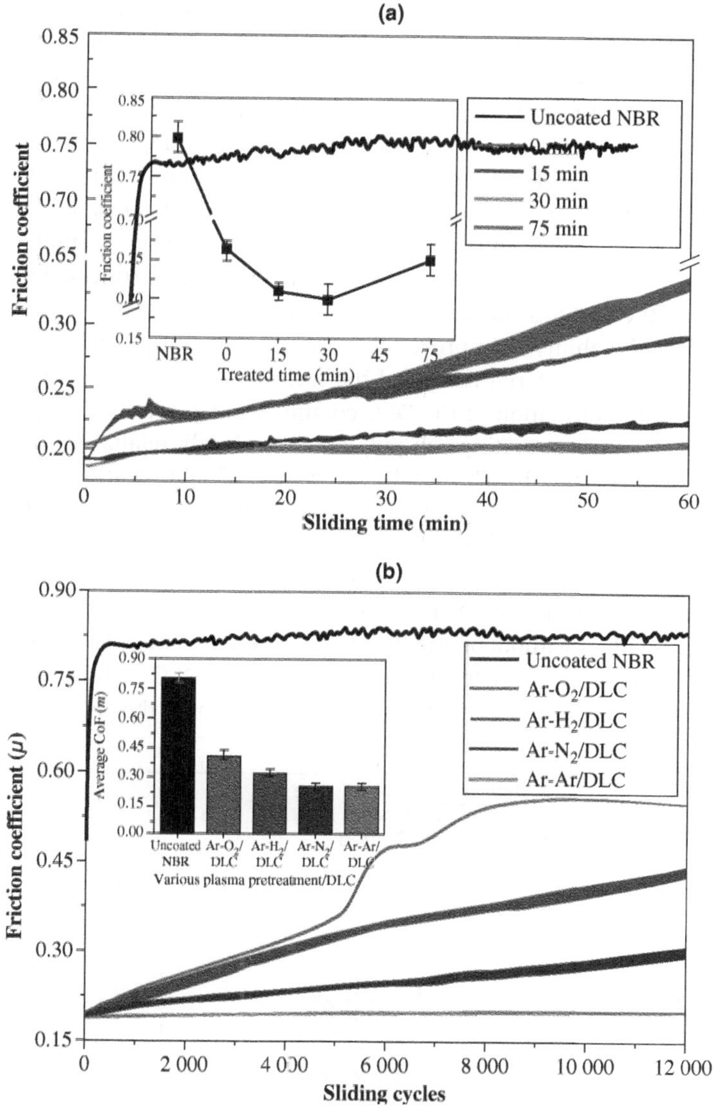

Figure 11.5 Influence on the COF: (a) duration of pretreatment processes of NBR substrates in argon atmosphere. *Source:* Reproduced from [47] / with permission of ELSEVIER. (b) Change in the gas atmosphere of plasma pretreatment processes [46] / Springer Nature / CC BY 4.0.

mixtures with other gases (hydrogen, oxygen, nitrogen) for plasma pretreatments of rubber substrates. As shown in Figure 11.5a, simply changing the duration of pretreatment processes for NBR substrates in an argon atmosphere affects the courses of friction coefficient of DLC coatings produced on such modified substrates [47].

Additionally, as shown in Figure 11.5b, a change in the gaseous atmosphere of the plasma pretreatments will influence the friction coefficient values obtained [46].

DLC coatings' good wear resistance properties may be due, among other things, to their low state of residual stress. The study by Ozkaya et al. demonstrates that by using microwave plasma on PET substrates, DLC coatings can be produced even with a zero stress state [69]. However, despite the ability to control the stress state of DLC coatings and their adhesion to polymer substrates, examples of adhesion retention problems can be found in the literature. Fisher attempted to use DLC coatings to modify flexible polyurethane tubes for artificial feeding of patients, but the coatings cracked and delaminated over time [32].

Ultimately, the ability to improve the wear resistance of polymeric materials will depend on the material itself and the technological treatments performed on it prior to the formation of the DLC coating. Changes in the friction coefficient values of carbon coatings on polymers are closely related to the type of substrate [35]. A study by Thirumalai of the use of DLC coatings for sealing systems (Figure 11.6) showed that by using an identical coating manufacturing process, depending on the polymer substrate, it was possible to reduce the averaged friction coefficient values from 28.8 to 56.4% [35].

The COF values obtained are also influenced by the way the processes are carried out (e.g. with or without premodification of the substrates). The course of the interlayer formation process during the manufacture of DLC coatings on polymer substrates and its properties are in close correlation to both the polymer material

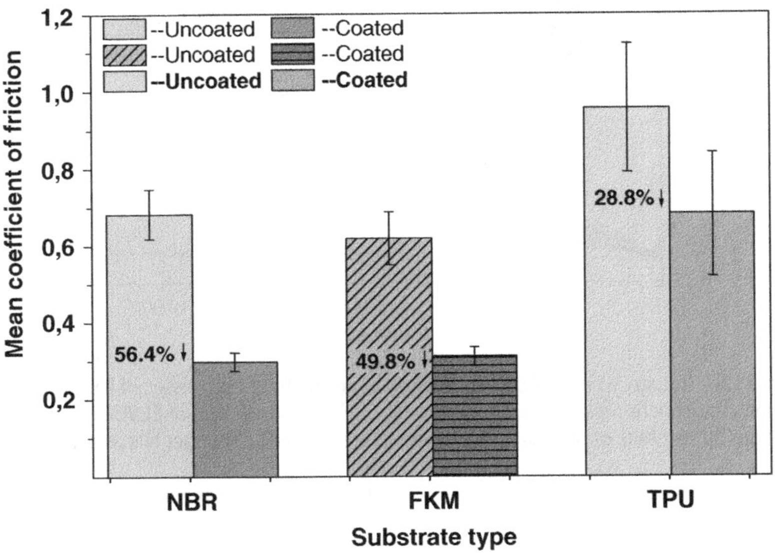

Figure 11.6 Average values of friction coefficients of elastomers before and after fabrication of DLC coatings. *Source:* Reproduced from [35] / with permission of ELSEVIER.

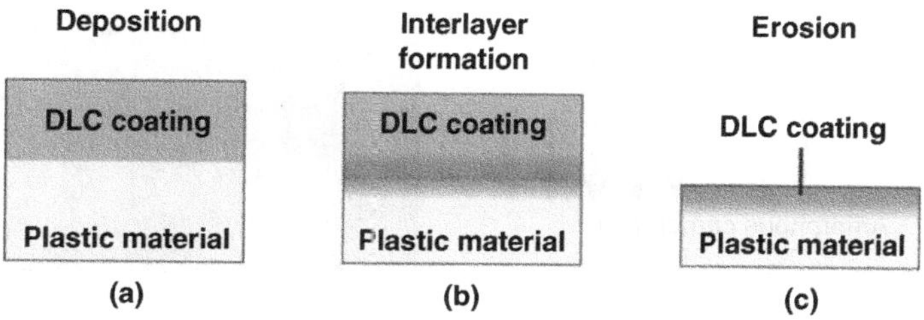

Figure 11.7 Three possible interactions of plasma with polymeric materials during the production of carbon coatings on them: (a) formation of a carbon coating directly on the polymer material; (b) formation of a carbon coating with interlayer formation; (c) formation of a carbon coating with erosion of the polymer material. *Source:* Reproduced from [30] / with permission of ELSEVIER.

used and the modification steps and their execution conditions [1, 15, 30, 70]. According to Fischer and his team, three types of plasma-polymer material interactions are possible during the production of DLC coatings (Figure 11.7), determining the production of a carbon coating without interlayer (Figure 11.7a) or with interlayer participation without erosion of the polymer material (Figure 11.7b) or with erosion of the polymer material (Figure 11.7c) [30]. Each of the coatings obtained as a result of the mentioned interactions will have different mechanical properties, particularly the wear resistance.

Surface modification processes of polymeric materials to produce DLC coatings on them result in changes in surface morphology and topography. Typically, amorphous or grain-like structures are formed on the modified polymer surfaces, the size of which increases with the thickness of the coating [71]. An interesting property of selected polymeric materials is that wrinkles can be formed on their surface. Their formation occurs, for example, with substrates such as PDMS [13–17]. The formation of wrinkles on PDMS surfaces is the result of a structural mismatch between the elastic modulus of the coating and the polymer substrate [13, 72]. This phenomenon is explained as the effect of a mismatch resulting in the deformation of the coating when the compressive stresses exceed a threshold value [73, 74]. The geometric parameters of the resulting wrinkles are particularly influenced by the modification conditions used, among others, the applied polarization potential (with RF-PACVD technologies) [16] or, as shown in Figure 11.8, the duration of the processes [75]. On the surface of DLC coatings produced on PDMS substrates, wrinkles can also be shaped using additional tensile stresses [76].

Among other important properties of DLC coatings on polymeric materials, mention should be made of their ability to alter surface wettability. The use of such modifications most often helps to achieve hydrophobic surface properties [3, 77–79]. Caschera, in his studies on the production of DLC coatings on cotton

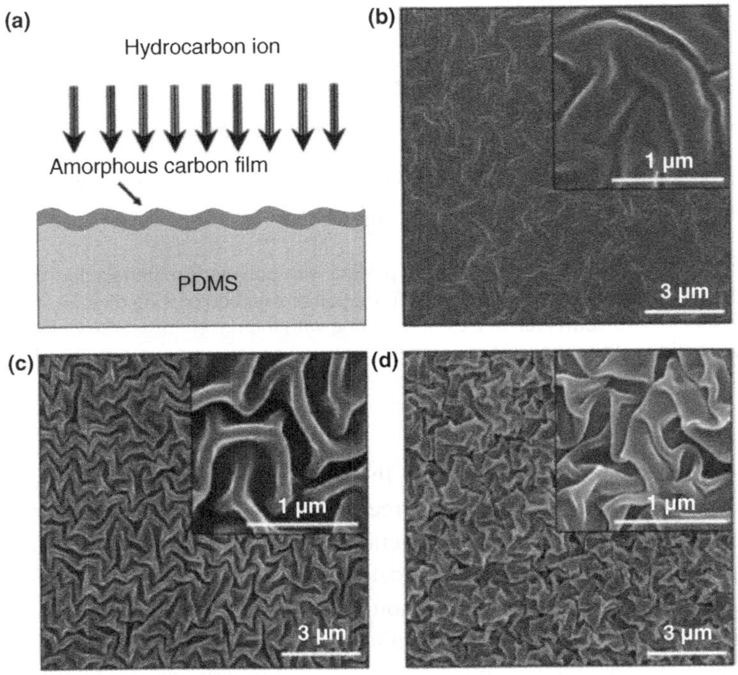

Figure 11.8 (a) SEM images of wrinkles on the surface of PDMS coated with carbon coatings produced using different times: (b) 30 seconds; (c) 5 minutes; (d) 30 minutes. *Source:* Reproduced from [75] / with permission of ELSEVIER.

fibers, shows that high surface wettability angles of up to 146° for deionized water can be obtained using this type of modification. In addition, as Cortese shows that the obtained wetting angle value of carbon coatings produced on cotton fibers is significantly influenced by the plasma pretreatment time [80]. Her study showed that DLC coatings obtained after pretreatment of fibers in oxygen plasma for a time of 60 minutes can reach a wetting angle for deionized water of up to 165°. The hydrophobicity of polymer fabrics coated with DLC coatings is also demonstrated in other publications. Wang shows that it is characterized by a-C coatings produced on terylene fabric, for which a wetting angle of about 131° was obtained [3]. Nevertheless, the wetting angles of DLC coatings on polymers can take values below 90°, i.e. characterized by hydrophilicity [2, 11, 19, 66, 70, 78, 81, 82].

An important feature of DLC coatings on polymers is that their wetting angle does not change after fabrication, as shown in Figure 11.9 [43]. Such results are a confirmation of the barrier properties of the carbon coatings produced. When other gases, such as oxygen, argon, or nitrogen, are used for plasma modification of polymer substrates, the effect of a change in wetting angles, over time, from a value indicative of their strong hydrophilicity toward hydrophobic properties is always observed [16].

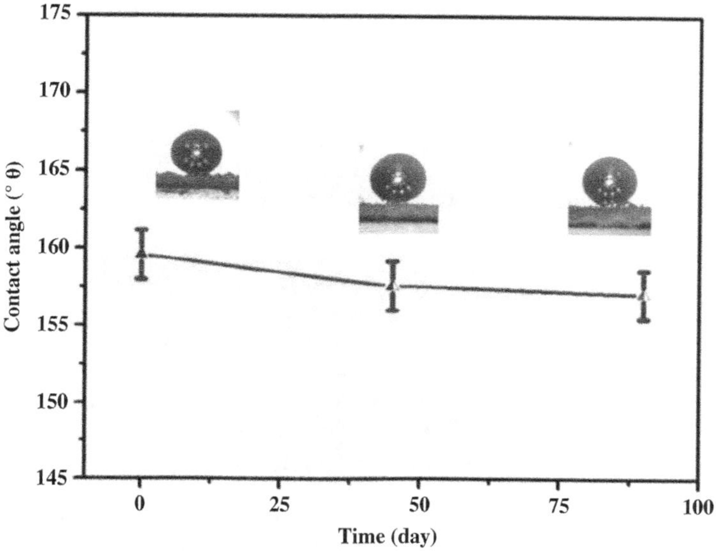

Figure 11.9 Aging analysis of DLC coatings on cotton fibers. *Source:* Reproduced from [43] / with permission of ELSEVIER.

Carbon coatings are also an irreplaceable way of modifying polymer surfaces to improve their biological properties. Carbon coatings make the surfaces of polymeric materials more amenable to various cells. Hubacek demonstrated improved vascular smooth muscle cell (VSMC) proliferation on PET and PTFE polymer surfaces modified with DLC coatings (Figure 11.10) [11]. Modifications of polymers with carbon coatings improve adhesion and proliferation to surfaces and other cells, such as fibroblasts [42], osteoblasts [81, 83], or endothelial cells [84].

Carbon coatings produced on polymeric materials can also inhibit bacterial growth. DLC-coated cotton fabrics show antibacterial properties against bacteria (*Staphylococcus aureus* and *Klebsiella pneumoniae*). Kitahara and her team showed that after 18 hours of incubation of unmodified cotton fabrics, a 100-fold increase in the number of bacteria on the surface is was observed, while analogous studies using modifications with carbon coatings show no active bacteria on the surfaces tested [39]. Similar results were also obtained by Kuwada and co-workers investigating the effect of DLC coatings on polyurethane substrates [85]. An essential feature of DLC coatings produced on polymeric materials is the improved hemocompatibility that can be achieved with them [84, 86]. Such modifications can reduce platelet adhesion to the surface. Alanazi showed that DLC coatings on PC exhibit lower levels of platelet adhesion compared to unmodified or heparin-modified surfaces [87]. In addition to those mentioned earlier, most studied biological aspects, DLC coatings may find their application, for example, in the construction of components

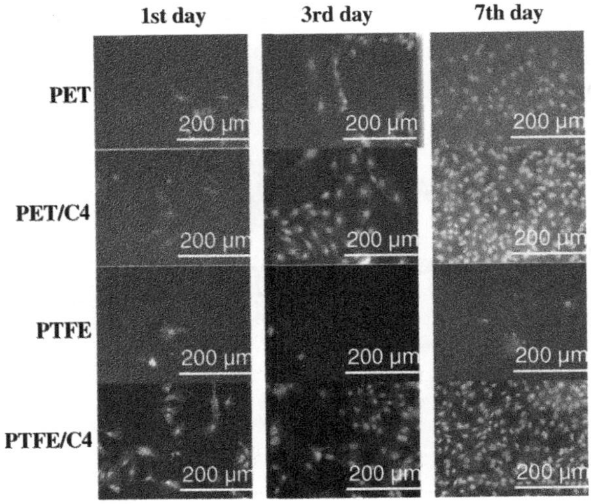

Figure 11.10 Adhesion and proliferation analysis of VSMC cells on PET and PTFE polymer substrates without and with carbon coatings. *Source:* Reprinted from Ref. [11] with permission from Elsevier (Copyright 2013).

in permanent contact with urine. Mozzocchi showed that modifications of PDMS substrates guarantee less formation of urine encrustations on the surface [88]. Sometimes, in order to further improve the biological properties, the carbon coating is given further functionalization processes involving, for example, the formation of amine groups. In this way, Mo has shown that it is possible to achieve better osteointegration of implants made of PEEK coated with functionalized DLC coatings in in-vivo tests [19]. Improved osteointegration of PEEK implants can also be achieved using the functionalization of the carbon coatings produced on them with biomolecules. In-vitro and in-vivo studies conducted on the use of osteoprotegerin and alendronate for this purpose show the great potential of such solutions [38].

Of course, the biological properties of DLC coatings can be improved by introducing other additional elements into their structure. One of these may be fluorine (F), which determines the achievement of good surface antithrombogenicity [22]. Pandiyaraj has shown that DLC coatings doped with titanium on PET substrates can inhibit platelet adhesion and activation [89]. Lackner analyzed various carbon coatings produced on polyurethane substrates for their use in the construction of an artificial heart. He showed that of the coatings used: DLC, DLC doped with titanium, silicon and nitrogen, the latter showed the best antithrombotic behavior [90]. More common, however, is the doping of DLC coatings with elements that improve resistance to microbial colonization. Juknius and his team modified synthetic silk with silver-doped DLC coatings, demonstrating its 99% effectiveness in constructing a dressing material against all tested bacterial strains including *Staphylococcus aureus*. It is also important to note that the amount of silver doping was chosen at a level below that, likely to affect toxicity to cells [91, 92]. Similar results were obtained by the Antunes group, which analyzed the antibacterial properties of silver-doped coatings with an atomic content of 0–12% produced on three different fabrics: cotton, jersey, and non-woven fabric (TNT) [21, 93]. In order to obtain the antimicrobial properties of DLC coatings, they are also doped with zinc (Zn) [94]. Santos et al. applied DLC coatings doped with a bioactive compound synthesized by plants, camphor, on polyurethane substrates in order to improve the surface resistance of polymeric materials against the growth of *Candida albicans* biofilms [34].

DLC coatings on polymer surfaces are used as a barrier to reduce the permeability of water molecules. To this end, the water vapor transmission rate (WVTR), which expresses the number of water molecules passing through the surface under test per unit time, is used to evaluate the effectiveness of carbon coatings. As shown in Figure 11.11a, DLC coatings with a thickness of more than 1 μm can be used to modify printed acrylonitrile butadiene styrene (ABS) and Verogray substrates, making it possible to reduce the WVTR rate by up to 70% in selected cases [12]. DLC coatings can be produced on PET carbonated beverage bottles, improving the barrier properties of the polymers by 5–10 times [5]. These are, of course, influenced not only by the density of the DLC coatings [95] but also their thickness, as shown in Figure 11.11b [96]. Among the applications mentioned here, carbon coatings are of particular importance as a barrier against the ingress of carbon dioxide, but also oxygen or water through the packaging [5, 20, 22, 31, 33, 95–98].

Another feature for which DLC coatings find use in polymer modification is their ability to impart chemical resistance to surfaces. By using carbon coatings on a

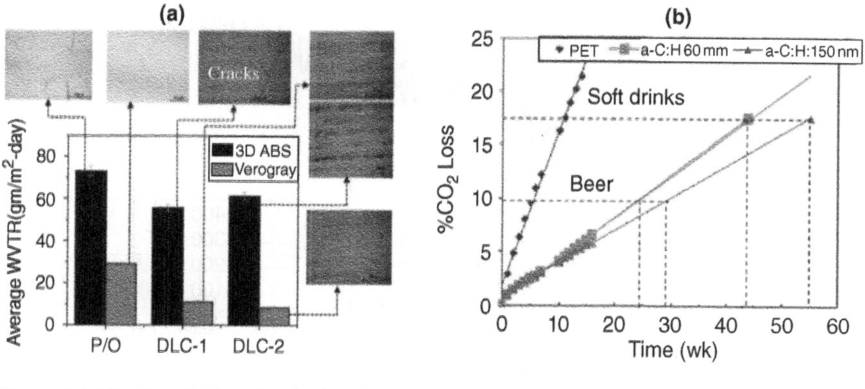

Figure 11.11 Possibility of reducing the permeability of polymeric materials through the use of carbon coatings: (a) analysis of changes in the permeability of water molecules using ABS and Verogray substrates. *Source:* Reproduced from [12] / with permission of ELSEVIER. (b) Analysis of changes in CO_2 permeability using PET substrates. *Source:* Reproduced from [96] / with permission of ELSEVIER.

PMMA substrate, it is possible to protect surfaces from dissolution in acetone [24]. Eshaghi showed that DLC coatings produced on PC substrates increase their chemical resistance to alkaline environments (acetone and sodium hydroxide) [25]. These modifications also perform well in acidic solutions. Honda showed that DLC-coated PMMA powders do not change surface morphology after immersion tests in a 30% HNO_3 acid solution at room temperature with additional stirring (1000 rpm) [99]. Carbon coatings can also withstand such demanding conditions as those found in the stomach. Two-week tests in a solution resembling gastric acids showed that DLC coatings still kept their elasticity and mechanical properties [26].

Carbon coatings are also used on polymers due to their specific properties such as optical or electrical properties. For example, Lee and co-workers analyzing the optical emission spectra of PI substrates without and with a DLC coating on them (Figure 11.12) concluded that the permeability of modified surfaces depends on its homogeneity, roughness, and the content of the C–C sp^3 phase in the coatings [4]. Ban, in turn, describes the possibility of using DLC coatings on PMMA microfluidic devices, for which there is an acceptable few percent decrease in transmittance ranging from 400 to 600 nm of the wavelength under test [24].

Carbon coatings can be useful in stretchable conductors deposited on flexible polymer substrates such as PDMS. Compared to solutions without an interlayer, a 280-times reduction in resistance can be achieved by using these coatings as interlayers between the polymer and the gold coating [29]. According to Nakahigashi, DLC coatings produced on silicon rubber are excellent insulators achieving a resistance of $1–2 \times 10^{12}\ \Omega$ cm [53].

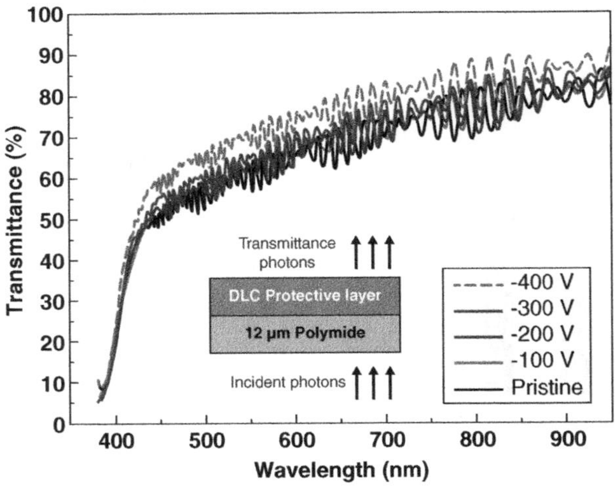

Figure 11.12 Optical transmission spectra of the DLC/polyamide modified by the magnetron method using a negative sample holder potential from −100 to −400 V. *Source:* Reproduced from [4] / with permission of ELSEVIER.

11.4 Application Areas for Carbon Coatings for Polymers and Textiles

DLC coatings are applied to polymeric materials because of their specific properties mentioned in the earlier section. In the simplest terms, we can apply carbon coatings to polymeric materials to increase their wear resistance. In the literature, we can find a number of examples describing the benefits of coatings in different tribological junctions and environments, starting with biomedical joints, e.g. polymers used in hip joint construction [63, 100] or artificial heart components [58, 90], and increasing the durability of equipment and industrial workstations that use polymer-based parts in their construction [55]. Very often, a significant decrease in friction coefficients on modified surfaces is achieved along with a reduction in wear and tear through the production of DLC coatings, resulting in a reduction in the energy efficiency of processes. DLC coatings can also be found in modern solutions for the space industry. Multilayer systems based on them provide good protection against the degradation of polymeric materials on satellites and space installations under the influence of atomic oxygen and ultraviolet [101].

Another interesting application of polymeric materials modified with carbon coatings is their shaping of surfaces for cell culture. As shown in Figure 11.13, the

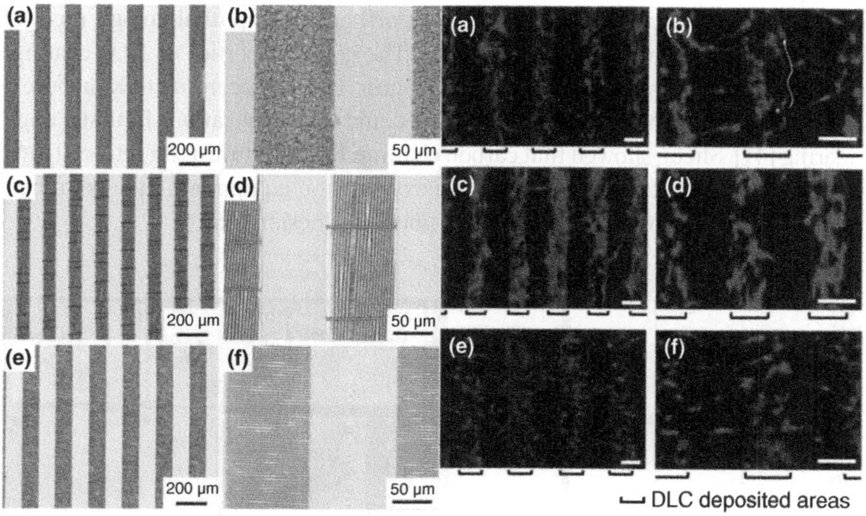

Figure 11.13 Confocal laser microscope images of wrinkles on PDMS substrate (left) and human neuroblastoma cells, SH-SY5Y, cultured on them in fluorescent microscope images. Axons and cell bodies stained in red were merged with the nuclei in blue. The cell marked with an asterisk expanded the axon without any departure from the DLC deposited areas through meandering. The scale bar corresponds to 100 μm [56].

use of different ways of creating wrinkles on the surface of PDMS modified with DLC coatings and masking part of the surface makes it possible to study the axon behaviors of neuronal cells in vitro [56].

An important reason for using DLC coatings on polymeric materials is the surface wettability they can achieve, particularly their superhydrophobicity. Carbon coatings on fabrics result in a change in wetting angles and free surface energy, enabling them to become water-repellent. Work in this direction shows that they may have a future in the production of advanced self-cleaning fabrics [102]. Wetting angles of cotton fibers modified with DLC coatings can remain high above 140° even after 90 days following their manufacturing process or 20–50 wash cycles of coated fabrics [43, 45, 80]. Cotton fabrics with manufactured carbon coatings can be used to separate diesel oil/water mixtures, as seen in Figure 11.14 [43].

A change in wettability toward hydrophobic properties was also observed with DLC coating modifications of polymer nanotissues with polyvinyl alcohol (PVA) [103]. An additional effect of producing coatings on such substrates was also a reduction in sodium chloride aerosol's permeability with an average size of 0.6 μm from 97% for the unmodified PVA nanotissue to about 64% after the modification process performed [104].

Some of the most popular applications of DLC coatings on polymers are examples of their use in biomedicine. Diamond-like coatings are produced on implants for cardiac surgery, neurosurgery, orthopedics, ophthalmology, medical instrumentation, and biomedical devices. The good properties of DLC coatings can also be used for blood vessel fabrics (Figure 11.15). According to Koucourek, they represent an interesting material that could find applications in tissue engineering [41]. Sheng showed that carbon coatings have a future as an artificial lung membrane in extracorporeal membrane oxygenation equipment, thanks to their ability to reduce protein adsorption and platelet adhesion [86].

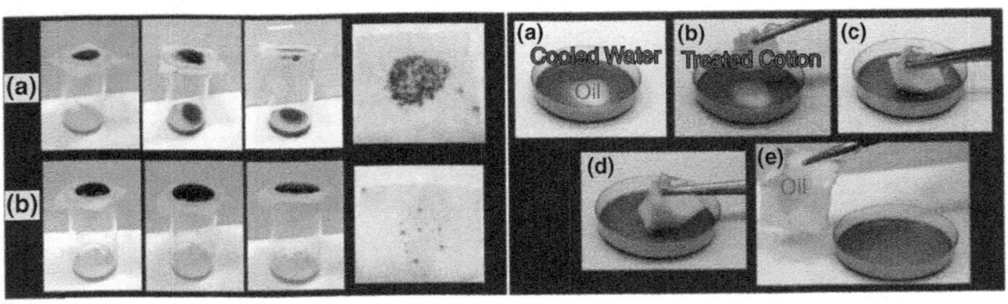

Figure 11.14 From left – separation process of oil/water with (a) unmodified cotton fabric and (b) carbon-coated cotton fabric. Right – the oil removal process from the oil/water mixture using a cotton fabric with a carbon coating. *Source:* Reprinted from Ref. [43] with permission from Elsevier (Copyright 2019).

Figure 11.15 DLC-coated textile ARTEKOR blood vessel. *Source:* Reprinted from Ref. [41] with permission from Springer Nature (Copyright 2008).

Junknius' research shows that silver-doped DLC coatings can be a valuable source of silver ions released from dressings (Figure 11.16) [91, 92]. By coating an artificial silk fabric with Ag-DLC coatings, activating them in an oxygen plasma, and producing smart bandages with them, a product capable of killing 99.9% of bacteria (even *Staphylococcus aureus*) after 320 minutes can be obtained.

There are other applications for DLC coatings on polymeric materials, which are related to both their properties and aspects of their manufacturing technology. For example, due to the impermeability of the coatings and their ability to reflect the modified surface, such coatings can be used as a modification of packaging materials to create a barrier against various agents, such as water or oxygen (Figure 11.17). Solutions for DLC coating technology on PET bottles are available on an industrial scale, with companies such as Mitsubishi Heavy Industries, Inc. and Sidel, Inc. having developed technology lines for their production [22]. Ongoing research shows that the barrier properties of DLC coatings are also maintained at elevated temperatures, which is why carbon coatings are used on packaging for hot tea drinks [18].

Figure 11.16 Disk diffusion method study of antibacterial properties with the *S. aureus* LTSaM01 Ag-DLC coatings with three different silver contents produced on artificial silk fabrics before (a, b, c) and after 20 seconds of oxygen plasma treatment (d, e, f) [91].

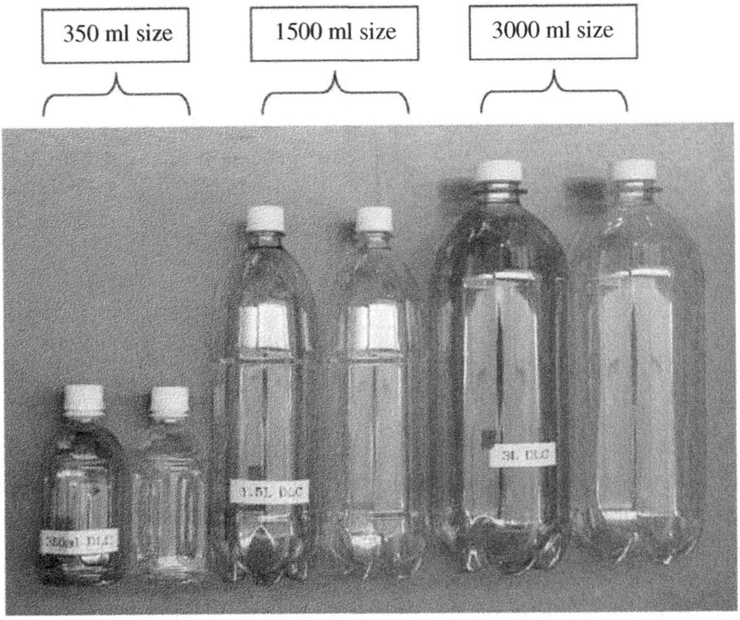

Figure 11.17 View of PET bottles without and with DLC coatings inside them. *Source:* Reprinted from Ref. [22] with permission from Elsevier (Copyright 2006).

Carbon coatings are also used to modify polymers used in drug delivery systems. In this case, the proposed solutions involve obscuring parts of the active surface with a fabricated DLC coating. With the use of micropattern techniques, the coatings in question can be used for the controlled release of drugs from the modified surface [20, 105]. Bito shows the use of this technique to modify poly(2-methacryloyloxyethyl phosphorylcholine) MPC nanofibers with curcumin to produce stents [105]. The potential to control the amount of drug release from DLC-modified surfaces has also been confirmed for other polymeric materials, but for their industrial application, in-depth studies are needed to reduce the risk of loss of continuity of the applied coatings under real conditions of use [20].

When analyzing the benefits of DLC coatings on fabrics, it is worth noting that they can delay aging processes and the damage they cause to fiber surfaces [3]. Another possible application is the modification of fibers exposed to salt water. As shown by Wang, DLC coatings produced on terylene held for two months in 1 M NaCl solution effectively blocked the modified material from corrosion [3]. Carbon coatings can find application on PMMA substrates used in micro-total analysis systems (μ-TAS). They have been shown to increase the chemical resistance of polymers, allowing their use in aggressive environments and only slightly reducing the transparency of the polymers used [24].

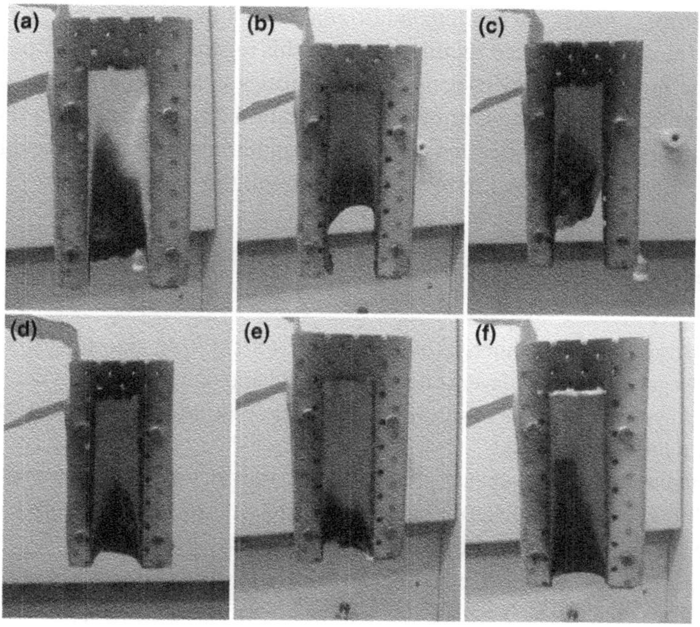

Figure 11.18 Optical images of the burning test results after five seconds for (a) pristine cotton, (b) DLC-COT, (c) DLC-O_2-60, (d) DLC-H_2-10, (e) DLC-H_2- 30, and (f) DLC-H_2-60. *Source:* Reprinted from Ref. [18] with permission from Springer Nature (Copyright 2015).

An interesting direction for the application of DLC coatings on fabrics was shown by Caschera, who demonstrated the effect of DLC coatings produced on cotton fabrics on reducing their flammability [18, 102]. In this case, the carbon coating delays the degradation processes of the fibers and results in an increase in ignition time (Figure 11.18).

Due to their optical properties, carbon coatings have also been proposed for the modification of polymer contact lenses. Studies performed in this direction show that the transmittance (transmission) of polymeric materials with DLC coatings increases with the content of the C–C sp^3 phase in the coatings [28, 106]. Such coatings can successfully reduce the formation of biofilm on the contact lens surface [26].

11.5 Conclusion

Carbon coatings have an extensive range of applications in the modification of polymeric materials and fabrics. Thanks to their specific properties (e.g. physicochemical, tribological, biological), which are dependent on the conditions of their manufacture (process parameters, additional cross-linking, or doping), they can meet the most sophisticated requirements. In this regard, it is probably possible to shape the individual properties of carbon coatings to suit a particular substrate and its application. The large number of publications within the area of the use of DLC coatings in the modification of polymeric materials and fabrics is evidence of the topicality of this subject in many teams around the world.

References

1 Schlebrowski, T., Wehner, S., and Fischer, C. (2023). Thin amorphous hydrogenated carbon layers on polylactic acid and polyhydroxybutyrate biopolymers-influence of substrate chemistry on interlayer formation. *Thin Solid Films* 780: 139943.

2 Schlebrowski, T., Beucher, L., Bazzi, H. et al. (2019). Prediction of a-C:H layer failure on industrial relevant biopolymer polylactide acide (PLA) foils based on the sp^2/sp^3 ratio. *Surface & Coatings Technology* 368: 79–87.

3 Wang, Y., Li, Y., Wang, Y. et al. (2022). Fabrication and performance of flexible aC films on terylene. *Vacuum* 196: 110766.

4 Lee, J.-A., Lin, C.-R., Pan, P.-C. et al. (2020). Dramatically enhanced mechanical properties of diamond-like carbon films on polymer substrate for flexible display devices via argon plasma pretreatment. *Chemical Physics* 529 (2020): 110551. doi: 10.1016/j.chemphys.2019.110551.

5 Ray, S., Mukherjee, D., Sarma, S. et al. (2017). Functional diamond like carbon (DLC) coatings on polymer for improved gas barrier performance. *Diamond and Related Materials* 80: 59–63.

6 Dufils, J., Faverjon, F., Héau, C. et al. (2017). Evaluation of a variety of aC: H coatings on PEEK for biomedical implants. *Surface & Coatings Technology* 313: 96–106.

7 Kaczorowski, W., Batory, D., Szymanski, W. et al. (2015). Evaluation of the surface properties of PEEK substrate after two-step plasma modification: etching and deposition of DLC coatings. *Surface & Coatings Technology* 265: 92–98.

8 Kaczorowski, W., Gajewski, K., Szymanski, W. et al. (2018). Evaluation of mechanical properties of carbon coatings synthesised in radio frequency plasma on PDMS. *Surface & Coatings Technology* 333: 220–228.

9 Liu, J., Yang, T., Cao, H. et al. (2022). Diamond-like carbon films for tribological modification of rubber. *Nanotechnology Reviews* 11 (1): 2839–2856.

10 Tsubone, D., Hasebe, T., Kamijo, A. et al. (2007). Fracture mechanics of diamond-like carbon (DLC) films coated on flexible polymer substrates. *Surface & Coatings Technology* 201 (14): 6423–6430.

11 Hubacek, T., Siegel, J., Khalili, R. et al. (2013). Carbon coatings on polymers and their biocompatibility. *Applied Surface Science* 275: 43–48.

12 Dangnan, F., Espejo, C., Liskiewicz, T. et al. (2021). Water barrier performance of additively manufactured polymers coated with diamond-like carbon films. *Diamond and Related Materials* 119 (1): 108541.

13 Teixeira, F., Araújo, W., Gushiken, N. et al. (2017). On the influence of PDMS (polydimethylsiloxane) substrate surface energy in wrinkling of DLC (diamond-like carbon) thin films. *Journal of Applied Physics* 122 (13): 135308.

14 Ban, M., Tobe, S., and Takeuchi, L. (2018). Effects of diamond-like carbon thin film and wrinkle microstructure on cell proliferation. *Diamond and Related Materials* 90: 194–201.

15 Kaczorowski, W, Swiatek, H., Luczak, K. et al. (2021). Impact of plasma pre-treatment on the tribological properties of DLC coatings on PDMS substrates. *Materials* 14 (2): 433.

16 Kaczorowski, W., Batory, D., Szymanski, W. et al. (2022). Barrier diamond-like carbon coatings on polydimethylsiloxane substrate. *Materials* 15 (11): 3883.

17 Kaczorowski, W., Szymanski, W., Batory, D. et al. (2014). Effect of plasma treatment on the surface properties of polydimethylsiloxane. *Journal of Applied Polymer Science* 132: 41635.

18 Caschera, D., Toro, R., Federici, F. et al. (2015). Flame retardant properties of plasma pre-treated/diamond-like carbon (DLC) coated cotton fabrics. *Cellulose* 22 (4): 2797–2809.

19 Mo, S., Zhao, F., Gao, A. et al. (2021). Simultaneous application of diamond-like carbon coating and surface amination on polyether ether ketone: towards superior mechanical performance and osseointegration. *Smart Materials in Medicine* 2 (2021): 219–228. doi: 10.1016/j.smaim.2021.07.004.

20 Atsushi, H. and Terumitsu, H. (2013). Diamond-like carbon coated on polymers for biomedical applications (Chapter 6). In: *Thin Films and Coatings in Biology* (ed. S. Nazarpour), 171–228. Dordrecht Heidelberg New York London: Springer.

21 Antunes, J., Matos, K., Carvalho, S. et al. (2021). Carbon-based coatings in medical textiles surface functionalisation: an overview. *Processes* 9 (11): 1997.

22 Shirakura, A., Nakaya, M., Koga, Y. et al. (2006). Diamond-like carbon films for PET bottles and medical applications. *Thin Solid Films* 494 (1): 84–91.

23 Yamamoto, S., Kodama, H., Hasebe, T. et al. (2005). Oxygen transmission of transparent diamond-like carbon films. *Diamond and Related Materials* 14 (3): 1112–1115.

24 Ban, M. and Yuhara, T. (2009). Chemical resistance of DLC thin film deposited PMMA substrates. *Surface & Coatings Technology* 203 (17): 2587–2590.

25 Eshaghi, A. and Salehi, M. (2018). Fabrication and characterization of optical, mechanical and chemical properties of diamond-like carbon thin film deposited on polymer substrate. *Optical and Quantum Electronics* 50 (12): 431.

26 Elinson, V., Sleptsov, V., Laymin, A. et al. (1999). Barrier properties of carbon films deposited on polymer-based devices in aggressive environments. *Diamond and Related Materials* 8 (12): 2103–2109.

27 Baek, S., Shirafuji, T., Saito, N. et al. (2011). Fabrication of transparent protective diamond-like carbon films on polymer. *Japanese Journal of Applied Physics* 50 (8): 08JD08.

28 Pisarik, P. (2022). Biocompatible carbon nanolayers for coating lenses. *Lekar a Technika – Clinician and Technology* 52 (2): 42–47.

29 Manzoor, M.U., Tuinea-Bobe, C.L., McKavanagh, F. et al. (2011). Amorphous carbon interlayers for gold on elastomer stretchable conductors. *Journal of Physics D-Applied Physics* 44 (24): 245301.

30 Fischer, C.B., Rohrbeck, M., Wehner, S. et al. (2013). Interlayer formation of diamond-like carbon coatings on industrial polyethylene: thickness dependent surface characterization by SEM, AFM and NEXAFS. *Applied Surface Science* 271: 381–389.

31 Zhang, Z., Song, R., Li, G. et al. (2013). Improving Barrier Properties of PET by Depositing a Layer of DLC Films on Surface. *Advances in Materials Science and Engineering* 6: https://doi.org/10.1155/2013/861804.

32 Fischer, C.B., Rohrbeck, M., Zentgraf, S. et al. (2014). Diamond-like carbon coatings on medically relevant polyurethane tubing with a follow-up aging study. *Biointerface Research in Applied Chemistry* 4: 837–842.

33 Riley, P., Joshi, P., Machekposhti, S. et al. (2021). Enhanced vapor transmission barrier properties via silicon-incorporated diamond-like carbon coating. *Polymers* 13 (20): 3543.

34 Santos, T., Vieira, A., Paula, L. et al. (2017). Flexible camphor diamond-like carbon coating on polyurethane to prevent *Candida albicans* biofilm growth. *Journal of the Mechanical Behavior of Biomedical Materials* 68: 239–246.

35 Thirumalai, S., Hausberger, A., Lackner, J. et al. (2016). Effect of the type of elastomeric substrate on the microstructural, surface and tribological characteristics of diamond-like carbon (DLC) coatings. *Surface & Coatings Technology* 302: 244–254.

36 Marian, M., Zambrano, D., Rothammer, B. et al. (2023). Combining multi-scale surface texturing and DLC coatings for improved tribological performance of 3D printed polymers. *Surface & Coatings Technology* 466: 129682.

37 Yamamoto, G., Okabe, T., and Ikenaga, N. (2018). Fracture behavior of diamond-like carbon films deposited on polymer substrates. *Journal of Vacuum Science & Technology A* 36 (2): 02C101.

38 Chen, J., Xie, L., Ruan, Q. et al. (2022). Diamond-like carbon coating and surface grafting of osteoprotegerin and alendronate on polyetheretherketone to ameliorate the mechanical performance and osseointegration simultaneously. *Composites Part B-Engineering* 236: 109815.

39 Kitahara, N., Sato, T., Isogawa, H. et al. (2010). Antibacterial property of DLC film coated on textile material. *Diamond and Related Materials* 19 (7–9): 690–694.

40 Niemczyk, A., Moszynski, D., Goszczynska, A. et al. (2022). Understanding the DLC film–polyamide 12 substrate interrelation during pulsed laser deposition. *Applied Surface Science* 576: 151872.

41 Kocourek, T., Jelinek, M., Vorlícek, V. et al. (2008). DLC coating of textile blood vessels using PLD. *Applied Physics A-Materials Science & Processing* 93: 627–632.

42 Jelínek, M., Kocourek, T., Remsa, J. et al. (2010). Diamond/graphite content and biocompatibility of DLC films fabricated by PLD. *Applied Physics A-Materials Science & Processing* 101 (4): 579–583.

43 Sohbatzadeh, F., Farhadi, M., and Shakerinasab, E. (2019). A new DBD apparatus for super-hydrophobic coating deposition on cotton fabric. *Surface & Coatings Technology* 374: 944–956.

44 Morozov, I., Kamenetskikh, A., Beliaev, A. et al. (2022). Carbon deposition and argon post-treatment of polyurethane surface: structural-mechanical and fracture properties. *Surface & Coatings Technology* 437: 128372.

45 Caschera, D., Cortese, B., Mezzi, A. et al. (2013). Ultra hydrophobic/superhydrophilic modified cotton textiles through functionalized diamond-like carbon coatings for self-cleaning applications. *Langmuir* 29 (8): 2775–2783.

46 Bai, C., Gong, Z., An, L. et al. (2021). Adhesion and friction performance of DLC/rubber: the influence of plasma pretreatment. *Friction* 9: 627–641.

47 Bai, C., Liang, A., Cao, Z. et al. (2018). Achieving a high adhesion and excellent wear resistance diamond-like carbon film coated on NBR rubber by Ar plasma pretreatment. *Diamond and Related Materials* 89 (2018): 84–93. doi: 10.1016/j.diamond.2018.08.013.

48 Lubwama, M., Corcoran, B., Rajani, K. et al. (2013). Raman analysis of DLC and Si-DLC films deposited on nitrile rubber. *Surface and Coatings Technology* 232 (2013): 521–527. doi: 10.1016/j.surfcoat.2013.06.013.

49 Foong, Y.M., Koh, A.T.T., Ng, H.Y. et al. (2011). Mechanism behind the surface evolution and microstructure changes of laser fabricated nanostructured carbon composite. *Journal of Applied Physics* 110 (5): 054904.

50 Batory, D., Jedrzejczak, A., Kaczorowski, W. et al. (2015). Influence of the process parameters on the characteristics of silicon-incorporated a-C:H:SiO_x coatings. *Surface & Coatings Technology* 271: 112–118.

51 Kaczorowski, W., Batory, D., Szymanski, W. et al. (2016). Frictional behavior of polyurethane modified by carbon coatings synthesized in dual-frequency plasma. *Tribology Transactions* 59 (3): 530–537.

52 Rothammer, B., Marian, M., Neusser, K. et al. (2021). Amorphous carbon coatings for total knee replacements—part II: tribological behavior. *Polymers* 13 (11): 1880.

53 Nakahigashi, T., Tanaka, Y., Miyake, K. et al. (2004). Properties of flexible DLC film deposited by amplitude-modulated RF P-CVD. *Tribology International* 37 (11–12): 907–912.

54 Kaczorowski, W., Szymanski, W., Batory, D. et al. (2014). Tribological properties and characterization of diamond like carbon coatings deposited by MW/RF and RF plasma-enhanced CVD method on poly (ether-ether-ketone). *Plasma Processes and Polymers* 11 (9): 878–887.

55 Bobzin, K., Kalscheuer, C., Thiex, M. et al. (2023). DLC-coated thermoplastics: tribological analyses under dry and lubricated sliding conditions. *Tribology Letters* 71: 2.

56 Ban, M. and Chen, J. (2023). Fabrication of plane-type axon guidance substrates by applying diamond-like carbon thin film deposition. *Scientific Reports* 13 (1): 8489.

57 Alanazi, Ali, S. (2018). Medical application of diamond-like carbon (DLC) coating – a review. *International Journal of Clinical & Medical Informatics* 1: 74–82.

58 Szuber, A., Koscielniak-Ziemniak, M., Janiczak, K. et al. (2013). Polymeric and carbon materials applied in the construction of polish mechanical valve disk – Evaluation of selected physical and biological properties. *Engineering of Biomaterials* 16 (121): 25–32.

59 Bai, C., Qiang, L., Zhang, B. et al. (2022). Optimizing the tribological performance of DLC-coated NBR rubber: the role of hydrogen in films. *Friction* 10 (6): 866–877.

60 Bui, X., Pei, Y., and De Hosson, J. (2008). Magnetron reactively sputtered Ti-DLC coatings on HNBR rubber: the influence of substrate bias. *Surface & Coatings Technology* 202 (20): 4939–4944.

61 Lubwama, M., Corcoran, B., McDonnell, K. et al. (2014). Flexibility and frictional behaviour of DLC and Si-DLC films deposited on nitrile rubber. *Surface & Coatings Technology* 239: 84–94.

62 Lubwama, M., McDonnell, K., Kirabira, J. et al. (2012). Characteristics and tribological performance of DLC and Si-DLC films deposited on nitrile rubber. *Surface & Coatings Technology* 206 (22): 4585–4593.

63 Zhang, L., Zong, X., Guo, F. et al. (2020). Effect of fluorine incorporation on dlc films deposited by pulsed cathodic arc deposition on nitrile butadiene rubber and polyurethane rubber substrates. *Coatings* 10 (9): 878.

64 Pei, Y.T., Bui, X.L., Zhou, X.B. et al. (2008). Tribological behavior of W-DLC coated rubber seals. *Surface & Coatings Technology* 202 (9): 1869–1875.

65 Wang, C., Xu, C., Wu, Z. et al. (2023). The study of surface structure and the tribological property of DLC-modified NBR elastomers using DC-MS. *Coatings* 13 (2): 468.

66 Rothammer, B., Neusser, K., Marian, M. et al. (2021). Amorphous carbon coatings for total knee replacements—part I: deposition, cytocompatibility, chemical and mechanical properties. *Polymers* 13 (12): 1952.

67 Bayrak, S., Paulkowski, D., Stöckelhuber, K. et al. (2020). A comprehensive study about the role of crosslink density on the tribological behavior of DLC coated rubber. *Materials* 13 (23): 5460.

68 Matsumoto, R., Sato, K., Ozeki, K. et al. (2008). Cytotoxicity and tribological property of DLC films deposited on polymeric materials. *Diamond and Related Materials* 17 (7): 1680–1684.

69 Ozkaya, B., Bahre, H., Boeke, M. et al. (2014). Electrochemical analysis of strain-induced crack formation of bilayer barrier plasma polymer films on metal and polymer substrates. *Surface & Coatings Technology* 244: 173–179.

70 Catena, A., Guo, Q., Kunze, M. et al. (2016). Morphological and chemical evolution of gradually deposited diamond-like carbon films on polyethylene terephthalate: from subplantation processes to structural reorganization by intrinsic stress release phenomena. *ACS Applied Materials & Interfaces* 8 (16): 10636–10646.

71 Catena, A., Kunze, M., Agnello, S.et al. (2016). Amorphous hydrogenated carbon (a-C:H) depositions on polyoxymethylene: substrate influence on the characteristics of the developing coatings. *Surface & Coatings Technology* 307: 658–665.

72 Alam, M.S., Mukherjee, N., and Ahmed, S.K.F. (2018). Electron field emission property of nanostructure wrinkle thin film induced by amorphous diamond like carbon. *Materials Today: Proceedings* 5: 2082–2088. doi: 10.1016/j.matpr.2017.09.204.

73 Wang, Y. and Xiao, J. (2017). Programmable, reversible and repeatable wrinkling of shape memory polymer thin films on elastomeric substrates for smart adhesion. *Soft Matter* 13: 5317–5323.

74 Li, S., Wu, K., Yuan, H. et al. (2019). Formation of wrinkled patterns in metal films deposited on elastic substrates: tunability and wettability. *Surface & Coatings Technology* 362: 35–43.

75 Ahmed, S.F., Nagashima, S., Lee, J.Y. et al. (2014). Self-assembled folding of a biaxially compressed film on a compliant substrate. *Carbon* 76: 105–112.

76 Ban, M., Hagiwara, T., and Masumoto, Y. (2017). Partial formation of linear concavo-convex microstructure onto microwells by diamond-like carbon thin film deposition. *Diamond and Related Materials* 74: 138–144.

77 Zia, A. and Birkett, M. (2021). Deposition of diamond-like carbon coatings: conventional to non-conventional approaches for emerging markets. *Ceramics International* 47 (20): 28075–28085.

78 Zia, A., Shah, A., Lee, S. et al. (2015). Development of diamond-like-carbon coated abaca-reinforced polyester composites for hydrophobic and outdoor structural applications. *Polymer Bulletin* 72: 2797–2808.

79 Kyziol, K., Jablonski, P., Niemiec, W. et al. (2020). Deposition, morphology and functional properties of layers based on DLC:Si and DLC:N on polyurethane. *Applied Physics A-Materials Science & Processing* 126: 751. doi: 10.1007/s00339-020-03939-y.

80 Cortese, B., Caschera, D., Federici, F. et al. (2014). Superhydrophobic fabrics for oil/water separation through a diamond like carbon (DLC) coating. *Journal of Materials Chemistry A* 2 (19): 6781–6789.

81 Wang, H., Xu, M., Zhang, W. et al. (2010). Mechanical and biological characteristics of diamond-like carbon coated poly aryl-ether-ether-ketone. *Biomaterials* 31 (32): 8181–8187.

82 He, F.F., Bai, W.Q., Li, L.L. et al. (2016). Enhancement of adhesion by a transition layer: deposition of aC film on ultrahigh molecular weight polyethylene (UHMWPE) by magnetron sputtering. *Applied Surface Science* 364: 280–287.

83 Kyziol, K., Oczkowska, J., Kottfer, D. et al. (2018). Physicochemical and biological activity analysis of low-density polyethylene substrate modified by multi-layer coatings based on DLC structures, obtained using RF CVD method. Coatings 8: 135. doi: 10.3390/coatings8040135.

84 Kaczorowski, W., Batory, D., Jakubowski, W. et al. (2016). Physicochemical and biological investigation of different structures of carbon coatings deposited onto polyurethane. *Brazilian Archives of Biology and Technology* 59: e16150305.

85 Kuwada, N., Fujii, Y., Nakatani, T. et al. (2023). Diamond-like carbon coating to inner surface of polyurethane tube reduces *Staphylococcus aureus* bacterial adhesion and biofilm formation. *Journal of Artificial Organs* 27 (2): 108–116.

86 Sheng, D., Zhang, L., Shang, H. et al. (2023). The surface of a pmp hollow fiber membrane was modified with a diamond-like carbon film to enhance the blood compatibility of an artificial lung membrane. *Langmuir* 39 (37): 13258–13266.

87 Alanazi, A., Nojiri, C., Kido, T. et al. (2000). Engineering analysis of diamond-like carbon coated polymeric materials for biomedical applications. *Artificial Organs* 24 (8): 624–627.

88 Mazzocchi, T., Lucarini, G., Roehrer, I. et al. (2022). PDMS and DLC-coated unidirectional valves for artificial urinary sphincters: opening performance after 126 days of immersion in urin. *Journal of Biomedical Materials Research Part B-Applied Biomaterials* 110 (6): 817–827.

89 Pandiyaraj, K.N., Heeg, J., Lampka, A. et al. (2012). In vitro cyto and blood compatibility of titanium containing diamond-like carbon prepared by hybrid sputtering method. *Plasma Science & Technology* 14: 829.

90 Lackner, J.M., Waldhauser, W., Hartmann, P. et al. (2012). Hemocompatibility of inorganic physical vapor deposition (PVD) coatings on thermoplastic polyurethane polymers. *Journal of Functional Biomaterials* 3 (2): 283–297.

91 Juknius, T., Ruzauskas, M., Tamulevicius, T. et al. (2016). Antimicrobial properties of diamond-like carbon/silver nanocomposite thin films deposited on textiles: towards smart bandages. *Materials* 9 (5): 371.

92 Juknius, T., Jukniene, I., Tamulevicius, T. et al. (2020). Preclinical study of a multi-layered antimicrobial patch based on thin nanocomposite amorphous diamond like carbon films with embedded silver nanoparticles. *Materials* 13 (14): 3180.

93 Antunes, J., Matos, K., Carvalho, I. et al. (2022). Physical vapor deposition technology in personal protective equipment production: improved antibacterial and hydrophobic character of textiles. *Coatings* 12 (10): 1399.

94 Correia, R., Sampaio, A., Milhan, N. et al. (2023). Zn:DLC films via PECVD-HIPIMS: evaluation of antimicrobial activity and cytotoxicity to mammalian cells. *Journal of Vacuum Science & Technology A* 41 (3): 033103.

95 Abbas, G., Roy, S., Papakonstantinou, P. et al. (2005). Structural investigation and gas barrier performance of diamond-like carbon based films on polymer substrates. *Carbon* 43 (2): 303–309.

96 Casiraghi, C., Robertson, J., and Ferrari, A. (2007). Diamond-like carbon for data and beer storage. *Materials Today* 10 (1–2): 44–53.

97 Boutroy, N., Pernel, Y., Rius, J. et al. (2006). Hydrogenated amorphous carbon film coating of PET bottles for gas diffusion barriers. *Diamond and Related Materials* 15 (4): 921–927.

98 Mostafavi, A., Mishra, A., Gallucci, F. et al. (2023). Advances in surface modification and functionalization for tailoring the characteristics of thin films and membranes via chemical vapor deposition techniques. *Journal of Applied Polymer Science* 140 (15): e53720.

99 Honda, Y., Akamaru, S., Inoue, M. et al. (2012). Development of a polygonal barrel-plasma enhanced chemical vapor deposition method for preparing powdered materials with a diamond-like carbon film. *Chemical Engineering Journal* 209: 616–622.

100 Song, Z., Liu, Y., Liu, Y. et al. (2023). Improving the protective properties of an artificial joint friction couple by using TiN and DLC films. *Surface & Coatings Technology* 472: 129944. doi: 10.1016/j.surfcoat.2023.129944.

101 Smith, C., Delkowki, M., Anguita, J. et al. (2021). Complete atomic oxygen and UV protection for polymer and composite materials in a low earth orbit. *ACS Applied Materials & Interfaces* 13 (5): 6670–6677.

102 Caschera, D., Toro, R.G., Cortese, B. et al. (2019). AAPP Atti della Accademia Peloritana dei Pericolanti, Classe di Scienze Fisiche, Matematiche e Naturali 97 C7 – AAPP97S2A27.

103 Rozek, Z., Kaczorowski, W., Lukáš, D. et al. (2008). Potential applications of nanofiber textile covered by carbon coatings. *Journal of Achievements in Materials and Manufacturing Engineering* 27: 35–38.

104 Kaczorowski, W. (2016). Plazmowa modyfikacja biomateriałów polimerowych z wykorzystaniem powłok węglowych, Wydawnictwo Politechniki Łódzkiej, łodz.

105 Bito, K., Maeda, T., Hagiwara, K. et al. (2015). Poly (2-methacryloyloxyethyl phosphorylcholine) (MPC) nanofibers coated with micro-patterned diamond-like carbon (DLC) for the controlled drug release. *Journal of Biorheology* 29 (1): 51–55.

106 Písarík, P., Jelínek, M., Smetana, K. et al. (2013). Study of optical properties and biocompatibility of DLC films characterized by sp^3 bonds. *Applied Physics A-Materials Science & Processing* 112: 143–148.

12

Diamond-like Carbon Coating Applications in MEMS

Muhammad Usman[1], *Hafiz Mahmood Ilahi*[1], *Muhammad Umar Farooq*[2], *Mudassar Rauf*[3], *Muhammad Asim*[4], *Sarmad Ali Khan*[5]

[1] *Department of Mechanical Engineering, City University of Hong Kong, Hong Kong, SAR China*
[2] *The Sargent Centre for Process Systems Engineering, Department of Chemical Engineering, University College London, London, UK*
[3] *School of Mechanical and Electrical Engineering, Wenzhou University, Wenzhou, China*
[4] *School of Professional Education & Executive Development, The Hong Kong Polytechnic University, Kowloon, Hong Kong.*
[5] *Department of Industrial and Manufacturing Engineering, University of Engineering and Technology Lahore, Pakistan*

CONTENTS

12.1 Introduction, 346
12.2 Mechanical Properties of DLC Coatings, 348
12.3 DLC Applications for Sensors, 351
 12.3.1 DLC Applications for Force Sensors, 351
 12.3.2 DLC Applications for Temperature Sensors, 354
 12.3.3 DLC Applications for Piezoresistive Sensor, 355
 12.3.4 DLC Applications for Strain Sensors, 355
 12.3.5 DLC Applications for Pressure Sensors, 356
12.4 DLC Applications for Nanoactuators, 357
12.5 Industrial processing, 358
12.6 Applications of DLC Coatings Structural Integrity, 359
12.7 DLC Applications for MEMS Tribology, 359
 12.7.1 Nanofretting, 360
 12.7.2 Nanodot Contacts, 361
 12.7.3 Lubricant Life Assessment, 363
12.8 DLC Coatings System and Performance, 363
 12.8.1 Hydrogenated and Non-hydrogenated DLC Coating, 363
 12.8.2 Role of Substrate Material, 364
 12.8.3 Wear at Microscopic Scale, 365
 12.8.4 Static Friction, 366
12.9 Conclusion, 366

Applications of Diamond-like Carbon Coatings, First Edition. Abdul Wasy Zia.

12.1 Introduction

Micro-electromechanical systems (MEMS) are micro- to nanometer-scale devices in their simplest form. These systems typically include sensors, actuators, microprocessors, micropower supplies, and antenna structures, all packaged together in a single electronic packaging. They always include a sensor or an actuator, but other components are not guaranteed to be part of the system. A MEMS device with all key components on board can sense the environment, process the acquired data, and react accordingly to make a desired adjustment. These microsystems improve the performance of engineering systems by enhancing their robustness, reliability, and resourcefulness [1]. Examples of such MEMS devices are accelerometers, optical switches, and microvalves [2–4]. MEMS devices play a crucial role in various industries, such as biomedical, aerospace, automotive, and entertainment, among others [20, 21]. In the past few decades, MEMS has revolutionized many fields through breakthroughs, e.g. biotechnology, information technology, energy, materials, etc., representing the next technology frontiers in the new millennium [5–7].

MEMS technology is dominated by silicon (Si)-based materials due to their mechanical and electrical properties [8]. However, high surface energy, small energy bandgap, high coefficient of friction, high wear rate, and low toughness limit its use [9–12]. Therefore, researchers have tried many new materials for new applications. Diamond-like carbon (DLC) coatings are among the most promising candidates as they generally introduce anti-stiction and hydrophobic properties in microelectromechanical components, protect mechanical components from wear and friction, provide structural stability, and improve biocompatibility in medical implants [13, 14]. Further, its thermal stability up to 600 °C degrees and chemical inertness can benefit in high-temperature applications [15]. Therefore, it is usually applied in MEMS technology in two ways: (1) DLC coating for improved surface properties, and (2) structural components of MEMS are made by DLC films fabricated using the usual wafer fabrication process.

Surface forces, such as van der Wall's and capillary forces, become considerably dominant at the micro- or submicro-scale, and engineering solutions are required to deal with these forces. Moreover, surface adhesion and friction of moving components can critically hinder the manufacturability and performance of MEMS devices. Silicon-based MEMS devices are generally produced with a standard fabrication process using SiO_2 as sacrificial material; adhesion forces due to capillary attraction pull the freestanding components, which then stick to the surface of the substrate, causing a considerable fabrication loss. Although newer methods like freeze-releasing and dry ion etching address this problem, the freestanding components may fail due to the same adhesion forces. DLC coatings, with their low surface energy and contact angle larger than 90 degrees, assist in overcoming the adhesion forces and induce superior anti-stiction properties that are paramount in MEMS fabrication and operation [15]. Another critical problem for MEMS devices is the wear and friction of moving components, such as in micromotors and microengines. MEMS devices are traditionally made with Si materials, and a very high coefficient of friction (COF) between Si and Si (0.6) causes a critically high wear rate [11, 12]. At the microlevel moving components in MEMS, where the liquid lubricants are not a feasible solution due to the dominance of viscous shear forces, DLC coating is an excellent choice as a solid lubricant with an achievable coefficient of friction of 0.25 or even below. In a few experiments, moving components coated with a 5–15 nm thick DLC prolonged the life cycle of components to 80 million cycles, 300 times higher than that of Si-to-Si contact. A very low coefficient of friction and wear rate for DLC coating has already been reported [16]. Therefore, tribological properties induced through DLC coatings justify their application in MEMS.

Apart from DLC usage as surface coatings, their diamond-like characteristics, such as high Young's modulus, high hardness, low wear rates, and superior thermal conductivity, in addition to a low coefficient of thermal expansion (CTE), promote their applications to produce MEMS mechanical structures and moving components [15, 17]. DLC material MEMS components such as cantilever beams, comb drive structures, fixed-end bimorph beams, as well as microgears and linear motors can be fabricated using the traditional MEMS fabrication process [18, 19]. DLC MEMS structures, owing to their high Young's modulus to density ratio, are being used in high-frequency resonators with their high resonant frequencies reaching 640 MHz [20]. In bilayer thermal MEMS actuators, where two materials have large coefficients of thermal expansion difference, DLC structures show a low CTE (~10^{-6} K^{-1}) material layer combined with other high CTE material, producing a large out-of-plane motion. Similarly, DLC MEMS structures command their superiority in other MEMS applications, including but not limited to bistable beams, piezoresistive sensors, field emission tips, accelerometers, and microfluidic channels [15]. Moreover, because of biocompatibility (does not trigger the body's immune system), chemical resistance, and thermal stability, DLC coatings have become a go-to choice for coating implantable structural materials such as titanium for implantable joint prosthetics and dental implants. These coatings are also employed in implantable applications such as stents, heart valves, etc. In other biomedical applications, DLC coating is used for biosensing because of its bioactivity, such as in polymerase chain reaction devices. DLC offers resistance to nonspecific DNA and protein surface binding, a critical requirement for biosensor detection [21–25].

The DLC coatings' excellent mechanical properties, better tribological performance, biocompatibility, better surface properties, chemical inertness, and structural and thermal stability make them suitable for MEMS sensors, nanoactuators, biosensors, MEMS tribology, and structural elements. This chapter discusses DLC applications in MEMS for different purposes in addition to the commonly used fabrication process.

12.2 Mechanical Properties of DLC Coatings

The requirements ascertained on the use of DLC in MEMS are quite stringent in comparison to their usage as macroscale mechanical components surface coating. These requirements include but are not limited to tensile strength, hardness, fracture toughness, and bending stress. Past studies demonstrate the behavior of DLC prepared for MEMS under varying preparation parameters. Figure 12.1 is one of many attempts that characterize the relationship between DLC thickness and bending stiffness, bending stress, and fracture toughness. Figure 12.1a shows the central deflection vs loading force required to decrease the curvature of the produced DLC films of various thicknesses. A very thin 0.5-micron film showed nearly no bending comparatively. Figure 12.1b demonstrates that by increasing the DLC film thickness, both the bending stiffness and force to remove curvature increase. A similar trend was observed for bending stress and bending stresses in terms of energy release rate (Figure 12.1c). Figure 12.1d shows that with an increasing film thickness of DLC films, the fracture toughness decreases while the ratio of bending force at curvature removal to force at crack initiation increases, suggesting that thicker DLC films have a larger proportion of bending forces than those of thinner DLC films. Such behavior of DLC films can be observed in Table 12.1, wherein the tensile strength of various samples with varying lengths and thicknesses is shown. For a fixed length of DLC film, the tensile strength decreases with an increase in thickness. Figure 12.2 emphasizes this phenomenon where ta-C DLC film typically used in MEMS structures demonstrates that increasing the thickness of the film reduces its tensile strength, and a similar trend is observable when the volume of the film is increased [26, 27].

The process parameters for fabricating DLC films for MEMS are of key importance apart from thickness. These process parameters control the sp^3/sp^2 ratio in a DLC film and, in turn, decide the hardness, coefficient of friction, and wear rate of the fabricated films. Table 12.2 shows the effect of process parameters on the mechanical properties of DLC films. A more hydrogenated DLC film demonstrates improved lubrication because the hydrogen atoms passivate the free bonds on the surface. Thus a more hydrogenated DLC film possesses lower wear rates and a lower coefficient of friction.

A further comparison of DLC coatings prepared under various process conditions and their sp^3 and sp^2 bonding and hydrogen content dominance is demonstrated through a ternary phase diagram in Figure 12.3. Diamond is at the peak with complete dominance of sp^3 bonding and no hydrogen content, while pure carbon has complete dominance of sp^2 bonding. Several types of DLC materials can be tailor-made with varying sp^3/sp^2 ratios together with hydrogenation for various applications. Generally, ta-C is a material of interest here due to its

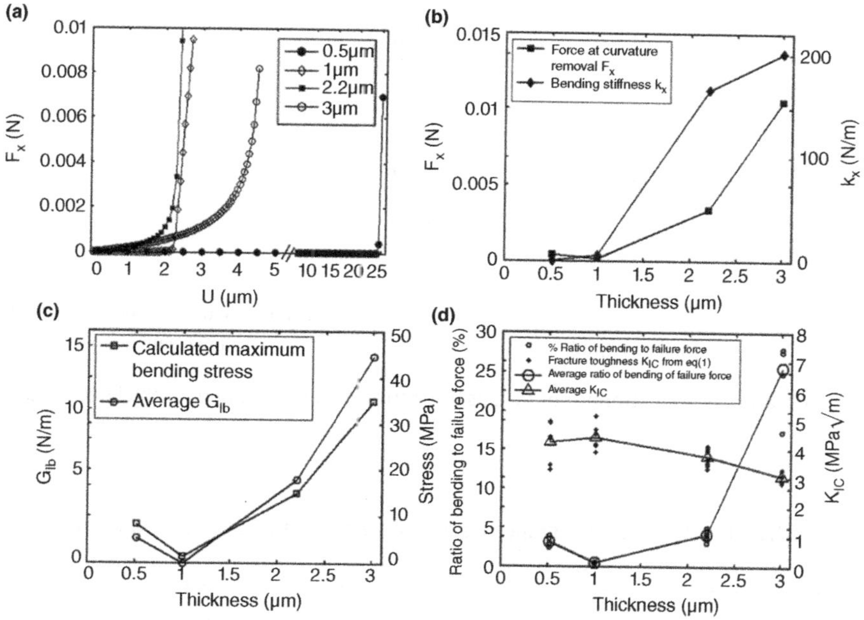

Figure 12.1 (a) In-plane force as a function of horizontal crosshead displacement, (b) bending stiffness and applied in-plane force at curvature removal as a function of sample thickness, (c) calculated maximum bending stress and energy release rate due to bending, and (d) ratio of the applied in-plane force at curvature removal (termed bending force in the plot) to the applied force at fracture, and apparent fracture toughness. Reproduced from [26] / with permission of ELSEVIER.

Table 12.1 Tensile strength of ta-C as a function of specimen dimensions [27] / John Wiley & Sons / CC BY 4.0.

Thickness (μm)	Length (μm)	Tensile strength (GPa)
0.5	500	5.4 ± 1.2
	1000	5.7 ± 0.3
1.0	500	5.5 ± 0.9
	1000	3.9 ± 0.7
2.2	500	4.3 ± 1.4
	1000	3.4 ± 0.9
3.0	500	2.2 ± 0.5
	1000	2.3 ± 0.5

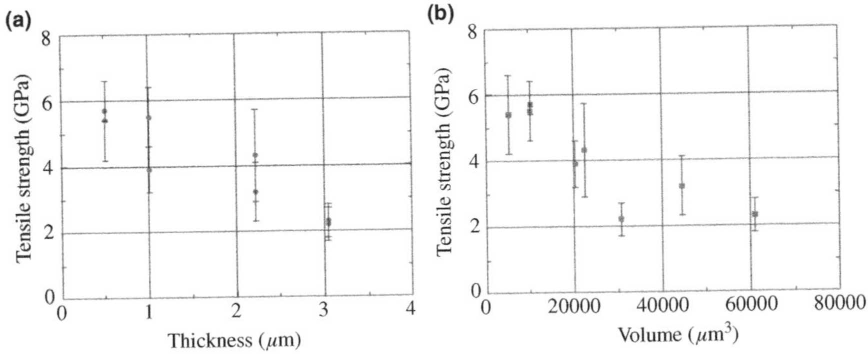

Figure 12.2 Tensile strength of ta-C as a function of the specimen (a) thickness and (b) volume. Adapted from [3, 28].

Table 12.2 Friction and wear properties of DLC. Reproduced from [15] / with permission of IOP Publishing.

Carbon film grown in	COF	K_w mm³ N⁻¹ m⁻¹
100% CH_4	0.015	9×10^{-9}
25% H_2 + 75% CH_4	0.01	7.36×10^{-9}
50% H_2 + 50% CH_4	0.004	1.23×10^{-8}
90% H_2 + 10% CH_4	0.004	2.8×10^{-10}
100% C_2H_2	0.27	7.5×10^{-7}
Hydrogen-free DLC	0.25	2.8×10^{-7}
Diamond	0.07	n/a

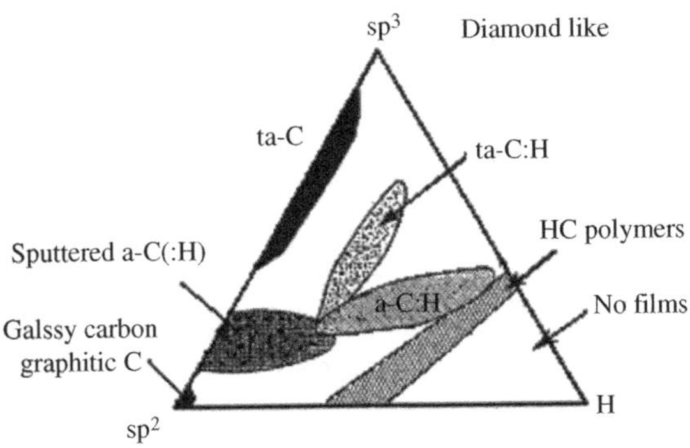

Figure 12.3 Ternary phase diagram defining the different regions of DLC materials. Reproduced from [15] / with permission of IOP Publishing.

crucial application in MEMS structures as it possesses diamond properties with no hydrogenation and approximately 80% sp^3 bonding [15, 29, 30]. Various deposition methods also heavily influence the hydrogen content, sp^3 fraction, and mechanical properties of DLC material. As the hydrogen content decreases, the material becomes denser, tougher, and harder, with density approaching that of a diamond [13, 14].

12.3 DLC Applications for Sensors

Amorphous (a-C) and hydrogenated amorphous carbon films (a-C:H) have high-stress sensitivity and excellent electrical properties besides chemical inertness, high hardness, and low COF. Their gauge factor (GF: 50–1200) has been reported to be one order of magnitude higher than silicon-based piezoresistive material, depending on the preparation process and component. The cluster size of the sp^2 and sp^3 matrix is the reason for different GF values in a-C coatings. It increases and decreases with a high sp^3/sp^2 ratio and low sp^2 cluster size, respectively [31]. Additionally, third-element addition can exhibit a weak dependence of resistivity on temperature in a certain range [32, 33]. These properties make carbon coatings a suitable candidate for MEMS sensors. Moreover, a-C preparation processes are safe, green, flexible, mature, and compatible with MEMS sensors. It can be easily deposited on various substrates compared to other promising piezoresistive options, e.g. silicon nanowires, graphene, and carbon nanotubes (CNTs) [34–37]. Therefore, DLC coatings are widely employed in various types of sensors, e.g. force, temperature, piezoresistive, strain, and pressure. A detailed discussion can be read in the following sections.

12.3.1 DLC Applications for Force Sensors

A detailed fabrication process of an a-C-based MEMS sensor fabricated using conventional MEMS fabrication techniques is shown in Figure 12.4a–d [38]. In this example, an integrated MEMS force sensor is fabricated on a 4-inch n-type silicon substrate after growing a SiO_2 layer on the surface for better electrical insulation. The a-C films are deposited using DC magnetron sputtering. This deposition process is preferable because the deposition temperature is not high, excluding the need to use a heat-resistant mask. A photoresist can be directly used as a mask for the piezoresistors in this setting, as shown in Figure 12.4a. Figure 12.4b displays the deposition of a-C film on the substrate with the designed mask. Figure 12.4c demonstrates the final a-c piezoresistor shape after removing the excess deposited materials using the lift-off process, and the photoresist mask is removed using the acetone ultrasonic process. To develop electrical connections of thus developed

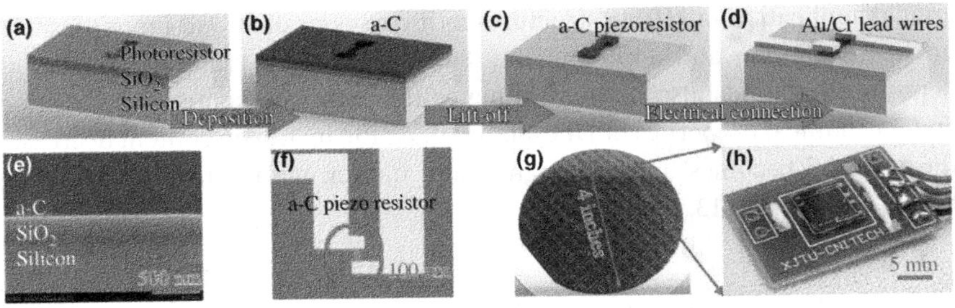

Figure 12.4 Schematic of the fabrication process of the a-C diaphragm type force sensor, (a) photolithography, (b) a-C deposition, (c) lift-off, (d) electrical connection, (e) the cross-sectional SEM image of a-C and the substrate, (f) the optical microscopy image of the a-C piezoresistor, (g) the image of a 4-inch substrate, and (h) the packaged force sensor. Reproduced from [38] / with permission of AIP Publishing.

a-C piezoresistor, 20 nm Cr and 200 nm Au are deposited again using magnetron sputtering, and the final electrical connections resulting after the off process are shown in Figure 12.4d. The backside structure is fabricated using an inductively coupled plasma etching process with silicon diaphragm thickness maintained at 170 ± 10 μm. Figure 12.4e shows the SEM image of the cross-section of the fabricated sensor, and a smooth interface between the SiO_2 and a-C surface can be seen. Similarly, an intact piezoresistor can be seen in the optical micrograph of the sensor cross-section in Figure 12.4f. The images of the 4-inch sensor on a silicon substrate are shown in Figure 12.4g and the final a-C MEMS force sensor is shown in Figure 12.4h. In this example, the a-C film showed good adhesion and a high value of spallation in the groove L_{c3} was achieved, which is a key factor in the a-C sensor fabrication. The methods used in this example of fabrication are flexible and can be replaced with other deposition and preparation methods to vary the characteristics of the force sensor. The discussion on these methods is beyond the scope of this chapter.

The a-C film-based microforce sensor, thus prepared under variable input fabrication conditions, is characterized using Raman spectra to evaluate the critical parameters governing sensor performance, such as I_D/I_G and sp^2 content. These characterizations from the previously discussed example are shown in Figure 12.5, where Raman spectra results are detailed for the sensors prepared under varying process conditions [38]. The Raman spectra of this force sensor a-C films deposited under varying bias voltage is shown in Figure 12.5a, and a typical broadband wavelength in the range of 800–2000 cm^{-1} characteristics can be seen. Gaussian function-based curve fitting is employed to identify D and G peaks in the spectrum, and the ratio of I_D/I_G is plotted against negative bias voltage in Figure 12.5b. The ratio did not change much by changing the bias voltage from 0 to −200 V but increased rapidly to 2.48 when the bias voltage was increased to −300 V

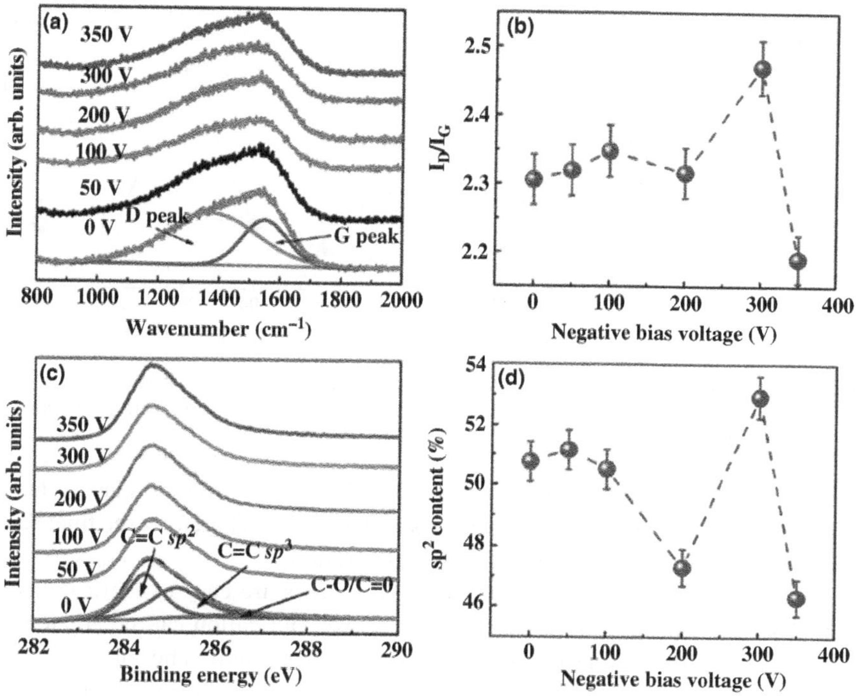

Figure 12.5 (a) Raman spectra of the films, (b) the correlation between fitted values of I_D/I_G and negative bias voltage, (c) XPS C1s figure of the a-C films, and (d) the correlation between the fitted sp^2 content and negative bias voltage. Reproduced from [38] / with permission of AIP Publishing.

and decayed even faster to 2.19 with the further increase in negative bias voltage. A larger I_D/I_G ratio is related to the larger sp^2 cluster size. To further analyze the sp^2 content in a-C films, the results from X-ray Photoelectron Spectroscopy (XPS) C 1s spectra are depicted in Figure 12.5c. The curve with peak position 284.6 eV, 285.4 eV, and 286.5 eV corresponds to sp^2 hybridization, sp^3 hybridization, and a combination of sp^2/sp^3 hybridization, respectively. The sp^2 content percentage derived from these curves is plotted against negative bias voltage in Figure 12.5d. The sp^2 content varies by varying negative bias voltage following the I_D/I_G ratio trend. The sp^2 content drops to 47.3% from 51% when bias voltage is increased from −50 V to −200 V and the peak sp^2 content of 52.9% is achieved at −300 V, which drops again with a further increase in bias voltage. The minimum sp^2 content, which is a key factor in a-C MEMS fabrication, is achieved at −350 V.

The sensors are also subjected to performance testing under loading conditions as shown in Figure 12.6 to evaluate their GF [31]. A testing strip supported at two ends is subjected to concentrated load and the deflection, and changes in resistance are measured. The GF of a-C films is noted and plotted against fabrication

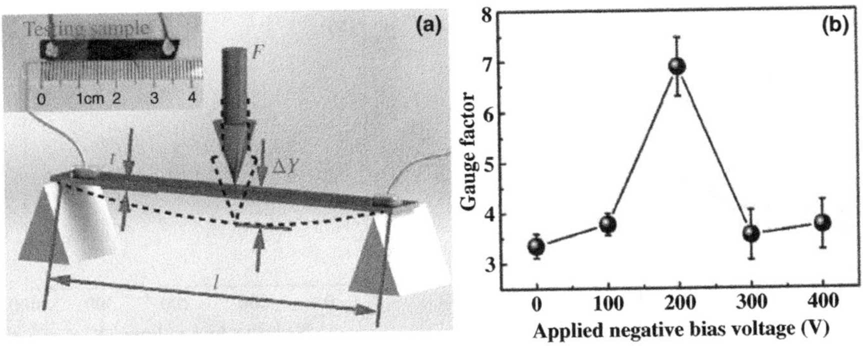

Figure 12.6 (a) The 3D schematic of the homemade gauge factor testing device and the photograph of the testing sample, and (b) GF testing result. Reproduced from [31] / with permission of ELSEVIER.

bias voltage. The highest gauge factor is achieved at −200 V bias voltage, producing low sp^2 content and optimum sp^2 cluster size. Therefore, the final sensor is fabricated at this bias voltage for a-C deposition.

To conclude, a-C film-based MEMS force sensors are easy to fabricate as a conventional deposition for a-C films and sensor fabrication can be efficiently employed and the whole fabrication process is compliant. The fabrication is safer as no toxic gases are required during the deposition process. Strict process control, efficient performance measure methods, superior GF, sensitivity, linearity, and endurance to harsh environments make a-C MEMS force sensor suitable for current and future applications.

12.3.2 DLC Applications for Temperature Sensors

DLC films are an excellent choice for fabricating a temperature sensor on a silicon substrate due to their exceptional thermal and chemical stability, special electrical properties, mechanical strength, and negative temperature coefficient of resistance (TCR). Due to the ease and controllability of the fabrication process, DLC film-based temperature sensors can also be integrated on a single Si substrate with other sensors. Such integration is advantageous as the chip temperature can be monitored in real time, and this temperature data can improve the accuracy and reliability of temperature compensation to cover the temperature-induced sensor drift. The test data of one such sensor prepared by depositing two types of DLC films (hydrogenated and hydrogen-free) using ion vapor deposition technology and integrated with an accelerometer on the same chip is found in the literature. The data shows the linear relationship between resistance and temperature for hydrogen-free DLC and an exponential relationship between the two for hydrogenated DLC films, suggesting the feasibility of using hydrogen-free DLC films in temperature sensors integrated with other sensors [39].

12.3.3 DLC Applications for Piezoresistive Sensor

Harsh environments can be explained in terms of voltage, contaminations, vibrations, corrosive gases, moisture, radiations, shock, vibrations, pressure, heat exposure, and operating temperature range fluctuations. Every material cannot perform under such conditions therefore, researchers are searching for suitable candidates. Elastic modulus, thermal conductivity, coefficient of thermal expansion, resistance to creep, fatigue, and yield strength are the desirable properties, as all these are temperature-dependent. Generally, degradation of structural materials, loss in elasticity, and plastic deformations cause problems. The most popular silicon-based MEMS is an inappropriate choice for the aforementioned conditions, particularly in high temperatures, as their piezoresistive coefficients change. Moreover, wide bandgap semiconductor materials have inherent advantages, i.e. high thermal conductivity, high breakdown field, and high saturated electron velocity. Among other wide bandgap semiconductor materials such as SiC and TiO_2, which are compatible with conventional MEMS fabrication techniques, DLC has presented itself as a good choice. It can be seen in Figure 12.7 that DLC outperforms the other suitable candidates in GF under a strain and is a function of activation energy [40]. The next section also presents an example of a strain sensor for high-temperature applications.

12.3.4 DLC Applications for Strain Sensors

Many applications like aeronautical, aerospace, petrochemical, and automotive are subjected to harsh environments, i.e. high pressure, temperature, corrosive, etc. The low performance of silicon sensors restricts their usage in such conditions. Therefore, multiple thin films have been tried so far in search of good candidates

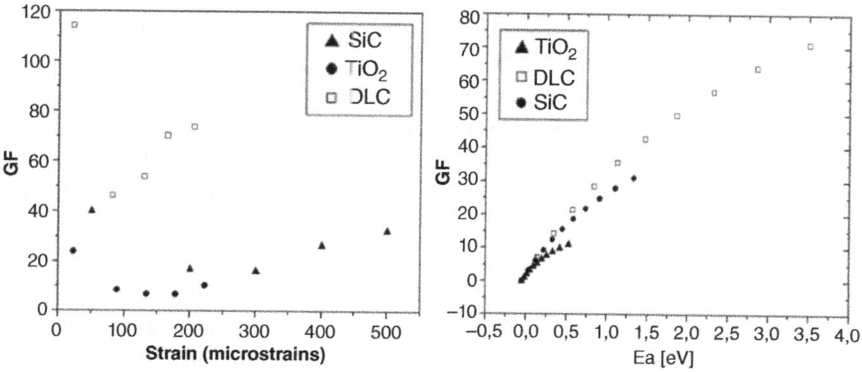

Figure 12.7 (a) Gauge factor of sputtered SiC, DLC, and TiO_2 thin films as a function of strain. (b) Gauge factor of sputtered SiC, DLC, and TiO_2 thin films as a function of activation energy. Reproduced from [40] / with permission of Springer Nature.

with mechanical and chemical stability, compatibility with integrated circuits, in addition to a large piezoresistive effect. DLC-based thin film strain sensors are promising due to their superior performance, especially in harsh environments.

A comparison was made for the TCR, change in resistance as a function of applied force (in terms of stress), and as a function of strain for strain sensors prepared with various thin films [41]. The DLC-based thin film sensor resistance change is the highest among the other candidates unless stress values are too high. Similarly, the resistance changes with the applied strain in DLC film-based sensors are also highest, proving the higher sensitivity of DLC film-based thin sensors. DLC thin film sensor exhibits the lowest TCR values when a temperature is applied, suggesting the strong candidacy of these sensors in high-temperature applications of strain sensors. Moreover, the DLC thin film sensors have a very high GF value of 69 as compared to SiC's 31 and TiO_2's 9. These comparisons suggest the superiority of using DLC films in piezoresistive strain sensors in harsh environments.

12.3.5 DLC Applications for Pressure Sensors

MEMS piezoresistive microsensors are already being used in medical and industrial applications wherein there is a need to monitor tiny pressure changes, i.e. intraocular, intracranial, aviation, etc. New applications, including space exploration, clean nuclear energy systems, and deep sea submersibles, demand the development of new smart structures and sensitive materials for harsh environments. However, it is not easy to fabricate uniform ultra-thin films due to complexity, availability, and cost. DLC has been identified as a promising material for carbon materials after trials on nanocrystalline diamond and graphene-BN. This can be attributed to its physiochemical and piezoresistive behavior with gauge factor ranging from −3200 to 1200. Moreover, high-temperature stability, facile tailoring on thickness, and compatibility with MEMS processes make DLC an ultra-sensitive functional material candidate for new-generation micropressure sensors.

The performance results of a classical submicron thickness self-supporting flat film structure micro pressure sensor fabricated using a conventional MEMS fabrication process are discussed in the literature [42]. A good linearity of the thermal compensation resistor at a TCR value of −1247.3 ppm/K in a temperature range of −20 to 100 degrees is found. The TCR value of DLC films is in the negative range; however, it can be easily adjusted to zero or positive values with doping. The observed sensitivity of the DLC thin film pressure sensor is 5.3×10^{-5} kPa^{-1} and within a 0–60 kPa pressure range. It shows good linearity and hysteresis behavior with nonlinearity and hysteresis were only 5.7 % and 0.8% full scale, respectively. The sensitivity only varied ±7% within the −20 to 100 degrees temperature range, with sensitivity at 20 degrees being the benchmark for comparison. Overall, the sensor offers good repeatability, sensitivity, linearity, and hysteresis with DLC offering rather easily adjustable TCR values and more convenient and uniformly controllable thin film fabrication. DLC-based micropressure sensor only lags in sensitivity to nanocrystalline diamonds however, DLC is more feasible due to the simplicity and process controllability of the film deposition process.

12.4 DLC Applications for Nanoactuators

Nanoactuators can play an important role in the advancement of nanotechnologies. Previously, tungsten and silicon dioxide-based bimorph, hybrid platinum/SWCNT, cell wall cutting tools, and nanorotors have been fabricated using the focused ion beam chemical vapor deposition (FIB-CVD) technique. The Bimorph segment was made of the tungsten-based conductor (TBC) and DLC in this versatile fabrication. Five segments, as shown in Figure 12.8 at different angles (30, 60 degrees), were made before the identification pillar, and only TBC

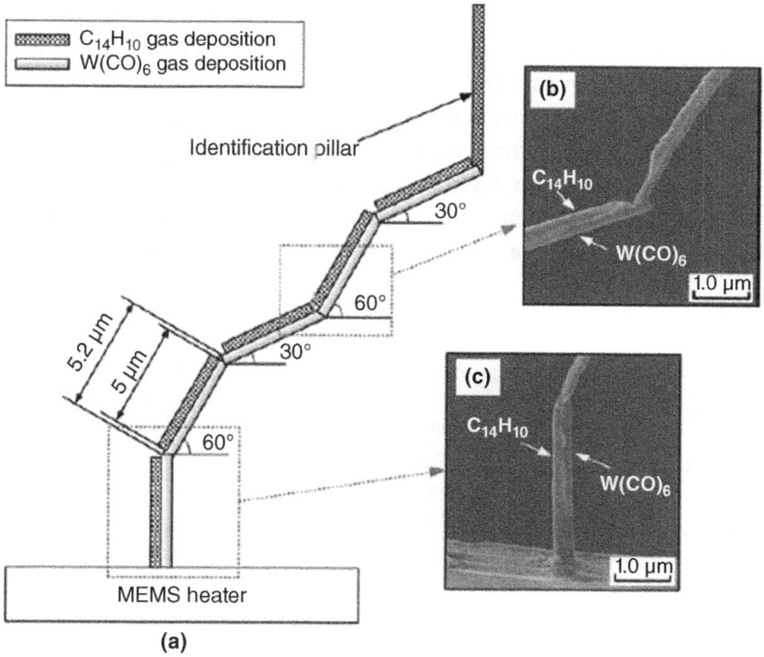

Figure 12.8 (a) Side view of the nanoactuator. The two SIM images in (b) and (c) show two different areas of the structure and relatively poor dimensional control of the manufacturing process. Reproduced from [43] / with permission of ELSEVIER.

structures are linked to avoid strain accumulation in the joint area (for maximum bending). Due to easy fabrication and actuation, a thermal actuator was used herein. However, the other options include biomorph-type micro-actuators and electromagnetically and electrostatically driven nanoactuators. Projection displacement in the above-mentioned (100 μm × 10 μm × 50 μm) structure was 600 ± 60 nm by input power of 0.16 A. Hence, the bimorph segment-based nanoactuator provides guidelines for future designs [43].

12.5 Industrial Processing

Industry 4.0, the fourth revolution, highlighted the need for sensor systems for load, temperature, and pressure control in production for customer-oriented products. Moreover, lower parts rejection and cost efficiency are also by-products. Forming and deep drawing real-time process applications are discussed herein [44, 45]. Using Plasma-assisted chemical vapor deposition (PACVD), a hydrogenated carbon coating with excellent tribological and piezoresistive behavior was deposited. The thin film piezoresistive hydrogenated carbon layer sensor system (6 μm thick deposited using PACVD) designed on the insert is shown in Figure 12.9, wherein the array of sensor structures is visible in the load region. These structures have better wear resistance due to silicon doping in hydrogenated carbon as it was required due to its placement in the drawing cushion during the forming process. A multifunctional thin film system (PACVD) for temperature and pressure measurement on a blank holder tool surface was also developed for the drawing process, wherein the load measurement area, conductive path, and contact areas were visible. Both case studies proved the efficacy of thin carbon film applications at the industrial level.

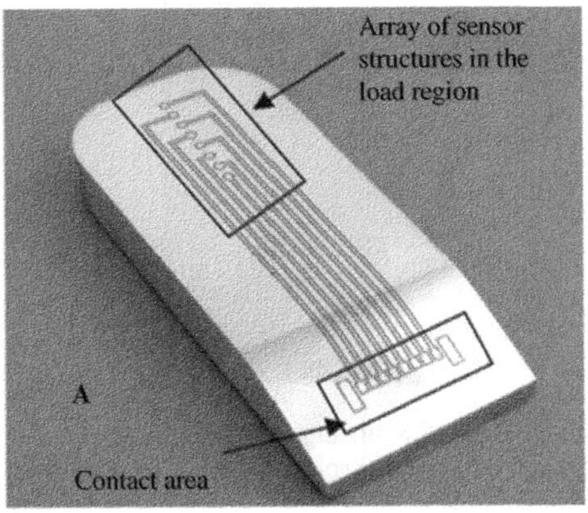

Figure 12.9 Piezoresistive carbon layer coated on steel insert for forming process [44].

12.6 Applications of DLC Coatings Structural Integrity

The high hardness, chemical inertness, electrical resistance, and high wear resistance of DLC make it useful for being widely employed in magnetic hard drives, bio-coatings, optical windows, electrical components, protective coatings, and MEMS. In MEMS, elastic contact between components and a large surface-to-volume ratio plays an important role in surface damage. Therefore, it is critical to determine its mechanical properties to support design and functionality, i.e. Young's modulus, Poisson's ratio, fracture strength, etc. The nanoindentation technique is generally applied to measure mechanical properties; however, the substrate effect may not be ruled out for thin coatings (< 100 nm) even though the standard of 10% depth is followed. For that reason, different values of Young's modulus are reported. Moreover, due to Poisson ratio assumptions, surface acoustic waves, a non-destructive method didn't provide an accurate enough value for modulus measurement. Tensile and deflection experiments have been developed to characterize the mechanical properties accurately. These studies also simultaneously present the structural stability of DLC to be employed in MEMS applications. The fracture strength and Young's modulus of submicron-thick 8 DLC samples (−100 V: 4, −300 V: 4) were determined on 19 μm single-crystal silicon (SCS) deposited using plasma-enhanced chemical vapor deposition. The fracture strength (0.82–1.36 GPa) changes with the thickness of DLC (characterized using atomic force microscopy (AFM)), and the maximum was found for the minimum thickness at both biased voltages. Young's modulus (−100 V: 101.8 GPa, −300 V: 120.9 GPa) of DLC/SCS was approximately the same for samples deposited at the same biased voltage across all thicknesses [46]. The membrane deflection experiment in another study also demonstrates that fracture strength increases with a decrease in membrane width under nanoindenter and a reduction in film thickness [47]. DLC's excellent properties also proved its suitability for high-performance infrared emitters at the commercial level [48]. This means DLC can be tailored according to desired properties, and the advantage of better fracture strength at a small size makes it further attractive for MEMS.

12.7 DLC Applications for MEMS Tribology

MEMS devices can be classified into two categories: sensors and actuators. Sensor-based MEMS are commercially available; however, actuator-based MEMS are not found in the market. This is because, unlike sensor-based MEMS, actuator-based MEMS elements undergo relative motion. This is where tribology comes into play as there is no reliable tribological solution available currently for micro/nanoscale. For machine elements at the micro/nanolevel (microgears, micromotors, microturbines, etc.), contact lubrication is essential however due to size constraints, it is not feasible. Moreover, the surface area to volume ratio becomes very large, which in turn signifies the role of surface forces, including friction and adhesion over gravitational and inertial forces. Therefore, liquid due to liquid-mediated adhesion is not an appropriate option in addition to solid lubricants (MoS_2, graphite) having a size equivalent to MEMS element [49]. This poses a challenge to the smooth functioning of actuator-based MEMS.

Conventionally, MEMS are made from silicon due to well-established design and fabrication methods for IC technology. Nonetheless, it exhibits poor tribological performance. To resolve this issue, researchers followed two directions; topographical and chemical modifications [50]. Topographical modifications include surface patterning/texturing and roughening whereas thin coatings (DLC, self-assembled monolayers) are named as chemical. DLC coatings are employed in MEMS due to stable chemical behavior, excellent mechanical properties, low friction, and outstanding wear resistance. The following studies highlight the importance of DLC as a promising candidate to develop and expand its applications in MEMS/NEMS actuators.

12.7.1 Nanofretting

Cyclic motion of contact interfaces is referred to as nano fretting, wherein non-uniform distribution in the contact area is revealed by relative displacement and displacement amplitude is in nanometer scale [51]. This phenomenon may occur at contacts of silicon in MEMS/NEMS due to mechanical vibration or temperature variation. This was first proposed in 2003 and extensively studied on monocrystalline silicon. However, this study revealed that DLC can minimize nanofretting damage of silicon contacts in MEMS/NEMS. Ultrathin DLC (2 nm) was deposited on p-Si (100) using the physical vapor deposition technique. Root mean square values of the wafer were 0.06 nm and 0.11 nm before and after deposition, respectively, measured using AFM. The hardness and modulus of DLC were approximately 70 GPa and 720 GPa, respectively. Figure 12.10a shows the SiO_2 tip (1 μm radius) with the relative displacement of 0.5-100 nm (D) under 2 μN applied load (F_n).

Due to its hydrophobic surface, DLC may reduce the adhesion force. Figure 12.10b, c shows the adhesion force results for Si/SiO_2 and DLC/SiO_2 pairs before and after the nanofretting test (vacuum, air). It shows that DLC can reduce up to 14% of the adhesion forces compared to the Si/SiO_2 pair. The adhesion forces on both pairs are almost the same after nanofretting, perhaps due to weak nanofretting damage and the absence of vapors. However, the strong capillary action (vapor abundance) induces larger adhesion forces in the air. The adhesion forces

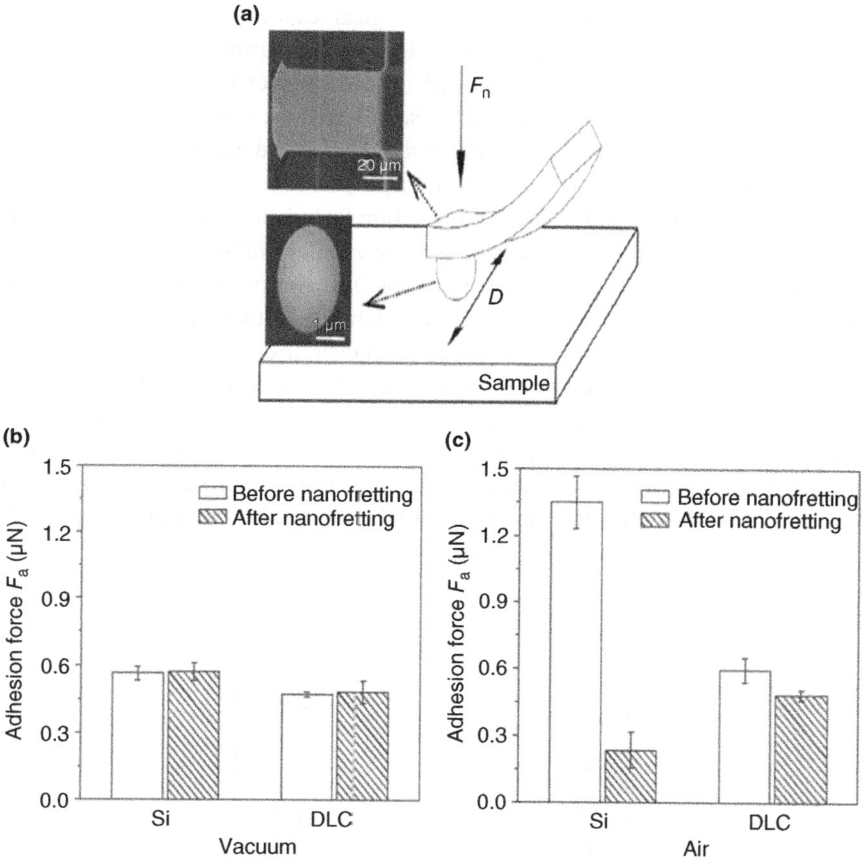

Figure 12.10 (a) Schematic of nanofretting test, adhesion force F_a of Si/SiO_2 and DLC/SiO_2 pairs before and after nanofretting tests (b) in vacuum and (c) in air. Reproduced from [52] / with permission of ELSEVIER.

decreased by 82% and 19% for Si and DLC, respectively. This significant variation on the Si surface may be attributed to SiO_2 sphere chemical modifications during the nanofretting process. Superior mechanical properties (prevent wear) and chemical inertness (avoid tribochemical damage) are the reasons for resisting nanofretting damage on silicon substrates [52].

12.7.2 Nanodot Contacts

Tribological properties of DLC and Si were studied at micro- and nanolevels for possible applications in miniaturized devices. For that purpose, nanodots were created by DLC deposition on nickel nanodots (fabricated by annealing ~9 μm Ni thin films at 800 degrees for 15 min). Tribological tests were conducted on DLC, nanodot DLC, and Si flat surface using AFM. A borosilicate ball with ~1.25 μm diameter was employed as tip mounted on a cantilever for nano-testing. The parameters were different for the micro (soda lime ball: 1 mm diameter, sliding speed: ~1mm/s, load: ~3000 μN, scan length: ~3 mm) and nano (sliding speed: ~2 μm/s, load: ~0–120 nN, scan size: ~5 μm × 5 μm) scales.

Adhesion, coefficient of friction, and friction force values have been plotted in Figure 12.11a–c. Adhesion values (caused by van der Walls, capillary, and electrostatic forces) were the lowest for nanodot DLC compared to DLC and Si flat surfaces. This was due to the hydrophobic nature (contact angle > 90 degrees) and reduction in the real area of contact. Moreover, friction forces are plotted as a function of applied load. Interestingly, friction forces even exist at zero applied load due to inherent adhesive force (dominated by capillary force). Si surfaces are hydrophilic and therefore produce adhesion (inherent), which is the reason for high friction. The coefficient of friction at the microscale is shown in Figure 12.11c. It can be seen that nanodot DLC has the lowest coefficient of friction influenced by a smaller contact area [53].

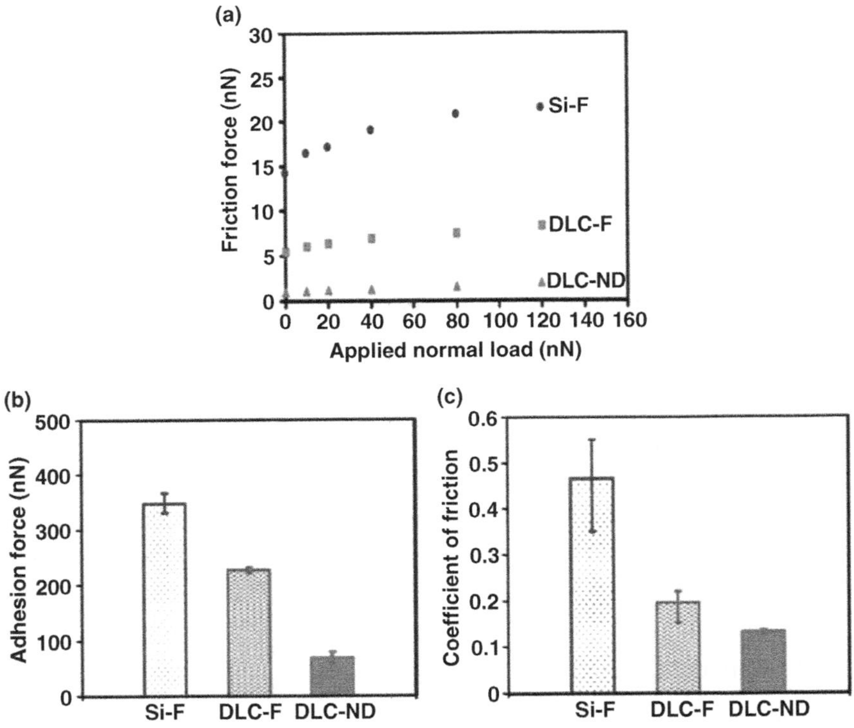

Figure 12.11 (a) Friction force comparison. (b) Adhesion force. (c) Coefficient of friction. Reproduced from [53] / with permission of ELSEVIER.

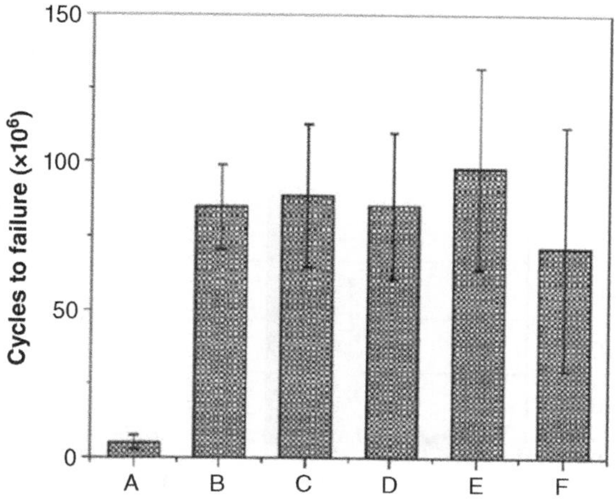

Figure 12.12 Coated and uncoated motor performance (A) uncoated (B) ~15 nm DLC plus plasma etching (C) B+0.1% Zdol cured at 150 degrees for an hour (D) C after one year (E) coated with 0.1% Zdol on polysilicon devices cured at 150 degrees for an hour (F) after one usage. Reproduced from [54] / with permission of ELSEVIER.

12.7.3 Lubricant Life Assessment

The electrostatic lateral output motor was used to investigate the fluorinated lubricant aging effect. Fomblin Zdol coated device performance is quite good; however, it deteriorates with storage in the air. This is possibly due to the lubricant decomposing more slowly in contact with the polysilicon. The polysilicon wafers were plasma etched after coating with DLC (~15 nm). Cycles to failure were compared for uncoated and coated samples tested under different conditions (lubricated and unlubricated). The uncoated polysilicon sample failed early, and all other counterparts, including the DLC-coated one, performed excellently, as shown in Figure 12.12. The DLC+Zdol survived 8.56×10^7 cycles even after 12 months on polysilicon motors, exploring another aspect of DLC usage [54].

12.8 DLC Coatings System and Performance

12.8.1 Hydrogenated and Non-hydrogenated DLC Coating

DLC coatings have been investigated extensively in search of suitable wear-resistance candidates. Generally, ta-C DLC is preferred for wear resistance-related applications; however, other types have also been investigated. Hydrogenated and hydrogen-free DLC were deposited using plasma-assisted chemical vapor deposition and DC sputtering, respectively. For tribological characterization, oscillating scratch tests and single scratches under single asperity were used. The counter

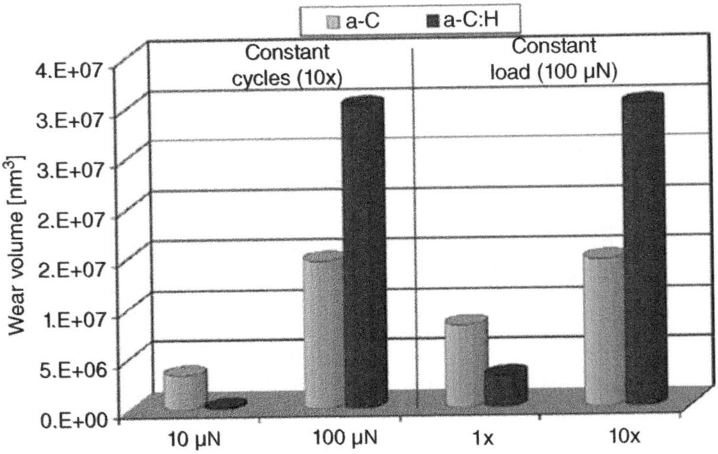

Figure 12.13 a-C and a-C:H 250 nm thick DLC wear comparison with constant cycles and loading. Reproduced from [55] / with permission of ELSEVIER.

body was a 1 μm radius conical diamond tip. Figure 12.13 shows the a-C and a-C:H (~250 nm thickness) wear results. The left and right sides present data after a constant number of 10 cycles and a fixed load of 100 μN, respectively. It can be seen that for one cycle and low load a-C:H showed better wear resistance whereas under high load and more cycles, a-C performed well [55]. It depicts the applicability of both DLC coatings for different conditions.

12.8.2 Role of Substrate Material

For the applications of microtribological coatings for micro-actuators and MEMS, DLC coatings' wear and microfiction behavior were investigated with different substrate materials with different Young's moduli. Plasma-enhanced chemical vapor deposition was used for coating deposition. Multiple materials, including steel (100Cr6), silicon, SU 8 resin on silicon, and BK7 glass, were employed as substrates. Hysitron nanoindenter, AFM, and pin on disc (10 mm ball diameter, 3 N load) were utilized for characterization. Figure 12.14a demonstrates that friction coefficient values decrease with a decrease in Young's modulus due to the plate bending effect [56]. The coefficient of friction values decreases by more than 20% with a change in Young's modulus (5, 6, 70, 165 GPa). The wear of DLC also gets affected by different Young's modulus. A band length to grind an 80 nm deep crater into the surface was set for wear assessment. Wear resistance was found to be maximum with the soft substrate (SU 8) with a 10 nm titanium interface (Figure 12.14b). Load application dispensed over the larger area reduces the wear as it fits closer to the counter body.

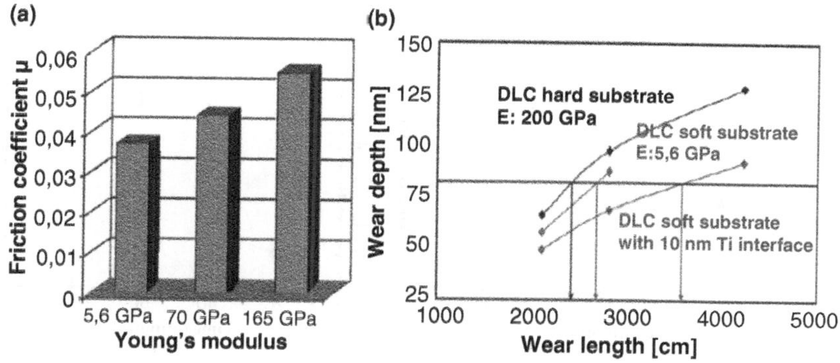

Figure 12.14 Dependence and influence of substrate material on (a) coefficient of friction and (b) abrasive wear. Reproduced from [56] / with permission of ELSEVIER.

12.8.3 Wear at Microscopic Scale

This study reported a detailed investigation of microscopic wear for MEMS applications. Triangular macroscopic specimens were coated with a range of materials (DLC, SCS, Si_3N_4, SiO_2). The contact pressure and area of contact were 10^{-6}–2 MPa and 20 μm^2–1 mm^2, respectively. A well-known photolithography process and reactive ion etching were used for specimen fabrication whereas these were coated by photoresist. A specific specimen was employed for the comparison of all materials (Figure 12.15) at 170 Pa contact pressure. Wear was plotted against sliding distance at a load of 30 μN. The graph demonstrates that

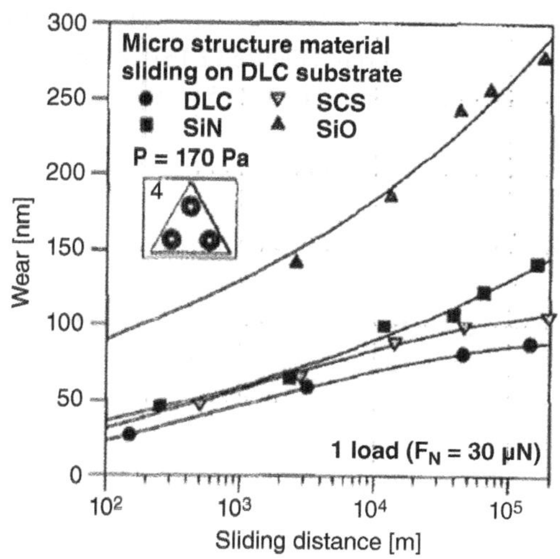

Figure 12.15 Wear rate for different materials coupled on DLC at 170 Pa contact pressure. Reproduced from [57] / with permission of ELSEVIER.

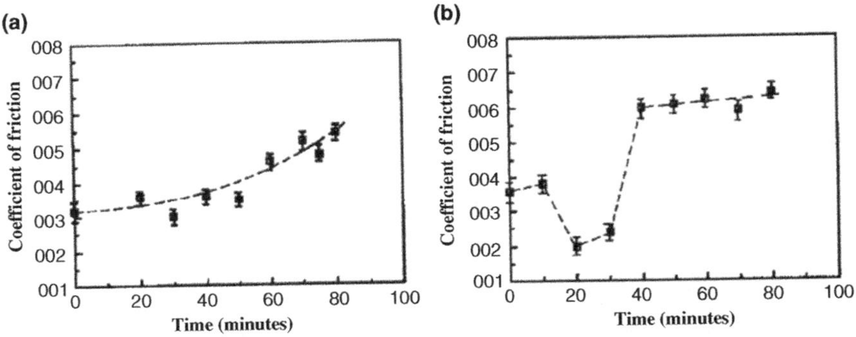

Figure 12.16 Coefficient of friction of DLC/DLC contacts under exposure: (a) oxygen 10^{-7} Torr and (b) carbon monoxide 5×10^{-6} Torr. Reproduced from [58] / with permission of ELSEVIER.

DLC/DLC produced the minimum wear at all sliding distances undisputedly. Two wear mechanisms were explored: (i) asperity fracture and (ii) asperity deformation. All the above-mentioned materials fell into the former category except polysilicon (the latter one) [57].

12.8.4 Static Friction

The gas environment's effect on the coefficient of static friction of a 3 mm radius aluminum bullet coated with DLC (directly and on silicon nitride-coated using plasma-enhanced chemical vapor deposition) and a flat DLC substrate was studied. In an ultra-high vacuum 10^{-7} Torr, increasing the duration of oxygen exposure (in DLC/DLC contact) to 8 minutes increased the coefficient of friction by approximately 50% (0.04–0.06). However, the value is still less than that of silicon-based materials (Figure 12.16). On the other hand, carbon monoxide exposure presents a different trend. At 5×10^{-6} Torr, the value decreases for 2–3 minutes and increases onwards [58].

12.9 Conclusion

DLC coatings find widespread applications in MEMS due to various reasons. The summary is presented as follows:

1. The fabrication processes are safe, green, flexible, mature, and compatible with MEMS.
2. Low surface energy and a larger contact angle (>90°) assist in overcoming the adhesion forces and induce superior anti-stick properties that are helpful in MEMS fabrication and operation.

3. High-stress sensitivity and excellent electrical properties, a higher gauge factor, and weak dependence of resistivity on the temperature in a certain range make carbon coatings a suitable candidate for MEMS sensors.
4. High hardness and Young's modulus, tensile strength, low friction and wear rates, superior thermal conductivity, and low coefficient of thermal expansion promote their applications to produce MEMS mechanical structures and moving tribological components.

References

1 Varadan, V.K. (2003). Nanotechnology, MEMS and NEMS and their applications to smart systems and devices, (in English). *Proceedings of the Society of Photo-Optical Instrumentation* 5062: 20–43. doi: 10.1117/12.514830.

2 Qian, J.Y., Hou, C.W., Li, X.J. et al. (2020). Actuation mechanism of microvalves: A review. *Micromachines (Basel)* 11 (2): 172. doi: 10.3390/mi11020172.

3 Zhang, Z., Zhang, H., Hao, Y. et al. (2024). A review on MEMS silicon resonant accelerometers, (in English). *Journal of Microelectromechanical Systems* 33 (2): 174–208. doi: 10.1109/Jmems.2024.3354235.

4 Chen, S.S., Zhang, Y.Y., Hong, X.R. et al. (2022). Technologies and applications of silicon-based micro-optical electromechanical systems: A brief review, (in English). *Journal of Semiconductors* 43 (8): 081301. doi: 10.1088/1674-4926/43/8/081301.

5 Forouzanfar, S., Pala, N., Madou, M. et al. (2021). Perspectives on C-MEMS and C-NEMS biotech applications. *Biosensors and Bioelectronics Biosensors and Bioelectronics* 180: 113119. doi: 10.1016/j.bios.2021.113119.

6 Le, X., Shi, Q., Vachon, P. et al. (2022). Piezoelectric MEMS-evolution from sensing technology to diversified applications in the 5G/Internet of Things (IoT) era, (in English). *Journal of Micromechanics and Microengineering* 32 (1): 014005. doi: 10.1088/1361-6439/ac3ab9.

7 Hossain, M.I., Zahid, M.S., Chowdhury, M.A. et al. (2023). MEMS-based energy harvesting devices for low-power applications-a review, (in English). *Results in Engineering* 19: 101264. doi: 10.1016/j.rineng.2023.101264.

8 Tilli, M. and Haapalinna, A. (2020). Chapter 1 - Properties of silicon. In: *Handbook of Silicon Based MEMS Materials and Technologies, 3e* (eds. M. Tilli, M. Paulasto-Krockel, M. Petzold et al.), 3–17. Amsterdam: Elsevier.

9 World's first diamond micromachines created at Sandia. https://www.sandia.gov/media/NewsRel/NR2000/diamond.htm (accessed 10 August 2024).

10 Van Spengen, W.M., Puers, R., and De Wolf, I. (2003). On the physics of stiction and its impact on the reliability of microstructures, (in English). *Journal of Adhesion Science and Technology* 17 (4): 563–582. doi: 10.1163/15685610360554410.

11 Tanner, D.M., Miller, W.M., Peterson, K.A. et al. (1999). Frequency dependence of the lifetime of a surface micromachined micro engine driving a load, (in English). *Microelectronics Reliability* 39 (3): 401–414. doi: 10.1016/S0026-2714(98)00248-0.

12 Tanner, D.M., Walraven, J.A., Helgesen, K.S. et al. (2000). MEMS reliability in a vibration environment. In: *2000 IEEE International Reliability Physics Symposium Proceedings. 38th Annual (Cat. No.00CH37059)*, 10-13 139–145. doi: 10.1109/RELPHY.2000.843904.

13 Grill, A. (1997). Tribology of diamondlike carbon and related materials: an updated review, (in English). *Surface and Coatings Technology* 94: 507–513. doi: 10.1016/S0257-8972(97)00458-1.

14 Robertson, J. (2002). Diamond-like amorphous carbon. *Materials Science and Engineering: R: Reports* 37 (4): 129–281. doi: https: 10.1016/S0927-796X(02)00005-0.

15 Luo, J.K., Fu, Y.Q., Le, H.R. et al. (2007). Diamond and diamond-like carbon MEMS. *Journal of Micromechanics and Microengineering* 17 (7): S147. doi: 10.1088/0960-1317/17/7/S12.

16 Smallwood, S.A., Eapen, K.C., Patton, S.T. et al. (2006). Performance results of MEMS coated with a conformal DLC, (in English). *Wear* 260 (11–12): 1179–1189. doi: 10.1016/j.wear.2005.07.019.

17 Yu, W., Wang, J., Huang, W. et al. (2020). Improving high temperature tribological performances of Si doped diamond-like carbon by using W interlayer, (in English). *Tribology International* 146: 106241. doi: 10.1016/j.triboint.2020.106241.

18 Chua, D.H., Milne, W.I., Sheeja, D. et al. (2004). Fabrication of diamond-like amorphous carbon cantilever resonators, (in English). *Journal of Vacuum Science & Technology B* 22 (6): 2680–2684. doi: 10.1116/1.1824042.

19 Milne, W.I. (2003). Electronic devices from diamond-like carbon. *Semiconductor Science and Technology* 18 (3): S81.

20 Sekaric, L., Parpia, J.M., Craighead, H.G. et al. (2002). Nanomechanical resonant structures in nanocrystalline diamond, (in English). *Applied Physics Letters* 81 (23): 4455–4457. doi: 10.1063/1.1526941.

21 Birkett, M., Zia, A.W., Devarajan, D.K. et al. (2023). Multi-functional bioactive silver- and copper-doped diamond-like carbon coatings for medical implants. *Acta Biomaterialia* 16: 54–68. doi: 10.1016/j.actbio.2023.06.037.

22 Zia, A.W. and Birkett, M. (2021). Deposition of diamond-like carbon coatings: Conventional to non-conventional approaches for emerging markets, (in English). *Ceramics International* 47 (20): 28075–28085 doi: 10.1016/j.ceramint.2021.07.005.

23 Hauert, R. (2003). A review of modified DLC coatings for biological applications, (in English). *Diamond and Related Materials* 12 (3–7): 583–589 doi: 10.1016/S0925-9635(03)00081-5.

24 Joshi, P., Mishra, R., and Narayan, R.J. (2021). Biosensing applications of carbon-based materials, (in English). *Current Opinion in Biomedical Engineering* 18: 100274. doi: 10.1016/j.cobme.2021.100274.

25 Khanna, P., Villagra, A., Kim, S. et al. (2006). Use of nanocrystalline diamond for microfluidic lab-on-a-chip, (in English). *Diamond and Related Materials* 15 (11–12): 2073–2077. doi: 10.1016/j.diamond.2006.05.014.

26 Jonnalagadda, K., Cho, S.W., Chasiotis, I. et al. (2008). Effect of intrinsic stress gradient on the effective mode-I fracture toughness of amorphous diamond-like carbon films for MEMS, (in English). *Journal of the Mechanics and Physics of Solids* 56 (2): 388–401. doi: 10.1016/j.jmps.2007.05.013.

27 Jonnalagadda, K.N. and Chasiotis, I. (2009). Strength and fracture resistance of amorphous diamond-like carbon films for MEMS, (in English). *Journal of Nanomaterials* 2009(1): 204281. doi: 10.1155/2009/204281.

28 McCarty, A. and Chasiotis, I. (2007). Description of brittle failure of non-uniform MEMS geometries, (in English). *Thin Solid Films* 515 (6):3267–3276. doi: 10.1016/j.tsf.2006.01.039.

29 Erdemir, A., Eryilmaz, O.L., Nilufer, I.B. et al. (2000). Synthesis of superlow-friction carbon films from highly hydrogenated methane plasmas. *Surface and Coatings Technology* 133: 448–454. doi: 10.1016/S0257-8972(00)00968-3.

30 Enke, K. (1981). Some new results on the fabrication of and the mechanical, electrical and optical properties of i-carbon layers. *Thin Solid Films* 80 (1): 227–234. doi: 10.1016/0040-6090(81)90226-1.

31 Ma, X., Tong, X., Guo, P. et al. (2020). MEMS piezo-resistive force sensor based on DC sputtering deposited amorphous carbon films, (in English). *Sensor and Actuators A-Physical* 303: 111700. doi: 10.1016/j.sna.2019.111700.

32 Tibrewala, A., Peiner, E., Bandorf, R. et al. (2006). Transport and optical properties of amorphous carbon and hydrogenated amorphous carbon films, (in English). *Applied Surface Science* 252 (15): 5387–5390. doi: 10.1016/j.apsusc.2005.12.046.

33 Petersen, M., Bandorf, R., Bräuer, G. et al. (2012). Diamond-like carbon films as piezoresistors in highly sensitive force sensors, (in English). *Diamond and Related Materials* 26: 50–54. doi: 10.1016/j.diamond.2012.04.004.

34 Smith, A.D., Niklaus, F., Paussa, A. et al. (2016). Piezoresistive properties of suspended graphene membranes under uniaxial and biaxial strain in nanoelectromechanical pressure sensors. *ACS Nano* 10 (11): 9879–9886. doi: 10.1021/acsnano.6b02533.

35 Khalid, H.R., Nam, I.W., Choudhry, I. et al. (2018). Piezoresistive characteristics of CNT fiber-incorporated GFRP composites prepared with diversified fabrication schemes, (in English). *Composite Structures* 203: 835–843. doi: 10.1016/j.compstruct.2018.08.003.

36 He, R. and Yang, P. (2006). Giant piezoresistance effect in silicon nanowires. *Nature Nanotechnology* 1 (1): 42–46. doi: 10.1038/nnano.2006.53.

37 Winkler, K., Bertagnolli, E., and Lugstein, A. Origin of anomalous piezoresistive effects in VLS grown Si nanowires. *Nano Letters* 15 (3): 1780–1785. Mar 11 2015. doi: 10.1021/nl5044743.

38 Ma, X., Guo, P., Tong, X. et al. Piezoresistive behavior of amorphous carbon films for high performance MEMS force sensors, (in English). *Applied Physics Letters* 114 (25). doi: 10.1063/1.5096225.

39 Li, B., Zhao, Y., Ma, X. et al. (2017). Design of a resonant accelerometer integrated with a diamond like carbon film temperature sensor, (in English). *IEEE Sensor* 804–806. [Online]. Available: <Go to ISI>://WOS:000427677500275.

40 Fraga, M.A., Furlan, H., Pessoa, R.S. et al. (2012). Studies on SiC, DLC and TiO thin films as piezoresistive sensor materials for high temperature application, (in English). *Microsystem Technologies* 18 (7–8): 1027–1033. doi: 10.1007/s00542-012-1435-y.

41 Fraga, M.A., Furlan, H., and Pessoa, R.S. (2011). Comparison among performance of strain sensors based on different semiconductor thin films, (in English). *Proceedings of SPIE* 8066: 742–750. doi: 10.1117/12.886624.

42 Ma, X., Zhang, Q., Guo, P. et al. (2022). Self-supporting ultrathin DLC/SiN/SiO for micro-pressure sensor, (in English). *IEEE Sensors Journal* 22: 5 3937–3944. doi: 10.1109/Jsen.2021.3133935.

43 Chang, J.Y., Min, B.K., Kim, J. et al. (2009). Bimorph nano actuators synthesized by focused ion beam chemical vapor deposition, (in English). *Microelectronic Engineering* 86 (11): 2364–2368. doi: 10.1016/j.mee.2009.04.022.

44 Biehl, S., Staufenbiel, S., Hauschild, F. et al. (2010). Novel measurement and monitoring system for forming processes based on piezoresistive thin film systems. *Microsystem Technologies* 16 (5): 879–883. doi: 10.1007/s00542-010-1058-0.

45 Biehl, S., Paetsch, N., and Meyer-Kornblum, E. (2017). Thin film system with integrated load and temperature sensors for the technical application in the deep drawing process. In: *Smart Sensors, Actuators, and MEMS VIII*. 10246: 154–161.

46 Isono, Y., Namazu, T., and Terayama, N. (2006). Development of AFM tensile test technique for evaluating mechanical properties of sub-micron thick DLC films. *Journal of Microelectromechanical System* 15 (1): 169–180.

47 Zhou, B. Wang, L., Mehta, N. et al. The mechanical properties of freestanding near-frictionless carbon films relevant to MEMS. *Journal of Micromechanics and Microengineering* 16 (7): 1374.

48 Ohlckers, P., Skotheim, T., Dmitriev, V. et al. (2008). Advantages and limitations of diamond-like carbon as a MEMS thin film material. In: *Technical Proceedings of the 2008 NSTI Nanotechnology Conference and Trade Show*. 1: 63–66.

49 Henck, S.A. (1997). Lubrication of digital micromirrordevicesTM. *Tribology Letters* 3 (3): 239–247.

50 Singh R.A. and. Yoon E.-S. (2007). Friction of chemically and topographically modified Si (1 0 0) surfaces. *Wear* 263 (7–12): 912–919.

51 Qian, L., Zhou, Z., Sun, Q. et al. (2007). Nanofretting behaviors of NiTi shape memory alloy. *Wear* 263 (1–6): 501–507.

52 Chen, L., Yang, M., Yu, J. et al. (2011). Nanofretting behaviorsbehaviors of ultrathin DLC coating on Si (1 0 0) substrate. *Wear* 271 (9–10): 1980–1986.

53 Singh, R.A., Na, K., Yi, J.W. et al. (2011). HHDLC nano-dot surfaces for tribological applications in MEMS devices. *Applied Surface Science* 257 (8): 3153–3157.

54 Eapen, K.C., Patton, S.T., Smallwood, S.A. et al. (2005). Aging of a fluorinated lubricant on bare and DLC-coated silicon-based MEMS. *Surface and Coatings Technology* 197 (2–3): 270–277.

55 Bandorf, R., Lüthje, H., Henke, C. et al. (2005). Different carbon-based thin films and their microtribological behaviour in MEMS applications. *Surface and Coatings Technology* 200 (5–6): 1777–1782.

56 Bandorf, R., Lüthje, H., and. Staedler, T. (2004). Influencing factors on microtribology of DLC films for MEMS and microactuators. *Diamond and Related Materials* 13 (4–8): 1491–1493.

57 Beerschwinger, U., Albrecht, T., Mathieson, D. et al. (1995). Wear at microscopic scales and light loads for MEMS applications. *Wear* 181: 426–435.

58 Deng, K. and. Ko, W.H. (1992). Static friction of diamond-like carbon film in MEMS. *Sensors and Actuators A: Physical* 35 (1): 45–50.

13

Data-centric Process, Property, and Performance of Carbon Coatings

Muhammad Ikhlaq[1,2], *Abdul Wasy Zia*[3]

[1] *School of Engineering, Newcastle University, Newcastle upon Tyne, UK*
[2] *Dyson Institute of Engineering and Technology, Malmesbury, UK*
[3] *Institute of Mechanical Process and Energy Engineering (IMPEE), School of Engineering and Physical Sciences, Heriot-Watt University, Edinburgh, UK*

CONTENTS

13.1 Introduction, 373
13.2 Data-centric Manufacturing Process of Carbon Coatings, 375
13.3 Data-centric Virtual Synthesis and Structural Design of Carbon Coatings, 380
13.4 Understanding the Properties of Carbon Coatings with Data-centric Approaches, 383
13.5 Data-centric Performance Assessment of Carbon Coatings, 387
13.6 Applications of Carbon Coating in Enhancing the Heat Transfer Processes, 397
13.6.1 Condensation Heat Transfer, 398
13.6.2 Boiling Heat Transfer, 402
13.7 Conclusion, 405

13.1 Introduction

Industrial revolutions are categorized from Industry 1.0 to Industry 5.0 along a transformative timeline. Industry 1.0 began around 1784 with mechanization. Industry 2.0 emerged around 1870 with electrification. Industry 3.0 started around 1970, characterized by automation. Industry 4.0 began around 2011, marked by digitalization, and Industry 5.0 commenced around 2021, focusing on resilient, sustainable, and human-centric industrialization [1]. An analogous progression is observed in the materials field. The Bronze Age, starting around 3500 BC, transitioned to Materials 1.0 during the Iron and Steel Age before the twentieth century. Materials 2.0 began in the twentieth century, emphasizing prototyping, testing, life cycle assessments, and

Applications of Diamond-like Carbon Coatings, First Edition. Abdul Wasy Zia.

mass manufacturing. Materials 3.0 emerged around 1970 with advancements in information technology, fostering hypothetical materials research using conventional methodologies. Materials 4.0, arriving in the twenty-first century, ushered in the era of computation and digitalization, where big data drives materials informatics. This includes multiscale, multiagent modeling, virtual synthesis and testing of materials, virtual optimizations, and redesigning manufacturing procedures following virtual validations, leading to savings in materials, energy, and costs.

Materials 4.0 [2] is promoting data-driven materials development and sustainable manufacturing, leveraging multidisciplinary areas to enhance product life and performance while supporting sustainability goals. In the modern age, data-centric engineering for materials innovation connects data science with engineering by integrating mathematical modeling, big data, and artificial intelligence protocols to innovate efficient, robust, sustainable, and advanced engineering solutions. Applications with significant financial, legal, or hazardous risks benefit the most from integrating data-driven methods [3] with simulations, highlighting the value of data-centric engineering approaches [4]. The data-centric approaches have emerged in the material and manufacturing industries; however, extensive data availability in a wider engineering context is an ongoing challenge [5]. Artificial intelligence tools like machine learning models warrant larger amounts of data to provide useful outcomes. Therefore, along with large-scale data collection, physics-informed computational simulations are performed when data is insufficient to be processed by machine learning tools. Sufficient availability of data not only helped to develop a good benchmark but also provide reliable predictions. In this context, material informatics can enhance the design, manufacturing process, and performance of carbon coatings to meet Industry 5.0 demands. Optimizing carbon coatings for durability, low friction coefficients, and wear rates against high contact pressures can extend their usable life, providing significant technological, economic, and environmental benefits by reducing recoating costs, recycling needs, and waste management, and promoting a circular economy of coatings [6].

Carbon coatings are primarily deposited using physical vapor deposition (PVD) [7] and plasma-enhanced chemical vapor deposition (PECVD) methods [8], which encompass over 10 different technologies, including sputtering [9], arc methods [10], pulsed laser deposition [11], and various types of plasma techniques [12]. Each of these methods involves multiple parameters, such as electric power inputs, dimensional aspects, and working pressures, which impact the growth and performance of diamond-like carbon (DLC) coatings. Consequently, the optimal bias voltage, typically around 100 eV [13], varies depending on the system configuration and deposition method. Optimizing the properties and performance of DLC is crucial and involves more than just adjusting a single parameter like bias voltage. For instance, while hard DLC coatings exhibit low friction and reduced wear rates, they often have higher residual stresses. Similarly, hardness and toughness in DLC coatings typically exhibit inverse relationships, which can limit their effectiveness in high-contact load applications. Research indicates that DLC coatings achieve the highest toughness and lowest hardness at a 40 V bias, while a 100 V bias results in the highest hardness but lower toughness [14]. To refine DLC coatings for optimal properties, such as hardness, toughness, wear resistance, surface roughness, and residual stress, there are numerous parameter combinations to consider [15]. These combinations are often more complex than single-variable adjustments, involving over 50 different factors that need to be balanced.

The application of computational tools to develop process–property–performance (PPP) relationships is a growing area to develop efficient material designs for new industrial applications. The high-throughput computational materials science [16] is innovating new materials design and synthesis methods for a broad range of applications. Recent technological advancements in data acquisition and storage, microstructure characterization and reconstruction (MCR), materials modeling and simulation, data processing and machine learning, and deep learning-based optimization in combination with physical manufacturing have greatly enhanced researchers' capabilities in developing PPP relationships and inverse material design [17]. This chapter explores these advancements from a design research perspective, introducing a data-centric approach that encompasses three fundamental aspects of carbon coatings: design representation, design synthesis, and design evaluation. Developments in each aspect are guided by and benefit from domain knowledge. Accordingly, we present a diverse array of computational methods for each aspect, illustrating how their integration enables data-centric materials discovery and design.

In recent years, data analytics has played a significant role in optimizing DLC coatings. Approaches like the Taguchi method and its combinations with fuzzy logic have emerged to enhance performance. For example, researchers have improved the tribological performance of multilayered DLC coatings using the orthogonal array L18 test [18]. Others have applied adaptive fuzzy inference systems to study tribological behavior, finding good agreement with experimental results. More recently, hybrid models combining gray and fuzzy reasoning have been used to optimize the conditions for multilayer DLC coatings on AISI 52100 steel [19]. Additionally, metaheuristic models have been employed to study DLC coating hardness, with some models showing better accuracy than traditional genetic algorithms [20].

13.2 Data-centric Manufacturing Process of Carbon Coatings

As discussed in the earlier chapters, DLC holds very good erosion resistance (tribological) and has low friction and high wear strength. Albeit the adhesive properties of DLC become very poor when they are subjected to high residual stresses, this restricts them from being utilized for many applications where sliding or movement against surfaces is needed. Understanding the complex behavior of the process of adhesion is quite difficult experimentally due to the thickness of coatings and limited access to the measurement equipment. Computational simulations allow us to understand the complex physical phenomenon of adhesion along with their fundamental understanding. To simulate the surface coating process, predominantly, two different approaches of numerical simulations are adopted: molecular dynamics (MD) simulations and probabilistic simulations.

Numerous techniques are being utilized to simulate molecular dynamics; the two most widely accepted empirical interatomic potential methods in the coatings are reactive force field (ReaxFF) [21] and Tersoff [22]. ReaxFF and Tersoff potential are being used to investigate the systems that involve covalent bonds for carbon-based materials. To mimic chemical reactions, combustion processes, catalysis, and other situations where bond breaking and creation are important, ReaxFF is frequently employed. It is particularly well suited for research on systems with functionalized surfaces and intricate molecular interactions. Tersoff potential is frequently used to investigate the mechanical, thermal, and crystallographic properties of materials. It works effectively for modeling crystalline systems when the emphasis is on material characteristics rather than chemical processes. Although all physical phenomena that can be simulated using Tersoff potential can be simulated using ReaxFF, it is impractical to utilize it for material properties as ReaxFF is significantly computationally expensive (approximately 100 times more [23]) than the Tersoff potential method.

Kametani et al. [24] looked at the evolution of residual stresses during the process of deposition for DLC coatings by the utilization of MD simulations using the Tersoff potential method. They unravelled the mechanism behind residual stress formation in DLC films through MD simulations (Tersoff) of carbon atom deposition on diamond substrates. It is observed that the deposition process established a surface dominated by sp^2 carbon bonding (about 70%), tapering linearly toward the film's interior. Conversely, the sp^3 fraction was minimal at the surface (around 10%), gradually intensifying deeper within, and the sp^1 fraction remained consistently at roughly 20% across the film thickness. The film exhibited compressive stresses in both in-plane and thickness orientations, escalating with growing film thickness. Notably, the maximum in-plane stress reached approximately 0.6 GPa, significantly surpassing the stress along the thickness. Compressive stress stemmed from sp^3 and sp^1 bonding, while sp^2 bonding induced tensile stress. Particularly, the stress in sp^3 greatly exceeded that in sp^2 and sp^1. Calculations indicated bond lengths in stress-free conditions at 500 K, measuring 1.54 Å for sp^3, 1.42 Å for sp^2, and 1.48 Å for sp^1. Film average bond lengths deviated from these values, reflecting compressive and tensile stresses in sp^3 or sp^1 and sp^2, respectively. Examining from film surface to interior revealed a reduction in sp^3

clusters coupled with their size enlargement. This growth was prompted by the transition from sp^2 to sp^3 bonding at the surface of sp^3 clusters. Given sp^3's larger bond length compared to sp^2, this transition under constrained conditions engendered compressive stress within sp^3 clusters. The ensuing expansion of sp^3 clusters further amplified compressive stress, thus constituting the core mechanism behind residual compressive stress in the film.

Figure 13.1 presents the distribution of sp^3 clusters for the xz cross-section at t = 6000 ps. The y-direction thickness is kept at 50 Å. Each cluster is represented by distinct colors to differentiate between clusters. Z = 0 is represented by the dotted red line in Figure 13.1 (left). The visualization of clusters is constructed using surface mesh using The Open Visualization Tool 'OVITO' by OVITO GmbH [25]. Figure 13.1 (right) shows the distributions in the z-direction of the average diameter and the number of sp^3 clusters at t = 6000 ps, calculated by the cluster analysis of OVITO. The number of sp^3 clusters monotonically decreased, and the average volume increased from the surface to approximately z = 15 Å inside the film (region I). In the region with $5 \leq z \leq 15$ Å, the number and volume of sp^3

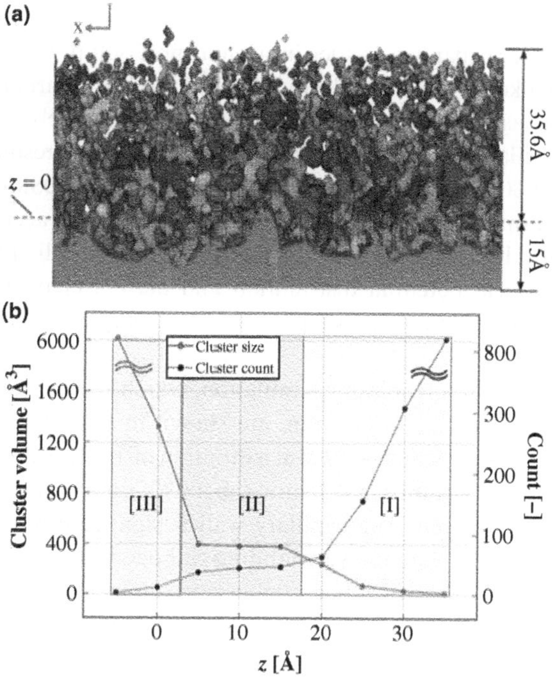

Figure 13.1 (a) Arrangement of sp^3 clusters (t = 6000 ps) in 50 Å thin slice. Each color presents an individual sp^3 cluster. (b) The relationship between cluster average volume and number of sp^3 clusters in growth direction. *Source:* Reproduced from [24] / with permission of ELSEVIER.

clusters are almost constant (region II), and in the region z < 5 Å, the number of sp^3 clusters decreased sharply and the volume increased (region III). From the results, it can be predicted: in region I, the sp^3 clusters nucleate in the sp^2 structure near the film surface and grow with time. During growth, some sp^3 clusters shrink and disappear, or fuse and grow larger. Region II corresponds to a condition in which more time has passed. In this region, relatively large sp^3 clusters were found to exist stably, and this region corresponds to the region with high constant stress. Region III is in the original substrate, where sp^3 clusters exist in the fluctuating substrate, as shown in Figure 13.1.

Probabilistic methods use a statistical approach to solve the equations using a particle-based method, and most of them utilize the Monte Carlo Method. For the kinetic behavior of the gases, the Boltzmann transport equation can be solved statistically to model the behavior. Hofmann et al. [26] simulated the plasma-assisted surface treatment for a hollow cylinder. Due to the shape of the geometry, it is quite difficult to control coating parameters, which is why probabilistic simulation is suitable to study the effect of different parameters. The aspect ratio of the hollow cylinder (geometry used for the simulation) >>1. Particle-based simulation methods were used to model plasma and flow in high-pressure systems, revealing an under-expanded gas jet in a blind hole setup. Unlike continuum-based methods like computational fluid dynamics (CFD), particle methods did not require extra assumptions or complex mathematical models. The simulations showed that capillary position and gas flow rate affected the resulting flow, with stable plasma at 300 sccm. Direct Simulation Monte Carlo (DSMC) simulation's agreement with impact pressure measurements validated its accuracy for determining pressure distribution in the geometry. Particle-in-cell (PIC) simulation highlighted electrical potential distribution and plasma characteristics, aiding process parameter optimization and phenomenon comprehension in blind hole cathode discharge.

The outcomes of the PIC plasma simulation, which was performed at a position (capillary position, z_k) of 10 mm, are shown in Figure 13.2. The electric potential between −271 V and −283 V at a duration of 10^{-5} s is depicted in Figure 13.2. In the region of 11 mm ≤ z ≤ 26 mm (about 7 V), a marginally greater positive potential can be seen. The capillary wall is significantly more negatively charged (around −3 V) than the surrounding cathode wall along its flow region (0 mm ≤ z ≤ 10 mm). Electrical potential perturbations in the inflow region are brought on by a brief charge carrier compression. Potential densifications resembling shock patterns can be found along the inflow region, which may be related to the geometry, background gas, static potential distribution, and dynamic potential distribution. Looking at the cross-sections of Figure 13.2b makes the charge carrier dispersion more obvious. The electrical potential distribution and charge carrier density along the z = 12 mm and z = 18 mm cross-sections are

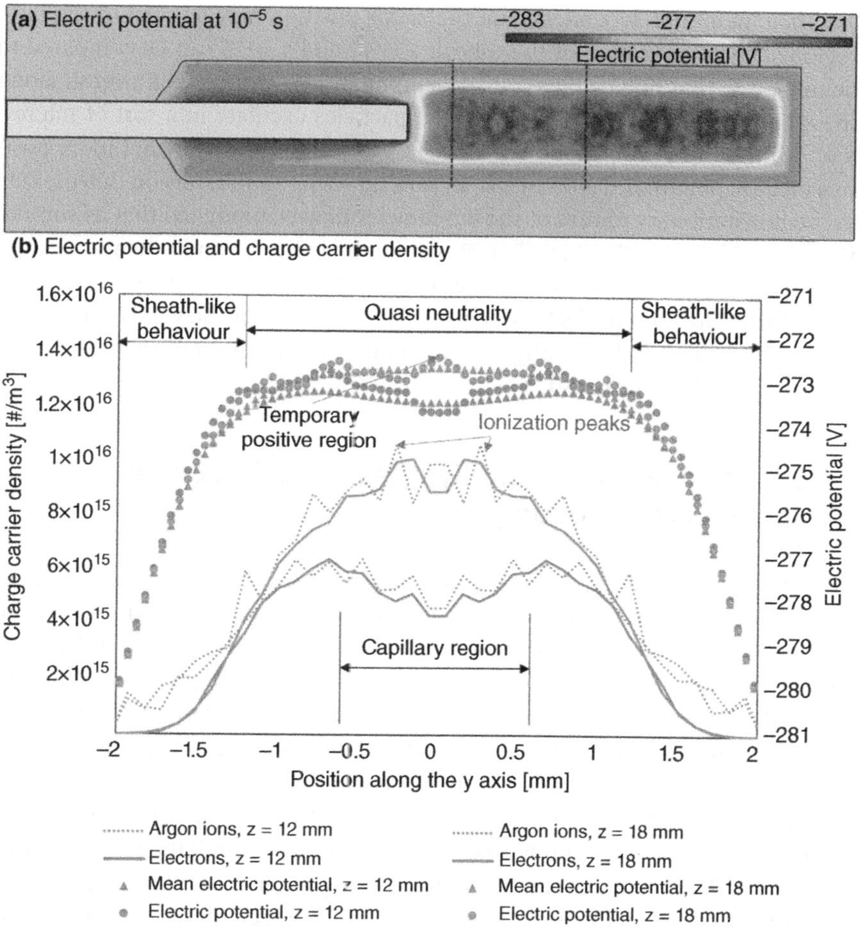

Figure 13.2 Computational mapping of (a) electric potentials at an instant of $t = 10^{-5}$ s with an increment of 10^{-12} s and (b) the variance in electric positional and charge carrier density along the blind hole and developing sheath-like and quasi-neutral zones. *Source:* Reproduced from [26] / with permission of ELSEVIER.

displayed here for a single instant in time. There are two zones in the observed charge carrier dispersion. The electron density falls more quickly than the ion density in the region close to the cathode wall, exhibiting sheath-like behavior.

The electrons emitted by the ion impact are most highly accelerated and produce fast electrons in the sheath-like zone. The bulk region, which makes up the middle, is roughly quasi-neutral but deviates significantly due to the varied charge carrier mobility. The impact ionization-produced electrons in this situation are not accelerated, resulting in a collection of sluggish electrons. An electrical potential dependence on the z-axis position may be seen in the bulk region. The cross-section of the tube at z = 12 and z = 18 can be compared to demonstrate that no virtual anode development [27] can be seen overall along the whole tube length. Instead, charged particles oscillate in a sort of microfield, creating tiny changes in the electric potential. The somewhat more positive electric potential region in z = 18 mm serves as an illustration of this. Due to the nonstationary nature of the outcome, it may be predicted that as simulation time advances, the disturbance will be made up for and a temporary virtual anode will form in a new place. An oscillation of charge carriers results from this. Ionization peaks between −0.2 and 0.2 mm for the y-axis can also be found [28]. It is possible to align the qualitative distribution of the charge carriers with the literature [29].

13.3 Data-centric Virtual Synthesis and Structural Design of Carbon Coatings

Data-centric engineering is significantly helping to understand and correspondingly fabricate the DLC with the desired structural design. DLC coatings are gaining popularity because of their superior, reliable structure and enhanced chemical, electronic, and mechanical properties. During the fabrication process of semiconductor devices, masking is a very important process that allows to perform doping or metallization in the selected areas. DLC coatings are one of the potential candidates for the etch hard mask, but their usage is limited due to low-temperature stability, high internal stresses, and the naturally weak shield against the attachment of fluorine (F) to sub-layers. To overcome these limitations, the atomic structure of DLC coatings can be modified using different methods, and one of the most effective methods is the influence of hydrogen atoms (H) during the deposition process. The hydrogen atoms can modify the DLC coatings' atomic structure, especially in the case of amorphous carbon (a-C) films.

By employing hydrocarbon blended with precursor gases like CH_4 or C_2H_2 diluted in H_2, the a-C films are primarily produced on large surfaces at relatively low temperatures by PECVD [30, 31]. The extremely high reactivity of F atoms with other atoms with an electron-donating nature, such as B, Al, and Ga, causes the bond breaking of sublayer materials to form F-containing alloys. However, it is well known that the F atoms, which originate from the injection of etchant gases like CF_x during the device fabrication process, generally attack the sub-layers and significantly degrade the device's performance. Future integrated semiconductor device development is currently hampered by this issue [32, 33]. Due to the experimentally limited observations on the atomistic scale, a theoretical understanding of the a-C films is necessary for the device construction process to overcome this issue.

The sp^3-to-sp^2 hybridization ratio and the H content affect the characteristics of a-C films. These films display a variety of qualities, including hardness, electrical resistivity, optical clarity, and chemical inertness, depending on the composition. In brief, Mikami et al. [34] observed that when DLC films are deposited by radiofrequency (RF) magnetron sputtering, an increase in the H_2 flow rate causes an increase in the sp^3 fraction and disorder. Using mass spectrometry, Vietzke et al. [35] looked into the chemical degradation of C atoms from a-C films caused by hydrogen. They noted that the reaction between the heated H atoms in the a-C films produced CH_3 radicals as well as a variety of high hydrocarbons. Few other studies claimed that hydrogen had a significant impact on the growth rate, film structure, mechanical characteristics, optical gaps, and field emission [36, 37]. Despite numerous experimental investigations into the impact of an a-C film's hydrogen content, no recommendations exist for enhancing the a-C film as an etch hard mask since the connection between the structure of the a-C film and its etching properties is still not evident. Density functional theory (DFT) and *ab initio* molecular dynamics (AIMD) calculations are useful tools for figuring out the atomic and electronic specifics and giving a fuller understanding of how the a-C films reduce stresses [38, 39].

Figure 13.3 depicts the atomic structures of 64 C atom-containing a-C films with increasing hydrogen concentrations (15.6–31.2 at%) and atomic densities (3.2–4.0 $g.cm^{-3}$). Figure 13.4(a) shows how the computed bulk modulus depends on the atomic density. The bulk modulus of the reference a-C film (Ref. a-C) is predicted to be around 212 GPa. The bulk modulus sharply rose with increasing atomic density; at an atomic density of 4.0 $g\ cm^{-3}$, a maximum bulk modulus of around 617 GPa is reached, which is roughly three times greater than that of the reference a-C film. The bulk modulus, on the other hand, increased as a function of the H concentration upon the introduction of H atoms into the a-C films, as shown in Figure 13.4(b), and then slightly increased with the subsequent rise of H concentration. An observed peak bulk modulus of roughly 248 GPa is reached at an H concentration of 31.2 at%, which represents a modest increase of 17% in comparison to the reference a-C film. The results of DFT and AIMD are well-validated against published experimental data. The amorphous phase is distinguished by a short-range order, but the atoms are bound in arbitrarily disordered positions because of a mechanism that prevents the establishment of a regular arrangement; therefore, there is some disparity with experimental findings.

As a final observation from DFT and AIMD simulations, it is being established that by increasing the atomic density of the a-C films, a sharp increase in bulk modulus will be observed, which indicates that it would be challenging for the films that are hardened by high atomic density to relieve the stress of the individual layers within the overall stack, especially in integrated semiconductor

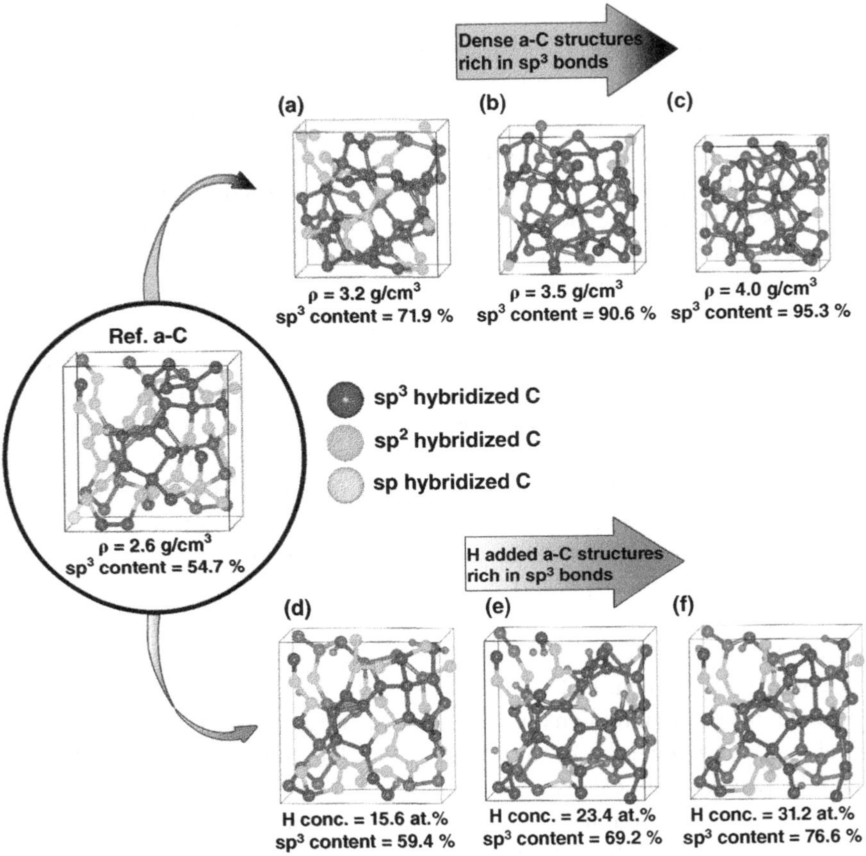

Figure 13.3 Illustration of sp, sp^2, and sp^3 bond arrangements in hydrogen-free and hydrogenated carbon structures. The density increases with an increase in sp^3 composition. The sp^3 bonds increase with the addition of hydrogen attached to available active sites and dangling bonds. It is worth mentioning that the sp bonds are highly unstable; therefore, they are not counted for DLC structures. *Source:* Reproduced from [40] / with permission of Royal Society of Chemistry.

devices. Even though the sp^3 content rose, the addition of hydrogen to the a-C films had no impact on raising the bulk modulus. These results indicate that adding hydrogen rather than increasing atomic density should be used in the process design for high-sp^3 content in a-C films. The importance of these requirements rises along with the difficulty of meeting them as device integrity and reliability improve. A change in the sp^3 concentration from the standpoint of the F-blocking nature raises the F atom's diffusion barrier and suppresses it, which means that the F atom would follow the diffusion path that passes the sp^2 carbon and not the sp^3 carbon because of the noticeably high barrier. This new simple method for

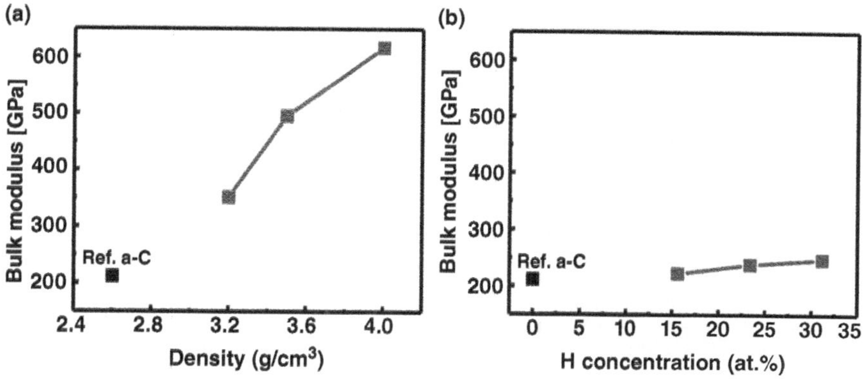

Figure 13.4 The increase in sp^3 which increases the (a) density of hydrogen-free carbon coatings or (b) with the addition of hydrogen changes coating properties, such as bulk modules. This is perceived to directly impact the mechanical and tribological performance of carbon coatings. *Source:* Reproduced from [40] / with permission of Royal Society of Chemistry.

designing a-C films with acceptable doping characteristics can be tailored for superior mechanical and other innovative physical and chemical features, and this becomes possible because of DFT and AIMD simulation.

13.4 Understanding the Properties of Carbon Coatings With Data-centric Approaches

Besides structure, the data-centric approaches are helping to understand the properties of carbon coatings. As discussed earlier, computational simulation can provide insight into how the molecular structure can affect the properties of DLC coatings, as it is difficult to run a parametric experimental study for a single parameter and isolate all other variables. Similarly, the change in the properties (e.g. mechanical/structural) of the DLC coatings can also be studied using a computation/numerical method. To measure the mechanical properties of the DLC coatings, the nanoindentation (experimental/simulation) method can be utilized. Valencia et al. [41] studied the plastic deformation mechanical response of the nonhydrogenated films using the nanoindentation approach for the different concentrations of sp^3. Initially, the simulation results were compared with experimental results to establish confidence in the methodology utilized for the numerical simulation. The molecular dynamics simulation is conducted using an open-source code (LAMMPS [42]), which is widely adopted by researchers and developed by the researchers at Sandia National Laboratory. Fatigue and cyclic exposure to elevated temperatures also affect the performance of DLC coatings,

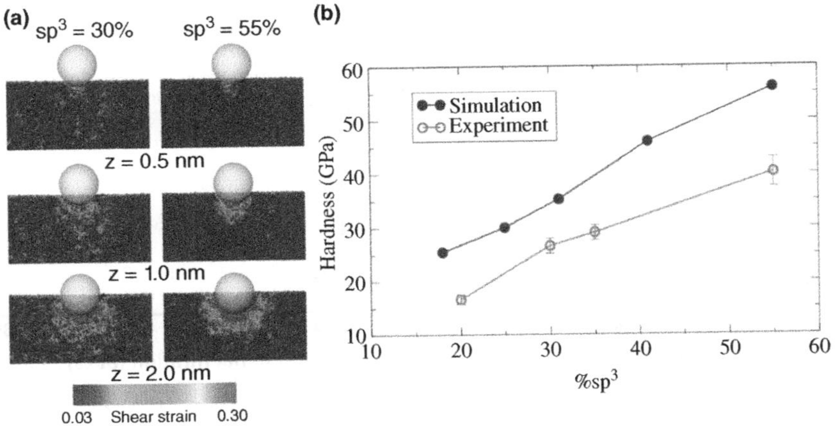

Figure 13.5 Nanoindentation studies of DLC coatings. (a) Shear strain distribution at an atomistic level for DLC coatings made with varying amounts of sp^3 bonds. (b) The relationship between hardness and sp^3 fraction. The growth trends of MD simulation and experimental studies are in good agreement. The higher values of computational results are possibly due to empirical hardness, where differences mostly exist due to instrumental accuracies and experimental protocols. *Source:* Reproduced from [41] / with permission of ELSEVIER.

which are being studied using ANSYS software along with nanoindentation experiments for up to 400 °C [43].

The relationship between hardness at full penetration and sp^3 content is depicted in Figure 13.5 (right), utilizing the contact area determined by OVITO. The simulated hardness values closely align with the conducted experimental results, and when considering higher sp^3 values, the findings corroborate with the DLC experiments in the literature [44]. Regardless of the absolute values, which may be greatly influenced by sample preparation, the indentation size effect, and the penetration rate, the results exhibit the same trends as experimental data for hardness in relation to sp^3 content. An alternative approach by Ziegenhaim [45] yields a slightly smaller contact area, resulting in higher hardness values. The geometric contact area typically exceeds the atomistic contact area obtained from OVITO, and the MD hardness results decrease by 30–40% and exhibit a much closer agreement with experimental outcomes. A comprehensive analysis of the nanomechanical characterization of carbon films is presented by Beake and Liskiewicz [46], wherein a section is dedicated to discussing the factors that influence the reliable and comparable determination of hardness and elastic modulus. They highlight ultra-thin thickness, indenter geometry, and surface roughness as three major challenges in hardness and elastic modulus estimations during experiments. Furthermore, atomistic simulations of indentation generally yield higher computed hardness values compared to experimental counterparts. The differences in these values often arise from disparities in timescales (where MD simulations typically involve large indentation rates) and length scales (such as the tip diameter effect), or a combination of both, as discussed by Ruestes and colleagues [47]. An atomistic perspective of the nanoindentation simulations is illustrated in Figure 13.5 (left). Each carbon atom is color-coded based on its shear strain, and the panels depict films with sp^3 percentages of 30 and 55 at varying indenter depths. In samples with low sp^3 content, the shear response extends well beyond the contact region, indicating a long-range collective mechanical response for sp^2-rich samples. Conversely, the sample with 55% sp^3 content exhibits a concentrated shear strain for atoms in contact with the indenter, and negligible shear strain is observed far from the indentation region. This disparity is attributed to the rigid tetrahedral arrangement of sp^3 bonds, which restrict atomic displacements far from the tip.

The results obtained through the Finite Element Method (FEM) supported simulation of the impact test are presented in Figure 13.6a. In this figure, the stress fields that developed in both the coating and its substrate during the loading and relaxation phases of the impact test are illustrated. These stress fields correspond to an impact load of 100 daN and a temperature of 25 °C. The coating deforms in a manner that resembles a spherical surface in response to the substrate deformation. The maximum stresses in the coating structure occur near the crater, where bending happens and the coating failure initiates. If the difference between the maximum stresses during loading and relaxation exceeds the corresponding fatigue endurance safe stress range, the coating may fail. Figure 13.6b, shows the equivalent stress fields developed in the coating during loading and relaxation at a critical impact force of 180 daN and an ambient temperature of 25 °C are depicted. According to these stress fields, the maximum equivalent stresses during loading and relaxation amount to 4.28 GPa and 1.68 GPa, respectively. These data, along with the mechanical properties of the coating, are necessary for the construction of the DLC coating Smith-like diagram.

Table 13.1 shows the maximum equivalent stresses during the loading and relaxation phases of the impact test at critical fatigue endurance forces and elevated temperatures. In Case 5, where the critical impact force is low, the substrate behaves elastically, resulting in no remaining stresses in the coating during relaxation. Furthermore, when the impact force remains constant, the coating stresses increase due to the more significant deformation of the substrate at higher temperatures. Therefore, although the coating retains its strength properties up to a temperature of 150°C, it may fail at a lower load compared to ambient conditions due to the deterioration of the substrate's mechanical properties and the resulting larger deformation of the coating.

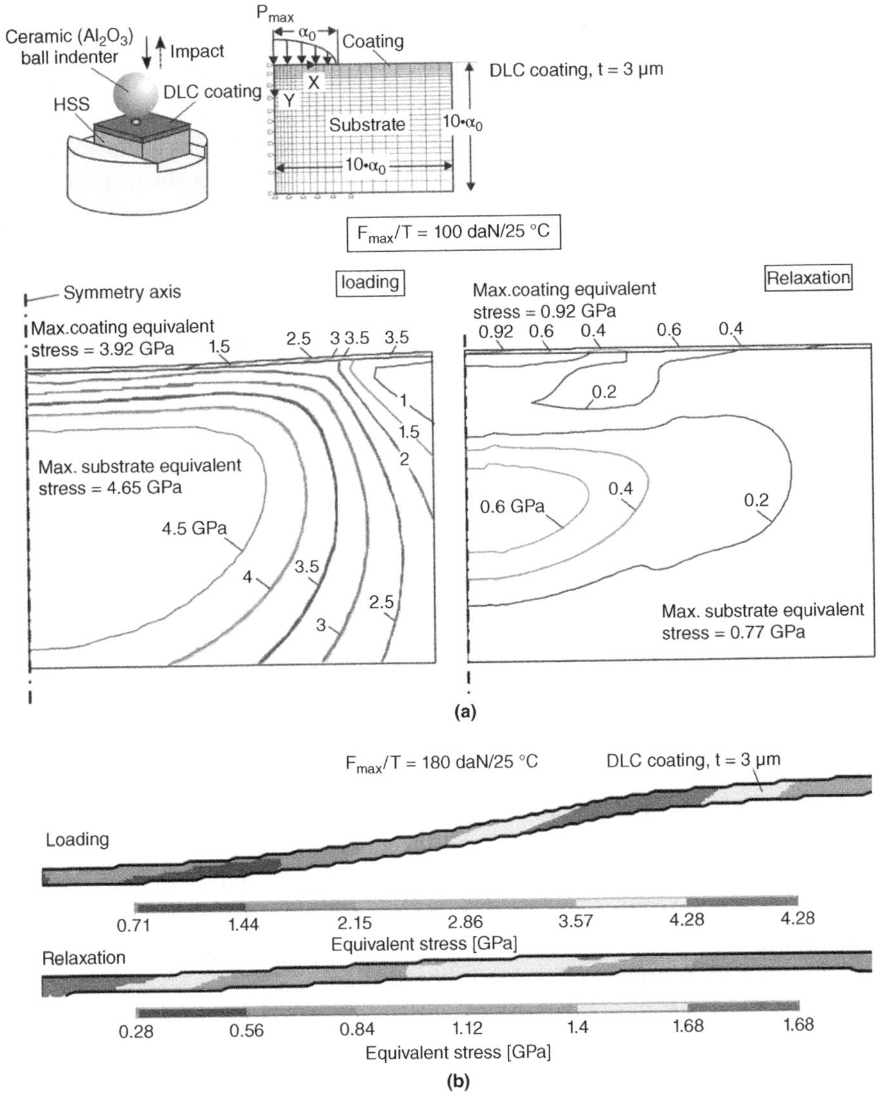

Figure 13.6 (a) The maximum and retaining stress contours in coating and substrate and (b) their mapping under impact and relaxation modes studied through finite elemental analysis. The stress-strain distributions help to design a mechanically safe coating-to-substrate interface with good interface adhesion and interfacial toughness. *Source:* [43] / MDPI / CC BY 4.0.

Table 13.1 Maximum equivalent stresses during loading and relaxation at the fatigue endurance critical forces at various temperatures. [43] / MDPI / CC BY 4.0.

Case no.	Temperature	Critical impact force F [daN]	Maximum equivalent stress [GP]	
			Loading	Relaxation
1	25	180	4.28	1.68
2	80	80	3.84	1.02
3	150	42	3.48	0.66
4	200	25	2.88	0.48
5	300	3	0.88	0
6	350	0	0	0

13.5 Data-centric Performance Assessment of Carbon Coatings

As discussed in the earlier sections, molecular dynamics simulations are quite helpful in understanding the process of developing the DLC coatings and evaluating the structure and properties of DLC coatings. Similarly, MD simulations are being used to study the performance of DLC coatings under different tribological conditions. Large-scale Atomic/Molecular Massively Parallel Simulator (LAMMPS) is being adopted by many researchers for MD simulations either to study the structure, properties, or performance of DLC coatings. LAMMPS is maintained and distributed by the Sandia National Laboratory and Temple University. Figure 13.7 presents the results from the MD simulation conducted by LAMMPS. The simulation is conducted to look at the contact interfaces/configurations between the diamond slider and DLC film. Figure 13.7 presents the results of different slider loading at two different positions of the slider: L = 0 (Figure 13.7a,b) and 50 Å (Figure 13.7c–e).

When subjected to a substantial Load (F_n = 588 nN), the slider's lower part infiltrates the DLC film, resulting in the emergence of a chip in front of the slider during the sliding (Figure 13.7d or e). This occurrence is accompanied by a significant rise in worn atoms (N_w), which can be attributed to the fact that a large F_n profoundly alters the contact configuration, thereby influencing the wear pattern between the slider and the film. The modification of the wear pattern due to changes in the load has been validated in numerous experiments [49, 50, 51].

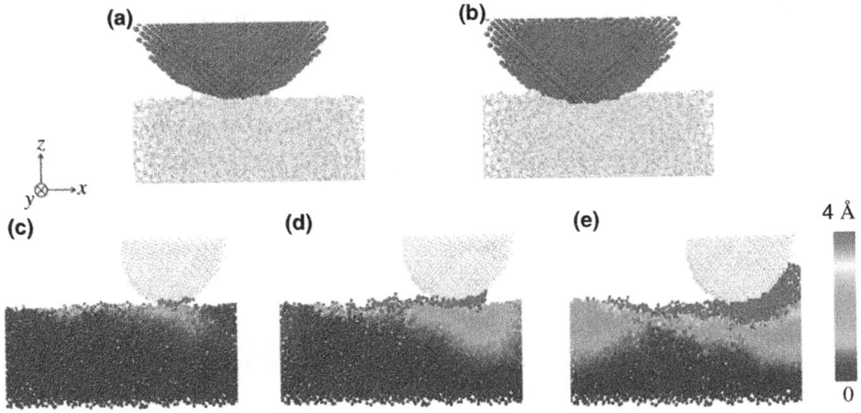

Figure 13.7 DLC coating behavior against diamond slide in stationary contact with varying normal loads of (a) 78 nN and (b) 588 nN. The atomic mobility and corresponding atomic contours can be observed when slider travels on DLC surfaces with increasing normal loads of (c) 78 nN, (d) 392 nN, and (e) 588 nN. The data-centric approach helps to understand the tribological performance (friction and wear mechanisms) of DLC coatings. *Source:* [48] / MDPI / CC BY 4.0.

According to Yoon et al. [52], the friction force of the DLC film is primarily influenced by adhesion at low loads, which leads to elastic deformations in the film. Smerdova et al. have suggested that as the worn area expands, other wear mechanisms come into play [49]. Yang et al., through MD simulations, have shown that wear can manifest in two forms: local shedding of atoms and large-scale plastic wear [53]. At a lower F_n, a portion of the atoms is shed from the DLC film and slides along with the slider (Figure 13.7d). However, in the case of $F_n = 588$ nN, the wear pattern transitions from localized atom shedding to large-scale plastic wear, resulting in a substantial increase in N_w (Figure 13.7e). Consequently, the average contact stress is determined by dividing the applied load F_n by the average contact area between the slider and the DLC film after the sliding process reaches a stabilized state. The findings reveal that F_n values of 78, 196, 392, and 588 nN correspond to average contact stresses of approximately 13.2, 30.6, 46.7, and 63.9 GPa, respectively. The hardness limit of this film was determined to be around 60.8 GPa in the room temperature nanoindentation test. Furthermore, the increase in temperature at the contact interface can significantly diminish the strength of the DLC film during the sliding process, thereby resulting in a further reduction in the hardness limit of this film. Consequently, for $F_n = 588$ nN, such a reduction leads to the structural deterioration of the DLC films and a substantial increase in N_w. The evolutionary patterns of the atomic bond structures in DLC coatings, which serve as representatives, as well as the changes in the concentration of sp^3 hybridized bonds during the sliding process, have been visually represented in Figure 13.8.

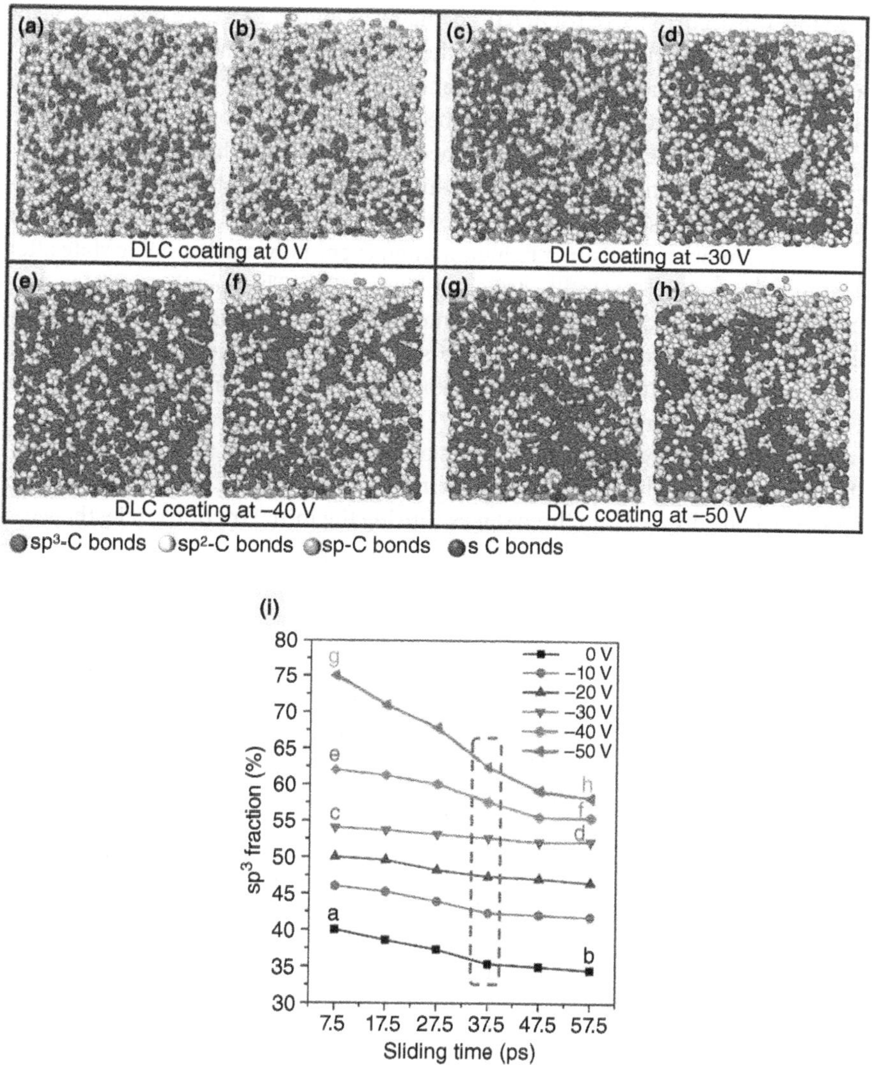

Figure 13.8 Correlation between carbon atomic bonds and their variance during frictional motion. (a–h) The composition of sp, sp^2, and sp^3 carbon bonds in DLC coatings at instants of 7.5 ps and 57.5 ps, virtually produced as a function of bias voltage. (i) The transformation of carbon atomic structure during sliding motion, where sp^3 bonds continue to reduce as a function of sliding time. The data-centric approach assists in selecting a suitable bias voltage to deposit DLC coatings that can meet application-specific tribological performance. *Source:* Reproduced with permission from Ref. [54], Copyright (2020) American Chemical Society.

Before the occurrence of friction (as depicted in Figure 13.8), the presence of sp^3 bonds is evenly distributed throughout the DLC film, while the surface layer, with a thickness of a few atoms, exhibits a higher occurrence of sp^2 bonds. The fraction of sp^3 carbon hybridization bonds has been determined through comprehensive calculations of the entire film. It has been observed that the proportion of sp^3 carbon hybridization bonds decreases while the proportion of sp^2 carbon hybridization bonds increases across all DLC coatings. Furthermore, the sp^3 fractions for each case demonstrate distinct reductions and eventually reach a stable state after a sliding time of 37.5 ps, as illustrated in Figure 13.8i. The energy generated during the sliding process disrupts the internal bonding structure, consequently inducing a transformation from sp^3 to sp^2. Additionally, the elevated internal stress experienced during sliding can expedite the conversion from sp^3 to sp^2 for DLC coatings with high SBVs.

Figure 13.9 displays the morphologies of the wear scars on a-C films with varying sp^3/sp^2 ratios. These films cover the range of 1st to 40th circles. The friction process can be divided into three stages. During the initial friction circle, the wear

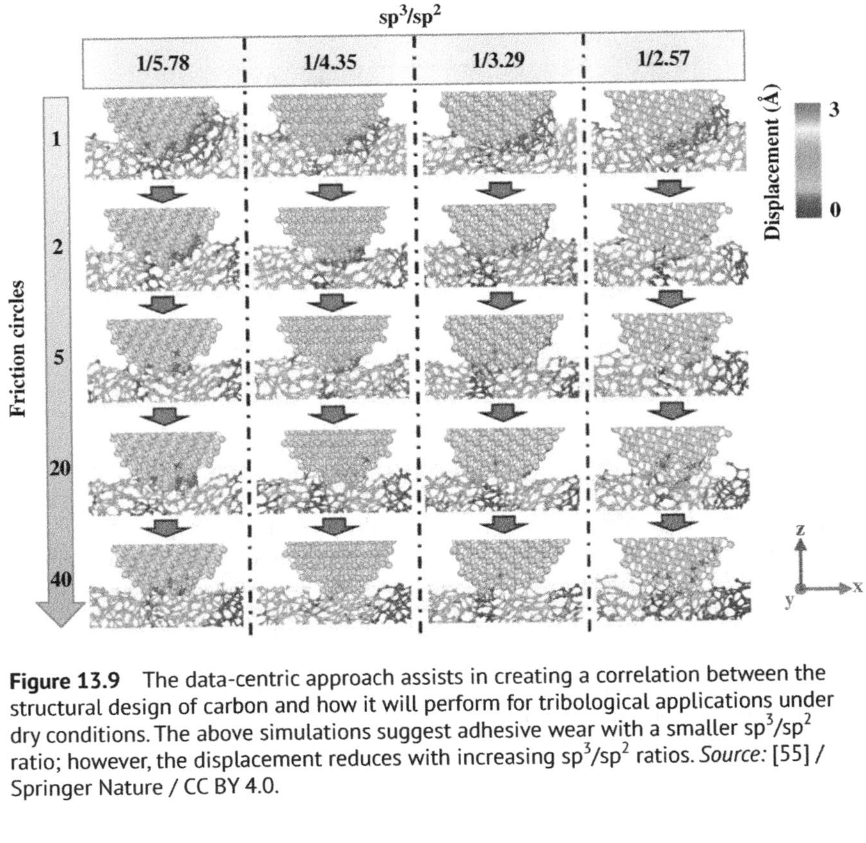

Figure 13.9 The data-centric approach assists in creating a correlation between the structural design of carbon and how it will perform for tribological applications under dry conditions. The above simulations suggest adhesive wear with a smaller sp^3/sp^2 ratio; however, the displacement reduces with increasing sp^3/sp^2 ratios. *Source:* [55] / Springer Nature / CC BY 4.0.

scar morphology changes significantly due to chip formation. Subsequent friction circles (2nd–5th) show relatively small changes in wear morphology. After the 5th friction circle, the change in wear scar morphology becomes less evident. Therefore, they selected the wear morphologies of the 1st, 2nd, 5th, 20th, and 40th friction circles as representative of the wear track on the film for the three stages.

In the initial stage of friction (first friction circle), the displacement of C atoms and their accumulation in front of the counterpart indicate bond breakage and chip formation during abrasive wear. In the early stage of the friction/wear process, the worn surface of the a-C films experiences plowing. However, as the number of friction circles increases, C atoms primarily follow the α-Fe counterpart, suggesting a shift from abrasive wear to a combination of abrasion and adhesive wear for the a-C films. Figure 13.9 also demonstrates that the displacement decreases as the sp^3/sp^2 ratio increases (more blue sites), indicating that adhesive wear becomes more prominent when carbon atoms in the film have a low sp^3/sp^2 ratio. While the presented MD simulation is conducted on small systems (nanoscale) to explore underlying mechanisms rather than to obtain quantitatively accurate data comparable to realistic situations, the changes in coefficient of friction (CoF) curves, wear rate, and wear morphology obtained from the experiments align with those from the MD simulations [55]. The ease with which the bonds at the contact interface between the films and their counterparts are broken is demonstrated by both the results in Figure 13.9 and the studies mentioned in the literature [56]. However, the process of metastable transformation in the hybridized a-C remains unclear and needs further investigation. The ultralow friction using a-C films can be achieved when mixed with lubricating oil. Du et al. [57] looked at the different textures in lubricating oil conditions, showing that textured surfaces can reduce friction, but the effectiveness depends on the depth and width of the texture. Deeper textures initially decrease friction, while wider textures lead to increased friction due to lubricant atom aggregation at the friction interface. Figure 13.10 presents the displacement of oil atoms for different heights and widths, and the results are also compared with the untextured surface as well.

The representations of the dispersion of lubricant atoms at varying displacements can be observed in the insets of each depicted image in Figure 13.10. Initially, in the case of the untextured system, the outcome indicates a scarcity of atoms exhibiting high mobility. Conversely, most atoms only exhibit minimal displacement, thus corresponding to a low diffusion coefficient. However, because of the texturing treatment, the flow of atoms showcases two distinct extremes: atoms with low mobility and atoms with high mobility, which constitute the majority of lubricating oil atoms. There is a limited presence of atoms with intermediate mobility. Notably, the low-mobility atoms encompass those that are ensnared within the textured layer and those that are affixed to the lower surface platform.

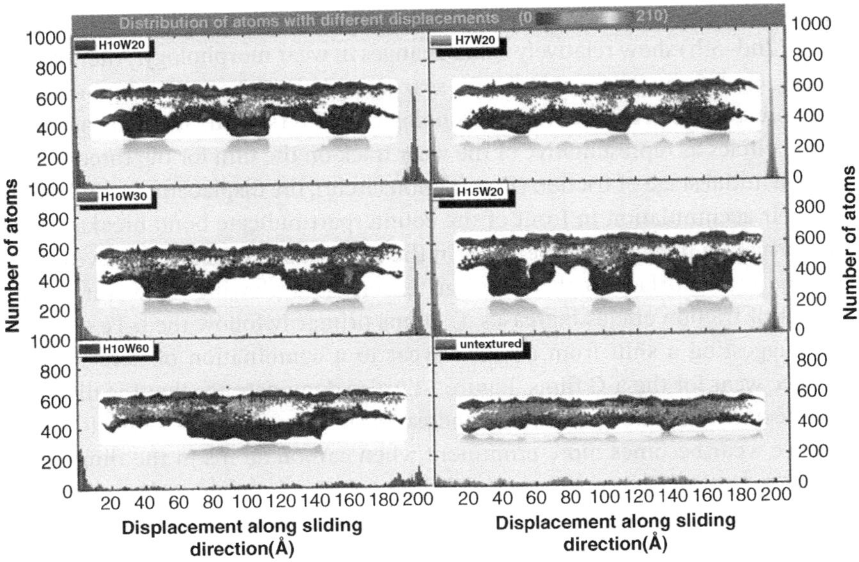

Figure 13.10 Atomic and molecular-scale mobility of lubricating oil at carbon coatings of varying textures by width (W) and depths (H). The data-centric approach assists in visualizing the surface roughness and its impact on the tribological performance of carbon coatings under wet conditions. *Source:* Reproduced from [57] / with permission of ELSEVIER.

The atoms with high mobility, on the other hand, can be attributed to the significant displacement caused by adsorption on the upper surface layer during sliding, which is indicative of the formation of the lubricating film. A small fraction of atoms produces minor displacements due to their connection to both the upper and lower a-C surfaces. The process of sliding induces shearing forces, ultimately resulting in adhesive failure and subsequent capture of lubricating oil molecules by the textured layer. It is also plausible that some of these atoms originate from the captured lubricating oil. During the process of sliding, they re-establish contact with the lubricating oil at the upper a-C surface.

Sliding friction simulations (Figures 13.7 and 13.9) give quite a good understanding of the performance of the DLC coatings at low temperatures, but to predict the behavior of the coating at elevated temperatures, there is a need to perform simulations at high temperatures as shown in Figure 13.11. The entire process of simulation comprises three distinct components, namely the relaxation process, the loading process, and the sliding friction process. The relaxation process is depicted in Figure 13.11a. Initially, the α-Fe ball was positioned above the a-C film surface at a vertical distance of 15 Å so as to ensure the absence of any interaction between the a-C film and the α-Fe ball. Subsequently, the entire simulation system underwent relaxation for a duration of 30 ps, ensuring the attainment of a stable state for

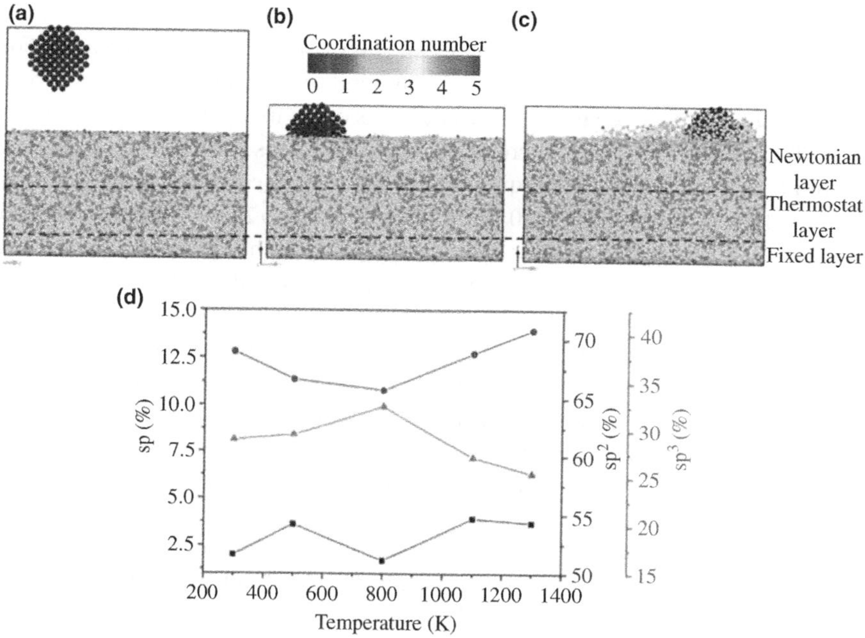

Figure 13.11 A simulation model presenting atomistic response during (a) relaxation, (b) loading, and (c) sliding friction processes in a Newtonian layer. Likewise, (d) presents the change in carbon atomic composition in a Newtonian layer as a function of stimulated temperature between 250 and 1300 K. *Source:* Reproduced from [58] / with permission of American Chemical Society.

the a-C/α-Fe system. The loading process is illustrated in Figure 13.11b. A constant normal load of 500 nN was exerted on all atoms within the α-Fe ball, causing it to move along the negative Z-axis and make contact with the a-C film. The loading duration spanned 50 ps, guaranteeing complete contact between the α-Fe ball and the a-C film, as well as the attainment of the desired normal load. During the loading process, an exploration was conducted on the associations between loading time and various parameters, including the total energy of the system (E_{tot}), system temperature, normal load/friction force, and thickness in the Z-direction at different simulated temperatures (STs). Finally, the sliding friction process is depicted in Figure 13.11c. A constant tangential velocity of v = 50 m/s was imposed on the α-Fe ball, moving along the positive Y-axis. This value was deemed reasonable within the given context. The sliding duration extended to 100 ps.

According to the data presented in Figure 13.11d, the plotted curves between the percentages of sp, sp^2, and sp^3 in the a-C film Newtonian layer. Prior to sliding, the values for sp, sp^2, and sp^3 were 1.24%, 65.73%, and 32.18%, respectively. Comparing these values to those before sliding, it is observed that sp remains relatively unchanged, while sp^2 increases and sp^3 decreases. Moreover, when the ST is below 800 K, sp^2 decreases and sp^3 increases as the ST increases. For instance, at ST = 300 K, sp^2 is equal to 68.96% and sp^3 is equal to 29.01%, whereas at ST = 800 K, sp^2 is equal to 64.89% and sp^3 is equal to 33.41%. However, when the ST surpasses 1100 K, sp^2 increases, and sp^3 decreases as the ST increases. For example, at ST = 1100 K, sp^2 is equal to 68.86% and sp^3 is equal to 27.09%, whereas at ST = 1300 K, sp^2 is equal to 70.07% and sp^3 is equal to 26.79%. In addition, Cui et al. [4] also discovered in their experiments that a higher friction coefficient is accompanied by a higher sp^2 percentage. When ST exceeds 1100 K, the percentages of sp^2 and sp^3 fluctuate, indicating that excessively high temperatures disrupt the stability of the structure and result in phase transition 46. The literature indicates that 4,47 an increase in temperature and local tensile stress facilitates the conversion of sp^3 into sp^2. Consequently, when the ST is below 800 K, the transformation of sp^3 into sp^2 is primarily governed by local tensile stress, whereas at ST = 1100 K, the transformation of sp^3 into sp^2 is primarily governed by temperature increase.

From the aforementioned results, it is evident that when the ST is below 800 K, the percentage of sp^2 decreases, and the percentage of sp^3 increases with an increase in temperature. This leads to a gradual rise in the friction coefficient of the a-C film. However, when the ST reaches 1100 K, the excessively high temperature disrupts the stability of the structure, resulting in an increase in the friction coefficient.

Figure 13.12 on the left side presents the temperature ϑ in coated polymer-steel contacts for various coatings with thermal effusivities of $e_c = (578, 1156, 2312)$ J/(Ks$^{-0.5}$ m^2) for a constant coating thickness of $t_c = 1$ μm. A coating with $e_c = 578$ J/(Ks$^{-0.5}$ m^2)

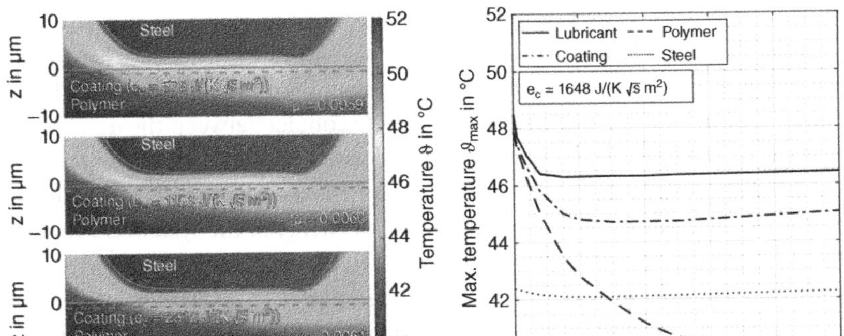

Figure 13.12 Temperature distribution at steel and carbon-coated polymer interface (left) for a coating thickness between 6 ± 1 nm. The right image presents the trends of temperature change in polymer, steel, and lubricant as a function of coating thickness. *Source:* [59] / Springer Nature / CC BY 4.0.

represents a coating that possesses thermophysical properties similar to the polymer. On the right side, Figure 13.12 (right) illustrates the impact of coating thickness t_c on the maximum temperatures in the lubricant, coating, polymer, and steel for the coating properties. As the thermal effusivity e_c increases, the temperature ϑ of the lubricant decreases, while the coefficient of friction μ experiences a slight increase as shown in Figure 13.12 (left). The coating demonstrates a decrease in the depth of temperature penetration as e_c increases. Consequently, lower maximum temperatures are observed at the interface between the coating and the polymer. Therefore, even very thin coatings with $t_c = 1$ μm and higher thermal effusivity compared to the polymer can serve as a thermal barrier to protect the polymer. Generally, as the coating thickness (t_c) increases to approximately 30 μm, the maximum temperatures in the lubricant, coating, and steel decrease. This decrease is most noticeable when the coating thickness is approximately 8 μm. However, for coating thicknesses greater than approximately 30 μm, these temperatures increase slightly. The coatings considered here contribute to improved heat removal due to their higher thermal effusivity compared to the polymer (as illustrated in Figure 13.12). On the contrary, as the coating thickness (tc) increases, the pressure, shear stress, and frictional heat also increase slightly. Both influences have an inverse effect on temperature. The effect of the coating stiffness on the Thermo elasto-hydrodynamically lubricated contact seems to be significant when the coating thickness is approximately greater than 50 μm, whereas the influence of the thermophysical properties of the coatings is already noticeable at very low coating thicknesses less than 8 μm. A transition point is observed at approximately 30 μm.

Figure 13.13 presents the simulated coefficient of friction μ_{sim}, demonstrating an acceptable agreement with the measured coefficient of friction μ_{exp}. Within this context, the simulation results also illustrate the impact of thermal insulation effects on the reduction of fluid friction. The heat transport into the disks is impeded to a greater extent as the thermal inertia of the solid materials decreases, resulting in a higher EHL temperature. Consequently, the effective lubricant viscosity, which exhibits an exponential dependence on temperature, is reduced. As a result, the shear stress of the lubricant and fluid friction experience a decline. These mechanisms are thoroughly discussed in the works of Siegltrum et al. [61] and Habchi and Bair [62]. Although the experimental studies on the performance and structure are few, they give insights into different performance parameters. Huang et al., assessed the cutting performance of a-C and hydrogenated amorphous carbon (a-C:H) coatings on machining 2A50 aluminum alloy, concluding that a-C coatings outperformed a-C:H coatings in terms of service life and workpiece surface quality due to reduced adhesion of aluminum alloy. They also provided behavior with respect to the work of adhesion (W_{ad}) as shown in Figure 13.14.

To provide further elucidation on the extensive behavior of the a-C:H/Al model, a comparison of W_{ad} is made between the model before and after optimization, as depicted in Figure 13.14. Additionally, the bonding strength of C-H

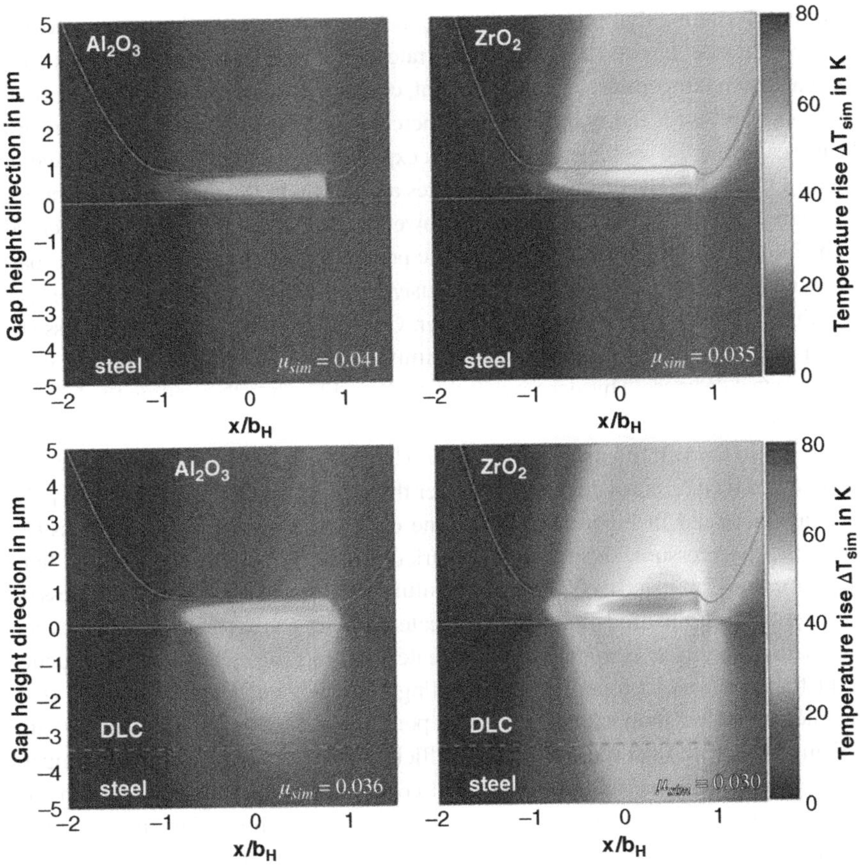

Figure 13.13 Simulated elastohydrodynamically lubricated (EHL) temperature rise of Al_2O_3/steel, ZrO_2/steel (test) and their comparison with Al_2O_3/DLC and ZrO_2/DLC. The data-centric simulation helps to understand thermal inertia associated with compression or shearing behavior of lubricants in the frictional process. *Source:* Reproduced from [60] / with permission of ELSEVIER.

bonds is evaluated through population analysis. The assessment of the covalent or ionic nature of a bond can be determined by the overlap population. A high value of the bond population signifies a covalent bond, while a low value indicates an ionic interaction [64, 65]. The observations from Figure 13.14 reveal that the distance and angle of the C–H bonds increase from 1.14 to 1.348 Å and from 90 to 123.456 after optimization. Moreover, the proximity between the C atoms and Al atoms near the interface decreases from 2.95 to 2.482 Å due to the repositioning of H atoms. Furthermore, at the second layer, the corresponding distance between carbon atoms increased from 1.348 to 1.632 Å at the third

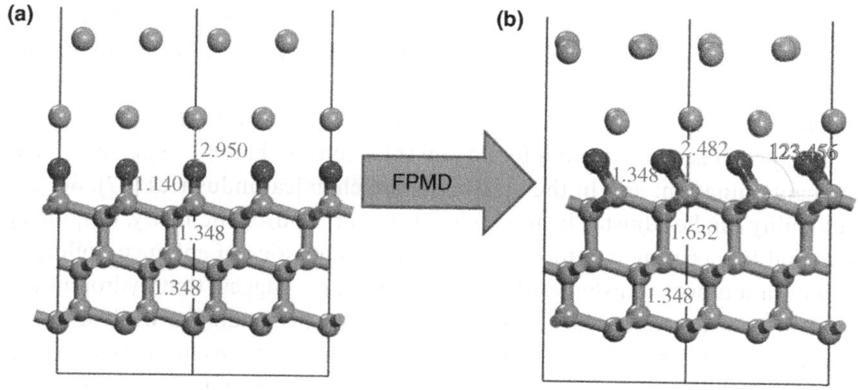

Figure 13.14 The data-centric engineering assists in underpinning DLC behavior for machining and cutting applications. *Source:* [63] / MDPI / CC BY 4.0.

layer. Simultaneously, the number of C–H bonds decreases from 0.63 to 0.54. Consequently, it can be deduced that H atoms in the a-C:H coating exhibit transverse motion below 300 °C, which subsequently alters the charge distribution of C atoms at the interface and intensifies the interaction between C and Al. The adhesion of aluminum alloys to a-C:H-coated tools in the machining process is facilitated compared to a-C-coated tools. Simultaneously, the positioning of other carbon atoms in the coating undergoes changes. However, the emergence of suspended bonds exhibits a stronger interaction with the Al alloy upon the breakage of C–H bonds. All the aforementioned data provide an explanation for the superior adhesion resistance exhibited by the a-C coating in comparison to the a-C:H coating when applied to aluminum alloy cutting.

13.6 Applications of Carbon Coating in Enhancing the Heat Transfer Processes

Phase change heat transfer is frequently linked with elevated heat fluxes in comparison to alternative conventional methods of heating and cooling, such as forced convection of single-phase gases and liquids. The occurrence of liquid-vapor phase change phenomena, including boiling and condensation, is extensively utilized in various industrial systems like power plants, refrigeration and air conditioning systems, desalination plants, water processing facilities, and thermal management devices due to their superior heat transfer capacity when contrasted with single-phase processes. Particularly, the existence of surface alterations has the potential to modify surface chemistry and significantly impact solid-liquid-vapor interfacial dynamics, thereby substantially facilitating phase transition [66]. Wettability denotes the behavior of a liquid spreading on a surface. The manipulation of wettability is crucial, for example, in the automotive sector to repel water from windshields or paints, prevent fouling, and decrease friction in moving components; in the marine industry to avert fouling; in the electric power sector to inhibit fouling on solar cells; in biology to prevent bacterial contamination; and in the electronic and chemical industries [67]. Surface wettability can be adjusted by modifying its texture and/or chemistry. Employing diamond-like coating can modify surface characteristics to enhance boiling or condensation heat transfer. Multiple research works suggest that hydrophilicity can boost surface wettability and enhance boiling heat transfer, while a hydrophobic surface is beneficial in enhancing drop-wise condensation (DWC, [68]). The subsequent sections elaborate on the advancement in boiling and condensation heat transfer in depth.

13.6.1 Condensation Heat Transfer

DWC provides a significantly higher heat transfer efficiency compared to filmwise condensation (FWC), amounting to approximately 10 times more, a well-established fact for over six decades. This enhancement is greatly influenced by the wettability of the condensation surface, typically assessed through the contact angle formed by the condensing fluid with the surface, correlating with the surface energy. A higher surface energy translates to better wettability. Metals generally exhibit high surface energies conducive to FWC, although reducing surface energy is achievable through methods such as surface coatings. For instance, a copper disc partially coated with a film, as depicted in Figure 13.15a, promotes DWC due to its lower surface energy. Regions uncoated undergo FWC, while the coated sections prominently display DWC. Achieving DWC on a metallic surface necessitates diminishing its surface energy, commonly done by coating it with a material of lower surface energy. These coatings may include organic compounds with hydrophobic attributes, inorganic compounds, polymers, or surface alloys. Despite their benefits, many of these coatings suffer from long-term mechanical instability, hindering their practical application in technical processes. Recent innovations involve modifying amorphous hydrogenated carbon films (a-C:H) with elements like silicon or fluorine to lower their surface energy. Koch et al. [70] and Leipertz and Paul-Forba [69] delve into the heat transfer characteristics of these modified coatings, a discussion that will be touched upon later in this section.

The fundamental prerequisite for DWC formation is the incomplete wettability of the surface, quantifiable through the contact angle size, which forms between the tangent to a droplet's contour and the solid surface's horizontal plane, as illustrated in Figure 13.15b. This equilibrium contact angle is an outcome of the

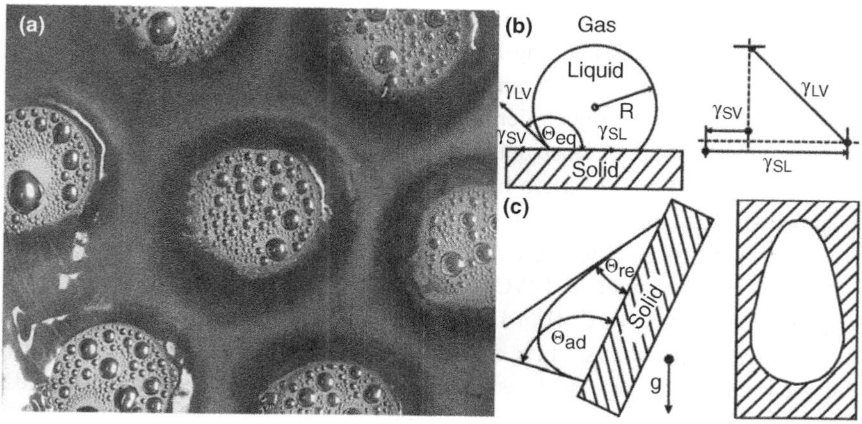

Figure 13.15 (a) Filmwise and dropwise condensation of steam on a partially carbon-coated copper plate. (b) Equilibrium contact angle of a droplet and vector diagram presenting surface tension for incomplete surface wettability. (c) A droplet response on inclined surfaces making two contact angles. *Source:* Reproduced from [69] / with permission of Taylor & Francis Group.

interplay of surface tensions between liquid/vapor (γ_{LV}), solid/vapor (γ_{SV}), and solid/liquid (γ_{SL}) at thermodynamic equilibrium, encapsulated in the following Young's equation.

$$\cos\theta_{eq} = \frac{\gamma_{SV} - \gamma_{SL}}{\gamma_{LV}} \tag{13.1}$$

The surface energy of a solid can be reduced to increase the contact angle of a specific liquid. In order to achieve significant DWC (greater than 60° [40], greater than 90° [41]), it is necessary to attain large equilibrium contact angles. In the case of an inclined surface, two angles are used to describe the behavior of the condensed liquid [64] as depicted in Figure 13.15c. To ensure that the liquid drains off in the form of droplets, the advance contact angle must not be too small. Typically, it is quite similar to the value, while the recede contact angle can be significantly smaller. The contact angle hysteresis is influenced by the roughness of the surface and is minimized for homogeneous and extremely smooth surfaces. This minimum level can be achieved for Ra < 0.1 μm, necessitating meticulous processing of the base material before applying a coating [65].

If the prerequisites for stable DWC are not met, the initial DWC will swiftly transition to FWC, and usually the form of mix condensation has been observed. Erb [71] conducted a study over several days on various systems; initially, 60–80% of the surface was exhibiting DWC, which eventually transformed entirely into FWC by the end of the testing period. This transformation is linked to inadequate corrosion protection and the development of oxide layers.

A typical schematic diagram in Figure 13.16 illustrates the production and assessment techniques for long-lasting superhydrophobic surfaces on Al6061 for vapor condensation heat transfer. Initially, varied microstructures are created on substrates through microblasting based on the duration of blasting. Subsequently, micro/nanostructured diamond-like coatings/films (Florine base) are applied to the micro-blasted Al6061. The condensation modes and heat transfer efficiency can be assessed using a steam/vapor condensation heat transfer system as outlined by Ryu et al. [72].

For the enhancement of DWC, coatings consisting of amorphous layers of hydrogenated carbon (a-C: H or DLC) have exhibited unique properties suitable for industrial applications. These coatings display exceptional mechanical and tribological characteristics, such as chemical inertness, high resistance to acids,

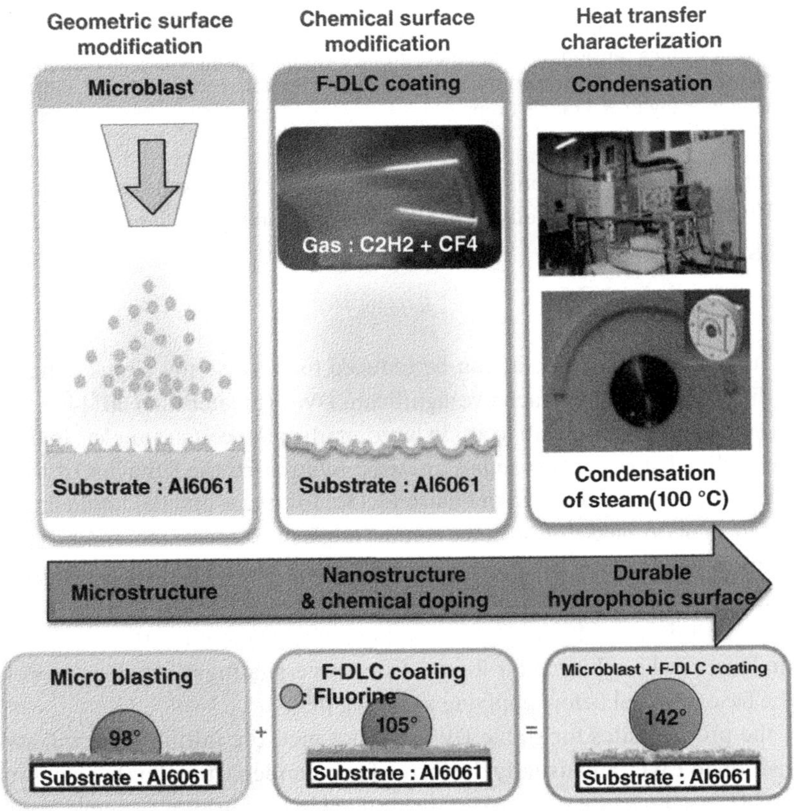

Figure 13.16 Process flow for coating superhydrophobic surfaces carbon film on Al6061surface for vapor condensation heat transfer. *Source:* Reproduced from [72] / with permission of ELSEVIER.

alkalis, and solvents, as well as limited wettability with a contact angle of equilibrium of around 70° when water is the condensing fluid. By introducing elements like fluorine, silicon, or silicon combined with oxygen during the PECVD coating process, larger contact angles for water can be achieved, thus reducing the surface energy of the pure DLC layer. The decrease in surface energy is contingent upon the specific modifying elements utilized (F, Si, and Si/O) and the operational parameters during the production process.

The heat transfer coefficient in DWC systems is influenced by the surface subcooling temperature and the magnitude of the contact angle. Coatings of DLC with modifying elements exhibit superior heat transfer coefficients compared to DLC lacking these elements owing to their higher surface energy. As the surface subcooling temperature increases, the heat transfer coefficient decreases for every DWC surface. This phenomenon is attributed to the accumulation of condensate on the condenser wall with elevated thermal flow, resulting in additional heat conduction resistance. Consequently, at high subcooling temperatures, the heat transfer coefficients approach those observed in laminar filmwise condensation. The efficient removal of condensates from the surface while maintaining the DWC mode can significantly alter the heat transfer performance of a steam condensation system.

In Figure 13.17, the relationship between heat transfer coefficient, heat flux, subcooling temperature, and surface conditions is depicted. On polished and microblasted surfaces of Al6061, the heat transfer coefficient and heat flux closely adhere to Nusselt theory, as filmwise condensation mode predominates on these surfaces. The relatively low heat transfer coefficient and heat flux on these surfaces are due

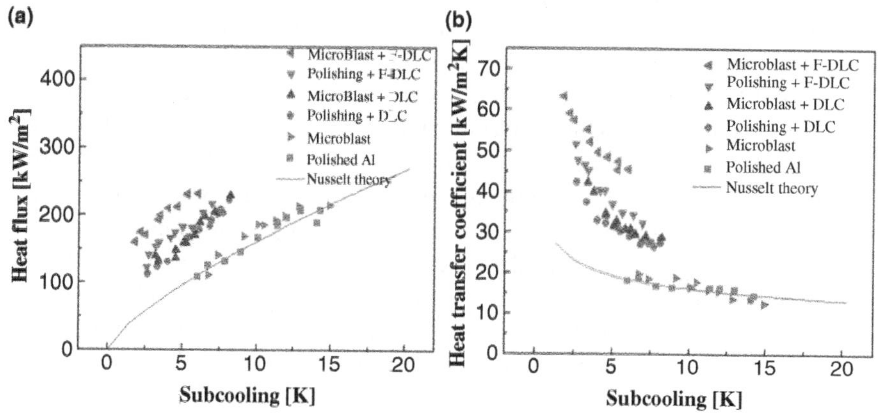

Figure 13.17 DLC coatings have shown superior (a) heat flux and (b) heat transfer coefficient as functions for a range of subcooling temperatures. *Source:* Reproduced from [72] / with permission of ELSEVIER.

to the continuous liquid film covering the condensing interface, leading to increased thermal resistance. Following the nucleation of condensates, subsequent vapor clusters typically exhibit a lower contact angle than the static contact angle (Young angle) due to the high surface energy of the liquid. The surface energy at the condensing interface rises as condensation progresses, making it challenging for filmwise condensates to detach under gravitational force. In contrast, the heat transfer coefficient and heat flux on DLC- and F-DLC-coated surfaces show significant improvement irrespective of the microblasting treatment. This highlights the transition from filmwise condensation mode to dropwise condensation mode as a key factor in enhancing heat transfer performance in steam condensation systems. Moreover, the adoption of microblasting prior to DLC and/or F-DLC coatings further enhances the heat transfer performance.

13.6.2 Boiling Heat Transfer

Power systems, such as electronic components or fuel cells, are experiencing an increase in heat dissipation due to the rising power densities associated with continuous advancements in their miniaturization processes. This escalating heat generation necessitates efficient heat removal to prevent any damage to the components. To achieve this, various cooling methods can be implemented, including air convection, liquid cooling, and boiling flows. Among these options, air convection stands out as the most commonly used due to its ease of implementation. However, this method demonstrates subpar heat transfer performance, and it has already reached its operational limits. Consequently, there has been a growing focus on the development of cooling systems based on liquid and boiling flows, with flow boiling emerging as the most effective mode offering superior heat transfer efficiencies. In this cooling process, a portion of the heat is transferred through latent heat during the liquid-vapor phase change, enhancing convective effects [73].

Kew and Cornwell [74] established criteria specific to microchannels, where capillary effects become significant compared to inertia and viscous effects, unlike in macro-size channels. The role of surface wettability in two-phase pressure drop has been extensively explored in the literature, with various works discussing the utilization of DLC coating to alter the contact angle. The fabrication process for microchannels with coating, as depicted in Figure 13.18 [73], involves physical and chemical vapor deposition techniques to achieve a contact angle of 63° for DLC coating. The impact of surface wettability on single-phase pressure drop is minimal, but in boiling conditions, it becomes crucial due to the surface tension force at the fluid-wall interface.

The surface wettability has a negligible impact on the single-phase flow pressure drop. However, in boiling conditions, the surface wettability can play a significant role because of the surface tension force generated at the fluid-wall interface. This

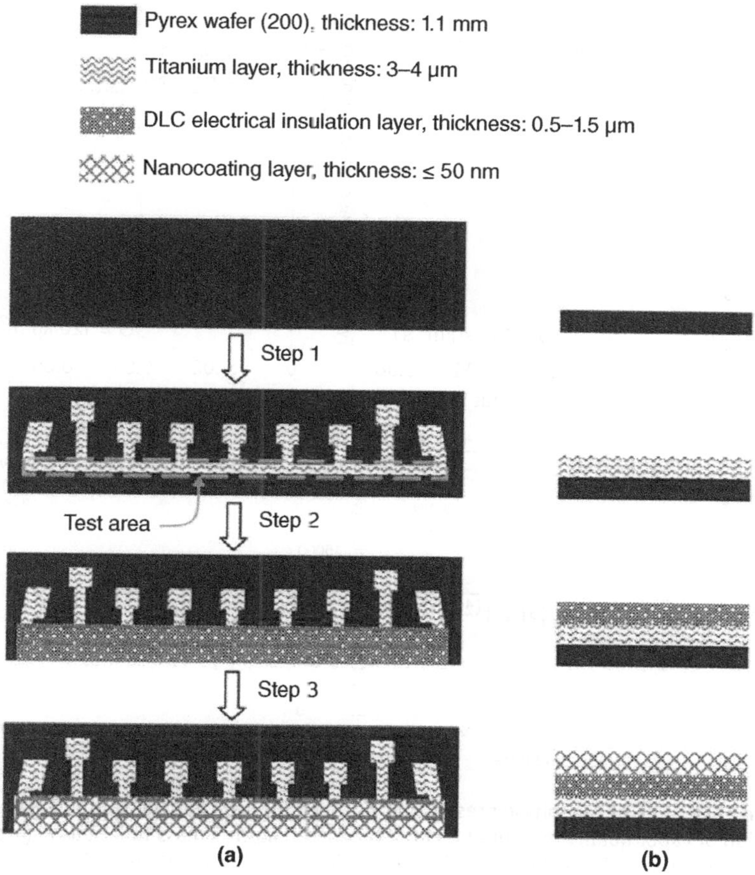

Figure 13.18 Process flow for fabricating DLC-coated microchannels. The process steps involve the deposition of a thick titanium layer on a Pyrex wafer, then the deposition of DLC film for electrical insulation, and the deposition of functional nanocoating such as controlling surface wettability. Sample fabrication processes: (a) top view and (b) side view. Reproduced from [73] / with permission of ELSEVIER.

remark is highlighted by experimental measurement for two different mass fluxes: 100 kg/m^{2}s and 120 100 kg/m^{2}s as shown in Figure 13.19a,c, respectively [73]. It can be seen that the higher the contact angle, the higher the pressure drop for that particular coating. Experimental measurements have highlighted the significant role of surface wettability in boiling conditions, particularly in generating surface tension forces at the fluid-wall interface. This effect is demonstrated through experiments conducted at two different mass fluxes: 100 and 120 kg/m^{2}s, as illustrated in Figure 13.19a,c [75]. The results show that a higher contact angle corresponds to a higher pressure drop for the specific coating applied.

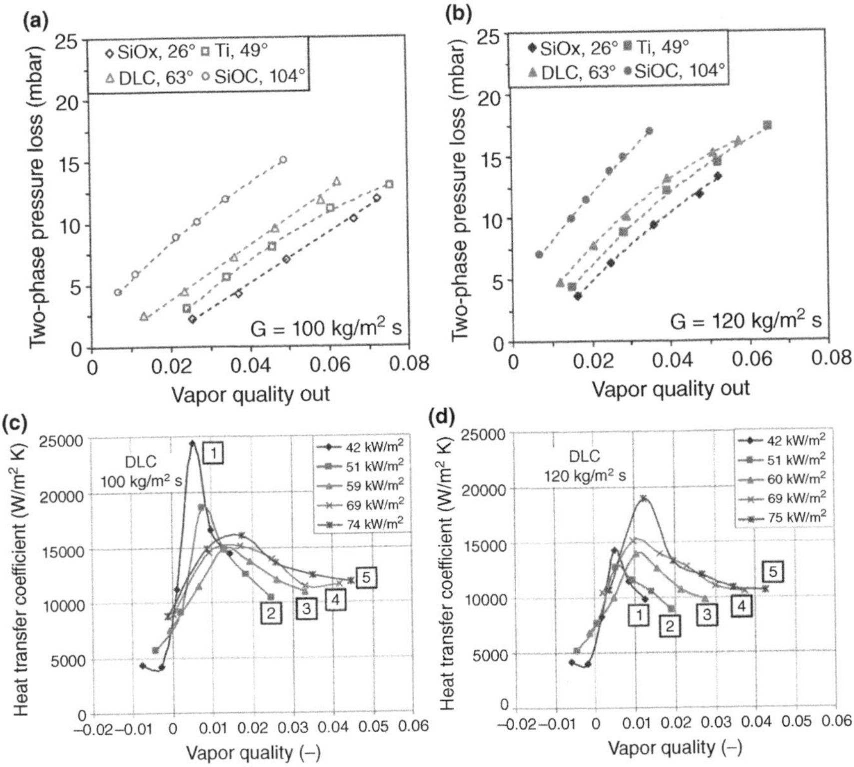

Figure 13.19 (a, b) Two-phase pressure drops and (c, d) heat transfer coefficient as a function of vapor quality measured (a) at 100 and 120 kg/m^2s mass flex. *Source:* (a, b) Reproduced from [73] / with permission of ELSEVIER. (c, d) [75] / ASME / CC BY.

Figures 13.19b,d illustrate the progression of the heat transfer coefficient concerning the vapor quality across various mass fluxes. The heat transfer coefficient, under boiling conditions, ranges from 5000 to 30000 W/m^2 K, a finding that aligns with Sobierska et al.'s [76] observations. As the vapor quality increases, it is noted that the heat transfer coefficient rises when boiling commences under a given heat flux. Within a vapor quality spectrum of 0.005–0.015, the heat transfer coefficient peaks before stabilizing at a constant level. An escalation in heat flux triggers a rise in this consistent value. The impact of mass flux on the heat transfer coefficient is observed to be minimal; at 120 kg/m^2s, the optimum heat transfer coefficient is achieved with the highest operational heat flux, approximately 80 kW/m^2. Conversely, at 100 kg/m^2s, lower heat flux levels can yield superior values for the maximum heat transfer coefficient for DLC coating. The heat transfer coefficient peaks at 25000 W/m^2 K when the heat flux equals 42 kW/m^2. This fluctuation in mass flux's effects could be attributed to

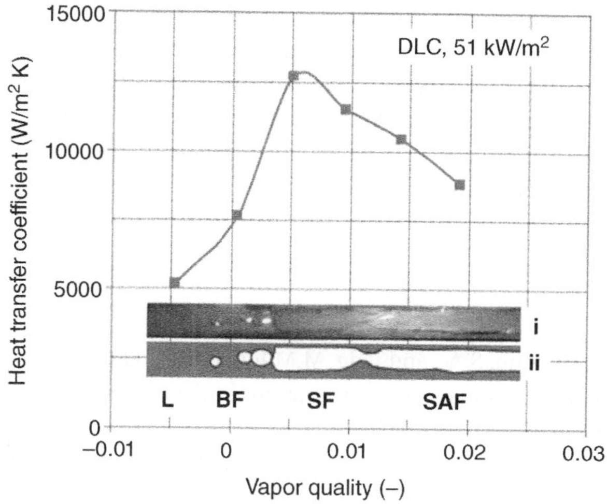

Figure 13.20 (i) Original and (ii) contour sharpening of flow patterns at DLC coating surfaces at mass flux of 120 kg/m^{2}s. *Source:* [75] / ASME / CC BY.

nucleate boiling's influence at low vapor qualities, which often proves more pronounced at lower mass flux levels.

For a more comprehensive understanding of the boiling mechanisms, Figure 13.20 captures an image taken from a high-speed camera, showcasing the flow patterns on the DLC surface [54]. The visual representation in Figure 13.20 delineates three distinct boiling flow patterns: bubbly flow (BF), slug flow (SF), and semi-annular flow (SAF). An intermittent dry-out phenomenon is recognized through the condensation of vapor on the top wall of the test channel, particularly during the transition from slug to semi-annular flow [75].

13.7 Conclusion

In conclusion, the integration of data-centric approaches into the realm of carbon coatings represents a significant advancement in both their development and application. Through data-driven manufacturing processes, the production of carbon coatings is likely to become more efficient and precise, ensuring higher quality outcomes. Virtual synthesis and structural design, powered by computational tools, allow for the pre-emptive evaluation and optimization of coating structures, reducing the need for extensive physical prototyping. Additionally, understanding the properties of carbon coatings through data analysis provides a deeper insight into their behavior for potential applications, enabling tailored solutions for specific needs. Performance assessments, driven by real-time data and analytics, further validate the effectiveness and durability of these coatings in various environments. Finally, the application of carbon coatings in enhancing heat transfer processes, particularly through condensation and boiling mechanisms, underscores their practical value in improving thermal system efficiency. Together, these advancements highlight the pivotal role of data-centric methodologies in pushing the boundaries of carbon coating technology.

References

1 Zia, A.W., Hussain, S.A., and Baig, M.M.F.A. (2022). Optimizing diamond-like carbon coatings-From experimental era to artificial intelligence. *Ceramics International* 48 (24): 36000–36011.

2 Pascoal-Faria, P., da Silva, D.P., Mateus, A. et al. (2023). Digitalisation of material science–Improving product design in the context of Industry 4.0. *Materials Today: Proceedings*, In Press.

3 Wang, J., Li, Y., Gao, R.X. et al. (2022). Hybrid physics-based and data-driven models for smart manufacturing: Modeling, simulation, and explainability. *Journal of Manufacturing Systems* 63: 381–391.

4 Pan, I., Mason, L.R., and Matar, O.K. (2022). Data-centric engineering: Integrating simulation, machine learning and statistics. Challenges and opportunities. *Chemical Engineering Science* 249: 117271.

5 Bauer, S., Benner, P., Bereau, T. et al. (2024). Roadmap on data-centric materials science. *Modeling and Simulation in Materials Science and Engineering* 32 (6): 063301.

6 Cherrington, R., Marshall, J., Alexander, A.T. et al. (2022). Exploring the circular economy through coatings in transport. *Sustainable Production and Consumption* 32: 136–146.

7 Ichou, H., Arrousse, N., Berdimurodov, E. et al. (2024). Exploring the advancements in physical vapor deposition coating: A review. *Journal of Bio-and Tribo-Corrosion* 10 (1): 3.

8 Zheng, J., Mei, S., Chao, Y. et al. (2024). Bias voltage-induced structural change in PECVD-deposited diamond-like carbon coatings for optimizing the tribological performance. *Surface and Coatings Technology* 484: 130861.

9 Garg, R., Gonuguntla, S., Sk, S. et al. (2024). Sputtering thin films: Materials, applications, challenges and future directions. *Advances in Colloid and Interface Science* 330: 103203. doi: 10.1016/j.cis.2024.103203.

10 Roy, A., Wang, S., and Komvopoulos, K. (2023). A review of plasma-assisted deposition methods for amorphous carbon thin and ultrathin films with a focus on the cathodic vacuum arc technique. *Journal of Materials Research* 38 (3): 586–616. doi: 10.1557/s43578-022-00868-9.

11 Yu, J., Han, W., Suleiman, A.A. et al. (2024). Recent advances on pulsed laser deposition of large-scale thin films. *Small Methods* 8 (7): 2301282.

12 Zia, A.W. and Birkett, M. (2021). Deposition of diamond-like carbon coatings: Conventional to non-conventional approaches for emerging markets. *Ceramics International* 47 (20): 28075–28085. doi: 10.1016/j.ceramint.2021.07.005.

13 Hofsäss, H., Feldermann, H., Merk, R. et al. (1998). Cylindrical spike model for the formation of diamondlike thin films by ion deposition. *Applied Physics A* 66: 153–181.

14 Zia, A.W., Zhou, Z., Shum, W. et al. (2017). The effect of two-step heat treatment on hardness, fracture toughness, and wear of different biased diamond-like carbon coatings. *Surface and Coatings Technology* 320: 118–125.

15 Kalita, K., Ghadai, R.K., and Chakraborty, S. (2022). Parametric optimization of CVD process for DLC Thin film coatings: A comparative analysis. *Sādhanā* 47 (2): 57.

16 Ludwig, A. (2019). Discovery of new materials using combinatorial synthesis and high-throughput characterization of thin-film materials libraries combined with computational methods. *NPJ Computational Materials* 5 (1): 70.

17 Pei, Z., Rozman, K.A., Doğan, Č.N. et al. (2021). Machine learning microstructure for inverse material design. *Advanced Science* 8 (23): 2101207.

18 Jean, M.D., Liu, C.D., Wang, S.F. et al. (2013). Design and optimization of surface properties for diamond-like carbon films by sputtering depositions. *Applied Mechanics and Materials* 401: 762–766.

19 Solis-Romero, J., Rodríguez-Molina, A., Solis-Cordova, J.J. et al. (2022). Parametric optimisation of friction and wear of a multi-layered aC: H coating on AISI 52100 steel. *Materials Letters* 318: 132166.

20 Kalita, K. and Ghadai, R.K. (2022). Optimization of plasma enhanced chemical vapor deposition process parameters for hardness improvement of diamond-like carbon coatings. *Scientia Iranica* 29 (4): 1795–1805.

21 Kański, M., Hrabar, S., van Duin, A.C.T. et al. (2022). Development of a charge-implicit ReaxFF for C/H/O systems. *The Journal of Physical Chemistry Letters* 13 (2): 628–633. doi: 10.1021/acs.jpclett.1c03867.

22 Tersoff, J. (1988). New empirical approach for the structure and energy of covalent systems. *Physical review B* 37 (12): 6991–6991.

23 Senftle, T.P., Hong, S., Islam, M.M. et al. (2016). The ReaxFF reactive force-field: development, applications and future directions. *npj Computational Materials* 2 (1): 1–14.

24 Kametani, N., Nakamura, M., Yashiro, K. et al. (2022). Investigating residual stress evolution in the deposition process of diamond-like carbon film through molecular dynamics. *Computational Materials Science* 209: 111420. doi: 10.1016/j.commatsci.2022.

25 Stukowski, A. (2009). Visualization and analysis of atomistic simulation data with OVITO–the open visualization tool. *Modeling and Simulation in Materials Science and Engineering* 18 (1): 015012–015012.

26 Hofmann, P., Gryga, R., Müller, M. et al. (2022). Multiscale simulation of hollow cathode assisted internal plasma treatment process. *Surface and Coatings Technology* 442: 128422. doi: 10.1016/j.surfcoat.2022.

27 Bárdoš, L. and Dušek V. (1988). High rate jet plasma-assisted chemical vapour deposition. *Thin Solid Films* 158(2) 265-270. doi: 10.1016/0040-6090(88)90029-6.

28 Caetano, R.F., Hoyer, Y.D.U., Oliveira, C. et al. (2014). Study of microhollow cathode glow discharge to improve the wettability on surface of polypropylene film. *researchgate.netRF, J. Condensed Matter Physics, 2014•researchgate.net* 2014(3A): 32–39. doi: 10.5923/s.ajcmp.201401.05.

29 Muhl, S. and Pérez, A. (2015). The use of hollow cathodes in deposition processes: A critical review. *Thin Solid Flims* 579:174–198. https://www.sciencedirect.com/science/article/pii/S0040609015001947.

30 Kim, Y., Cho, S., Choi, W. et al. (2003). Dependence of the bonding structure of DLC thin films on the deposition conditions of PECVD method. *Surface and Coatings Technology* 169–170: 291–294. https://www.sciencedirect.com/science/article/pii/S0257897203000690.

31 Cicala, G., Bruno, P., and Losacco, A.M. (2004). PECVD of hydrogenated diamond-like carbon films from CH4–Ar mixtures: Growth chemistry and material characteristics. *Diamond and Related Materials* 13: 1361–1365. https://www.sciencedirect.com/science/article/pii/S0925963503004485.

32 Dammann, M., Chertouk, M., Jantz, W. et al. (2000). Reliability of InAlAs/InGaAs HEMTs grown on GaAs substrate with metamorphic buffer. *Microelectronics Reliability* 40: 1709–1713. https://www.sciencedirect.com/science/article/pii/S0026271400001645.

33 Yang, H., Brown, G.A., Hu, J.C. et al. (1997). A comparison of TiN processes for CVD W/TiN gate electrode on 3 nm gate oxide. *ieeexplore.ieee.org, International Electron Devices Meeting. IEDM Technical Digest, 1997•ieeexplore.ieee.org*, [Online]. Available: https://ieeexplore.ieee.org/abstract/document/650423/.

34 Mikami, T., Nakazawa, H., Kudo, M. et al. (2005). Effects of hydrogen on film properties of diamond-like carbon films prepared by reactive radio-frequency magnetron sputtering using hydrogen gas. *Thin Solid Films* 488: 87–92. https://www.sciencedirect.com/science/article/pii/S0040609005004323.

35 Vietzke, E., Flaskamp, K., Philipps, V. et al. (1987). Chemical erosion of amorphous hydrogenated carbon films by atomic and energetic hydrogen. *Journal of Nuclear Materials* 145–147: 443–447. https://www.sciencedirect.com/science/article/pii/0022311587903783.

36 Silinskas, M., Grigonis, A., Kulikauskas, V. et al.-T. S. Films, and undefined (2008). Hydrogen influence on the structure and properties of amorphous hydrogenated carbon films deposited by direct ion beam. *Thin Solid Films* 516: 1683–1692. https://www.sciencedirect.com/science/article/pii/S0040609007008024.

37 Wei, C., Peng, K. S., and Hung, M. S. (2016). The effect of hydrogen and acetylene mixing ratios on the surface, mechanical and biocompatible properties of diamond-like carbon films. *Diamond and Related Materials* 63: 108–114. https://www.sciencedirect.com/science/article/pii/S0925963515300790.

38 Park, H., Lee, S., Kim, H.J. et al. (2019). Effects of nitrogen doping in amorphous carbon layers on the diffusion of fluorine atoms: A first-principles study. *pubs.aip.org, Journal of Applied Physics* 125: 155701–155701. doi: 10.1063/1.5064437.

39 Li, X., Ke, P., and Wang, A. (2015). Probing the stress reduction mechanism of diamond-like carbon films by incorporating Ti, Cr, or W carbide-forming metals: Ab initio molecular dynamics simulation. *Journal of Physical Chemistry C* 119 (11): 6086–6093. doi: 10.1021/ACS.JPCC.5B00058.

40 Park, H., Woo, D., Lee, J.M. et al. (2020). The influence of hydrogen concentration in amorphous carbon films on mechanical properties and fluorine penetration: A density functional theory and: Ab initio molecular dynamics study. *RSC Advances* 10 (12): 6822–6830. doi: 10.1039/c9ra09328a.

41 Valencia, F.J., Santiago, J., González, R.I. et al. (2021). Nanoindentation of Amorphous Carbon: A combined experimental and simulation approach. *Acta Materialia* 203: 116485. doi: 10.1016/j.actamat.2020.116485.

42 Plimpton, S. (1995). Fast parallel algorithms for short-range molecular dynamics. *Journal of Computational Physics* 117 (1): 1–19. https://www.sciencedirect.com/science/article/pii/S002199918571039X.

43 Bouzakis, E. (2020). Fatigue endurance assessment of DLC coatings on high-speed steels at ambient and elevated temperatures by repetitive impact tests. *Coatings* 10 (6): 547. doi: 10.3390/COATINGS10060547.

44 Charitidis, C.A. (2010). Nanomechanical and nanotribological properties of carbon-based thin films: A review. *International Journal of Refractory Metals and Hard Materials* 28 (1): 51–70. https://www.sciencedirect.com/science/article/pii/S0263436809001139.

45 Ziegenhain, G., Urbassek, H. M., and Hartmaier, A. (2010). Influence of crystal anisotropy on elastic deformation and onset of plasticity in nanoindentation: A simulational study. *Journal of Applied Physics* 107 (6). doi: 10.1063/1.3340523.

46 Beake, B.D. and Liskiewicz, T.W. (2017). Nanomechanical characterization of carbon films. IN: Tiwari, A., and Natarajan, S., (Eds.) *Applied Nanoindentation in Advanced Materials*, 19–68. Wiley, USA. doi: 10.1002/9781119084501.CH2.

47 Ruestes, C.J., Alhafez, A., Urbassek, H.M. (2017). Atomistic studies of nanoindentation—A review of recent advances. *Crystals* 7 (10): 293. doi: 10.3390/cryst7100293.

48 Yin, Z., Wu, H., Zhang, G. et al. (2022). Wear estimation of dlc films based on energy-dissipation analysis: A molecular dynamics study. *Materials* 15 (3): 893. doi: 10.3390/ma15030893.

49 Smerdova, O., Mazuyer, D., and Cayer-Barrioz, J. (2014). Links between energy dissipation and wear mechanisms in solid epoxy/epoxy sliding contact. *Tribology International* 77: 148–159. doi: 10.1016/J.TRIBOINT.2014.04.009.

50 Ma, X.G., Komvopoulos, K., Wan, D. et al. (2003). Effects of film thickness and contact load on nanotribological properties of sputtered amorphous carbon thin films. *Wear* 254 (10): 1010–1018. doi: 10.1016/S0043-1648(03)00307-7.

51 Zhang, H.S., Endrino, J.L., and Anders, A. (2008). Comparative surface and nano-tribological characteristics of nanocomposite diamond-like carbon thin films doped by silver. *Applied Surface Science* 255 (5): 2551–2556. doi: 10.1016/J.APSUSC.2008.07.193.

52 Yoon, E.S., Singh, R.A., Oh, H.J. et al. (2005). The effect of contact area on nano/micro-scale friction. *Wear* 259 (7–12): 1424–1431. doi: 10.1016/J.WEAR.2005.01.033.

53 Yang, Y., Huang, L., and Shi, Y. (2016). Adhesion suppresses atomic wear in single-asperity sliding. *Wear* 352–353: 31–41. doi: 10.1016/J.WEAR.2016.02.002.

54 Wang, L., Zhang, Z., Chen, H. et al. (2023). Friction behavior of diamond-like carbon coatings with different sp3 contents by atomistic-scale friction dynamics. *Surface and Coatings Technology* 464: 129580. doi: 10.1016/j.surfcoat.2023.129580.

55 Wang, L., Zhang, Z., Chen, H. et al. (2023). Metastable hybridized structure transformation in amorphous carbon films during friction—A study combining experiments and MD simulation. *Friction* 11 (9): 1708–1723. doi: 10.1007/s40544-022-0690-x.

56 Kunze, T., Posselt, M., Gemming, S. et al. (2014). Wear, plasticity, and rehybridization in tetrahedral amorphous carbon. *Tribology Letters* 53 (1): 119–126. doi: 10.1007/s11249-013-0250-7.

57 Du, N., Feng, C., Chen, K. et al. (2023). Friction dependence on the textured structure of an amorphous carbon surface: A reactive molecular dynamics study. *Applied Surface Science* 610: 155584. doi: 10.1016/j.apsusc.2022.155584.

58 Shao, W., Shi, Z., Rao, L. et al. (2020). High-temperature sliding friction behavior of amorphous carbon films: Molecular dynamics simulation. *Langmuir* 36 (50): 15319–15330. doi: 10.1021/acs.langmuir.0c02765.

59 Ziegltrum, A., Maier, E., Lohner, T. et al. (2020). A numerical study on thermal elastohydrodynamic lubrication of coated polymers. *Tribology Letters* 68 (2): 71. doi: 10.1007/s11249-020-01309-6.

60 Ebner, M., Ziegltrum, A., Lohner, T. et al. (2020). Measurement of EHL temperature by thin film sensors – Thermal insulation effects. *Tribology International* 149: 105515. doi: 10.1016/j.triboint.2018.12.015.

61 Ziegltrum, A. Lohner, T., and Stahl, K. (2018). TEHL Simulation on the Influence of lubricants on the frictional losses of DLC coated gears. *Lubricants* 6: 17; 6 (1): 17–17. doi: 10.3390/LUBRICANTS6010017.

62 Habchi, W. and Bair, S. (2016). Effect of lubricant rheology on friction in coated elastohydrodynamic lubricated contacts. *Proceedings of the Institution of Mechanical Engineers, Part J: Journal of Engineering Tribology* 231 (8): 975–985. doi: 10.1177/1350650116684657.

63 Huang, B., Zhang, E.G., Zhou, Q. et al. (2021). Research on the performance of diamond-like carbon coatings on cutting aluminum alloy: Cutting experiments and first-principles calculations. *Coatings* 11 (1): 63. doi: 10.3390/coatings.

64 Neumann, A.W., Abdelmessih, A.H., and Hameed, A. (1978). The role of contact angles and contact angle hysteresis in dropwise condensation heat transfer. *International Journal of Heat and Mass Transfer* 21 (7): 947–953. doi: 10.1016/0017-9310(78)90185-2.

65 Busscher, H.J., van Pelt, A.W.J. de Boer, P. et al. (1984). The effect of surface roughening of polymers on measured contact angles of liquids. *Colloids and Surfaces* 9 (4): 319–331. doi: 10.1016/0166-6622(84)80175-4.

66 Upot, N.V., Rabbi, K.F., Khodakarami, S. et al. (2023). Advances in micro and nanoengineered surfaces for enhancing boiling and condensation heat transfer: A review. *Nanoscale Advances* 5 (5): 1232–1270. doi: 10.1039/D2NA00669C.

67 Attinger, D., Frankiewicz, C., Betz, A.R. et al. (2014). Surface engineering for phase change heat transfer: A review. *MRS Energy & Sustainability* 1 (1): E4–E4. doi: 10.1557/MRE.2014.9.

68 Qin, S., Ji, R., Miao, C. et al. (2024). Review of enhancing boiling and condensation heat transfer: Surface modification. *Renewable and Sustainable Energy Reviews* 189: 113882–113882. doi: 10.1016/J.RSER.2023.113882.

69 Leipertz, A. and Fröba, A.P. (2008). Improvement of condensation heat transfer by surface modifications. *Heat Transfer Engineering* 29 (4): 343–356. doi: 10.1080/01457630701821563.

70 Koch, G., Zhang, D.C., Leipertz, A. et al. (1998). Study on plasma enhanced CVD coated material to promote dropwise condensation of steam. *International Journal of Heat and Mass Transfer* 41 (13): 1899–1906. doi: 10.1016/S0017-9310(97)00356-6.

71 Erb, R.A. (1965). Wettability of metals under continuous condensing conditions. *Journal of Physical Chemistry* 69 (4): 1306–1309. doi: 10.1021/J100888A035.

72 Ryu, H., Kim, J., Kim, J. et al. (2020). Enhancement of a heat transfer performance on the Al6061 surface using microstructures and fluorine-doped diamond-like carbon (F-DLC) coating. *International Journal of Heat and Mass Transfer* 148: 119108–119108. doi: 10.1016/J.IJHEATMASSTRANSFER.2019.119108.

73 Phan, H.T., Caney, N., Marty. P. et al. (2011). Flow boiling of water in a minichannel: The effects of surface wettability on two-phase pressure drop. *Applied Thermal Engineering* 31 (11–12): 1894–1905. doi: 10.1016/J.APPLTHERMALENG.2011.02.036.

74 Kew, P.A. and Cornwell, K. (1997). Correlations for the prediction of boiling heat transfer in small-diameter channels. *Applied Thermal Engineering* 17 (8–10): 705–715. doi: 10.1016/S1359-4311(96)00071-3.

75 Phan, H.T., Caney, N., Marty, P. et al. (2011). Flow boiling of water on titanium and diamond-like carbon coated surfaces in a microchannel. *2010 14th International Heat Transfer Conference, IHTC 14*. 1: 127–132. doi: 10.1115/IHTC14-22024.

76 Sobierska, E., Kulenovic, R., and Mertz, R. (2007). Heat transfer mechanism and flow pattern during flow boiling of water in a vertical narrow channel—experimental results. *International Journal of Thermal Sciences* 46 (11): 1172–1181. doi: 10.1016/J.IJTHERMALSCI.2007.06.011.

14

Circularity and Techno-economic Analysis of Carbon-coated Products

Ashmal James, Abdul Wasy Zia

Institute of Mechanical Process and Energy Engineering (IMPEE), School of Engineering and Physical Sciences, Heriot-Watt University, Edinburgh, UK

CONTENTS

14.1 Introduction, 413
14.2 Potential Reprocessing Methods for Carbon De-coating, 415
14.3 Decision Matrix of Carbon De-coating Methods, 417
14.3.1 Process Robustness for De-coating, 417
14.3.2 Process Efficiency to De-coat Carbon-coated Components, 418
14.3.3 De-coating Processes and Base Material Safety, 418
14.3.4 Sustainability Aspects of De-coating Processes, 418
14.3.5 Technology Readiness Level of De-coating Processes, 419
14.3.6 Cost, 419
14.4 Technical Analysis of Carbon De-coating Methods, 419
14.4.1 Robustness of De-coating Processes, 420
14.4.2 Circularity Index for De-coating Processes, 421
14.4.3 Safety Factor Index for De-coating Processes, 421
14.4.4 Technology Readiness Levels of De-coating Processes, 422
14.5 Economic Analysis of Carbon De-coating Methods, 422
14.5.1 Capital Cost, 422
14.5.2 Running Cost, 422
14.6 Case Study – Carbon De-coating from Cutting Tools, 423
14.7 Conclusion, 426

14.1 Introduction

Carbon coatings are widely used for dies and molds, stamping and cutting tools, fuel engine components, optical filters, solar panels and battery anodes, and orthodontics and orthopedic implants. Their applications are expanding for

Applications of Diamond-like Carbon Coatings, First Edition. Abdul Wasy Zia.

plastics, textiles, and electronics. Carbon capture from gaseous phases like CO_2 received wider attention and there is no clear framework for green recycling or circular usage of (solid) carbon-coated products (CCPs). The global revenue of carbon coating was estimated at £1.5 billion in 2020, which is increasing at a compound annual growth rate of 2.6% and is expected to reach £2.2 billion by 2027. The wastage from CCPs is also increasing proportionally because carbon-coated components are retired with degradation of the top carbon coat as product efficiency declines. Some components such as cutting tools are processed to remove the remainder coatings with physical and chemical methods and harvested with fresh coatings. The carbon released into the environment through liquid, gaseous, or particulate forms threatens the green ecology.

The circular economy is gaining prominence in modern techno-economic systems, focusing on the reuse and regeneration of products to extend their life cycle and minimize waste, thereby supporting green ecology and environmental restoration [1]. Unlike the linear economy [2], which follows a produce-use-dispose model and fails to prioritize recycling or reuse, leading to resource depletion and environmental degradation, the circular economy offers a sustainable alternative [3]. This shift is crucial for conserving nature, protecting the environment, and promoting sustainability, while also delivering economic benefits through waste reduction and enhanced energy efficiency.

Carbon coatings, which involve applying a thin carbon skin to engineering components and products, play a vital role in enhancing product performance and functionality. Despite acting as thin skin, these coatings significantly improve the material properties to meet engineering requirements across various applications [4]. Pure carbon coatings contain significant amounts of carbon nanomaterial while doped carbon coatings contain valuable compounds such as titanium, silver, gold, or chromium, tungsten, and molybdenum depending on the application. Therefore, CCPs contribute to carbon emissions, which exacerbate the greenhouse effect and global warming; large consumption of precious and critical materials extends economic and supply chain burdens; and leaching of hazardous compounds increases climate challenges. Therefore, adopting a circular approach for CCP could help mitigate these environmental impacts by reducing carbon waste, energy consumption, and emissions, thereby making CCP more sustainable and contributing to economic growth.

Carbon coatings are widely applied across various industries, such as automotive components, cutting tools, space technology, medical instruments and products, and electronics, due to their ability to enhance durability, reduce friction, prevent corrosion, and increase wear resistance. Among numerous industries, the cutting tool sector is a major contributor to the carbon footprint globally, driven by the widespread use of carbon-coated tools. In 2021, the global market for cutting tools was valued at $65.49 billion, rising to $78.89 billion by 2023, with projections of reaching $120.44 billion by 2030. Sandvik Coromant Ltd. reported that recycled tools maintain the same high standards while consuming 70% less energy and generating 40% fewer CO_2 emissions [5]. Given the rapidly increasing carbon footprint associated with cutting tools, there is a pressing need for sustainable solutions. This extends to other CCPs, such as diamond-like carbon (DLC) coatings used in medical implants. For instance, DLC coatings on knee implants reduce wear, extending the longevity and reliability of the implants. In the United Kingdom alone, over 100 000 knee implants are performed annually. Globally, the knee implant market was valued at $11.1 billion in 2022 and is expected to reach $10.85 billion by 2030. This growth will significantly increase the global carbon footprint, emphasizing the need to de-coat such products before disposal or recycling. Repurposing or recycling de-coated carbon can help minimize waste and reduce environmental impact.

De-coating CCP offers numerous benefits for manufacturers and environmental sustainability. For products made from expensive or limited resources, de-coating allows for the recovery and reuse of the primary carbon and other doped materials, thereby reducing waste. Since carbon can significantly harm the environment, removing carbon coatings through de-coating provides a pathway to recycle the underlying material more effectively. This process not only cuts down on waste but also lowers replacement costs, as de-coated components could be injected back into the product cradle-to-cradle cycle. Additionally, in industries where CCPs are widely used, de-coating is crucial for repairing and maintaining parts. Removing old coatings before applying new ones can also reduce the product's weight, further enhancing efficiency [6]. This work briefly describes potential reprocessing methods and their working principles, develops a techno-economic analysis (TEA), and demonstrates the tribology as suitable carbon de-coating method particularly attractive for micro- and small to medium enterprises and in low- and middle-income country climates.

14.2 Potential Reprocessing Methods for Carbon De-coating

This section briefly introduces common methods suitable to de-coat CCPs such as chemical and electrochemical stripping or etching, thermal decomposition, mechanical abrasion, laser ablation, plasma processing, ultrasonic cleaning, etc.

The process of chemical stripping or etching involves choosing the right type of chemical solvent. The CCPs could be immersed into this solvent while it weakens the coating and finally dissolves it. The underlying substrate material is left intact after the stripping process is completed [7]. Whereas electrochemical stripping could use fuel cell phenomena. The CCP to be de-coated is placed in an electrochemical cell as an electrode connected to a positive terminal. This cell includes two electrodes, and the other counter electrode is connected to the negative terminal. The electric current is applied, and electrolysis takes place resulting in the dissolution of carbon coating. The coating particles are converted into gases and soluble compounds. This de-coating process with ultrasonic cleaning also consists of immersing CCP in solvents. This tank is equipped with ultrasonic transducers that produce high-frequency sound waves. The frequency used in the industry ranges from 20 to 120 kHz. Normally 20–40 kHz is the most used in industry, whereas 80–120 kHz is available at customization. The energy waves pass through the solution causing an effect called cavitation, which creates microscopic bubbles that implode creating microscopic waves. These waves result in scrubbing, which removes the carbon coating of the substrate. The ultrasonic cleaning could be enhanced with a choice of solvents and solvent temperature. The carbon residue could be filtered, collected, and processed for circular usage.

Pyrolysis could be a suitable process for thermal decomposition of carbon coatings. This method also sometimes referred to as fire separating process. CCPs are placed in a heating chamber and the carbon coating disintegrates into two byproducts as a result of pyrolysis process. One byproduct is hydrocarbons in the form of gaseous mixtures and the other is solid carbon residue that settles on the underlying substrate. This is removed by the combustion process resulting in carbon dioxide and heat being released. The carbon footprints could be directly or indirectly captured by a high efficiency particulate air (HEPA) filter and processed for circular usage.

Mechanical abrasion is a physical technique where the carbon coating is de-coated by mechanical abrasive methods. The process includes the physical removal of carbon from the substrate, layer by layer. The product to be de-coated is worn down by abrasive techniques such as sandblasting, grinding, blasting media, scraping, or using sandpaper. The friction, pressure, and motion forces that are applied to the CCP, by the abrasive technique action, result in the removal of the carbon coating, and the substrate is brushed off to remove the remaining carbon residue [8]. Tribology also falls under the family of mechanical abrasion. It uses friction, wear, and controlled mechanical force to de-coat CCP for optimal efficiency and minimal damage to the underlying materials [9]. The carbon residue could be collected and processed for circular usage.

For laser ablation-based de-coating, a laser system of high intensity such as an infrared or ultraviolet laser is focused on the CCP for the ablation process. Ultraviolet laser is the most common laser used in the industry, with a wavelength of 355 nm and a characteristic pulse duration of 10 ps. The coating absorbs this high energy of the laser, which results in a rapid heating of the carbon material. This selectively vaporizes the carbon coating of the product. It is a very controlled method of de-coating as you can tune the intensity of the laser and is also very efficient as the intensity can specifically be focused. A HEPA filter could collect carbon residue from the gas flue. The lasers ablate material from small areas while plasma can etch materials for large areas. For plasma processing, the product to be de-coated is placed in a chamber that is vacuumed to create a low-pressure environment. Then, argon gas could be introduced to produce plasma in the chamber. The argon ions are ionized by introducing electrical energy in the chamber. These ionized atoms bombarded at carbon coating, on the CCPs causing them to break down. As a result of this breakdown, carbon monoxide and carbon dioxide are produced as byproducts [10].

14.3 Decision Matrix of Carbon De-coating Methods

This work considers the robustness, efficiency, base material safety, sustainability, technology readiness levels (TRLs), and cost to develop a decision matrix. The learning outcomes of the decision matrix are carried forward to develop a TEA. Table 14.1 presents the decision matrix of carbon de-coating methods, and the indices criterion is discussed in the following section.

14.3.1 Process Robustness for De-coating

This is defined as the amount of time taken to de-coat the carbon coating. The quickest method will be a high energy process (HEP), making it 5-rated [11]. Sandblasting and tribology are also a fast-stripping method. They have a very high turnaround time earning a high rate of 4 [12, 13]. Electrochemical stripping de-coats the CCP's coating at a higher and steadier rate than chemical stripping, hence, it is rated 4 [14]. Ultrasonic has comparatively higher rates of de-coating than chemical stripping but will still consume a considerable time to de-coat the CCP, as it requires immersion of the product; therefore, it is 3-rated [15]. Chemical stripping or etching is rated 2, as it requires some hours of immersion of the CCP in the solvent solution for de-coating [14].

Table 14.1 Decision matrix of the de-coating methods.

Methods	Criteria 1	Criteria 2	Criteria 3	Criteria 4	Criteria 5	Criteria 6	Total of criteria
	Speed	Efficiency	Sustainability	Minimum effect on substrate	Technology readiness	Cost	Total
1. Chemical stripping or etching	2	4	2	4	5	5	22
2. Sand blasting	4	2	4	2	4	4	20
3. High energy processes	5	5	3	3	2	2	20
4. Ultrasonic cleaning	3	4	5	5	4	4	25
5. Electrochemical methods	4	3	2	4	3	3	19
6. Tribology method	4	4	5	3	5	4	25

14.3.2 Process Efficiency to De-coat Carbon-coated Components

This is defined as what percentage of carbon coating is successfully de-coated. Laser ablation, a high energy process (HEP) will have the most efficiency, hence rated as 5, as it changes the surface of the product at the microscopic level de-coating with a high efficiency [16]. Chemical etching and electrochemical stripping are rated to be 4. They have a very high-efficiency rate but not as high as laser ablation as the type of solvent used might not be as efficient as the laser that de-coats at the microscopic level. However, chemical stripping can de-coat the most durable and chemical-resistant coatings [12]. Ultrasonic cleaning is also rated a high 4 as it effectively de-coats complex areas and hard-to-reach surfaces [17]. The tribology method is rated 4 as it uses a controlled force that is exerted on the sample. The force can be manipulated by adding or removing weights, as well as by selecting different grit sandpapers [18]. Electrochemical stripping involves de-coating using electrolysis powered by electric current, so it is highly efficient at high current density surfaces but not very efficient at low current density surfaces such as corners, etc. Therefore, it is rated as 3 [14, 19]. Sandblasting is rated a low 2, as it involves the use of an uncontrolled mechanical force that is exerted on CCP [14].

14.3.3 De-coating Processes and Base Material Safety

This is one of the most important criteria to consider while choosing the most optimal de-coating method. Ultrasonic cleaning is rated the highest, i.e. 5, for this criterion as there is no mechanical force applied and only sound waves in a cleaning solution de-coat the product, leaving little to no effect on the substrate [15]. After that, chemical stripping is rated 4, as it does not have a massive effect on the substrate but depending on the solvent used it might damage it [12]. Electrochemical stripping will have the same effect on the substrate as chemical stripping [14]. Laser ablation is rated 3 due to its capacity to alter the substrate's microscopic characteristics but the conventional surface roughness (profile) of the product does not change significantly [11]. The tribology method will also have a considerable effect on the underlying substrate, but as it uses controlled force it will have less effect on the substrate compared to sandblasting [18]. Hence it is rated 3. Sandblasting will have the most effect on the substrate as it involves an uncontrolled physical force to be applied on the product to de-coat it. Therefore, it is rated 2 [8].

14.3.4 Sustainability Aspects of De-coating Processes

It is rated out of 5 and it is based on the environmental regulations of each method. Ultrasonic cleaning is the highest-rated method when it comes to preventing environmental damage and being green, as it uses only a cleaning solution and sound transducers powered by electricity. Hence, it is given a high 5 [17]. The tribology method is also highly green as it only uses sandpaper and electricity to power the machine. It de-coats with the help of friction and wear [20]. Hence its rating is 5. Sandblasting is rated a high 4 as the method is green. For instance, it only requires sand, which is natural and replenishable [13], and electricity to power the machine [12]. Laser ablation is rated a medium score of 3, making it environment-friendly. Chemical and electrochemical stripping are given a low 2 as both use corrosive materials, chlorinated solvents, and flammable solvents, which cause potential environmental problems [12].

14.3.5 Technology Readiness Level of De-coating Processes

This is rated based on the ease of obtaining equipment for each method. Chemical stripping equipment is easily accessible around the market, educational institutes, etc. Therefore, it is rated highly for this criterion and is given a 5. The tribology method only requires a machine and a few consumables, which are also readily available in the marketplace and educational institutes. It is also given a rating of 5. After that, ultrasonic cleaning is the easiest method for this criterion, making it a rating 4 for equipment availability, as it only requires a cleaning solution and an ultrasonic cleaning tank. Sandblasting is common process and it equipment are also readily available around the globe. Hence, it is given a 4. Electrochemical stripping is given a 3 because its equipment can be harder to secure as compared to other methods, whereas HEP such as laser ablation is rated a low 2 because comparatively, depending on the type of laser used, it can be a challenge to purchase laser ablation equipment [12, 14].

14.3.6 Cost

The cost will be the least for chemical stripping as it only needs a solvent such as an acid, which is very cheap. Hence it is rated a high 5. Sandblasting uses a cabinet that is very cost-efficient and hence rated 4. Ultrasonic cleaning is also rated 4 because it only requires a very economical cleaning tank. The tribology method uses sandpaper and a tribology machine, making the method also economical. Hence, it is rated 4. Electrochemical stripping is rated 3 as the assembly for it will not be cost-efficient and will not be as cheap as the aforementioned methods. Laser ablation is given a low 2 as it is a very costly method.

14.4 Technical Analysis of Carbon De-coating Methods

Figure 14.1a gives the overview of technical analysis of de-coating processes for CCP and details on indices are described in the following sections.

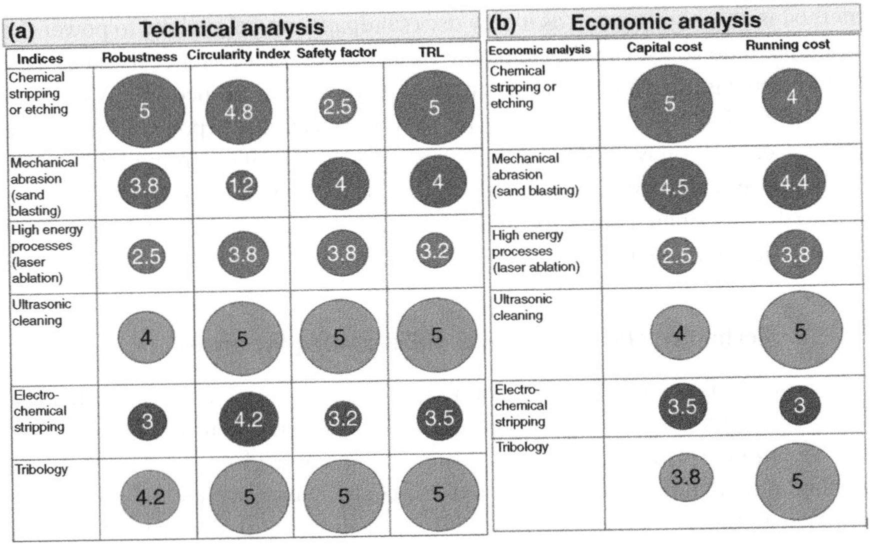

Figure 14.1 (a) Technical and (b) economic analysis of de-coating processes for CCPs.

14.4.1 Robustness of De-coating Processes

It refers to the consistency and effectiveness of a de-coating method in removing the carbon coatings from a product. Chemical stripping includes the complete emersion of CCP into a chemical solvent solution. Hence, it uniformly and effectively de-coats the carbon coating, earning a high 5 rating [12]. The tribology method includes a controlled force being homogenously applied on the CCP sample to be de-coated, making it comparatively more robust than other methods but still less robust than chemical stripping. Hence, it is rated 4.2 [18]. Ultrasonic cleaning follows the same principle as chemical stripping and provides good cleaning of carbon coatings between the smallest and tightest crevices. The only two differences would be the solvent used, as chemical etching uses a strong acid. As ultrasonic cleaning de-coats with a cleaning solution, it will comparatively be less effective, making it 4-rated. Sandblasting is rated 3.8, as it includes the process of forcing sand particles at high-velocity at the CCP at once, resulting in a consistent but not very efficient de-coating method [8]. In electrochemical stripping, the de-coating of the carbon happens consistently and effectively at high current density surfaces. Whereas, at lower current density areas, the removal of the coating might not be as effective, hence making the method not very robust, so it rated as 3 [14]. Laser ablation will be the least robust method comparatively as the laser is focused on a small surface area of CCP at a time, to de-coat the method. This may affect the consistency of de-coating [11].

14.4.2 Circularity Index for De-coating Processes

The circularity index mainly depends on two factors:

1. The amount of carbon coating removed from a product effectively, without harming the underlying substrate.
2. The ease of gathering the de-coated carbon of a product, after the de-coating process and recycling it.

Ultrasonic cleaning will have the least effect on the substrate and the carbon coating de-coated, will settle at the bottom of the tank in solid form, and can easily be extracted for recycling. Hence, it is rated 5 [15]. The tribology method removes the carbon coating of the CCP very effectively, but it might harm the underlying substrate. This drawback is overcome by the ease of gathering the de-coated carbon coating. It is normally gathered by using a vacuum and filtering it out using a HEPA filter [18]. The same principle as ultrasonic cleaning will apply to chemical stripping and as it does not have a significant effect on the substrate; hence it is rated 4.8 [12]. Electrochemical stripping will also follow the same principle. It has the same effect on the substrate as chemical stripping, but as it would not be as efficient as chemical stripping, hence it is rated 4.2 [14]. Laser ablation is rated 3.8 due to its capacity to alter the substrate's microscopic characteristics and to easily collect the de-coated carbon using a HEPA filter in the de-coating process chamber [11]. Sandblasting will have the most significant effect on the substrate because it is more aggressive, involving the propulsion of fine abrasive particles at high speeds toward the substrate. Moreover, collecting the de-coated carbon will not be straightforward, so it is rated 1.2 [8].

14.4.3 Safety Factor Index for De-coating Processes

The safety factor solely depends on how safe each de-coating method of CCP is. Ultrasonic cleaning will top this parameter by having a rating of 5 as it only uses cleaning solution and sound transducers, making it a very safe process [17]. The tribology method only uses sandpapers and a tribology machine. Sandblasting is rated 4 as it has a high advantage in meeting safety regulations because it only uses high-velocity sand for de-coating [12]. Laser ablation, i.e. HEP, does not pose a threat to safety, and they are safe methods. But they might be hazardous if a high-grade laser is used, making it a total of 3.8 rating. Electrochemical stripping is less rated than HEP as it involves the use of electric current and needs proper safety requirements, hence it is rated 3.2 [14]. Chemical stripping will have the lowest rating of 2.5, as it uses strong harmful acids that have potential health hazards [12].

14.4.4 Technology Readiness Levels of De-coating Processes

This parameter is defined as the framework to access the development and maturity of a method. TRL has nine different levels. It depends on how each method is adopted by the industry, educational institutes, etc., for the de-coating of CCPs. Ultrasonic cleaning, tribology method, and chemical stripping are mature and widely used methods, especially in industry. They are well-established methods, hence they both are rated 5 [14, 17, 18]. Sandblasting, a traditional method, is rated 4 as it is used in industry but is not as advanced as other methods [8]. Electrochemical stripping is also used in industries but is only used when chemical stripping becomes ineffective, unavailable, or too difficult to use [14]. Whereas laser ablation might be an advanced technique, but it is not very mature method and might be still under development in certain applications [16].

14.5 Economic Analysis of Carbon De-coating Methods

Figure 14.1b gives the overview of technical analysis of de-coating processes for CCP and details on indices are described in the following section.

14.5.1 Capital Cost

This is the cost of buying each unit of every method. Chemical stripping will have the least capital cost as it only requires a solvent and a tank, hence it is rated a high 5. Sandblasting will also have a very low capital cost as it only requires a sandblasting cabin, which will still be costly compared to chemical stripping, making it a 4.5-rated. Ultrasonic cleaning capital cost will be slightly higher, making it 4-rated. The tribology method will require a tribology machine and sandpapers. The sandpapers can be purchased cheaply but the machine can be costly, making its overall rating 3.8. Electrochemical stripping will have a higher capital cost, as the electrolysis assembly will be more expensive than the methods discussed above. Therefore, it is rated 3. Laser ablation will have the highest capital cost as it is an expensive method.

14.5.2 Running Cost

This is the cost of running each unit of every method. Ultrasonic cleaning will be 5-rated as it does not need any specific safety requirements. The cleaning machine also normally has a heating power of 600 W and a voltage of 110–220 V. Therefore, the running cost will be quite low. The tribology method is rated 5 as it only uses a few sandpapers and does not require any costly safety requirements. The tribology

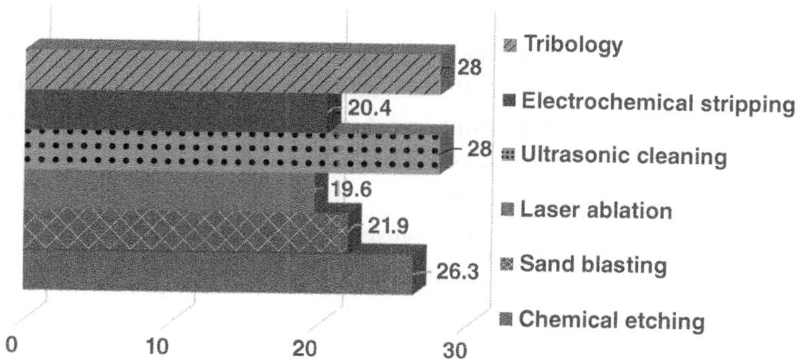

Figure 14.2 Techno-economic rating of de-coating processes for CCPs.

machine has a voltage of 240 V and a current of 2.4 A. Sandblasting will also only require electric power and some safety precautions to run. The sandblasting cabinet normally has a wattage of 1000 W and a voltage of 240 V. However, the sand cabinet requires maintenance, and the process has few safety requirements, making it 4.4-rated. Chemical stripping will be rated 4, as it does not need any power source to operate. It only includes immersion of CCP in an acid but because of the use of such strong acids, which can cause serious health hazards, proper government regulations need to be followed to ensure worker's health safety. This will include buying safety equipment and considering other precautions. The laser ablation machine has a wattage of 1000 W. As it involves the use of high-grade laser, it also needs proper safety equipment, making it 3.8-rated. Electrochemical stripping uses a high voltage of electric power and as it also involves the use of liquid in the form of solvent, extra safety precautions are needed, hence rated 3.

Figure 14.2 presents the overall techno-economic rating of potential de-coating processes for CCPs, particularly in low-resource environments.

14.6 Case Study – Carbon De-coating from Cutting Tools

After analysis, two methods emerged as the most optimal for de-coating: ultrasonic cleaning and tribology. The sample chosen for examination is a commercially used DLC-coated carbide cutting tool blade. This was selected because DLC is one of the hardest coatings, and to justify a de-coating method as the most optimal, it must be able to remove the hardest coatings. It can then be manipulated to de-coat less hard coatings. Initially, ultrasonic cleaning was attempted, but it did not affect the hard DLC coating, likely because it is a commercial-grade product. Consequently, tribology was selected and yielded the desired results.

Tribology-based de-coating process is demonstrated on 10 × 10 mm cutting blades and Figure 14.3 presents the DLC-coated and de-coated blades. The tribology experiment is carried out with four different sandpapers of grit P60, P80, P100, and P120 in dry sliding conditions. A customized 3D printed blade holder backed with a tensioner was used as contact material. Equations 14.1 and 14.2 are used to calculate the decorated carbon.

$$\%\,\text{of carbon coating decoated} = \left(\frac{\text{Mass of carbon removed from the sample (g)}}{\text{Total mass of carbon coating}}\right) \times 100 \tag{14.1}$$

$$= \left(\frac{\text{Initial mass of sample at an interval (g)} - \text{Final mass of sample at an interval (g)}}{\text{Total mass of carbon coating}}\right) \times 100 \tag{14.2}$$

P60 de-coats the sample, with the quickest taking 10 minutes. This is the coarsest grit among the sandpapers used for this experiment. The larger abrasive particles in P60 sandpaper allow it to remove material more aggressively. This aggressive removal means it can strip away coatings faster than finer grits, which is why it takes only 15 minutes to complete the de-coating. However, this

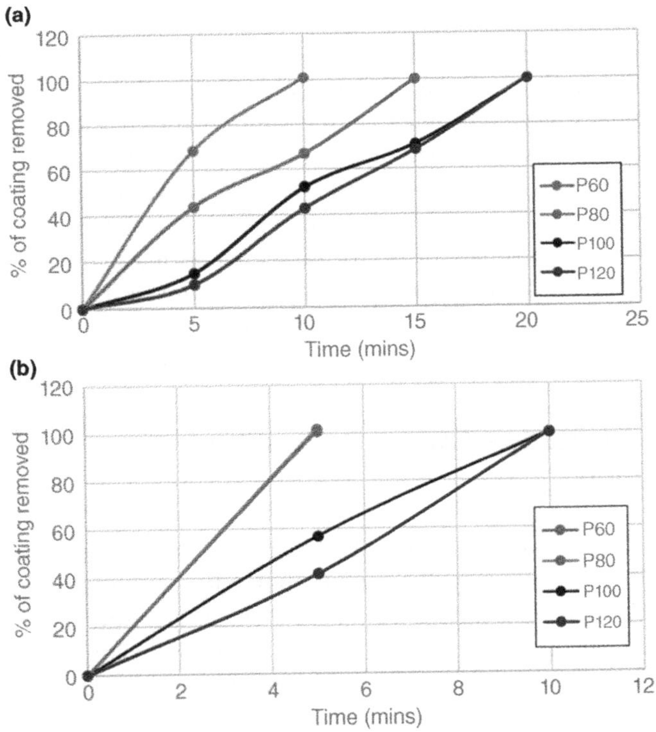

Figure 14.3 Percentage of coating removed as a function of time using sandpaper of different grit sizes in (a) incremental and (b) continuous dry sliding tribology test.

rapid removal can also cause more damage to the substrate. Therefore, for the last interval, the final mass of the sample is slightly less than the theoretical value, which is 2.405 g, indicating there is a small amount of material removal from the sample. P80 sandpaper, which is slightly finer than P60, has smaller abrasive particles, leading to somewhat less aggressive material removal. This reduction in aggressiveness results in a slightly longer time (20 minutes) needed to de-coat the same amount of material as P60. The trade-off gave a slightly smoother finish compared to P60. Coarser grits like P60 and P80 tend to generate more heat due to faster and more aggressive material removal. This can sometimes accelerate the de-coating process, as well as risks damaging the material due to heat stress. P100 and P120 are even finer grits, with P120 being the finest among the sandpapers used. Finer grits remove material less aggressively, which translates to slower coating removal rates. The smaller particles in P100 and P120 do not cut into the coating as deeply or quickly as coarser particles, resulting in the longest times (30 minutes) to achieve the same level of de-coating. The finer the abrasive, the more passes it generally takes to remove the same amount of material, leading to increased processing time. However, finer grits also minimize damage to the substrate and typically leave a smoother finish, which may be desired based on the application requirements. Finer grits like P100 and P120 generate less heat per pass, potentially preserving the material integrity better but requiring more time to remove the coating thoroughly. The thickness of the DLC coating is 70 μm, which is slightly higher than the average coating thickness. Hence, it was an extremely durable coating and weights were needed to be added to perform proper and homogeneous de-coating. Adding weight increases the downward pressure exerted by the sandpaper onto the surface of the DLC coating. Higher pressure enhances the abrasive action of the sandpaper against the surface leading to more effective abrasion per stroke or pass. The greater force can help break through tough coatings more efficiently, especially when dealing with high-hardness materials like DLC. It also allows for better control over the amount of material being removed. The extra weight also helped anchor the DLC sample in the sample holder, decreasing the likelihood that it would skid or slip during operation. This stability is crucial for achieving precise and safe material removal as it will reduce vibrations per stroke, improving control and enhancing safety. It is also observed the carbon coating is removed homogeneously as it de-coats the raised square first and then proceeds to remove the coating on the edges.

After experimentation, the de-coated surface area of each sample at each interval was calculated with ImageJ software. This is done by measuring the length and width of the de-coated sample at each interval and subtracting the area of the hole that is in the middle of the sample. This is also done to cross-validate and experiment with wear loss. Figure 14.3 presents the percentage of carbon coating

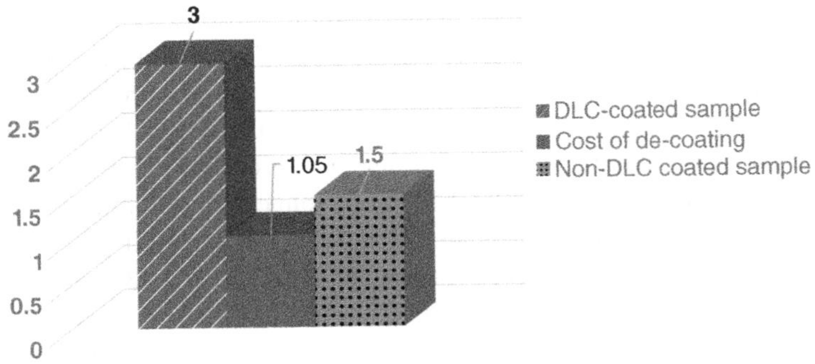

Figure 14.4 Cost estimation of pristine DLC-coated cutting blade, de-coating process, and uncoated cutting blade.

removed from cutting blades as a function of time using dry sliding tribological tests in incremental or continuous runs.

Figure 14.4 presents the cost savings if de-coated blades are recoated. Each DLC-coated cutting blade used has an estimated cost of £3. The same blade without the DLC coating has an estimated cost of around £1.5, about half the price of coated blades. The cost of de-coating for each sample is estimated at about £1.05056, which includes the cost of electricity and consumables. It is perceived that £0.50 per unit blade could be saved by returning de-coated blades in a cradle-to-cradle cycle. Besides cost, the de-coating will also complement energy saving, CO_2 emissions, logistic, and supply chain challenges throughout the cradle-to-cradle cycle.

14.7 Conclusion

Disposal of retired CCPs, especially in industry, has a high potential to release a large carbon footprint in ecology. Circularity and reducing the carbon footprint of such products are essential, and innovating new carbon de-coating processes is highly desirous to promote sustainable and green ecology. By qualitative and quantitative TEA of plasma, laser, chemical, electrochemical, ultrasonic, and sandblasting along with, experimentation tribology method using sandpaper of various grit sizes is demonstrated as a suitable de-coating method under low-resource environments, particularly making it attractive for small- and medium-sized enterprises and low-to-middle income countries. With increasing carbon waste, developing robust, economical, and safe carbon de-coating methods will lower the burden on mining, manufacturing, and supply chain sectors. The de-coated components could be returned to cradle-to-cradle cycle and safe recovery of carbon could be a circular pathway for secondary products like lubricant additives or graphitic pallets, etc.

References

1 Ruiz-Real, J.L., Uribe-Toril, J., De Pablo Valenciano, J. et al. (2018). Worldwide research on circular economy and environment: a bibliometric analysis, *International Journal of Environmental Research and Public Health* 15 (12): 2699.

2 Sariatli, F. (2017). Linear economy versus circular economy: a comparative and analyzer study for optimization of economy for sustainability. *Visegrad Journal on Bioeconomy and Sustainable Development* 6 (1): 31–34.

3 Korhonen, J., Honkasalo, A., and Seppälä, J. (2018). Circular economy: the concept and its limitations. *Ecological Economics* 143: 37–46.

4 Rajak, D.K., Kumar, A., Behera, A. et al. (2021). Diamond-like carbon (DLC) coatings: classification, properties, and applications. *Applied Sciences* 11 (10): 4445.

5 Tool regrinding and sustainability, M. Greene Ltd UK, 2021. https://mgreene.co.uk/tool-regrinding-and-sustainability (accessed 29 August 2024).

6 Mativenga, P., Rajemi, M., Marimuthu, S. et al. (2010). Establishing a basis for sustainable re-use of cutting tools through laser decoating. *Journal of Machine Engineering* 10 (2): 36–47.

7 Ahmed, J.K., Hamzah, A.F., and Hamed, A. (2017). Thermal and chemical etching of carbon fiber. *International Journal of Engineering and Technology* 7 (4) 519–526.

8 Rudawska, A. (2019). Mechanical treatment. *Surface Treatment in Bonding Technology* 1 (1) 87–128.

9 Choudhury, N., Kannan, A.G., and Dutta, N.K. (2013). Novel nanocomposites and hybrids for high-temperature lubricating coating applications (Chapter 21). *Tribology of Polymeric Nanocomposites* 2 (2): 717–778.

10 You, Y.B., Lee, Y.S., Kim, S.J. et al. (2022). Influence of additive N_2 on O_2 plasma ashing process in inductively coupled plasma. *National Library of Medicine* 12 (21): 3798.

11 Md. Shamsujjoha, S.R.A., Melia, M.A., Brooks, J.R. et al. (2015). Science direct. *Surface and Coatings Technology* 281 (1): 193–205.

12 El McKenzie (2013) Revolutionary pretreatment. Products Finishing. Gardner Business Media Inc. OH USA.

13 Different types of abrasive (sand) blasting. Swanton Welding Company, OH USA. 2022. https://swantonweld.com/different-types-of-sand-blasting/ (accessed 12 December 2024).

14 Altmayer, F. (2023). How to properly strip metallic coatings. Finishing and Coating, USA. https://finishingandcoating.com/index.php/plating/1609-how-to-properly-strip-metallic-coatings?highlight=WyJmcmFuayIsImFsdG1heWVyIl0= (accessed 12 December 2024).

15 Kim, S.H., Yuvaraj, H., Jeong, Y.T. et al. (2009). The effect of ultrasonic agitation on the stripping of photoresist using supercritical CO_2 and co-solvent formulation. *Microelectronic Engineering* 86 (2): 171–175.

16 Primus, T., Hlavinka, J., Zeman, P. et al. (2023). Experimental investigation of a method for selective and precise laser de-coating. *Lasers in Manufacturing and Materials Processing* 10 (1): 205–224.

17 Ultrasonic cleaning: exploring the efficiency and effectiveness. Precision Companies, FL USA. https://www.precgroup.com/exploring-the-benefits-of-ultrasonic-cleaning/ (accessed 12 December 2024).

18 Gammelgård, A., Tuuf, J., Strand, A. (2021). A method for tribological measurements of coated abrasives using a rheometer. *Tribology Letters* 69 (1): 62.

19 Economou, A. (2010). Recent developments in on-line electrochemical stripping analysis—an overview of the last 12 years. *Analytica Chimica Acta* 683 (1): 38–51.

20 Freschi, M., Paniz, A., Cerqueni, E. (2022). The twelve principles of green tribology: studies, research, and cases studies—a brief anthology. *Green Tribology* 10 (6): 129.

Index

a

ab initio 381
abrasion resistance 111, 132, 133
abutments 216, 218, 219
Actuators 346, 347, 357, 358, 359, 360, 364
additives 142, 144, 155, 157, 159, 160, 171, 176, 177, 182
adhesion 6, 10, 13, 22, 35, 71, 77, 83, 92, 96, 110, 114, 116, 120, 121, 130, 133, 139, 144, 158, 193, 195, 196, 197, 210, 216, 222, 223, 246, 148, 252, 263, 281, 286, 318, 321, 324, 327, 346, 360, 362, 366, 376, 388, 397,
adsorption 161, 176, 198, 251, 392
aerospace 34, 72, 100, 101, 121, 130, 138, 139, 144, 145, 146, 304, 346
AFM 50, 66, 143, 158, 166,177, 178, 202, 204, 294, 297, 298, 359, 360, 362, 364
AlCrN 121
AlTiN 121
aluminum 35, 73, 82, 85, 91, 92, 94, 111, 116, 216, 287, 306, 366, 395, 397
ambient air 92
antibacterial 35, 83, 120, 246, 258, 327, 329, 333
antimicrobial 3, 9, 10, 11, 17, 19, 262, 329
antireflective 12, 14, 56, 87, 262, 302, 304
anti-sticking 133
antithrombogenicity 328
artificial intelligence 3, 17, 18, 288, 374
artificial lung 86
atomic density 145, 381, 382
atomic force microscopy 56, 74, 75, 160, 202, 288, 294, 322, 359
atomic motion 135
Auger electron spectroscopy 54
automotive 6, 15, 22, 34, 72, 73, 100, 101, 118, 121, 130, 138, 144, 146, 158, 163, 304, 346, 355, 398, 414

b

bactericidal 83, 258, 259
bacteriostatic 258, 259
bandgap 12, 32, 71, 86, 299, 300, 301, 302, 303, 346, 355
batteries 12, 15, 21, 279, 280, 281, 284, 286, 287
battery 16, 279, 280, 281, 282, 287, 288, 413
big data 374
biocompatibility 9, 10, 11, 13, 19, 22, 33, 35, 54, 71, 73, 76, 77, 83, 84, 100, 130, 138, 170, 194, 195, 198, 199, 206, 210, 215, 216, 222, 246, 260, 263, 294, 346, 347
biodegradable 155, 158, 159, 160
biofilms 10, 19, 329
biomedical 4, 6, 8, 11, 15, 19, 22, 35, 50, 72, 83, 100, 101, 130, 136, 193, 196, 198, 201, 205, 245, 262, 263, 304, 315, 331, 332, 346, 347
biomedicine 317, 332

c

Candida albicans 10, 329
carbide 96, 97, 99, 110, 144, 146, 196, 248, 252, 253, 258, 423
carbon electrodes 16, 286
carbon footprint 414, 415, 416
carbon-coated products 413, 414
carboxyl groups 161
cardiac 247, 332
cardiovascular 3, 9, 11, 12, 100, 193, 194, 223, 259
chemical stripping 415, 417, 418, 419, 420, 421, 422, 423
chemical vapor deposition 4, 5, 31, 34, 48, 72, 109, 114, 120, 129, 130, 137, 138, 197, 245, 248, 280, 281, 294, 296, 316, 317, 357, 358, 359, 363, 364, 366, 402
circular economy 374, 414
circularity index 420, 421
CO_2 156, 199, 286, 329, 414, 415, 426
coated fabrics 332
coating failure 196, 197, 205, 225, 385
CoCrMo 206, 213, 214, 225, 226, 228
colonization 246, 258, 328
columnar 144, 145, 250, 253
combustion engines 156
computational fluid dynamics 378
condensation 37, 397, 398, 399, 400, 401, 402, 403, 405
contact design 180
copper 35, 111, 116, 258, 259, 287, 303, 398, 399
corrosion 4, 5, 6, 9, 10, 12, 15, 18, 22, 35, 77, 80, 81, 82, 83,100, 110, 114, 115, 120, 140, 144, 164, 166, 169, 195, 196, 198, 213, 216, 220, 221, 229, 232, 246, 247, 248, 249, 251, 253, 256, 257, 262, 263, 280, 306, 334, 399, 414
cotton 17, 325, 326, 327, 329, 332, 335
Cracks 74, 80, 111, 115, 204, 214, 251, 329
cradle-to-cradle 415, 426
critical load 92, 93, 114, 123, 206, 209
cross-linking 261, 315, 316, 321, 336
cutting tool 15, 34, 56, 72, 73, 100, 101, 109, 118, 136, 140, 145, 357, 413, 414, 415, 423
CVD 4, 5, 36, 37, 48, 72, 74, 77, 113, 114, 115, 116, 130, 137, 138, 245, 247, 249, 250, 251, 252, 254, 255, 256, 262, 280, 281, 282, 283, 297, 294, 302, 357
cyclic life 283, 284

d

data science 374
data-centric engineering 374, 380, 397
data-driven 15, 374, 405
decision matrix 413, 417
de-coating 46, 413–426
defrosting 319
delamination 35, 75, 90, 120, 196, 197, 198, 204, 205, 208, 213, 215, 228, 253, 319
dendrites 282, 286
density functional theory 381
detectors 305, 306
die tools 109, 110, 111, 112, 113, 115, 118, 123
digitalization 373, 374
doping elements 15, 73, 84, 97, 155, 160, 162, 171, 173, 176, 179, 182, 196, 197, 216, 247
doping material 47
droplet 77, 78, 79, 399
dropwise condensation 399, 402

e

elastohydrodynamic 162, 165, 172, 180
electric vehicles 18, 156, 280
electrical resistance 71, 77, 84, 359
electrochemical 82, 83, 167, 247, 279, 280, 281, 282, 283, 284, 286, 287, 288, 415, 416, 417, 418, 419, 421, 422, 423, 426
electrodes 16, 48, 49, 279, 280, 281, 282, 283, 284, 285, 287, 416
elevated temperatures 7, 19, 90, 114, 333, 383, 385, 392
endoprosthetic 198
energy consumption 155, 156, 157, 414
energy storage 3, 15, 16, 19, 22, 279, 280, 288
Escherichia coli 258

f

FCVAD 40, 41, 45, 48, 51
ferromagnetism 19
filmwise condensation 398, 401, 402
finite element method 385
fixtures 109, 216
flexible display 317
fluorine 11, 12, 35, 83, 133, 173, 221, 254, 258, 221, 254, 258, 259, 328, 380, 398, 400, 401
force sensors 351, 354
fourier transform infrared spectroscopy 294
friction 4, 6, 7, 10, 13, 22, 33, 35, 54, 56, 71, 72, 73, 75, 76, 87, 88, 89, 92, 100, 110, 114, 118, 121, 130, 132, 136, 138, 140, 141, 142, 145, 146, 156, 160, 167, 176, 179, 195, 199, 248, 254, 321, 346, 362, 374, 388, 391, 414, 419
fullerene 258, 280, 293,
functional groups 159, 166, 203
fuzzy reasoning 375

g

galvanic corrosion 83
gas flow 17, 137, 197, 201, 202, 378
Ge 10, 14, 44, 259, 303, 306
global warming 414
graphene oxide 166, 168
graphene quantum dots 166
graphite-like carbon 34, 122
greenhouse 156, 157, 414
growth 5, 7, 12, 16, 17, 22, 37, 41, 43, 45, 50, 52, 53, 57, 72, 83, 109, 197, 220, 250, 258, 282, 286, 327, 329, 374, 377, 378, 381, 384, 414, 415

h

haemocompatibility 273
hardness 6, 10, 19, 22, 32, 33, 35, 48, 53, 54, 56, 71, 72, 73, 75, 87, 90, 94, 95, 96, 97, 110, 111, 114, 117, 120, 125, 132, 135, 138, 143, 145, 146, 197, 201, 203, 205, 209, 210, 214, 218, 221, 252, 286, 293, 302, 316, 347, 359, 367, 375, 381, 384, 388, 425
hexagonal 55, 72, 76, 86, 218
hip joint 214, 215, 230, 248, 331
HiPIMS 114, 115, 118, 120, 317
hydrogen 4, 7, 14, 35, 43, 47, 74, 77, 80, 86, 117, 118, 129, 132, 144, 161, 171, 176, 202, 207, 248, 254, 256, 271, 302, 305, 321, 323, 348, 351, 354, 363, 380, 382
hydrogenation 4, 171, 302, 348, 351

i

IBAD 43, 51, 135, 206, 208, 213, 227, 229
ImageJ 78, 425
impedance 280, 282, 284, 287
impermeability 33, 164, 166, 333
implantable devices 215
implants 4, 9, 10, 15, 19, 73, 100, 130, 136, 140, 145, 193, 195, 198, 201, 213, 216, 222, 246, 262, 328, 332, 346, 347, 413, 415
in vitro 84, 198, 200, 206, 210, 328, 332
in vivo 197, 198, 199, 200, 213, 328
Industry 5.0 373, 374
inflammation 10
infrared light 84
initial capacity 284
interface 19, 79, 120, 155, 161, 178, 196, 215, 218, 280, 283, 284, 285, 288, 352, 364, 386, 388, 391, 394, 396, 397, 402, 403
ion beam deposition 4, 32, 34, 38, 39, 87, 135, 294, 296
ion energy 37, 38, 39, 40, 41, 44, 45, 47, 49, 51, 53, 57, 134, 197, 203
ionization 38, 40, 42, 47, 49, 50, 51, 57, 379, 380

k

kinetic behavior 378

l

LAMMPS 33, 383, 387
large-area 48
laser ablation 415, 416, 418, 419, 420, 421, 422, 423

ligament 199, 201
limb 261
limitations 5, 6, 18, 19, 22, 96, 97, 100, 116, 248, 263, 380
linear economy 414
lithium anodes 282
lithium-air batteries 286
low energy electron energy loss spectroscopy 54
low-to-middle income countries 426
lubricants 6, 89, 130, 155, 157, 158, 159, 160, 163, 166, 169, 171, 179, 182, 213, 347, 360, 396
lubrication 6, 75, 87, 96, 130, 131, 142, 157, 159, 161, 162, 163, 168, 178, 179, 182, 203, 206, 210, 213, 348, 360

m

machine learning 15, 374, 375
magnetic field 38, 41, 42, 48, 49, 205, 225
material informatics 374
materials 4.0 374
materials discovery 375
MD simulations 376, 385, 387, 388, 391
mechanical performance 94, 303
MEMS 12, 13, 22, 56, 130, 133, 345, 346, 347, 348, 351, 356, 359
metallosis 251
microblasting 400, 402
microelectronic 34, 300, 302, 304, 305, 306
microfluidic devices 330
microphage 84
molecular dynamics 33, 376, 381, 383, 387
molecular interactions 376
morphology 53, 56, 74, 76, 81, 98, 140, 144, 195, 196, 209, 245, 250, 261, 288, 317, 325, 330, 391
MoS_2 157, 163, 154, 168, 178, 360
molds 7, 109, 110, 111, 113, 117, 118, 121, 123, 413
MSIBD 38, 39, 45, 51
multiagent modeling 374
multilayer 47, 97, 98, 101, 144, 196, 214, 245, 247, 248, 331, 375
muscle cell 327

n

NaCl solution 115, 257, 334
nanoactuators 347, 357, 358
nanocomposite 7, 17, 116, 247
nanodot 345, 361, 362
nanofibers 284, 334
nanoindentation 94, 117, 118, 208, 210, 232, 254, 297, 298, 259, 383, 384, 385, 388
nanoparticles 155, 159, 162, 163, 164, 167, 168, 171, 178, 182, 259, 284
non-hydrogenated DLC 363
nonvolatile 13

o

oleic acid 161
oleophobic 176
Optical bandgap 12, 299, 300, 302, 303
optimization 15, 17, 18, 375, 378, 395, 396, 405
optoelectronic 21, 293, 295, 298, 299, 300, 303, 305, 307
orthodontic 194, 219, 220, 222
orthopedic 9, 10, 100, 145, 193, 194, 245, 249, 256, 413
oscillation 250, 380
osteointegration 194, 246, 328
osteoprotegerin 328
oxidation 79, 81, 99, 100, 115, 160, 164, 166, 247, 248
oxygen plasma 10, 11, 15, 322, 326, 333

p

Pseudomonas aeruginosa 10, 259
packaging 13
permeability 73, 317, 318, 319, 329, 333
photoelectronic 12, 34
physical vapor deposition 4, 31, 34, 37, 63, 74, 109, 113, 129, 130, 138, 139, 197, 245, 281, 295, 302, 316, 360, 374,

piezoresistive Sensor 345, 355
plasma discharge 35, 44, 48
plasma-enhanced chemical vapor deposition 34, 114, 197, 281, 295, 296, 316, 317, 359, 364, 366, 374
platelet 11, 83, 260, 327, 328, 322
polarization 76, 82, 252, 255, 257, 325
polyacetylene 36
polymer-like carbon 35, 117
precursor 34, 49, 50, 51, 74, 94, 96, 137, 203, 249, 317, 380
pressure sensors 345, 356
probabilistic simulations 376
profilometry 54, 215, 250
prostheses 56, 214, 229, 230, 261
protective coating 22, 282
pulsed laser deposition 4, 34, 44, 129, 136, 137, 219, 245, 281, 294, 316, 374
pyrolytic graphite 293

q

Q-carbon 3, 6, 14, 19, 20, 21, 22
quenched carbon 6

r

radial distribution function 54
Raman spectroscopy 20, 54, 74, 201, 202, 208, 254, 256, 257, 294, 297, 298, 316, 318
rate capability 16, 279, 280, 283, 284
recycling 374, 414, 415, 421
refractive 71, 76, 77, 84, 86, 87, 246, 250, 257, 299, 300, 301, 302, 303, 304
relative humidity 7, 88, 92
relaxation 52, 53, 253, 318, 385, 386, 387, 392
residual stress 7, 35, 98, 144, 208, 316, 324, 375, 376
RF power 48, 49
robustness 346, 413, 417, 420

s

sandpaper 416, 419, 424, 425, 426
sapphire 19, 21, 88, 205
scratch resistance 6, 33, 201, 297, 304, 318
semiconductor 84, 133, 303, 304, 355, 380, 381
semiconductors 6, 84, 300
sensors 34, 345, 346, 347, 351, 354, 355, 356, 359, 367
separator 279, 281, 286
shear strain 384, 385
silver 12, 17, 35, 196, 253, 258, 259, 260, 261, 262, 301, 303, 328, 329, 333, 414
small and medium-sized enterprises 426
smart bandages 333
soft materials 210
solar cells 87, 304, 305, 398
solid-electrolyte interface 280
solid-state batteries 280
sputtering 4, 31, 34, 37, 38, 42, 43, 51, 57, 74, 83, 91, 94, 99, 113, 114, 122, 129, 133, 134, 204, 208, 221, 232, 247, 257, 281, 284, 286, 287, 294, 296, 301, 316, 351, 363, 381
staphylococcus aureus 259, 327, 333
steel 10, 78, 79, 81, 88, 91, 93, 111, 112, 118, 143, 145, 203, 246, 255, 358, 364, 375, 394
strain energy 98
strain sensors 345, 355, 356
sulfur 157, 159, 160, 176, 254
superlubricity 160, 162, 168, 171
surface modification 120, 195, 252, 325, 400
surgical tools 145, 245, 262, 263
sustainable manufacturing 374

t

ta-C 4, 5, 7, 34, 36, 39, 44, 53, 55, 73, 77, 80, 86, 87, 99, 114, 117, 120, 121, 129, 131, 132, 143, 161, 171, 173, 175, 176, 205, 206, 212, 215, 217, 225, 231, 248, 250, 263, 348, 363
Taguchi 375
tantalum 216
techno-economic 413, 415, 423
technology readiness levels 413, 417, 422
TEM 54
temperature 4, 7, 8, 17, 19, 32, 34, 37, 52, 53, 80, 90, 94, 98, 99, 110, 111, 121, 122, 137, 169, 176, 251, 280, 283, 330, 351, 355, 358, 385, 393, 396, 416

temperature sensors 345, 354
Tetramethylsilane 50, 99, 118, 249, 256
textiles 3, 15, 17, 18, 19, 22, 11, 315, 331, 414
texture 74, 391, 398
thermal spike 52, 57
thermal strains 111
thermophysical 395
thickness 19, 55, 56, 74, 76, 81, 88, 91, 94, 110, 137, 142, 160, 161, 163, 176, 182, 202, 206, 211, 249, 253, 281, 287, 294, 298, 325, 348, 364, 376, 390, 395, 403, 425
thin films 10, 35, 44, 56, 57, 91, 110, 138, 242, 253, 283, 295, 300, 355, 362
TiAlN 79
TiN 44, 116, 118, 144, 206, 207, 214, 248, 305
tissues 198, 246, 247, 260, 263
TKR 199, 201, 222, 230
topography 226, 297, 298, 317, 325
total hip replacements 199
total knee replacements 199
toughness 10, 12, 17, 22, 111, 146, 298, 346, 348, 349, 374, 375
tribofilm 157, 162, 164, 165, 166, 168, 169, 170, 176, 178
tribology 87, 91, 129, 140, 145, 147, 155, 157, 158, 193, 345, 347, 259, 415, 416, 417, 418, 419, 420, 421, 422, 423, 426
tribopairs 171

u

UHMWPE 197, 199, 201, 202, 203, 204, 208, 209, 211, 212, 213, 215
undoped 6, 15, 34, 35, 72, 73, 101, 221, 222, 253, 257, 261, 287
uniform 4, 5, 44, 48, 136, 140, 171, 286, 287, 356

v

vacuum 38, 39, 40, 46 74, 77, 86, 88, 92, 114, 115, 133, 135, 136, 139, 205, 206, 294, 295, 316, 360, 361, 366, 421,
virtual optimizations 374
virtual synthesis 373, 374, 380, 405

w

wear rate 8, 10, 73, 118, 119, 123, 142, 203, 205, 211, 213, 255, 346, 347, 348, 365, 391
wettability 78, 79, 197, 203, 206, 224, 246, 257, 258, 317, 325, 326, 332, 398, 399, 401, 402, 403
wires 219, 220, 221, 352
WS_2 163

x

X-ray diffraction 54, 75, 207
X-ray photoelectron spectroscopy 54, 75, 171, 288, 294, 316, 353

z

zinc 77, 78, 82, 83, 157, 287, 301, 303, 306, 329

Printed and bound by CPI Group (UK) Ltd, Croydon, CR0 4YY
17/09/2026
14964309-0005